CONTEMPORARY MATHEMATICS

717

Mathematical Problems in Quantum Physics

QMATH13: Mathematical Results in Quantum Physics
October 8–11, 2016
Georgia Institute of Technology, Atlanta, Georgia

Federico Bonetto
David Borthwick
Evans Harrell
Michael Loss
Editors

AMERICAN MATHEMATICAL SOCIETY

2010 *Mathematics Subject Classification.* Primary 81V45, 81V70, 81V10, 82B28, 82B44, 82B20, 82B23, 82D77.

Library of Congress Cataloging-in-Publication Data

Names: QMath13 Conference (2016 : Atlanta, Georgia) | Bonetto, F. (Federico), editor. | Borthwick, David, editor. | Harrell, Evans M., editor. | Loss, Michael, 1954- editor.

Title: Mathematical problems in quantum physics / Federico Bonetto, David Borthwick, Evans Harrell, Michael Loss, editors.

Description: Providence, Rhode Island : American Mathematical Society, [2018] | Series: Contemporary mathematics ; volume 717 | Includes bibliographical references.

Identifiers: LCCN 2018023577 | ISBN 9781470436810 (alk. paper)

Subjects: LCSH: Mathematical physics–Congresses. | Quantum theory–Congresses.

Classification: LCC QC20.5 .Q28 2016 | DDC 530.13–dc23

LC record available at https://lccn.loc.gov/2018023577

Contemporary Mathematics ISSN: 0271-4132 (print); ISSN: 1098-3627 (online)

DOI: https://doi.org/10.1090/conm/717

∞ The paper used in this book is acid-free and falls within the guidelines
established to ensure permanence and durability.
Visit the AMS home page at `https://www.ams.org/`

10 9 8 7 6 5 4 3 2 1 23 22 21 20 19 18

Contents

Preface

In this volume are the proceedings of the 13th QMath conference, which was held at the Georgia Institute of Technology during October 8-11, 2016. The origins of this series date back to 1987 when Pavel Exner and Petr Šeba initiated the conference series "Mathematical Results in Quantum Theory" (or QMath). It is worth remembering that in those days Czechoslovakia, at that time the home country of Exner and Šeba, was still behind the Iron Curtain, and it took some courage to open scientific exchange between East and West.

The scientific aim of this conference series is not only to bring together people interested in the "quantum part" of mathematical physics, but also to stimulate a search for new quantum effects and a deeper understanding of quantum physics.

The proceedings can divided into essentially two topics. One is devoted to the many-body problem, in many ways the central problem in quantum physics and condensed matter physics. The mathematical questions there range from problems in atomic physics, e.g., the difficult problem of putting a bound on the excess charge an atom can hold, to the problem of deriving effective, i.e., simpler, equations for the behavior of large systems. A relatively new problem is 'Many-Body Localization'. Localization, i.e., the mathematical theory of insulators, is well understood for single particle models. The new problem is to understand this localization in connection with the many-body problem, and these proceedings contain some articles devoted to this very difficult subject.

The second main topic in these proceedings concerns quantum graphs. Because of the obvious connection to quantum networks, this field has received considerable attention in recent years. It also furnishes a host of new problems that allow a sharpening of the techniques used in the field of Schrödinger operators. Interesting connections appear, e.g., to the non-linear Schrödinger equation and to one-dimensional exactly solvable models.

It has been the hallmark of mathematical physics that many of its contributions do not fit neatly into traditional categories. Either new fields get created or problems from other fields of science are brought in. In this volume we have some examples of that, such as a mathematical discussion of resonances of open dimers, a problem in chemistry, and a discussion of the measurement process in quantum mechanics. Another theme is quantum information theory which, with the possibility of quantum computers, is a very active subject nowadays.

We hope with this collection to entice researchers to take up the challenges posed by some of these problems. We gratefully acknowledge partial support from National Science Foundation Grant DMS 1643086 as well as support from the American Institute of Physics, the International Union of Physics and Applied Physics, and the International Association of Mathematical Physics.

Contemporary Mathematics
Volume **717**, 2018
https://doi.org/10.1090/conm/717/14437

A short proof of the ionization conjecture in Müller theory

Rupert L. Frank, Phan Thành Nam, and Hanne Van Den Bosch

ABSTRACT. We prove that in Müller theory, a nucleus of charge Z can bind at most $Z + C$ electrons for a constant C independent of Z.

1. Introduction

In Müller theory [**12**], the energy of an atom is given by the functional

$$\mathcal{E}^{\mathrm{M}}(\gamma) = \operatorname{Tr}(-\Delta\gamma) - \int_{\mathbb{R}^3} \frac{Z\rho_\gamma(x)}{|x|}\,\mathrm{d}x + D(\rho_\gamma) - X(\gamma^{1/2}).$$

Here γ is the density matrix of the electrons and $\rho_\gamma(x) = \gamma(x,x)$ is its density. The Coulomb repulsion between the electrons is modeled by

$$D(\rho_\gamma) = \frac{1}{2}\iint_{\mathbb{R}^3\times\mathbb{R}^3} \frac{\rho_\gamma(x)\rho_\gamma(y)}{|x-y|}\,\mathrm{d}x\,\mathrm{d}y$$

and the exchange energy is described by

$$X(\gamma^{1/2}) = \frac{1}{2}\iint_{\mathbb{R}^3\times\mathbb{R}^3} \frac{|\gamma^{1/2}(x,y)|^2}{|x-y|}\,\mathrm{d}x\,\mathrm{d}y.$$

The ground state energy is then given by

$$E^{\mathrm{M}}(N) = \inf\left\{\mathcal{E}^{\mathrm{M}}(\gamma)\,|\,0 \le \gamma \le 1 \text{ on } L^2(\mathbb{R}^3), \operatorname{Tr}\gamma = N\right\}. \tag{1.1}$$

Here we ignore the electron spin for the sake of simplicity. Moreover, for our mathematical treatment we do not need to assume that the parameters $Z > 0$ (the nuclear charge) and $N > 0$ (the number of electrons) are integers.

Müller theory is a modification of Hartree–Fock theory, where the usual exchange energy $X(\gamma)$ is replaced by $X(\gamma^{1/2})$. On one hand, like Hartree–Fock theory [**2**], Müller theory correctly reproduces the Scott and Dirac–Schwinger corrections to Thomas–Fermi theory; see [**14**]. On the other hand, unlike the Hartree–Fock functional, the Müller functional is convex [**3**] and this leads to various mathematical simplifications. In particular, it follows from the discussion in [**3**, Subsection I.C] that the density of any minimizer (if it exists) is radially symmetric.

2010 *Mathematics Subject Classification.* Primary 81V45.

Key words and phrases. Maximal ionization, Müller density-matrix-functional theory.

The first author was supported in part by U.S. NSF Grant DMS-1363432.

The third author was supported in part by Conicyt (Chile) through CONICYT–PCHA/Doctorado Nacional/2014 Project # 116–0856 and Iniciativa Científica Milenio (Chile) through Millenium Nucleus RC–120002 "Física Matemática".

In [3], it was shown that the Müller functional has a minimizer if $N \le Z$, and it was conjectured that there is no minimizer if $N > N_c(Z)$ for a critical electron number $N_c(Z) < \infty$. As pointed out in [3], in Müller theory some electrons may form a nontrivial bound state at infinity, and therefore it is unclear how to apply the standard method of "multiplying the Euler-Lagrange equation by $|x|$" by Benguria and Lieb [1, 8, 9].

In [6], we used a different method to justify this conjecture and proved

THEOREM 1.1. *There is a constant $C > 0$ such that for all $Z > 0$, the Müller variational problem* (1.1) *has no miminizer if $N > Z + C$.*

The proof of Theorem 1.1 in [6] is adapted from our previous work on Thomas–Fermi–Dirac–von Weizsäcker theory [5]. It consists of two main ingredients. The first one is a new strategy to control the number of electrons far away from the nucleus, which is inspired by [13] and [4]. The second one is a comparison with Thomas–Fermi theory, following Solovej's fundamental work on Hartree–Fock theory [16]. In [6], we did not use the convexity of Müller funtional in order to illustrate the generality of our strategy. In fact, our proof has been generalized in [7] to cover a class of non-convex models between Müller and Hartree–Fock.

In this short note, we will provide a shorter proof of Theorem 1.1 by using the convexity of Müller functional and following Solovej's proof in reduced Hartree–Fock theory [15].

Acknowledgement. The first and second author are grateful to the organizers of the QMath 13 conference and for the invitation to speak there.

2. Exterior L^1-estimate

Throughout the paper we will assume that $N \ge Z$ and that the variational problem $E^{\rm M}(N)$ has a minimizer γ_0. As mentioned before we know that the density $\rho_0 = \rho_{\gamma_0}$ is radially symmetric. In many places we will use Newton's theorem

$$\int_{|y|<|x|} \frac{\rho_0(y)}{|x-y|}\, dy = \frac{1}{|x|}\int_{|y|<|x|} \rho_0(y)\, dy.$$

We start by proving a simple bound, which in particular verifies the conjecture in [3] that there is a critical electron number $N_c(Z) < \infty$.

LEMMA 2.1. $N \le 2Z + C(Z^{2/3} + 1)$.

PROOF. For any partition of unity $\chi_1^2 + \chi_2^2 = 1$, we have the binding inequality

$$\mathcal{E}^{\rm M}(\gamma_0) \le \mathcal{E}^{\rm M}(\chi_1\gamma_0\chi_1) + \mathcal{E}^{\rm M}_{Z=0}(\chi_2\gamma_0\chi_2). \tag{2.1}$$

We choose

$$\chi_j(x) = g_j\Big(\frac{\nu\cdot x - \ell}{s}\Big)$$

with $s > 0, \ell > 0, \nu \in \mathbb{S}^2$, and $g_j : \mathbb{R} \to \mathbb{R}^+$ satisfying

$$g_1^2 + g_2^2 = 1, \quad g_1(t) = 1 \text{ if } t \le 0, \quad g_1(t) = 0 \text{ if } t \ge 1, \quad |\nabla g_1| + |\nabla g_2| \le C.$$

By the IMS formula and the fact that

$$X(\chi_j\gamma_0^{1/2}\chi_j) \le X((\chi_j\gamma_0\chi_j)^{1/2})$$

(see [**3**, Lemma 3]), we can estimate

$$
\begin{aligned}
&\mathcal{E}^{\mathrm{M}}(\chi_1\gamma_0\chi_1)+\mathcal{E}^{\mathrm{M}}_{Z=0}(\chi_2\gamma_0\chi_2)-\mathcal{E}^{\mathrm{M}}(\gamma_0)\\
&\le \int\Big(|\nabla\chi_1(x)|^2+|\nabla\chi_2(x)|^2\Big)\rho_0(x)\,\mathrm{d}x+\int\frac{Z\chi_2^2(x)\rho_0(x)}{|x|}\,\mathrm{d}x\\
&\quad+\iint\frac{\chi_2^2(x)\Big(|\gamma_0^{1/2}(x,y)|^2-\rho_0(x)\rho_0(y)\Big)\chi_1^2(y)}{|x-y|}\,\mathrm{d}x\,\mathrm{d}y\\
&\le Cs^{-2}\int_{\nu\cdot x-s\le\ell\le\nu\cdot x}\rho_0(x)\,\mathrm{d}x+\int_{\ell\le\nu\cdot x}\frac{Z\rho_0(x)}{|x|}\,\mathrm{d}x\\
&\quad+\iint_{\nu\cdot y-s\le\ell\le\nu\cdot x}\frac{|\gamma_0^{1/2}(x,y)|^2}{|x-y|}\,\mathrm{d}x\,\mathrm{d}y-\iint_{\nu\cdot y\le\ell\le\nu\cdot x-s}\frac{\rho_0(x)\rho_0(y)}{|x-y|}\,\mathrm{d}x\,\mathrm{d}y.
\end{aligned}
$$

Thus from (2.1) it follows that for all $s>0,\ell>0$ and $\nu\in\mathbb{S}^2$,

$$
\begin{aligned}
\iint_{\nu\cdot y\le\ell\le\nu\cdot x-s}\frac{\rho_0(x)\rho_0(y)}{|x-y|}\,\mathrm{d}x\,\mathrm{d}y&\le Cs^{-2}\int_{\nu\cdot x-s\le\ell\le\nu\cdot x}\rho_0(x)\,\mathrm{d}x\\
&\quad+\int_{\ell\le\nu\cdot x}\frac{Z\rho_0(x)}{|x|}\,\mathrm{d}x+\iint_{\nu\cdot y-s\le\ell\le\nu\cdot x}\frac{|\gamma^{1/2}(x,y)|^2}{|x-y|}\,\mathrm{d}x\,\mathrm{d}y.
\end{aligned}
$$

Next, we integrate over $\ell\in(0,\infty)$, then average over $\nu\in\mathbb{S}^2$. We use Fubini's theorem and

$$\int_{\mathbb{S}^2}[\nu\cdot z]_+\,\frac{d\nu}{4\pi}=\frac{|z|}{4},\quad\forall z\in\mathbb{R}^3.$$

Moreover, we also use

$$\int_0^\infty \mathbb{1}(b-s\le\ell\le a)\,\mathrm{d}\ell\le[a-b]_++s$$

(for the right side) and

$$\int_0^\infty\Big(\mathbb{1}(b\le\ell\le a-s)+\mathbb{1}(-a\le\ell\le-b-s)\Big)\,\mathrm{d}\ell\ge\Big[[a-b]_+-2s\Big]_+$$

(for the left side) with $a=\nu\cdot x$, $b=\nu\cdot y$. All this leads to

$$N^2\le Cs^{-1}N+2ZN+2N+2s(D(\rho_0)+X(\gamma_0^{1/2})).$$

Optimizing over $s>0$ and using the a-priori estimate

$$D(\gamma_0)+X(\gamma_0^{1/2})\le C(Z^{7/3}+N)$$

(which follows by an easy energy comparison; see [**6**, Corollary 5]), we get $N\le 2Z+C(Z^{2/3}+1)$. □

In order to improve the bound in Lemma 2.1, we use the following observation. Heuristically, the electrons in the exterior region $|x|\ge r$ feel the rest of the system as an "effective nucleus" with the *screened nuclear charge*

$$Z_r=Z-\int_{|x|<r}\rho_0(x)\,\mathrm{d}x. \tag{2.2}$$

Therefore, by modifying the proof of Lemma 2.1 we can control the number of exterior electrons in terms of Z_r. We still lose a factor 2, but this is not a big problem because Z_r is much smaller than Z (if r is not too small).

Throughout the paper, we will use the cut-off functions

$$\chi_r^+(x) = \mathbb{1}(|x| \geq r), \quad \chi_r^+ \geq \eta_r \geq \chi_{(1+\lambda)r}^+, \quad |\nabla \eta_r| \leq C(\lambda r)^{-1}. \tag{2.3}$$

We have the following upgraded version of Lemma 2.1.

LEMMA 2.2 (Exterior L^1-estimate). *For all $r > 0, s > 0$ and $\lambda \in (0, 1/2]$,*

$$\int \chi_r^+ \rho_0 \leq C \int_{r<|x|<(1+\lambda)^2 r} \rho_0 + C\Big([Z_r]_+ + s + \lambda^{-2}s^{-1} + \lambda^{-1}\Big)$$
$$+ C\Big(s^2 \operatorname{Tr}(-\Delta \eta_r \gamma_0 \eta_r)\Big)^{3/5} + C\Big(s^2 \operatorname{Tr}(-\Delta \eta_r \gamma_0 \eta_r)\Big)^{1/3}.$$

PROOF. We use the binding inequality (2.1) with

$$\chi_j(x) = g_j\Big(\frac{\nu \cdot \theta(x) - \ell}{s}\Big)$$

where $\theta : \mathbb{R}^3 \to \mathbb{R}^3$ satisfies

$$|\theta(x)| \leq |x|, \quad \theta(x) = 0 \text{ if } |x| \leq r, \quad \theta(x) = x \text{ if } |x| \geq (1+\lambda)r, \quad |\nabla \theta| \leq C\lambda^{-1}$$

and proceed similarly as in Lemma 2.1. See [**6**, Lemma 7] for details. □

3. Comparison with Thomas–Fermi theory

In this section, we control the electron density in the exterior region $\{|x| \geq r\}$ in Müller theory by comparison with Thomas–Fermi (TF) theory. Recall that in usual TF theory, the ground state energy is obtained by minimizing the density functional

$$\mathcal{E}^{\rm TF}(\rho) = c^{\rm TF} \int_{\mathbb{R}^3} \rho^{5/3}(x)\,{\rm d}x - \int_{\mathbb{R}^3} \frac{Z\rho(x)}{|x|}\,{\rm d}x + D(\rho), \quad c^{\rm TF} = \frac{3}{5}(6\pi^2)^{2/3},$$

over all $0 \leq \rho \in L^1(\mathbb{R}^3) \cap L^{5/3}(\mathbb{R}^3)$. The TF minimizer $\rho^{\rm TF}$ is unique and has total mass $\int \rho^{\rm TF} = Z$ [**10**]. Here, as in [**15**], we will consider TF theory restricted to the exterior region $\{|x| \geq r\}$.

LEMMA 3.1 (Exterior TF theory). *Let $r > 0$ and $z \in \mathbb{R}$. Then the TF functional*

$$\mathcal{E}_r^{\rm TF}(\rho) = c^{\rm TF} \int_{\mathbb{R}^3} \rho(x)^{5/3}\,{\rm d}x - \int_{\mathbb{R}^3} \frac{z\rho(x)}{|x|}\,{\rm d}x + D(\rho)$$

has a unique minimizer $\rho_r^{\rm TF}$ among all densities satisfying

$$0 \leq \rho \in L^{5/3}(\mathbb{R}^3) \cap L^1(\mathbb{R}^3), \quad \operatorname{supp} \rho \subset \{|x| \geq r\}.$$

The minimizer $\rho_r^{\rm TF}$ is radially symmetric, has total mass $\int \rho_r^{\rm TF} = [z]_+$, has bounded kinetic energy

$$\int \left(\rho_r^{\rm TF}\right)^{5/3} \leq C[z]_+^{7/3} \tag{3.1}$$

and satisfies the TF equation

$$\frac{5c^{\rm TF}}{3}(\rho_r^{\rm TF})^{2/3} = [\varphi_r^{\rm TF}]_+ \qquad \text{in } \{|x| > r\} \tag{3.2}$$

with

$$\varphi_r^{\rm TF}(x) = \frac{z\chi_r^+(x)}{|x|} - \rho_r^{\rm TF} * |x|^{-1}.$$

Moreover, for every fixed $\kappa > 0$*, there is an* $\alpha(\kappa) > 0$ *such that if* $zr^3 \geq \kappa$ *and* $|x|r^{-1} \geq \alpha(\kappa)$*, then we have the Sommerfeld estimate*

$$\left|\rho_r^{\mathrm{TF}}(x) - \left(5\pi^{-1}c^{\mathrm{TF}}\right)^3|x|^{-6}\right| \leq C|x|^{-6}(|x|r^{-1})^{-\zeta} \tag{3.3}$$

with $\zeta = (\sqrt{73}-7)/2 \approx 0.77$*. For the full* ρ^{TF}*, for all* $x \neq 0$ *we have*

$$0 \geq \rho^{\mathrm{TF}}(x) - \left(5\pi^{-1}c^{\mathrm{TF}}\right)^3|x|^{-6} \geq -C|x|^{-6}(|x|z^{1/3})^{-\zeta}. \tag{3.4}$$

PROOF. See [**15**, Appendix B]. In fact, (3.3) is slightly stronger than [**15**, Theorem B3] and it is taken from [**16**, Lemma 4.4]. The bound (3.4) is taken from [**16**, Theorems 5.2, 5.4]. □

Convention. In what follows, ρ_r^{TF} denotes the minimizer from Lemma 3.1 with the choice $z = Z_r$ from (2.2).

The main result in this section is the following

LEMMA 3.2 (Comparison with TF). *For all* $r \geq s > 0$ *and* $\lambda \in (0, 1/2]$,

$$D(\eta_r^2\rho_0 - \rho_r^{\mathrm{TF}}) \leq Cs^{-2}\int \chi_r^+\rho_0 + C[Z_r]_+^{12/5}r^{-1/5}s^{2/5} + \mathcal{R} \tag{3.5}$$

where

$$\mathcal{R} = C\big(1 + (\lambda r)^{-2}\big)\int_{(1-\lambda)r\leq|x|\leq(1+\lambda)r}\rho_0 + C\lambda r^{1/2}[Z_{(1-\lambda)r}]_+^{5/2}$$
$$+ C\Big(\operatorname{Tr}(-\Delta\eta_r\gamma_0\eta_r)\Big)^{1/2}\Big(\int\chi_r^+\rho_0\Big)^{1/2}.$$

To prove this lemma we will use the following semi-classical estimates from [**16**, Lemma 8.2].

LEMMA 3.3 (Semi-classical analysis). *Let* $L_{\mathrm{sc}} = (15\pi^2)^{-1}$*. For every* $s > 0$*, fix a smooth function* $g_s : \mathbb{R}^3 \to [0, \infty)$ *such that*

$$\operatorname{supp} g_s \subset \{|x| \leq s\}, \quad \int g_s^2 = 1, \quad \int |\nabla g_s|^2 \leq Cs^{-2}.$$

(i) For all $V : \mathbb{R}^3 \to \mathbb{R}$ *such that* $[V]_+, [V - V * g_s^2]_+ \in L^{5/2}$ *and for all density matrices* $0 \leq \gamma \leq 1$*, we have*

$$\operatorname{Tr}((-\Delta - V)\gamma) \geq -L_{\mathrm{sc}}\int[V]_+^{5/2} - Cs^{-2}\operatorname{Tr}\gamma$$
$$- C\left(\int[V]_+^{5/2}\right)^{3/5}\left(\int[V - V * g_s^2]_+^{5/2}\right)^{2/5}. \tag{3.6}$$

(ii) If $V_+ \in L^{5/2}(\mathbb{R}^3) \cap L^{3/2}(\mathbb{R}^3)$*, then there is a density matrix* γ *such that*

$$\rho_\gamma = \frac{5}{2}L_{\mathrm{sc}}[V]_+^{3/2} * g_s^2$$

and

$$\operatorname{Tr}(-\Delta\gamma) \leq \frac{3}{2}L_{\mathrm{sc}}\int[V]_+^{5/2} + Cs^{-2}\int[V]_+^{3/2}. \tag{3.7}$$

PROOF OF LEMMA 3.2. **Step 1.** First, we show that the exterior density matrix $\eta_r\gamma_0\eta_r$ essentially minimizes the exterior reduced Hartree-Fock functional

$$\mathcal{E}_r^{\mathrm{RHF}}(\gamma) = \mathrm{Tr}(-\Delta\gamma) - \int_{\mathbb{R}^3} \frac{Z_r\rho_\gamma(x)}{|x|}\,\mathrm{d}x + D(\rho_\gamma)\,,$$

where Z_r is given by (2.2). Indeed, for all $r > 0$, $\lambda \in (0, 1/2]$ and for all density matrices

$$0 \le \gamma \le 1, \quad \mathrm{supp}(\rho_\gamma) \subset \{|x| \ge r\}, \quad \mathrm{Tr}\,\gamma \le \int \chi_r^+ \rho_0,$$

we have

$$\mathcal{E}_r^{\mathrm{RHF}}(\eta_r\gamma_0\eta_r) \le \mathcal{E}_r^{\mathrm{RHF}}(\gamma) + \mathcal{R} \tag{3.8}$$

The proof of (3.8) is straightforward, using a trial state argument. We refer to [**6**, Lemma 9] for details.

Step 2. Now we bound the right side of (3.8) by choosing γ as in Lemma 3.3 (ii) with $V = \chi_{r+s}^+\varphi_r^{\mathrm{TF}} \ge 0$. Note that $\rho_\gamma = \left(\chi_{r+s}^+\rho_r^{\mathrm{TF}}\right) * g_s^2$ by the TF equation, and hence

$$\mathrm{supp}\,\rho_\gamma \subset \{|x| \ge r\}, \quad \mathrm{Tr}\,\gamma = \int \rho_\gamma = \int \chi_{r+s}^+\rho_r^{\mathrm{TF}} \le \int \rho_r^{\mathrm{TF}} = Z_r \le \int \chi_r^+ \rho_0.$$

The last inequality here comes from our assumption $N = \int \rho_0 \ge Z$. On the other hand, by the semi-classical estimate (3.7),

$$\begin{aligned} \mathcal{E}_r^{\mathrm{RHF}}(\gamma) &\le \frac{3}{2}L_{\mathrm{sc}} \int [\varphi_r^{\mathrm{TF}}]^{5/2} - \int \frac{Z_r}{|x|}\rho_\gamma + D(\rho_\gamma) + Cs^{-2} \int [\varphi_r^{\mathrm{TF}}]_+^{3/2} \\ &\le \mathcal{E}_r^{\mathrm{TF}}(\rho_r^{\mathrm{TF}}) + \int_{r \le |x| \le r+s} \frac{Z_r}{|x|}\rho_r^{\mathrm{TF}} + Cs^{-2}\int \rho_r^{\mathrm{TF}}. \end{aligned} \tag{3.9}$$

Here we have used the TF equation, the convexity $D(\rho_\gamma) \le D(\rho_r^{\mathrm{TF}} * g_s^2) \le D(\rho_r^{\mathrm{TF}})$ and Newton's theorem

$$\int |x|^{-1}\rho_\gamma = \int (|x|^{-1} * g_s^2)(\chi_{r+s}^+\rho_r^{\mathrm{TF}}) = \int |x|^{-1}(\chi_{r+s}^+\rho_r^{\mathrm{TF}}).$$

Finally, we bound the error term by using (3.1) and Hölder's inequality:

$$\begin{aligned} \int_{r \le |x| \le r+s} \frac{Z_r}{|x|}\rho_r^{\mathrm{TF}} &\le C\frac{[Z_r]_+}{r}\Big(\int_{r \le |x| \le r+s} (\rho_r^{\mathrm{TF}}(x))^{5/3}\,\mathrm{d}x\Big)^{3/5}\Big(\int_{r \le |x| \le r+s} \mathrm{d}x\Big)^{2/5} \\ &\le C\frac{[Z_r]_+}{r}\Big([Z_r]_+^{7/3}\Big)^{3/5}\Big(r^2 s\Big)^{2/5} = C[Z_r]_+^{12/5} r^{-1/5} s^{2/5}. \end{aligned} \tag{3.10}$$

Step 3. To bound the left side of (3.8), we write

$$\mathcal{E}_r^{\mathrm{RHF}}(\eta_r\gamma_0\eta_r) = \mathrm{Tr}((-\Delta - \varphi_r^{\mathrm{TF}})\eta_r\gamma_0\eta_r) - D(\rho_r^{\mathrm{TF}}) + D(\eta_r^2\rho_0 - \rho_r^{\mathrm{TF}})$$

and use the semi-classical estimate (3.6) with $V = \varphi_r^{\mathrm{TF}}$. Note that by Newton's theorem,

$$-\rho_r^{\mathrm{TF}} * (|x|^{-1} - |x|^{-1} * g_s^2) \le 0$$

and

$$\begin{aligned} \left[\chi_r^+|\cdot|^{-1} - (\chi_r^+|\cdot|^{-1}) * g_s^2\right]_+(x) &\le \left[\chi_r^+|\cdot|^{-1} - \chi_{r+s}^+(|\cdot|^{-1} * g_s^2)\right]_+(x) \\ &= (\chi_r^+(x) - \chi_{r+s}^+(x))|x|^{-1}. \end{aligned}$$

Therefore, when $Z_r \geq 0$ and $r \geq s$ we can bound

$$[\varphi_r^{\rm TF} - \varphi_r^{\rm TF} * g_s^2]_+(x) \leq [Z_r]_+(\chi_r^+(x) - \chi_{r+s}^+(x))|x|^{-1}. \tag{3.11}$$

Using the TF equation (3.2) and the TF kinetic energy bound (3.1), we get, similarly to (3.10),

$$\begin{aligned}\left(\int [\varphi_r^{\rm TF}]_+^{5/2}\right)^{3/5}\left(\int [\varphi_r^{\rm TF} - \varphi_r^{\rm TF} * g_s^2]_+^{5/2}\right)^{2/5} &\leq C\left\|\rho_r^{\rm TF}\right\|_{L^{5/3}}[Z_r]_+ r^{4/5-1}s^{2/5}\\ &\leq C[Z_r]_+^{12/5} r^{-1/5}s^{2/5}.\end{aligned} \tag{3.12}$$

Note that (3.12) holds independently of the sign of Z_r since $[\varphi_r^{\rm TF}]_+ = 0$ if $Z_r \leq 0$. Thus,

$$\begin{aligned}\mathcal{E}_r^{\rm RHF}(\eta_r\gamma_0\eta_r) &= {\rm Tr}((-\Delta - \varphi_r^{\rm TF})\eta_r\gamma_0\eta_r) - D(\rho_r^{\rm TF}) + D(\eta_r^2\rho_0 - \rho_r^{\rm TF})\\ &\geq -L_{\rm sc}\int [\varphi_r^{\rm TF}]^{5/2} - Cs^{-2}\int \eta_r^2\rho_0 - C[Z_r]_+^{12/5}s^{2/5}r^{-1/5}\\ &\quad - D(\rho_r^{\rm TF}) + D(\eta_r^2\rho_0 - \rho_r^{\rm TF})\\ &= \mathcal{E}_r^{\rm TF}(\rho_r^{\rm TF}) + D(\eta_r^2\rho_0 - \rho_r^{\rm TF})\\ &\quad - Cs^{-2}\int \eta_r^2\rho_0 - C[Z_r]_+^{12/5}s^{2/5}r^{-1/5}.\end{aligned} \tag{3.13}$$

Putting together (3.8), (3.9), (3.10) and (3.13), we obtain (3.5). □

In order to translate (3.5) into an L^1-estimate, we will need

LEMMA 3.4. *For every $f \in L^{5/3}(\mathbb{R}^3) \cap L^1(\mathbb{R}^3)$ and $x \in \mathbb{R}^3$, we have*

$$\left|\int_{|y|<r} f(y)\,{\rm d}y\right| \leq C\|f\|_{L^{5/3}}^{5/6}D(f)^{1/12}r^{13/12}.$$

PROOF. From [**16**, Cor. 9.3] (see also [**5**, Lem. 18]) we have

$$\left|\int_{|y|<|x|} \frac{f(y)}{|x-y|}\,{\rm d}y\right| \leq C\|f\|_{L^{5/3}}^{5/6}(|x|D(f))^{1/12}, \quad \forall x \in \mathbb{R}^3.$$

Choosing $x = r\nu$ and averaging over $\nu \in \mathbb{S}^2$, we get the conclusion. □

We finish this section by proving some a-priori estimates for $\chi_r^+\rho_0$.

LEMMA 3.5 (A-priori estimates). *Assume that*

$$|Z_r| \leq Cr^{-3}, \quad \forall r \in (0, D] \tag{3.14}$$

for some $D \leq 1$. Then

$$\int \chi_r^+\rho_0 \leq Cr^{-3}, \quad \forall r \in (0, D], \tag{3.15}$$

$$\int \chi_r^+\rho_0^{5/3} \leq Cr^{-7}, \quad \forall r \in (0, D], \tag{3.16}$$

$${\rm Tr}(-\Delta\eta_r\gamma_0\eta_r) \leq Cr^{-7}, \quad \forall r \in (0, D]. \tag{3.17}$$

Here the constants are independent of D and the cut-off function η_r satisfies (2.3) with $\lambda \in [r/2, 1/2]$.

PROOF. We can choose $\gamma = 0$ in (3.8) to get $\mathcal{E}_r^{\rm RHF}(\eta_r\gamma_0\eta_r) \le \mathcal{R}$. Using the kinetic Lieb–Thirring inequality and TF lower bound, we find that

$$\begin{aligned}\operatorname{Tr}(-\Delta\eta_r\gamma_0\eta_r) \le C(\lambda r)^{-2}\Big(Z_{(1-\lambda)r} - Z_r + \int \chi_r^+\rho_0\Big) + C\lambda r^{1/2}[Z_{(1-\lambda)r}]_+^{5/2}\\ + C\Big(\operatorname{Tr}(-\Delta\eta_r\gamma_0\eta_r)\Big)^{1/2}\Big(\int \chi_r^+\rho_0\Big)^{1/2} + C[Z_r]_+^{7/3}.\end{aligned}$$

This bound, (3.14) and the choice $\lambda \ge r/2$ imply that

$$\operatorname{Tr}(-\Delta\eta_r\gamma_0\eta_r) \le C\Big(r^{-4}\int \chi_r^+\rho_0 + r^{-7}\Big), \quad \forall r \in (0,D]. \tag{3.18}$$

We recall that the estimate in Lemma 2.2 with $\lambda \ge r/2 = s/2$ gives

$$\begin{aligned}\int \chi_r^+\rho_0 \le C\int_{r<|x|<(1+\lambda)^2 r}\rho_0 + C\Big([Z_r]_+ + r^{-3}\Big)\\ + C\Big(r^2\operatorname{Tr}(-\Delta\eta_r\gamma_0\eta_r)\Big)^{3/5} + C\Big(r^2\operatorname{Tr}(-\Delta\eta_r\gamma_0\eta_r)\Big)^{1/3}.\end{aligned}$$

By inserting (3.14) and (3.18) into this bound we deduce that

$$\int \chi_r^+\rho_0 \le C(Z_r - Z_{(1+\lambda)^2 r}) + Cr^{-3}, \quad \forall r \in (0,D].$$

We can replace r by $(1+\lambda)^{-2}r$ and use (3.14) to get (3.15). Inserting (3.15) into (3.18) yields (3.17). Moreover, by (3.17) and the kinetic Lieb–Thirring inequality, we have

$$\int \chi_{2r}^+\rho_0^{5/3} \le \int (\eta_r^2\rho_0)^{5/3} \le C\operatorname{Tr}(-\Delta\eta_r\gamma_0\eta_r) \le Cr^{-7}, \quad \forall r \in (0,D].$$

Replacing r by $r/2$ we obtain (3.16). □

4. Proof of the main result

Now we prove Theorem 1.1. Since the (usual) TF minimizer $\rho^{\rm TF}$ has total mass Z [**10**], Theorem 1.1 is a direct consequence of the following

THEOREM 4.1 (Comparison with TF density). *There are universal constants $C > 0, \varepsilon > 0$ such that for all $N \ge Z \ge 1$ and $r > 0$,*

$$\left|\int_{|x|<r}\Big(\rho_0(x) - \rho^{\rm TF}(x)\Big)\,{\rm d}x\right| \le C(1 + r^{-3+\varepsilon}). \tag{4.1}$$

Note that the left side of (4.1) is $|Z_r - Z_r^{\rm TF}|$ where

$$Z_r^{\rm TF} := Z - \int_{|x|<r}\rho^{\rm TF}(x)\,{\rm d}x = \int_{|x|\ge r}\rho^{\rm TF}(x)\,{\rm d}x.$$

Recall that by the Sommerfeld estimate (3.4), for all $r > 0$ we have

$$Z_r^{\rm TF} = a^{\rm TF}r^{-3}\Big(1 + O((rZ^{1/3})^{-\zeta})\Big), \quad a^{\rm TF} := \frac{4(5c^{\rm TF})^3}{3\pi^2}. \tag{4.2}$$

Thus, (4.1) tells us that the screened nuclear charge Z_r can be approximated well by TF theory up to the distance $o(1)$, which is remarkably larger than the semi-classical distance $O(Z^{-1/3})$.

We will prove Theorem 4.1 using a bootstrap argument as in [**15**].

LEMMA 4.2 (Initial step). *There is a universal constant $C_1 > 0$ such that*

$$\left|\int_{|x|<r} (\rho_0(x) - \rho^{\mathrm{TF}}(x))\,\mathrm{d}x\right| \le C_1 (Z^{1/3} r)^{179/44} r^{-3+1/66}, \quad \forall r > 0. \tag{4.3}$$

PROOF. By writing $\mathcal{E}^{\mathrm{M}}(\gamma) = \mathcal{E}^{\mathrm{RHF}}(\gamma) - X(\gamma^{1/2})$, we obtain

$$\mathcal{E}^{\mathrm{RHF}}(\gamma_0) \le \inf\{\mathcal{E}^{\mathrm{RHF}}(\gamma) : 0 \le \gamma \le 1, \operatorname{Tr}\gamma \le N\} + X(\gamma_0^{1/2}).$$

Then we can use the semi-classical analysis as in the proof of Lemma 3.2 (now with $r = 0$ and $V(x) = \varphi^{\mathrm{TF}}(x) := Z|x|^{-1} - \rho^{\mathrm{TF}} * |x|^{-1} \ge 0$). The only difference is that instead of (3.11) we use

$$\int [\varphi^{\mathrm{TF}} - \varphi^{\mathrm{TF}} * g_s^2]_+^{5/2} \le Z^{5/2} \int \left(|x|^{-1} - |x|^{-1} * g_s^2\right)^{5/2} \le C Z^{5/2} s^{1/2}.$$

We thus obtain

$$D(\rho_0 - \rho^{\mathrm{TF}}) \le \int_{|x|\le s} \frac{Z}{|x|}\rho^{\mathrm{TF}} + Cs^{-2}N + C\|\rho^{\mathrm{TF}}\|_{L^{5/3}} Z s^{1/5} + X(\gamma_0^{1/2}).$$

Using the a-priori estimates

$$N \le CZ, \quad \int (\rho^{\mathrm{TF}})^{5/3} \le CZ^{7/3}, \quad X(\gamma_0^{1/2}) \le CZ^{5/3}$$

and optimizing over $s > 0$, we get $D(\rho_0 - \rho^{\mathrm{TF}}) \le CZ^{25/11}$. The desired estimate (4.3) then follows from Lemma 3.4. □

LEMMA 4.3 (Iterative step). *There are universal constants $C_2, \delta, \varepsilon > 0$ such that, if for some $D \le 1$*

$$\left|\int_{|x|<r} (\rho_0(x) - \rho^{\mathrm{TF}}(x))\,\mathrm{d}x\right| \le (a^{\mathrm{TF}}/2) r^{-3}, \quad \forall r \in (0, D], \tag{4.4}$$

then

$$\left|\int_{|x|<r} (\rho_0(x) - \rho^{\mathrm{TF}}(x))\,\mathrm{d}x\right| \le C_2 r^{-3+\varepsilon}, \quad \forall r \in [D, D^{1-\delta}]. \tag{4.5}$$

PROOF. Let $R \ge D \ge r$. Since $\int \rho_r^{\mathrm{TF}} = Z_r$ (see Lemma 3.1), we have the key identity

$$\begin{aligned} \int_{|x|<R} (\rho^{\mathrm{TF}} - \rho_0) &= Z_r - \int_{|x|<R} \chi_r^+ \rho_0 - \int_{|x|\ge R} \rho^{\mathrm{TF}} \\ &= \int_{|x|<R} (\rho_r^{\mathrm{TF}} - \chi_r^+ \rho_0) + \int_{|x|\ge R} (\rho_r^{\mathrm{TF}} - \rho^{\mathrm{TF}}). \end{aligned} \tag{4.6}$$

In order to estimate the first term on the right hand side, we start by establishing (3.14). If $r \ge Z^{-1/3}$ we estimate $|Z_r| \le |Z_r^{\mathrm{TF}}| + |Z_r - Z_r^{\mathrm{TF}}|$ and bound $|Z_r^{\mathrm{TF}}|$ by (4.2) (this requires $r \ge Z^{-1/3}$) and use the assumption (4.4). On the other hand, if $r \le Z^{-1/3}$, the bound $|Z_r| \le CZ \le Cr^{-3}$ follows by our a-priori bound in Lemma 2.1.This proves (3.14).

Therefore, we can apply Lemma 3.5 and we obtain the bounds (3.15), (3.16) and (3.17). We want to use these in Lemma 3.2 to bound $D(\eta_r^2 \rho_0 - \rho_r^{\mathrm{TF}})$. First, we use Lemma 3.5 to obtain

$$\mathcal{R} \le C(r^{-3} + \lambda^{-2} r^{-5} + \lambda r^{-7} + r^{-5}) \le C(\lambda^{-2} r^{-5} + \lambda r^{-7}).$$

For the other error terms in Lemma 3.2, we choose $s = r^{11/6}$, so that

$$C|Z_r|^{12/5} r^{-1/5} s^{2/5} + Cs^{-2} \int \chi_r^+ \rho_0 \le Cr^{-7+1/3}$$

and finally obtain

$$D(\eta_r^2 \rho_0 - \rho_r^{\mathrm{TF}}) \le Cr^{-7} \left(r^{1/3} + \lambda^{-2} r^2 + \lambda \right).$$

Inserting this bound, as well as (3.16), (3.1) and (3.14), into Lemma 3.4, we obtain

$$\left| \int_{|x| \le R} \left(\eta_r^2 \rho_0 - \rho_r^{\mathrm{TF}} \right) \right| \le Cr^{-3} \left(r^{1/3} + \lambda^{-2} r^2 + \lambda \right)^{1/12} (R/r)^{13/12}.$$

Moreover, from (3.16) and $0 \le \chi_r^+ - \eta_r^2 \le \mathbb{1}(r \le |x| \le (1+\lambda) r)$ it follows that

$$\int |\chi_r^+ \rho_0 - \eta_r^2 \rho_0| \le \|\chi_r^+ \rho_0\|_{L^{5/3}} \|\chi_r^+ - \eta_r^2\|_{L^{5/2}} \le Cr^{-3} \lambda^{2/5}.$$

Combining these estimates and choosing $\lambda = r^{1/3}$, we conclude that

$$\left| \int_{|x| \le R} \left(\chi_r^+ \rho_0 - \rho_r^{\mathrm{TF}} \right) \right| \le Cr^{-3+1/36} (R/r)^{13/12}. \tag{4.7}$$

Next, we use the Sommerfeld asymptotics to bound $(\rho_r^{\mathrm{TF}} - \rho^{\mathrm{TF}})$ on $\{|x| \ge R\}$. We will assume that $R \ge Lr \ge L^2 Z^{-1/3}$ for a universal constant $L > 0$ to be determined. From (4.2), by choosing $L > 0$ large enough, we have

$$Z_r^{\mathrm{TF}} \ge (3a^{\mathrm{TF}}/4) r^{-3}, \quad \forall r \ge LZ^{-1/3}.$$

Combining this bound with (4.4) we infer that

$$Z_r \ge (a^{\mathrm{TF}}/4) r^{-3}, \quad \forall r \ge LZ^{-1/3}.$$

Because of this we can, after increasing L if necessary, apply the Sommerfeld estimate (3.3) and deduce that

$$\left| \int_{|x| \ge R} \rho_r^{\mathrm{TF}}(x) \, \mathrm{d}x - a^{\mathrm{TF}} R^{-3} \right| \le CR^{-3} (R/r)^{-\zeta}, \quad \forall R \ge Lr.$$

Combining the latter estimate and (4.2) (with r replaced by R), we finally obtain

$$\left| \int_{|x| \ge R} \left(\rho_r^{\mathrm{TF}} - \rho^{\mathrm{TF}} \right) \right| \le CR^{-3} (R/r)^{-\zeta}, \quad \forall R \ge Lr. \tag{4.8}$$

Now let us conclude. From the bound (4.3) in the initial step, by choosing universal constants $\delta > 0$ and $\varepsilon > 0$ small enough we have

$$\left| \int_{|x| < r} (\rho_0 - \rho^{\mathrm{TF}}) \right| \le Cr^{-3+\varepsilon}, \quad \forall r \le (L^2 Z^{-1/3})^{(1-\delta)^2}.$$

Therefore, (4.5) holds true if $D \le (L^2 Z^{-1/3})^{1-\delta}$. It remains to consider the case $(L^2 Z^{-1/3})^{1-\delta} \le D \le 1$. We choose $r = L^{-1} D^{(1-\delta)^{-1}}$. Then we have $D \ge Lr \ge L^2 Z^{-1/3}$. Therefore, for all $R \in [D, D^{1-\delta}]$, by inserting (4.7) and (4.8) into (4.6), we get

$$\left| \int_{|x| < R} \left(\rho_0 - \rho^{\mathrm{TF}} \right) \right| \le Cr^{-3+1/36} (R/r)^{13/12} + CR^{-3} (R/r)^{-\zeta}.$$

Using $r \geq L^{-1}R^{(1-\delta)^{-2}}$ we deduce that

$$\left| \int_{|x|<R} \left(\rho_0 - \rho^{\mathrm{TF}}\right) \right| \leq CR^{-3+\varepsilon}, \quad \forall R \in [D, D^{1-\delta}]$$

if the universal constants $\delta > 0$ and $\varepsilon > 0$ are chosen small enough. □

Now we are ready to provide the

PROOF OF THEOREM 4.1. We use the notations in Lemma 4.2 and Lemma 4.3. Let $\beta = a^{\mathrm{TF}}/2$ and $C_0 = \max\{C_1, C_2, a^{\mathrm{TF}}\}$. The constant $\varepsilon > 0$ in Lemma 4.3 can be chosen to satisfy $\varepsilon \leq 1/66$. Let $D_n := Z^{-\frac{1}{3}(1-\delta)^n}$ for $n = 0, 1, 2, \ldots$

From Lemma 4.2, we have

$$\left| \int_{|x|<r} \left(\rho_0(x) - \rho^{\mathrm{TF}}(x)\right) \mathrm{d}x \right| \leq C_0 r^{-3+\varepsilon}, \quad \forall r \in (0, D_0]. \tag{4.9}$$

On the other hand, from Lemma 4.3, we deduce by induction that if

$$C_0(D_n)^{\varepsilon} \leq \beta,$$

then

$$\left| \int_{|x|<r} \left(\rho_0(x) - \rho^{\mathrm{TF}}(x)\right) \mathrm{d}x \right| \leq C_0 r^{-3+\varepsilon}, \quad \forall r \in (0, D_{n+1}].$$

Note that $D_n \to 1$ as $n \to \infty$ and that $C_0 > \beta$. Thus, there is a minimal $n_0 \in \{0, 1, 2, \ldots\}$ such that $C_0(D_{n_0})^{\varepsilon} > \beta$. If $n_0 \geq 1$, then $C_0(D_{n_0-1})^{\varepsilon} \leq \beta$ and therefore by the preceding argument

$$\left| \int_{|x|<r} \left(\rho_0(x) - \rho^{\mathrm{TF}}(x)\right) \mathrm{d}x \right| \leq C_0 r^{-3+\varepsilon}, \quad \forall r \in (0, D_{n_0}]. \tag{4.10}$$

If $n_0 = 0$, then (4.10) reduces to (4.9). Since $D_{n_0} \geq D := (C_0^{-1}\beta)^{1/\varepsilon}$, we have

$$\left| \int_{|x|<r} \left(\rho_0(x) - \rho^{\mathrm{TF}}(x)\right) \mathrm{d}x \right| \leq C_0 r^{-3+\varepsilon}, \quad \forall r \in (0, D]. \tag{4.11}$$

Note that $D \in (0, 1]$ is a universal constant. Using the exterior estimates (3.15) and (4.2) with $r = D$, we get

$$\int_{|x|\geq D} \left(\rho_0(x) + \rho^{\mathrm{TF}}(x)\right) \mathrm{d}x \leq C. \tag{4.12}$$

From (4.11) and (4.12), we obtain (4.1). □

References

[1] R. D. Benguria, *The von Weizsäcker and exchange corrections in the Thomas–Fermi theory*, Ph.D. Thesis, Princeton University, 1979.

[2] V. Bach, *Error bound for the Hartree-Fock energy of atoms and molecules*, Comm. Math. Phys. **147** (1992), no. 3, 527–548. MR1175492

[3] R. L. Frank, E. H. Lieb, R. Seiringer and H. Siedentop, *Müller's exchange-correlation energy in density-matrix-functional theory.* Phys. Rev. A 76 (2007), 052517.

[4] R. L. Frank, R. Killip, and P. T. Nam, *Nonexistence of large nuclei in the liquid drop model*, Lett. Math. Phys. **106** (2016), no. 8, 1033–1036, DOI 10.1007/s11005-016-0860-8. MR3520116

[5] R. L. Frank, P. T. Nam, and H. Van Den Bosch, *The ionization conjecture in Thomas-Fermi-Dirac–von Weizsäcker theory*, Comm. Pure Appl. Math. **71** (2018), no. 3, 577–614, DOI 10.1002/cpa.21717. MR3762278

[6] R. L. Frank, P. T. Nam and H. Van Den Bosch, *The maximal excess charge in Müller density-matrix-functional theory*, Ann. H. Poincaré (2018), to appear.

[7] C. Kehle, *The maximal excess charge for a family of density-matrix-functional theories including Hartree-Fock and Müller theories*, J. Math. Phys. **58** (2017), no. 1, 011901, 24, DOI 10.1063/1.4973850. MR3595176

[8] E. H. Lieb, *Thomas-Fermi and related theories of atoms and molecules*, Rev. Modern Phys. **53** (1981), no. 4, 603–641, DOI 10.1103/RevModPhys.53.603. MR629207

[9] E. H. Lieb, *Bound on the maximum negative ionization of atoms and molecules*, Phys. Rev. A 29 (1984), pp. 3018–3028.

[10] E. H. Lieb and B. Simon, *The Thomas-Fermi theory of atoms, molecules and solids*, Advances in Math. **23** (1977), no. 1, 22–116, DOI 10.1016/0001-8708(77)90108-6. MR0428944

[11] E. H. Lieb and W. E. Thirring. *Bound for the Kinetic Energy of Fermions Which Proves the Stability of Matter*, Phys. Rev. Lett. 35 (1975), 687.

[12] A. M. K. Müller, *Explicit approximate relation between reduced two- and one-particle density matrices*, Phys. Lett. A **105** (1984), no. 9, 446–452, DOI 10.1016/0375-9601(84)91034-X. MR767927

[13] P. T. Nam and H. Van Den Bosch, *Nonexistence in Thomas–Fermi–Dirac–von Weizsäcker theory with small nuclear charges*, Math. Phys. Anal. Geom. **20** (2017), no. 2, Art. 6, 32, DOI 10.1007/s11040-017-9238-0. MR3605165

[14] H. Siedentop, *Das asymptotische Verhalten der Grundzustandsenergie des Müllerfunktionals für schwere Atome* (German, with English and German summaries), J. Phys. A **42** (2009), no. 8, 085201, 9, DOI 10.1088/1751-8113/42/8/085201. MR2525482

[15] J. P. Solovej, *Proof of the ionization conjecture in a reduced Hartree-Fock model*, Invent. Math. **104** (1991), no. 2, 291–311, DOI 10.1007/BF01245077. MR1098611

[16] J. P. Solovej, *The ionization conjecture in Hartree-Fock theory*, Ann. of Math. (2) **158** (2003), no. 2, 509–576, DOI 10.4007/annals.2003.158.509. MR2018928

Rupert L. Frank, Mathematisches Institut, Ludwig-Maximilans Universität München, Theresienstr. 39, 80333 München, Germany—and—Mathematics 253-37, Caltech, Pasadena, California 91125

Email address: `rlfrank@caltech.edu`

Phan Thành Nam, Department of Mathematics and Statistics, Masaryk University, Kotlářská 2, 611 37 Brno, Czech Republic

Email address: `ptnam@math.muni.cz`

Hanne Van Den Bosch, Instituto de Física, Pontificia Universidad Católica de Chile, Av. Vicuña Mackenna 4860, Santiago, Chile

Email address: `hannevdbosch@fis.puc.cl`

Contemporary Mathematics
Volume **717**, 2018
https://doi.org/10.1090/conm/717/14438

Mean field dynamics of interacting fermionic systems

Marcello Porta

ABSTRACT. We consider the dynamics of interacting fermionic systems, in the mean-field regime. As the number of particles goes to infinity, the evolution of the system is expected to be well approximated by the time-dependent Hartree-Fock equation, a well-known example of effective evolution equation. We review some rigorous results about the validity of this approximation. We start by discussing the case of systems of particles interacting via bounded interaction potentials, at zero and at positive temperature. Under the assumption that a suitable semiclassical structure is propagated in time along the flow of the Hartree-Fock equation, the result can be extended to the case of Coulomb interactions.

1. Introduction

Effective theories are simplified models that are expected to provide an accurate description of complex physical systems in suitable scaling regimes. The complexity of the system arises, typically, from the large number of degrees of freedom involved in the description. Here we shall focus on effective theories for many-body quantum systems.

Thomas-Fermi theory is a famous example of effective theory, introduced independently by Thomas and Fermi to describe the physical properties of large atoms in a statistical way. Let us consider the Hamiltonian:

$$H^{\text{atom}}_{N,Z} := \sum_{j=1}^{N} \Big[-\Delta_j - \frac{Z}{|x_j|} \Big] + \sum_{i<j}^{N} \frac{1}{|x_i - x_j|}\,, \tag{1.1}$$

describing a non relativistic atom formed by a nucleus of charge $Z \in \mathbb{N}$ and by a cloud of N electrons with charge -1. The ground state energy of the system is:

$$E^{\text{GS}}_{N,Z} := \inf_{\psi_N \in L^2_{\text{a}}(\mathbb{R}^{3N})} \frac{\langle \psi_N, H^{\text{atom}}_{N,Z} \psi_N \rangle}{\langle \psi_N, \psi_N \rangle}\,, \tag{1.2}$$

where L^2_{a} is the space of fermionic wave functions, *i.e.* square integrable antisymmetric wave functions on $\mathbb{R}^{3N}$ (we neglect the spin of the particles for simplicity). As the number of particles N grows, the model becomes more and more difficult to study, even numerically. For this reason, one is interested in approximate descriptions, depending on much less degrees of freedom, that nevertheless allow to capture physically relevant macroscopic features of the system.

The author gratefully acknowledges financial support from the NCCR SwissMAP, and from the SNSF granf "Mathematical Aspects of Many-Body Quantum Systems".

Thomas-Fermi theory provides a description of the ground state properties of large atoms, taking as only input the *density* of the system. Given an N-particle fermionic wave function ψ, let us define the associated density as:

$$\rho_\psi(x) := N \int dx_2 \dots dx_N |\psi(x, x_2, \dots, x_N)|^2 , \tag{1.3}$$

where the normalization factor is chosen so that $\int dx\, \rho_\psi(x) = N$. The Thomas-Fermi energy of an atom with electron density $\rho_N \in \mathcal{F}_N := \{\rho \in L^1(\mathbb{R}^3) \cap L^{5/3}(\mathbb{R}^3) \mid \rho(x) \geq 0,\ \|\rho\|_1 = N\}$ is:

$$\mathcal{E}^{\mathrm{TF}}_{N,Z}(\rho_N) := \frac{3}{5}\gamma \|\rho_N\|^{5/3}_{5/3} - \int dx\, \frac{Z}{|x|}\rho_N(x) + \frac{1}{2}\int dx dy\, \frac{1}{|x-y|}\rho_N(x)\rho_N(y) , \tag{1.4}$$

where the first term describes the kinetic energy of the system, the second is the interaction of the electrons with the (static) nucleus and the last takes into account the electron-electron interactions. The Thomas-Fermi ground state energy is defined as:

$$E^{\mathrm{TF}}_{N,Z} := \inf_{\rho_N \in \mathcal{F}_N} \mathcal{E}^{\mathrm{TF}}_{N,Z}(\rho_N) . \tag{1.5}$$

It turns out that, for $Z = N + o(N)$, $E^{\mathrm{TF}}_{N,Z} = O(N^{7/3})$. In this regime, choosing appropriately the parameter γ, Lieb and Simon [**19**] proved that:

$$\big| E^{\mathrm{GS}}_{Z,N} - E^{\mathrm{TF}}_{Z,N} \big| \leq C N^{7/3 - \delta} , \tag{1.6}$$

for some explicit $\delta > 0$. That is, as $N \to \infty$, the leading contribution to the ground state energy is correctly described by Thomas-Fermi theory. The result has been recently extended in [**13**] to more general fermionic systems.

It is natural to ask whether it is possible to introduce more precise effective theories, depending on more information than just the density of the system, yielding a better approximation to physical quantities such as the ground state energy. *Hartree-Fock* theory is an effective theory which takes as input the *reduced 1-particle density matrix* of the system. In general, given a fermionic wave function $\psi_N \in L^2_{\mathrm{a}}(\mathbb{R}^{3N})$, $\|\psi_N\|_2 = 1$, its k-particle reduced density matrix is defined as:

$$\gamma^{(k)}_\psi := \binom{N}{k} \mathrm{Tr}_{k+1,\dots,N} |\psi_N\rangle\langle\psi_N| . \tag{1.7}$$

Reduced density matrices are important because they allow to compute averages of k-particle obervables. For instance, the energy of ψ_N can be written as:

$$\begin{aligned} E_{N,Z}(\psi) &:= \langle \psi_N, H^{\mathrm{atom}}_{N,Z}\psi_N\rangle \\ &= \mathrm{Tr}\Big[-\Delta - \frac{Z}{|x|}\Big]\gamma^{(1)}_\psi + \mathrm{Tr}\, \frac{1}{|x_1 - x_2|}\gamma^{(2)}_\psi . \end{aligned}$$

For a general large quantum system, it is essentially impossible to compute its reduced density matrices starting from first principles. *Slater determinants* are an example of fermionic states for which all reduced density matrices can be computed starting from the one-particle density matrix. Given a set of orthonormal functions $\{f_i\}_{i=1}^N$, $f_i \in L^2(\mathbb{R}^3)$, we define:

$$\psi_{\mathrm{Slater}}(x_1, \dots, x_N) := \frac{1}{\sqrt{N!}} \sum_\pi \mathrm{sgn}(\pi) f_{\pi(1)}(x_1) f_{\pi(2)}(x_2) \cdots f_{\pi(N)}(x_N) . \tag{1.8}$$

It is easy to see that:

$$\gamma^{(1)}_{\psi_{\text{Slater}}} \equiv \omega_N = \sum_{i=1}^{N} |f_i\rangle\langle f_i| \,, \tag{1.9}$$

where ω_N is a rank-N orthogonal projection: $\omega_N = \omega_N^2 = \omega_N^*$. For Slater determinants, it turns out that all reduced densities can be computed starting from ω_N via the fermionic Wick rule. For instance:

$$\gamma^{(2)}_{\psi_{\text{Slater}}}(x_1,x_2;y_1,y_2) = \frac{1}{2}[\omega_N(x_1;y_1)\omega_N(x_2;y_2) - \omega_N(x_2;y_1)\omega_N(x_1;y_2)] \,. \tag{1.10}$$

In particular, the energy of Slater determinant can be written as:

$$\begin{aligned}\langle\psi_{\text{Slater}}, H^{\text{atom}}_{N,Z}\psi_{\text{Slater}}\rangle &= \mathrm{Tr}\Big[-\Delta - \frac{Z}{|x|}\Big]\omega_N \\ &+\frac{1}{2}\int dxdy\,\frac{1}{|x-y|}[\rho_N(x)\rho_N(y) - |\omega_N(x;y)|^2] \equiv \mathcal{E}^{\text{HF}}_{N,Z}(\omega_N)\end{aligned}$$

with $\rho_N(x) \equiv \omega_N(x;x)$; $\mathcal{E}^{\text{HF}}_{N,Z}$ is called the *Hartree-Fock energy functional.* Notice that in Hartree-Fock theory the interaction is given by a sum of two terms, the *direct* term (also present in Thomas-Fermi theory) and the *exchange* term. The Hartree-Fock ground state energy is defined by restricting the minimization in the definition (1.2) to the set of Slater determinants. Equivalently,

$$E^{\text{HF}}_{N,Z} := \inf_{\omega_N\in\mathcal{P}_N} \mathcal{E}^{\text{HF}}_{N,Z}(\omega) \,, \tag{1.11}$$

where $\mathcal{P}_N$ is the set of rank-N orthogonal projections on $L^2(\mathbb{R}^3)$[1]. Bach [**3**], and afterwards Graf-Solovej [**16**], proved that Hartree-Fock theory gives a better approximation to the ground state energy of a large atom with respect to Thomas-Fermi theory. It follows that, as $N \to \infty$ and for $Z = N + o(N)$:

$$|E^{\text{GS}}_{N,Z} - E^{\text{HF}}_{N,Z}| \leq CN^{5/3-\delta} \,, \tag{1.12}$$

form some explicit $\delta > 0$. The error term is smaller than the smallest contribution to the Hartree-Fock energy, namely the exchange term, which is $O(N^{5/3})$.

It turns out that the regime $Z \simeq N \gg 1$, for which all these results hold, can be understood as a *mean-field* regime. Thomas-Fermi theory suggests that most of the electrons are localized at distances of order $N^{-1/3}$ from the nucleus. Therefore, let us rewrite $x_i = N^{-1/3}X_i$. After this transformation, we rewrite the Hamiltonian as:

$$\begin{aligned} H^{\text{atom}}_{N,N} &= N^{4/3}\Big[\sum_{j=1}^{N}\Big[-N^{-2/3}\Delta_{X_j} - \frac{1}{|X_j|}\Big] + \frac{1}{N}\sum_{i<j}^{N}\frac{1}{|X_i-X_j|}\Big] \\ &\equiv N^{4/3}\Big[\sum_{j=1}^{N}\Big[-\varepsilon^2\Delta_{X_j} - \frac{1}{|X_j|}\Big] + \frac{1}{N}\sum_{i<j}^{N}\frac{1}{|X_i-X_j|}\Big] \end{aligned} \tag{1.13}$$

where we defined $\varepsilon := N^{-1/3}$. That is, up to an overall factor $N^{4/3}$, the Hamiltonian of the atom can be understood as a *mean-field Hamiltonian*, in the presence of a joint *semiclassical scaling*. The question we are interested in discussing here is: how

[1]As proven by Lieb [**18**], for positive-definite interaction potentials the Hartree-Fock ground state energy can be obtained by minimizing over the larger set obtained by dropping the constraint $\omega_N = \omega_N^2$.

to describe in an effective way the *dynamics* of the system, after a perturbation is introduced in the Hamiltonian?

To understand this, we consider the dynamics generated by the Schrödinger equation:

$$\begin{aligned} i\varepsilon\partial_t\psi_{N,t} &= H_N\psi_{N,t} ,\\ H_N &= \sum_{j=1}^N -\varepsilon^2\Delta_j + \frac{1}{N}\sum_{i<j}^N V(x_i - x_j) , \end{aligned} \tag{1.14}$$

where V is now a general interaction potential, and suppose that the intial datum is close to a Slater determinant, in the sense that its reduced one-particle density matrix is close to the minimizer of the Hartree-Fock energy functional:

$$\mathcal{E}^{\rm HF}_N(\omega_N) = {\rm Tr}[-\varepsilon^2\Delta + V_{\rm ext}]\omega_N + \frac{1}{2N}\int dxdy\, V(x-y)[\rho_N(x)\rho_N(y) - |\omega_N(x;y)|^2] , \tag{1.15}$$

with $V_{\rm ext}$ a suitable confining potential. We assume that the system is initially confined in a region $\Lambda \subset \mathbb{R}^3$, with $|\Lambda| = O(1)$. The choice of the scaling in Eq. (1.15) ensures that kinetic and potential energy are both of order N. Moreover, the parameter ε is front of the time derivative implies that the time-scale for which the system performs a macroscopic change, that is for for which a given particle will travel a distance of order 1, is $t = O(1)$. Let $\psi_{N,t}$ be the solution of Eq. (1.14). For times of order 1, one expects that $\psi_{N,t}$ will still be close to a Slater determinant, in the sense that its reduced one-particle density matrix will be well approximated to the solution of the *time-dependent Hartree-Fock equation*:

$$\begin{aligned} &i\varepsilon\partial_t\omega_{N,t} = [h_{\rm HF}(t), \omega_{N,t}] , && h_{\rm HF}(t) := -\varepsilon^2\Delta + \rho_t * V - X_t ,\\ &\rho_t(x) := N^{-1}\omega_{N,t}(x;x) , && X_t(x;y) := N^{-1}V(x-y)\omega_{N,t}(x;y) . \end{aligned}$$

This is a *nonlinear* evolution equation, where the interaction is given by the self-consistent expression $\rho_t * V - X_t$. Notice that this equation still depends on N, and it is reasonable to ask what happens if $N \to \infty$. To understand this, let us define the *Wigner transform* of $\omega_{N,t}$ as:

$$W_{N,t}(x,p) := \frac{\varepsilon^3}{(2\pi)^3}\int dy\, e^{ipy}\omega_{N,t}(x + y/\varepsilon, x - y/\varepsilon) , \tag{1.16}$$

which might be thought as a "classical phase space density" (non-positive, in general). One expects that, as $N \to \infty$, $W_{N,t}$ will converge to the solution of the *Vlasov equation*:

$$\partial_t W_{\infty,t}(x,t) + p\cdot\nabla_x W_{\infty,t}(x,t) = (\nabla_x V * \rho_t)(x)\cdot\nabla_p W_{\infty,t}(x,p) , \tag{1.17}$$

a classical effective evolution equation.

Let us review some rigorous results about the derivation of these effective evolution equations. The first derivation of an effective evolution equation for mean-field fermionic systems, in the joint semiclassical/mean-field regime, is the work of [**22**], where the Vlasov equation was obtained starting from many-body quantum dynamics for analytic interaction potentials. The result has been extended in [**28**] to a larger class of interactions. In the same scaling regime, [**11**] gave a derivation of the time-dependent Hartree-Fock equation, for analytic interaction potentials and for short times. Here we will review the works [**6, 8, 9, 26**], which extended the result of [**11**] to a larger class of interactions, for all times, and allow to describe a larger class of initial data. More precisely, [**8**] considers the dynamics of pure states (zero temperature systems), with bounded interaction potentials. The result has been extended in [**9**] to the case of pseudo-relativistic fermions, and in [**6**] to non-relativistic mixes states (positive temperature systems). Then, [**26**] extended the result of [**8**] to the case of Coulomb interactions, under the assumption that the Hartree-Fock flow preserves a suitable semiclassical structure of the initial datum. Convergence to the Hartree-Fock equation in other scaling regimes, that do not involve a semiclassical scaling, have been obtained in [**4, 5, 14, 23, 24**].

Concerning the derivation of the Vlasov equation from the time-dependent Hartree-Fock equation, [**15, 20, 21**] proved convergence in an abstract sense, with no estimates on the rate; the analysis of [**20, 21**] also applies to the case of Coulomb potentials. Bounds on the rate of convergence have been established in [**2**], for mixed states and for bounded potentials. For smooth potentials, an expansion of the solution of the Hartree equation in powers of ε has been obtained in [**25**] (with no control on the remainder), and in [**1**]. The work [**7**] gave a quantitiative proof of convergence, in a similar spirit of [**2**]; with respect to [**2**], the method allows to consider a larger class of interaction potentials, and improves the rate of convergence. Moreover, [**7**] gave the first quantitative convergence result for the dynamics of pure states, such as Hartree-Fock ground states, at the level of expectations of semiclassical observables.

The article is organized as follows. In Section 2 we will introduce the fermionic Fock space, and some basic concepts of second quantization. In Section 3 we will spell out the results, while in Section 4 we will discuss some ideas of the proofs.

2. Second quantization

In this section we will map the many-body quantum dynamics in the fermionic Fock space, and rephrase the problem of proving convergence from the Schrödinger evolution to the Hartree-Fock evolution in the language of second quantization. We will first start by discussing the case of pure states, for which the initial state of the system can be represented by a vector in the Fock space. Then, we will consider the case of positive temperature states, for which the initial state (convex combinations of projections over pure states) can be represented as a vector in a doubled Fock space.

2.1. Pure states. The fermionic Fock space is defined as:

$$\mathcal{F}(L^2(\mathbb{R}^3)) \equiv \mathcal{F} = \mathbb{C} \oplus \bigoplus_{n\geq 1} L^2(\mathbb{R}^3)^{\wedge n} , \tag{2.1}$$

where $L^2(\mathbb{R}^3)^{\wedge n} \equiv L^2(\mathbb{R}^3) \wedge \cdots \wedge L^2(\mathbb{R}^3) \simeq L^2_{\mathrm{a}}(\mathbb{R}^{3N})$ is the n-fold antisymmetric tensor product of $L^2(\mathbb{R}^3)$. That is, a vector ψ in the fermionic Fock space consists of an infinite sequence of antisymmetric functions,

$$\mathcal{F} \ni \psi = (\psi^{(0)}, \psi^{(1)}, \ldots, \psi^{(n)}, \ldots,) , \qquad \psi^{(n)} \in L^2(\mathbb{R}^3)^{\wedge n} , \tag{2.2}$$

where each entry in the vector describes an n-particles system. An important example of vector in $\mathcal{F}$ is the *vacuum vector* $\Omega = (1, 0, 0, \ldots, 0, \ldots)$ corresponding to the physical situation in which no particles are present. Given $f \in L^2(\mathbb{R}^3)$ and $\psi \in \mathcal{F}$, it is convenient to introduce the fermionic creation and annihilation operators $a(f)$ and $a^*(f)$ as:

$$(a(f)\psi)^{(n)}(x_1,\ldots,x_n) = \sqrt{n+1}\int dx\overline{f(x)}\psi^{(n+1)}(x,x_1,\ldots,x_n)\,,$$

$$(a^*(f)\psi)^{(n)}(x_1,\ldots,x_n) = \frac{1}{\sqrt{n}}\sum_{j=1}^{n}(-1)^j f(x_j)\psi^{(n-1)}(x_1,\ldots,x_{j-1},x_{j+1},\ldots,x_n)\,,$$

which, respectively, annihilate and create a fermionic particle with wave function f. It is also convenient to introduce the operator valued distributions $a_x := a(\delta_x)$ and $a^*_x := a^*(\delta_x)$, with $\delta_x \equiv \delta(\cdot - x)$ the Dirac delta function at x. A straightforward consequence of the definition of the fermionic operators $a(f)$ and $a^*(f)$ are the canonical anticommutation relations:
(2.3)

$$\{a^*(f),a(g)\} = \langle g,f\rangle_{L^2}\,,\quad \{a(f),a(g)\} = \{a^*(f),a^*(g)\} = 0\,,\quad \forall f,g\in L^2(\mathbb{R}^3)\,.$$

A consequence of these relations is that $\|a(f)\| = \|f\|_2$.

Slater determinants can be conveniently represented in the fermionic Fock via *Bogoliubov transformations*, [**10**, **27**]. Given a rank-N orthogonal projection $\omega_N = \sum_{j=1}^N |f_j\rangle\langle f_j|$, a Bogoliubov transformation is implemented by a unitary operator $R_{\omega_N} : \mathcal{F}\to\mathcal{F}$ such that:

(i) $R_{\omega_N}\Omega = \varphi$, with $\varphi^{(k)} = 0$ for all $k\neq N$ and $\varphi^{(N)} = \psi_{\text{Slater}}$,
(ii) the map R_{ω_N} acts as a *particle-hole* transformation in $\mathcal{F}$. Let $\{f_i\}_{i=1}^\infty$ be a basis of $L^2(\mathbb{R}^3)$, where $f_1,\ldots,f_N$ are the orbitals of ψ_{Slater}. We have:

$$R^*_{\omega_N}a(f_i)R_{\omega_N} = \begin{cases} a^*(f_i) & \text{if } 1\le i\le N \\ a(f_i) & \text{if } i>N. \end{cases} \tag{2.4}$$

More generally, given any $g\in L^2(\mathbb{R}^3)$,

$$R^*_{\omega_N}a(g)R_{\omega_N} = a(ug) + a^*(\overline{v}\overline{g})\,, \tag{2.5}$$

where $u\equiv u_N = 1-\omega_N$ and $v\equiv v_N = \sum_{i=1}^N |\overline{f}_i\rangle\langle f_i|$. In terms of the operator valued distributions a^*_x, a_x, Eq. (2.5) reads:

$$R^*_{\omega_N}a_x R_{\omega_N} = a(u_x) + a^*(\overline{v}_x)\,, \tag{2.6}$$

where we introduced the short-hand notations $u_x(y)\equiv u_N(y;x)$ and $v_y\equiv v_N(y;x)$. It is important to notice that:

$$u_N v_N = 0\,,\qquad \overline{v}_N v_N = \omega_N\,. \tag{2.7}$$

In general, one can introduce a time-dependent Bogoliubov transformation $R_t\equiv R_{\omega_{N,t}}$, associated to the solution at the time t of the Hartree-Fock equation (1.16). In this case, $u_{N,t} = 1-\omega_{N,t}$ with $\omega_{N,t}\equiv\sum_{j=1}^N |f_{j,t}\rangle\langle f_{j,t}|$ the solution of (1.16), and $v_{N,t} = \sum_{j=1}^N |\overline{f}_{j,t}\rangle\langle f_{j,t}|$.

The dynamics in Fock space is generated by the second-quantized Hamiltonian:

$$\mathcal{H}_N = \int dx\,\varepsilon\nabla_x a^*_x \varepsilon\nabla_x a_x + \frac{1}{2N}\int dxdy\,V(x-y)\,a^*_x a^*_y a_y a_x\,, \tag{2.8}$$

in terms of which:

$$e^{-i\mathcal{H}_N t/\varepsilon}\psi = (e^{-iH_N^{(0)}t/\varepsilon}\psi^{(0)},\ldots,e^{-iH_N^{(n)}t/\varepsilon}\psi^{(n)},\ldots)\,,\qquad \forall\psi\in\mathcal{F}\,, \tag{2.9}$$

where:

$$H_N^{(n)} = \sum_{j=1}^{n} -\varepsilon^2\Delta_j + \frac{1}{N}\sum_{i<j}^{n} V(x_i - x_j)\,. \tag{2.10}$$

Thanks to these definitions, we can represent the kernel of the time-evolved reduced one-particle density matrix as:

$$\gamma^{(1)}_{N,t}(x;y) = \langle e^{-i\mathcal{H}_N t/\varepsilon} R_0\Omega, a^*_y a_x e^{-i\mathcal{H}_N t/\varepsilon} R_0\Omega\rangle\,, \tag{2.11}$$

while the kernel of the solution of the time-dependent Hartree-Fock equation (1.16) can be written as:

$$\omega_{N,t}(x;y) = \langle R_t\Omega, a^*_y a_x R_t\Omega\rangle \tag{2.12}$$

Intuitively, the two dynamics are "close" if the Fock space vector $R^*_t e^{-i\mathcal{H}_N t/\varepsilon} R_0\Omega \equiv \mathcal{U}_N(t;0)\Omega$ is "close" to the vacuum vector. The Fock-space unitary operator $\mathcal{U}_N(t;0) = R^*_t e^{-i\mathcal{H}_N t/\varepsilon} R_0$ is called the *fluctuation dynamics*: in describes the fluctuations between the initial many-body quantum evolution and the effective Hartree-Fock evolution. By using the properties of Bogoliubov transformations, we have:

$$\begin{aligned}\gamma^{(1)}_{N,t}(x;y) &= \langle \mathcal{U}_N(t;0)\Omega, R^*_t a^*_y a_x R_t \mathcal{U}_N(t;0)\Omega\rangle\\ &= \langle \mathcal{U}_N(t;0)\Omega, (a^*(u_y) + a(\overline{v}_y))(a(u_x) + a^*(\overline{v}_x))R_t\mathcal{U}_N(t;0)\Omega\rangle\\ &= \omega_{N,t}(x;y) + \langle \mathcal{U}_N(t;0)\Omega, \big[a^*(u_y)a(u_x) + a(\overline{v}_y)a(u_x)\\ &\qquad + a^*(u_y)a^*(\overline{v}_x) - a^*(\overline{v}_x)a(\overline{v}_y)\big]\mathcal{U}_N(t;0)\Omega\rangle\,,\end{aligned}$$

where we used the canonical anticommutation relations (2.3) together with the property $\overline{v}_{N,t}v_{N,t} = \omega_{N,t}$. Therefore, $\gamma^{(1)}_{N,t} = \omega_{N,t} + E_{N,t}$, for some explicit error term $E_{N,t}$. In order bound this error term, the following estimates are useful:

$$\begin{aligned}\Big\|\int dxdy\, O(x,y)a^*_x a_y\varphi\Big\| &\leq \|O\|_{\mathrm{op}}\|\mathcal{N}\varphi\|\,, \qquad \forall\varphi\in\mathcal{F}\\ \Big\|\int dxdy\, O(x;y)a^\sharp_x a_y\,\varphi\Big\| &\leq \min\big\{\|O\|_{\mathrm{HS}}\|\mathcal{N}^{1/2}\varphi\|, \|O\|_{\mathrm{tr}}\|\varphi\|\big\}\,,\end{aligned} \tag{2.13}$$

where: $a^\sharp_x = a^*_x, a_x$; $\|\cdot\|_{\mathrm{op}}$, $\|\cdot\|_{\mathrm{HS}}$, $\|\cdot\|_{\mathrm{tr}}$ are respectively the operator, Hilbert-Schmidt and trace norms; $\mathcal{N} = \int dx\, a^*_x a_x$ is the *number operator*. Using these bounds, one easily finds:

$$\|E_{N,t}\|_{\mathrm{HS}} \leq \langle \mathcal{U}_N(t;0)\Omega, (\mathcal{N}+1)\mathcal{U}_N(t;0)\Omega\rangle\,. \tag{2.14}$$

Similarly, one also finds:

$$\|E_{N,t}\|_{\mathrm{tr}} \leq N^{1/2}\langle \mathcal{U}_N(t;0)\Omega, (\mathcal{N}+1)\mathcal{U}_N(t;0)\Omega\rangle^{1/2}\,. \tag{2.15}$$

Hence, in order to control the distance between the two dynamics one has to control the expectation of the number operator over the fluctuation vector $\mathcal{U}_N(t;0)\Omega$. Notice that $\|\omega_{N,t}\|_{\mathrm{tr}} = N$, $\|\omega_{N,t}\|_{\mathrm{HS}} = N^{1/2}$; therefore, to prove convergence it is enough to show that:

$$\langle \mathcal{U}_N(t;0)\Omega, (\mathcal{N}+1)\mathcal{U}_N(t;0)\Omega\rangle \leq C(t) \qquad \forall t\in\mathbb{R}\,, \tag{2.16}$$

for some $C(t)$ independent of N.

2.2. Mixed states. Let us now discuss how to set up an analogous strategy for *mixed* states. Physically, mixed states describe fermionic system at positive temperature. Mathematically, a mixed state is represented by a *density matrix* $\rho : \mathcal{F} \to \mathcal{F}$:

$$\rho = \sum_j \lambda_j |\psi_j\rangle\langle\psi_j| \,, \qquad 0 \leq \lambda_j \leq 1 \,, \qquad \sum_j \lambda_j = 1 \,, \tag{2.17}$$

where $\{\psi_j\}$ is an orthonormal basis for $\mathcal{F}$. The conditions on $\{\lambda_j\}$ ensure that $\rho \geq 0$ and that $\mathrm{Tr}_{\mathcal{F}}\rho = 1$. A mixed state can be thought as an incoherent superposition of pure states, with λ_n the probability for the system to be in ψ_n.

Given a density matrix ρ as in Eq. (2.17), we define the operator $\widetilde{\kappa} : \mathcal{F} \to \mathcal{F}$ as:

$$\widetilde{\kappa} = \sum_j \varepsilon_j |\psi_j\rangle\langle\phi_j| \,, \tag{2.18}$$

where $\{\varepsilon_j\}$ is a sequence of complex number such that $\lambda_j = |\varepsilon_j|^2$ and $\{\phi_j\}$ is an orthonormal basis of $\mathcal{F}$. Clearly, $\widetilde{\kappa}\widetilde{\kappa}^* = \rho$. Now, we notice that the space of Hilbert-Schmidt operators on the Fock space, $\mathcal{L}^2(\mathcal{F})$, is isomorphic to the doubled Fock space $\mathcal{F} \otimes \mathcal{F}$, via the isomorphism $|\psi\rangle\langle\phi| \to \psi \otimes \overline{\phi}$, extended by linearity to the whole space $\mathcal{L}^2(\mathcal{F})$. Therefore, the operator $\widetilde{\kappa}$ is represented in $\mathcal{F} \otimes \mathcal{F}$ by the vector:

$$\kappa = \sum_j \varepsilon_j \psi_j \otimes \overline{\phi}_j \,. \tag{2.19}$$

It is well known that $\mathcal{F}(L^2(\mathbb{R}^3)) \otimes \mathcal{F}(L^2(\mathbb{R}^3))$ is isomorphic to $\mathcal{F}(L^2(\mathbb{R}^3) \oplus L^2(\mathbb{R}^3))$ (see [**10**] or any book in mathematical quantum field theory). The unitary operator U that implements this isomorphism is called the *exponential law*, and it is defined by the relations:

$$U(\Omega_{\mathcal{F}(L^2(\mathbb{R}^3))} \otimes \Omega_{\mathcal{F}(L^2(\mathbb{R}^3))}) = \Omega_{\mathcal{F}(L^2(\mathbb{R}^3)\oplus L^2(\mathbb{R}^3))}$$

$$U[a(f) \otimes 1]U^* = a(f \oplus 0) =: a_l(f) \,, \tag{2.20}$$

$$U[(-1)^{\mathcal{N}} \otimes a(f)]U^* = a(0 \oplus f) =: a_r(f) \,, \tag{2.21}$$

where $a_l(f)$ and $a_r(f)$ are the *left- or right- representations* of $a(f)$. Replacing f by δ_x, one obtains the operator valued distributions $a_{x,l}$, $a_{x,r}$. Finally, the time evolution in this new Fock space is generated by the *Liouvillian*:

$$\mathcal{L}_N = U(\mathcal{H}_N \otimes 1 - 1 \otimes \mathcal{H}_N)U^* \,. \tag{2.22}$$

In fact,

$$\Psi_t = U\kappa_t = Ue^{-i(\mathcal{H}_N\otimes 1 - 1\otimes\mathcal{H}_N)t/\varepsilon}\kappa \equiv e^{-i\mathcal{L}_N t/\varepsilon}\Psi \,. \tag{2.23}$$

The advantage of this formulation is that now mixed states can be represented as vectors on a "larger" space $\mathcal{F}(L^2(\mathbb{R}^3) \oplus L^2(\mathbb{R}^3))$. In particular, *mixed quasi free states* can be represented again via Bogoliubov transformations, acting on the vacuum vector $\Omega_{\mathcal{F}(L^2(\mathbb{R}^3)\oplus L^2(\mathbb{R}^3))}$. This is a particular example of a well-known construction in quantum statistical mechanics, called the *Araki-Wyss representation.* Let ω_N be the reduced one-particle density matrix of the quasi-free state, and let $u_N = 1 - \omega_N$, $v_N = \sqrt{\omega_N}$. The Bogoliubov transformation acts of the left- and right- fermionic operators as:

$$\begin{aligned} R^*_{\omega_N} a_{x,l} R_{\omega_N} &= a_l(u_x) - a^*_r(\overline{v}_x)\,, \\ R^*_{\omega_N} a_{x,r} R_{\omega_N} &= a_r(\overline{u}_x) + a^*_l(v_x)\,. \end{aligned} \tag{2.24}$$

The many-body evolution of a mixed quasi free state is represented by a vector $e^{-i\mathcal{L}_N t/\varepsilon} R_0 \Omega$, where we set $R_0 \equiv R_{\omega_N}$ and $\Omega \equiv \Omega_{\mathcal{F}(L^2(\mathbb{R}^3)\oplus L^2(\mathbb{R}^3))}$. Instead, the quasi-free state corresponding to the solution at time t of the time-dependent Hartree-Fock equation is represented by $R_t\Omega$, with $R_t \equiv R_{\omega_{N,t}}$. The kernels of the corresponding reduced densities can be written as:

$$\begin{aligned} \gamma^{(1)}_{N,t}(x;y) &= \langle e^{-i\mathcal{L}_N t/\varepsilon} R_0 \Omega, a^*_{x,l} a_{y,l} e^{-i\mathcal{L}_N t/\varepsilon} R_0 \Omega \rangle \\ \omega_{N,t}(x;y) &= \langle R_t \Omega, a^*_{x,l} a_{y,l} R_t \Omega \rangle\,. \end{aligned} \tag{2.25}$$

Proceeding in a way analogous to Eqs. (2.11)-(2.15), one finds out that in order to prove that the many-body evolution of the reduced one-particle density matrix converges to its Hartree-Fock counterpart, as $N \to \infty$, it is enough to show that:

$$\langle \mathcal{U}_N(t;0)\Omega, \mathcal{N}\mathcal{U}_N(t;0)\Omega \rangle \leq C(t) \qquad \forall t \in \mathbb{R}\,, \tag{2.26}$$

where now $\mathcal{N} = \sum_{\sigma=l,r} \int dx\, a^*_{x,\sigma} a_{x,\sigma}$, and $C(t)$ is bounded uniformly in N.

3. Results

In this section we will review the results obtained in [**6**, **8**, **26**]. Let us first review the result of [**8**], establishing convergence from the many-body Schrödinger dynamics to the Hartree-Fock dynamics, for pure states and bounded potentials.

THEOREM 3.1. *Let $V \in L^1(\mathbb{R}^3)$, such that $\int dp |\hat{V}(p)|(1+|p|^2) < \infty$. Let ω_N be a sequence of orthogonal projections such that* $\operatorname{Tr}\omega_N = N$ *and:*

$$\|[\omega_N, x]\|_{tr} \leq CN\varepsilon\,, \qquad \|[\omega_N, \varepsilon\nabla]\|_{tr} \leq CN\varepsilon\,. \tag{3.1}$$

Let $\psi_N \in L^2_a(\mathbb{R}^{3N})$ be such that its one-particle reduced density matrix $\gamma^{(1)}_N$ satisfies

$$\|\gamma^{(1)}_N - \omega_N\|_{tr} \leq CN^\alpha \tag{3.2}$$

for a constant $C > 0$ and an exponent $0 \leq \alpha < 1$. Let $\gamma_{N,t}$ be the reduced one-particle density matrix of $\psi_{N,t} = e^{-iH_N t/\varepsilon}\psi_N$ with Hamilton operator given by Eq. (1.14), and let $\omega_{N,t}$ be the solution of the time-dependent Hartree-Fock equation (1.16). Then, there exists $C, c > 0$ independent of N such that:

$$\begin{aligned} &\|\gamma^{(1)}_{N,t} - \omega_{N,t}\|_{HS} \leq CN^{\alpha/2} \exp c \exp c|t|\,, \\ &\|\gamma^{(1)} - \omega_{N,t}\|_{tr} \leq CN^{(\alpha+1)/2} \exp c \exp c|t| \qquad \forall t \in \mathbb{R}\,. \end{aligned} \tag{3.3}$$

REMARK 3.2. (i) The estimates in Eq. (3.3) imply convergence of the two dynamics, since $\|\omega_{N,t}\|_{\mathrm{tr}} = N$, $\|\omega_{N,t}\|_{\mathrm{HS}} = N^{1/2}$.

(ii) The assumptions in Eq. (3.1) encode the *semiclassical structure* of the initial data. They are expected to hold for the minimizers of the Hartree-Fock energy functional. It is easy to see that these bounds hold true for the semiclassical approximation of the minimizers, given by coherent states:

$$\omega^{\mathrm{c}}_N := \int_{p^2+\phi_{TF}(q)\leq\mu} |f_{pq}\rangle\langle f_{pq}|\,, \qquad f_{pq}(x) = e^{ipx/\varepsilon} g(x-q)\,, \tag{3.4}$$

with $g \in L^2(\mathbb{R}^3)$, $\phi_{\rm TF}(q) = V_{\rm ext}(q) + \rho_{\rm TF} * V$, where $\rho_{\rm TF}$ is the minimizer of the Thomas-Fermi energy functional. Roughly speaking, the commutator estimates (3.1) detect a separation of scales in $\omega_N(x;y)$, of the form $\varphi((x-y)/\varepsilon)\rho((x+y)/2)$ for some functions φ ("momentum distribution") and ρ ("density profile"). In [**8**] we prove that the bounds (3.1) are propagated along the flow of the time-dependent Hartree-Fock equation.

(iii) In [**8**] we actually prove convergence for all the k-particle reduced density matrices of $\psi_{N,t}$ to their Hartree-Fock counterparts, obtained from $\omega_{N,t}$ via the fermionic Wick rule.

(iv) It is well known that, for the class of bounded potentials we consider in this result, the exchange term is subleading with respect to the other terms in the Hartree-Fock equation. Our estimates do not deteriorate if we replace the Hartree-Fock dynamics with the Hartree dynamics.

(v) The theorem can be extended to the case of pseudorelativistic fermions. Consider a many-body Hamiltonian given by:

$$H_N = \sum_{j=1}^N \sqrt{-\varepsilon^2\Delta_j + m^2} + \frac{1}{N}\sum_{i<j} V(x_i - x_j)\,. \tag{3.5}$$

Under the same assumptions of Theorem 3.1, in [**9**] we prove the convergence of the many-body quantum dynamics to the pseudorelativistic Hartree-Fock equation:

$$i\varepsilon\partial_t \omega_{N,t} = \left[\sqrt{-\varepsilon^2\Delta + m^2} + \rho_t * V - X_t,\, \omega_{N,t}\right]\,. \tag{3.6}$$

The previous result describes the dynamics of zero temperature systems, corresponding to pure states. Physically, there is no reason to expect the validity of the time-dependent Hartree-Fock equation to depend on the temperature of the system. The next result proves that this intuition is indeed correct: we prove the convergence of the many-body quantum dynamics of mixed states to the Hartree-Fock dynamics. The initial datum is chosen to be close to a mixed quasi-free state, which is expected to describe the thermal state of an interacting fermionic system, in the mean-field regime. As for the previous theorem, the interaction potential is assumed to be bounded. The theorem has been proven in [**6**].

THEOREM 3.3. *Let $V \in L^1(\mathbb{R}^3)$, such that $\int dp\, |\hat V(p)|(1+p^2) < \infty$. Let ω_N be a sequence of operators on $L^2(\mathbb{R}^3)$ such that $0 \le \omega_N \le 1$, $\operatorname{Tr}\omega_N = N$ and*

$$\begin{aligned} &\|[\sqrt{\omega_N}, x]\|_{HS}^2 \le CN\varepsilon^2\,, \qquad &&\|[\sqrt{1-\omega_N}, x]\|_{HS}^2 \le CN\varepsilon^2 \\ &\|[\sqrt{\omega_N}, \varepsilon\nabla]\|_{HS}^2 \le CN\varepsilon^2\,, \qquad &&\|[\sqrt{1-\omega_N}, \varepsilon\nabla]\|_{HS}^2 \le CN\varepsilon^2\,. \end{aligned} \tag{3.7}$$

Let R_{ω_N} be the corresponding Bogoliubov transformation, acting on $\mathcal{F}(L^2(\mathbb{R}^3) \oplus L^2(\mathbb{R}^3))$. Let $\xi_N \in \mathcal{F}(L^2(\mathbb{R}^3) \oplus L^2(\mathbb{R}^3))$, such that $\langle \xi_N, \mathcal{N}^{10}\xi_N\rangle \le C$ uniformly in N, and such that $\xi_N = \chi(\mathcal{N} \le CN)\xi_N$, with $\chi(\cdot)$ the characteristic function. Let $\gamma^{(1)}_{N,t}$ be the operator with kernel:

$$\gamma^{(1)}_{N,t}(x;y) = \langle e^{-i\mathcal{L}_N t/\varepsilon} R_{\omega_N}\xi_N,\, a^*_{y,l}a_{x,l}\, e^{-i\mathcal{L}_N t/\varepsilon} R_{\omega_N}\xi_N\rangle\,, \tag{3.8}$$

with Liouvillian $\mathcal{L}_N$ defined in Eq. (2.22), and let $\omega_{N,t}$ be the solution of the time-dependent Hartree-Fock equation (1.16). Then, there exists constants $C, c > 0$

such that:

$$
\begin{aligned}
\|\gamma^{(1)}_{N,t} - \omega_{N,t}\|_{tr} &\leq CN^{1/2}\exp(c\exp c|t|)\,, \\
\|\gamma^{(1)}_{N,t} - \omega_{N,t}\|_{HS} &\leq C\exp(c\exp c|t|)\,, \qquad \forall t \in \mathbb{R}\,.
\end{aligned} \tag{3.9}
$$

REMARK 3.4.
(i) As in the statement of Theorem 3.1, the commutator estimates in Eq. (3.7) encode the semiclassical structure of the initial datum. Notice that the analogous estimates for pure states would be proportional to ε and not ε^2; the presence of the ε^2 factor is related to the smooth profile of the momentum distribution, compatible with a temperature $T = O(1)$.
(ii) The conditions on the vector ξ_N are needed in order to guarantee that the initial datum is close enough to a quasi-free state (the state would be exactly quasi-free if $\xi_N = \Omega$, for which all conditions are trivially true). The assumption on the expectation of the number operator can be relaxed to $\langle \xi_N, \mathcal{N}^3 \xi_N\rangle \leq C$, at the price of a worse rate of convergenge. We did not try to optimize on this conditions.
(iii) As discussed in [**6**], the result can be extended to prove convergence of all k-particle density matrices to their Hartree-Fock counterparts.

Theorems 3.1, 3.3 allow to study the dynamics of interacting systems with bounded potentials. However, if one is interested in the application to the dynamics of atoms and molecules, it is of great importance to be able to understand what happens if V is the Coulomb potential, namely $V(x-y) = 1/|x-y|$. This allows to describe, for instance, the electrostatic interactions between the electrons of large atoms. In this setting, the analogous problem of proving the validity of Hartree-Fock theory for the ground state of large Coulomb systems has been solved in [**3, 16**].

The next theorem provides a partial understanding of the dynamics of these important physical systems. It is a conditional result, holding under suitable assumptions on the solution of the time-dependent Hartree-Fock equation. The assumption holds for a special class of initial data, namely translation invariant systems (that is, electrons in $\mathbb{T}^3$). Nevertheless, it allows to reduce a statistical mechanics problem for an interacting many-body quantum system to a problem in nonlinear PDEs. The theorem has been proved in [**26**].

THEOREM 3.5. *Let $V(x-y) = 1/|x-y|$. Let ω_N be a sequence of orthogonal projections on $L^2(\mathbb{R}^3)$, with $\mathrm{Tr}\,\omega_N = N$ and such that $\mathrm{Tr}\,(-\varepsilon^2\Delta)\,\omega_N \leq CN$, for a constant $C > 0$. Let $\omega_{N,t}$ denote the solution of the Hartree-Fock equation (1.16) with initial data $\omega_{N,0} = \omega_N$. We assume that there exists a time $T > 0$, a $p > 5$ and a constant $C > 0$ such that*

$$
\sup_{t\in[0;T]} \left[\|\rho_{|[x,\omega_{N,t}]|}\|_1 + \|\rho_{|[x,\omega_{N,t}]|}\|_p\right] \leq CN\varepsilon\,. \tag{3.10}
$$

Let $\psi_N \in L^2_a(\mathbb{R}^{3N})$ be such that its one-particle reduced density matrix $\gamma^{(1)}_N$ satisfies

$$
\|\gamma^{(1)}_N - \omega_N\|_{tr} \leq CN^\alpha \tag{3.11}
$$

for a constant $C > 0$ and an exponent $0 \leq \alpha < 1$. Consider the evolution $\psi_{N,t} = e^{-iH_N t/\varepsilon}\psi_N$, and let $\gamma^{(1)}_{N,t}$ be the corresponding one-particle reduced density.

Then for every $\delta > 0$ there exists $C > 0$ such that

$$\sup_{t\in[0;T]} \|\gamma^{(1)}_{N,t} - \omega_{N,t}\|_{HS} \leq C\left[N^{\alpha/2} + N^{5/12+\delta}\right] \tag{3.12}$$

and

$$\sup_{t\in[0;T]} \|\gamma^{(1)}_{N,t} - \omega_{N,t}\|_{tr} \leq C\left[N^{\alpha} + N^{11/12+\delta}\right]. \tag{3.13}$$

REMARK 3.6.

(i) The bounds on the L^p norms of the function $\rho_{|[x,\omega_N]|}$ encode the semiclassical structure of the initial datum. With respect to the previous results, the assumption on $\|\rho_{|[x,\omega_{N,t}]|}\|_p$ for $p > 5$ is needed in order to control the singularity of the Coulomb potential. With respect to Theorems 3.1, 3.3, here we are assuming the validity of the semiclassical structure for a time interval $[0;T]$, instead of just assuming it at $t = 0$. It is an interesting open problem to replace the assumption Eq. (3.10) with an assumption on the initial data, and to prove that it is propagated in time along the Hartree-Fock flow. There is however one choice of intial datum for which this assumption is trivially true for all times, describing N free fermions in a periodic box. Consider the ground state of $\sum_{i=1}^{N} -\Delta_i$ on $L^2(\mathbb{T}^{3N})$,

$$\psi_N(x_1,\ldots,x_N) = \frac{1}{\sqrt{N!}}\sum_{\pi} \text{sgn}(\pi) f_{p_{\pi(1)}}(x_1)\cdots f_{p_{\pi(N)}}(x_N)$$

$$f_p(x) = e^{ip\cdot x} \quad \text{for } p \in (2\pi)\mathbb{Z}^3\,, \qquad p_i \neq p_j \iff i \neq j\,, \qquad |p_i| \leq cN^{1/3}\,, \tag{3.14}$$

for some $c > 0$ ensuring the normalization of the state. The corresponding reduced density matrix can be easily computed:

$$\omega_N(x;y) = \sum_{|p|\leq cN^{1/3}} e^{ip\cdot(x-y)} \simeq N\int_{|p|\leq c} dp\, e^{ip\cdot(x-y)/\varepsilon}\,. \tag{3.15}$$

Notice that in this case $\rho_{|[x,\omega_N]|}(x)$ is a constant, of order $N\varepsilon$. Moreover, since $\omega_N(x;y) \equiv \omega_N(x-y)$, the corresponding Hartree-Fock evolution is trivial: ω_N is a steady state of Eq. (1.16). Therefore, the assumption $\|\rho_{|[x,\omega_N]|}\|_p \leq CN\varepsilon$ propagates trivially on the flow of the time-dependent Hartree-Fock equation. Hence, Theorem 3.5 proves that the many-body evolution generated by the Hamiltonian H_N with Coulomb interactions leaves the Slater determinant in Eq. (3.14) approximately invariant, for all times $t = O(1)$.

(ii) The theorem remains true if one modifies the Hamiltonian H_N by adding an external potential, for instance describing the Coulomb attraction between the electrons and a static nucleus. However, the semiclassical assumption (3.10) might be more difficult to justify, in particular if the external potential is singular. Similarly, we stress that the theorem still holds if one replaces the interaction potential by the attractive Coulomb interaction $V(x-y) = -1/|x-y|$. Alse here, however, the assumption (3.10) might be more difficult to justify.

4. Proofs

Here we will discuss some aspects of the proofs of the results presented in Section 3. We shall focus on the proof of Theorem 3.5.

As pointed out already in Section 2, the proof of all these convergence results is based on a control of the growth of the average number of particles on the fluctuation vector $\mathcal{U}_N(t;0)\xi_N$, with $\xi_N \in \mathcal{F}$ such that $\langle\xi_N, \mathcal{N}\xi_N\rangle \leq C$. The goal is to get a bound like (2.16), for instance by using a Gronwall-type strategy. The starting point of our analysis is the following identity:

$$
\begin{aligned}
&i\varepsilon\frac{d}{dt}\left\langle \mathcal{U}_N(t)\xi_N, \mathcal{N}\,\mathcal{U}_N(t)\xi_N\right\rangle \\
&= \frac{4i}{N}\mathrm{Im}\int dx\,dy\,\frac{1}{|x-y|} \\
&\qquad\times\Big\{\left\langle \mathcal{U}_N(t;0)\xi_N, a^*(u_{t,x})\,a(\bar{v}_{t,y})\,a(u_{t,y})\,a(u_{t,x})\,\mathcal{U}_N(t;0)\xi_N\right\rangle \\
&\qquad+\left\langle \mathcal{U}_N(t;0)\xi_N, a^*(u_{t,y})\,a^*(\bar{v}_{t,y})\,a^*(\bar{v}_{t,x})\,a(\bar{v}_{t,x})\,\mathcal{U}_N(t;0)\xi_N\right\rangle \\
&\qquad+\left\langle \mathcal{U}_N(t;0)\xi_N, a(\bar{v}_{t,x})\,a(\bar{v}_{t,y})\,a(u_{t,y})\,a(u_{t,x})\,\mathcal{U}_N(t;0)\xi_N\right\rangle\Big\},
\end{aligned}
\tag{4.1}
$$

where, as in Section 2, we used the short hand notation $u_{t,x}(z) \equiv u_{N,t}(x;z)$, $v_{t,x}(z) \equiv v_{N,t}(x;z)$, with the operators $u_{N,t} = 1-\omega_{N,t}$ and $v_{N,t}$ such that $\bar{v}_{N,t}v_{N,t} = \omega_{N,t}$ (see discussion after Eq. (2.7)). The proof of (4.1) is a lengthy but straightforward computation that can be found in [**8**, Proof of Proposition 3.3].

The three terms in the right-hand side of Eq. (4.1) describe the fluctuation of the many-body quantum dynamics around the time-dependent Hartree-Fock dynamics. Let us show how to bound the largest of these terms, that turns out to be given by the third term appearing in the right-hand side of Eq. (4.1):

$$
\mathrm{III} = \frac{1}{N}\int dx\,dy\,\frac{1}{|x-y|}\left\langle \mathcal{U}_N(t;0)\xi_N, a(\bar{v}_{t,x})\,a(\bar{v}_{t,y})\,a(u_{t,y})\,a(u_{t,x})\,\mathcal{U}_N(t;0)\xi_N\right\rangle - \text{c.c.}
\tag{4.2}
$$

It is not difficult to prove that this term can be estimated proportionally to the quantity $\langle \mathcal{U}_N(t)\xi_N, \mathcal{N}\,\mathcal{U}_N(t)\xi_N\rangle$. To see this, recall that:

$$
\|a(\bar{v}_x)\| = \|\bar{v}_x\|_2 \equiv \omega_{N,t}(x;x)^{1/2}\,.
\tag{4.3}
$$

Using Eq. (4.3) one easily gets, inserting $1 = (\mathcal{N}+3)^{1/2}(\mathcal{N}+3)^{-1/2}$ right before $a(\bar{v}_{t,x})$ in Eq. (4.2):

$$
\begin{aligned}
&\frac{1}{N}\left|\int dx\,dy\,\frac{1}{|x-y|}\left\langle \mathcal{U}_N(t;0)\xi_N, a(\bar{v}_{t,x})\,a(\bar{v}_{t,y})\,a(u_{t,y})\,a(u_{t,x})\,\mathcal{U}_N(t;0)\xi_N\right\rangle\right| \\
&\le \frac{1}{N}\int dx\,dy\,\frac{1}{|x-y|}\omega_{N,t}(x;x)^{1/2}\omega_{N,t}(y;y)^{1/2} \\
&\qquad\times\left\|(\mathcal{N}+3)^{1/2}\mathcal{U}_N(t;0)\xi_N\right\|\left\|a(u_{t,y})\,a(u_{t,x})\,(\mathcal{N}+1)^{-1/2}\mathcal{U}_N(t;0)\xi_N\right\| \\
&\le \frac{C}{N}\left[\int dx\,dy\,\frac{1}{|x-y|^2}\omega_{N,t}(x;x)\omega_{N,t}(y;y)\right]^{1/2}\left\|(\mathcal{N}+1)^{1/2}\mathcal{U}_N(t;0)\xi_N\right\|^2,
\end{aligned}
\tag{4.4}
$$

where the second step follows from the bound (4.3) and from the Cauchy-Schwarz inequality, while the last follows from the Cauchy-Schwarz inequality and from the estimate:

$$
\int dxdy\,\|a(u_x)a(u_y)\varphi\|^2 \le \|\mathcal{N}\varphi\|^2 \qquad \forall\varphi\in\mathcal{F}\,.
\tag{4.5}
$$

By the Hardy-Littlewood-Sobolev inequality, the term in the square brackets in Eq. (4.4) can be estimated as:

$$
\int dx\,dy\,\frac{1}{|x-y|^2}\omega_{N,t}(x;x)\omega_{N,t}(y;y) \le C\|\rho_{N,t}\|_{3/2}^2\,,
\tag{4.6}
$$

where $\rho_{N,t}(x) = \omega_{N,t}(x;x)$. By interpolation, $\|\rho_{N,t}\|_{3/2} \leq \|\rho_{N,t}\|_1^{\lambda}\|\rho_{N,t}\|_{5/3}^{1-\lambda}$, with $\lambda = 1/10$. To bound the norm $\|\rho_{N,t}\|_{5/3}$ we use the Lieb-Thirring inequality, which implies

$$\|\rho_{N,t}\|_{5/3}^{5/3} \leq \mathrm{tr}\,(-\Delta)\omega_{N,t} \leq \varepsilon^{-2}\mathcal{E}_{\mathrm{HF}}(\omega_{N,t})\,, \tag{4.7}$$

with the Hartree-Fock energy

$$\begin{aligned}\mathcal{E}_{\mathrm{HF}}(\omega_{N,t}) &= \mathrm{tr}\,(-\varepsilon^2\Delta)\omega_{N,t}\\ &\quad + \frac{1}{2N}\int \frac{1}{|x-y|}\left[\omega_{N,t}(x;x)\omega_{N,t}(y;y) - |\omega_{N,t}(x;y)|^2\right] dxdy\,.\end{aligned}$$

By energy conservation, we have

$$\|\rho_{N,t}\|_{5/3}^{5/3} \leq \varepsilon^{-2}\mathcal{E}_{\mathrm{HF}}(\omega_N)\,. \tag{4.8}$$

Next, we remark that the potential part of $\mathcal{E}_{\mathrm{HF}}(\omega_N)$ can be bounded by its kinetic energy. In fact, applying the Hardy-Littlewood-Sobolev inequality and interpolation and using that $\|\rho_N\|_1 = N$, we find

$$\begin{aligned}\frac{1}{N}\int \frac{1}{|x-y|}\rho_N(x)\rho_N(y)dxdy &\leq \frac{C}{N}\|\rho_N\|_{6/5}^2\\ &\leq \frac{C}{N}\|\rho_N\|_1^{7/5}\|\rho_N\|_{5/3}^{3/5}\\ &= CN^{2/5}\|\rho_N\|_{5/3}^{3/5}\\ &\leq CN + CN^{-2/3}\|\rho_N\|_{5/3}^{5/3}\,,\end{aligned}$$

by Young's inequality. From the Lieb-Thirring inequality, we find

$$\frac{1}{N}\int \frac{1}{|x-y|}\rho_N(x)\rho_N(y)dxdy \leq CN + C\mathrm{tr}\,(-\varepsilon^2\Delta)\omega_N$$

and hence

$$\mathcal{E}_{\mathrm{HF}}(\omega_N) \leq CN + C\mathrm{tr}\,(-\varepsilon^2\Delta)\omega_N \leq CN \tag{4.9}$$

from the assumption $\mathrm{tr}\,(-\varepsilon^2\Delta)\omega_N \leq CN$ on the initial sequence of orthogonal projection ω_N. From (4.8), we conclude that $\|\rho_{N,t}\|_{5/3} \leq N$. Therefore, since $\|\rho_{N,t}\|_1 = N$, we can bound $\|\rho_{N,t}\|_{3/2}^2 \leq CN^2$. Plugging this bound into Eq. (4.6), Eq. (4.4) implies:

$$|\mathrm{III}| \leq C\langle \mathcal{U}_N(t;0)\xi_N, \mathcal{N}\mathcal{U}_N(t;0)\xi_N\rangle\,. \tag{4.10}$$

This, however, *does not* allow to follows the dynamics for times $t = O(1)$, since the time derivative in the left-hand side of Eq. (4.1) is multiplied by $\varepsilon = N^{-1/3}$. The bound we just found is compatible with a control of the dynamics for short times $t = O(\varepsilon)$. In order to obtain a better bound, which can be used to prove convergence of the many-body dynamics to the time-dependent Hartree-Fock dynamics on a physically relevant time scale, one has to use some extra information on the initial datum, consistent with the fact that the initial datum is close to the ground state of a mean-field Hamiltonian. This information turns out to be encoded in the bounds for the L^p norms of $\rho_{|[x,\omega_{N,t}]|}(x) = |[\omega_{N,t}, x]|(x;x)$ in Eq. (3.10).

The key remark, that allows to use these bounds to get a better estimate for the growth of the number of particles, is that the right-hand side of Eq. (4.1) would be exactly zero if the potential was constant, thanks to the orthogonality of $u_{N,t}$

and $v_{N,t}$. Of course, this is not the case here (the potential is actually unbounded); nevertheless, this information together with the semiclassical bounds Eq. (3.10) will allow to control the dynamics up to times of order 1.

It is convenient to rewrite the Coulomb potential using a smoothed version of the *Fefferman-de la Llave decomposition* [**12**]:

$$\frac{1}{|x-y|} = \frac{4}{\pi^2}\int_0^\infty dr\,\frac{1}{r^5}\int dz\,\chi_{(r,z)}(x)\chi_{(r,z)}(y)\,, \tag{4.11}$$

with $\chi_{(r,z)}(x) = \exp(-|x-z|^2/r^2)$. Let us plug the expression (4.11) into Eq. (4.2). We get:

$$\begin{aligned} \mathrm{III} = \frac{C}{N}\int dx\,dy\int_0^\infty \frac{dr}{r^5}\int dz\,\chi_{(r,z)}(x)\chi_{(r,z)}(y) \\ \times \langle \mathcal{U}_N(t;0)\xi_N, a(\bar{v}_{t,x})\,a(\bar{v}_{t,y})\,a(u_{t,y})\,a(u_{t,x})\mathcal{U}_N(t;0)\xi_N\rangle - \text{c.c.} \end{aligned}$$

We divide the r-intergral into two parts, setting $\mathrm{III} = \mathrm{III}_1 + \mathrm{III}_2 - \text{c.c.}$, with

$$\begin{aligned} \mathrm{III}_1 &= \frac{C}{N}\int \mathrm{d}x\,\mathrm{d}y\int_0^\kappa \frac{\mathrm{d}r}{r^5}\int \mathrm{d}z\;\chi_{(r,z)}(x)\chi_{(r,z)}(y) \\ &\quad\times \langle \mathcal{U}_N(t;0)\,\xi_N, a(\bar{v}_{t,x})\,a(\bar{v}_{t,y})\,a(u_{t,y})\,a(u_{t,x})\mathcal{U}_N(t;0)\,\xi_N\rangle\,, \\ \mathrm{III}_2 &= \frac{C}{N}\int \mathrm{d}x\,\mathrm{d}y\int_\kappa^\infty \frac{\mathrm{d}r}{r^5}\int \mathrm{d}z\;\chi_{(r,z)}(x)\chi_{(r,z)}(y) \\ &\quad\times \langle \mathcal{U}_N(t;0)\,\xi_N, a(\bar{v}_{t,x})\,a(\bar{v}_{t,y})\,a(u_{t,y})\,a(u_{t,x})\mathcal{U}_N(t;0)\,\xi_N\rangle\,. \end{aligned} \tag{4.12}$$

The parameter κ will be chosen later, in an optimal way. We start estimating III_1. Here, we start by integrating over z. Since

$$\int dz\,\chi_{(r,z)}(x)\chi_{(r,z)}(y) = r^3\chi_{(\sqrt{2}r,x)}(y)$$

we obtain:

$$\begin{aligned} \mathrm{III}_1 &= \frac{C}{N}\int_0^\kappa \frac{dr}{r^2}\int dxdy\,\chi_{(\sqrt{2}r,x)}(y) \\ &\quad\times \langle \mathcal{U}_N(t;0)\,\xi_N, a(\bar{v}_{t,x})\,a(\bar{v}_{t,y})\,a(u_{t,y})\,a(u_{t,x})\mathcal{U}_N(t;0)\,\xi_N\rangle \\ &\equiv \frac{C}{N}\int_0^\kappa \frac{dr}{r^2}\int dx\,\langle \mathcal{U}_N(t;0)\,\xi_N, B_{\sqrt{2}r,x}a(\bar{v}_{t,x})a(u_{t,x})\mathcal{U}_N(t;0)\,\xi_N\rangle\,, \end{aligned}$$

where we defined the Fock-space operator:

$$B_{r,z} = \int \mathrm{d}y\;a(\bar{v}_{t,y})\,\chi_{(r,z)}(y)a(u_{t,y}) = \int ds_1ds_2(\bar{v}_{N,t}\chi_{(r,z)}u_{N,t})(s_1;s_2)a_{s_1}a_{s_2}\,. \tag{4.13}$$

Since $\|\bar{v}_{N,x}\|^2 = \omega_{N,t}(x;x) =: \rho_{N,t}(x)$, we find

$$\begin{aligned} |\mathrm{III}_1| &\le \frac{C}{N}\int_0^\kappa \frac{dr}{r^2}\int dx\,\|B_{\sqrt{2}r,x}\|\,\rho_{N,t}^{1/2}(x)\,\|a(u_{t,x})\mathcal{U}_N(t;0)\,\xi_N\| \\ &\le \frac{C}{N}\int_0^\kappa \frac{dr}{r^2}\int dx\,\rho_{N,t}^{1/2}(x)\|[\chi_{(\sqrt{2}r,x)},\omega_{N,t}]\|_{\mathrm{tr}}\|a(u_{t,x})\mathcal{U}_N(t;0)\,\xi_N\|\,, \end{aligned} \tag{4.14}$$

where we used the trace norm bound in Eq. (2.13). Roughly speaking, we would like to use the smallness of the support of $\chi_{(r,z)}(x)$ for small r to control the r^{-5} singularity, and the semiclassical structure (3.10) of the initial datum in order to follow the dynamics up to times of order 1. To do so, the following lemma [**26**] will play a crucial role.

LEMMA 4.1. *Let $\chi_{r,z}(x) = \exp(-(x-z)^2/r^2)$. Then, for all $0 < \delta < 1/2$ there exists $C > 0$ such that the pointwise bound*

$$\left\| [\chi_{(r,z)}, \omega_{N,t}] \right\|_{tr} \leq C\, r^{\frac{3}{2}-3\delta} \sum_{i=1}^{3} \| \rho_{|[x_i,\omega_{N,t}]|} \|_1^{\frac{1}{6}+\delta} \left(\rho^*_{|[x_i,\omega_{N,t}]|}(z) \right)^{\frac{5}{6}-\delta} \tag{4.15}$$

holds true. Here $\varrho^*_{|[x_i,\omega_{N,t}]|}$ denotes the Hardy-Littlewood maximal function defined by

$$\rho^*_{|[x_i,\omega_{N,t}]|}(z) = \sup_{B: z\in B} \frac{1}{|B|} \int_B dx\, \rho_{|[x_i,\omega_{N,t}]|}(x) \tag{4.16}$$

with the supremum taken over all balls $B \in \mathbb{R}^3$ such that $z \in B$.

Using the pointwise bound (4.15) and the assumption (3.10), we obtain that, for all $\delta > 0$ sufficiently small, there exists a constant $C > 0$ such that

$$\begin{aligned} |\mathrm{III}_1| \leq \frac{C(N\varepsilon)^{\frac{1}{6}+\delta}}{N} \sum_{i=1}^{3} \int_0^{\kappa} \frac{dr}{r^{1/2+3\delta}} \int dx\, \rho_{N,t}^{1/2}(x) \left[\rho^*_{|[x_i,\omega_{N,t}]|}(x) \right]^{\frac{5}{6}-\delta} \\ \times \| a(u_{t,x}) \mathcal{U}_N(t;0)\, \xi_N \| . \end{aligned} \tag{4.17}$$

Applying Hölder's inequality, we conclude that

$$\begin{aligned} |\mathrm{III}_1| \leq \frac{C(N\varepsilon)^{\frac{1}{6}+\delta} \kappa^{1/2-3\delta}}{N} \sum_{i=1}^{3} \| \rho_{N,t} \|_{5/3}^{1/2} \| \rho^*_{|[x_i,\omega_{N,t}]|} \|_{\frac{25}{6}-5\delta}^{\frac{5}{6}-\delta} \\ \times \left[\int dx\, \| a(u_{t,x}) \mathcal{U}_N(t;0)\, \xi_N \|^2 \right]^{1/2} . \end{aligned} \tag{4.18}$$

By the Hardy-Littlewood maximal inequality and the assumption (3.10), we have

$$\| \rho^*_{|[x_i,\omega_{N,t}]|} \|_{\frac{25}{6}-5\delta} \leq \| \rho_{|[x_i,\omega_{N,t}]|} \|_{\frac{25}{6}-5\delta} \leq CN\varepsilon . \tag{4.19}$$

Furthermore, we have

$$\begin{aligned} \int dx\, \| a(u_{t,x}) \mathcal{U}_N(t;0)\, \xi_N \|^2 &= \langle \mathcal{U}_N(t;0)\, \xi_N, d\Gamma(u_{N,t}) \mathcal{U}_N(t;0)\, \xi_N \rangle \\ &\leq \langle \mathcal{U}_N(t;0)\, \xi_N, \mathcal{N} \mathcal{U}_N(t;0)\, \xi_N \rangle \\ &= \| \mathcal{N}^{1/2} \mathcal{U}_N(t;0)\, \xi_N \|^2 . \end{aligned} \tag{4.20}$$

Using that $\|\rho_{N,t}\|_{5/3} \leq CN$ (recall Eqs. (4.7)-(4.9)) and combining this estimate with (4.19) and (4.20), we obtain

$$|\mathrm{III}_1| \leq C\sqrt{N}\varepsilon\kappa^{1/2-3\delta} \| \mathcal{N}^{1/2} \mathcal{U}_N(t;0)\, \xi_N \| \leq \varepsilon \| \mathcal{N}^{1/2} \mathcal{U}_N(t;0)\, \xi_N \|^2 + CN\varepsilon\kappa^{1-6\delta}$$

for all $t \in [0;T]$.

Next, we estimate the second term in (4.12). With the definition (4.13), we have

$$|\mathrm{III}_2| \leq \frac{C}{N} \int_\kappa^\infty \frac{\mathrm{d}r}{r^5} \int \mathrm{d}z\, \|B_{r,z}\|^2 \leq \frac{C}{N} \int_\kappa^\infty \frac{\mathrm{d}r}{r^5} \int \mathrm{d}z\, \| [\chi_{(r,z)}, \omega_{N,t}] \|_{\mathrm{tr}}^2 .$$

With the bound (4.15) and the assumption (3.10), we obtain

$$|\mathrm{III}_2| \leq \frac{C(N\varepsilon)^2}{N} \int_\kappa^\infty \frac{\mathrm{d}r}{r^{2+6\delta}} \leq CN\varepsilon^2\kappa^{-1-6\delta} .$$

Hence,

$$|\mathrm{III}| \leq \varepsilon \| \mathcal{N}^{1/2} \mathcal{U}_N(t;0)\, \xi_N \|^2 + CN\varepsilon\kappa^{1-6\delta} + CN\varepsilon^2\kappa^{-1-6\delta} .$$

Minimizing over κ we find $\kappa = \varepsilon^{1/2}$ and we conclude

$$|\mathrm{III}| \leq \varepsilon \|\mathcal{N}^{1/2} \mathcal{U}_N(t;0)\, \xi_N\|^2 + CN\varepsilon^{3/2-6\delta}.$$

This concludes the bound of the largest term in the right-hand side of Eq. (4.1). It turns out, see [**26**], that all the other terms in the right-hand side Eq. (4.1) can be bounded by $C\varepsilon\langle \mathcal{U}_N(t;0)\,\xi_N, \mathcal{N}\mathcal{U}_N(t;0)\,\xi_N\rangle$, for some constant $C > 0$ independent of N. Therefore, we obtain from (4.1) that, for every $\delta > 0$ small enough, there is a constant $C > 0$ independent of N such that

$$\left|\frac{d}{dt}\langle \mathcal{U}_N(t;0)\,\xi_N, \mathcal{N}\mathcal{U}_N(t;0)\,\xi_N\rangle\right| \leq C\langle \mathcal{U}_N(t;0)\,\xi_N, \mathcal{N}\mathcal{U}_N(t;0)\,\xi_N\rangle + CN\varepsilon^{1/2-\delta}$$

for all $t \in [0;T]$. Therefore, Gronwall's lemma implies that there exists a constant $C > 0$ such that

$$\sup_{t\in[0;T]} \langle \mathcal{U}_N(t;0)\,\xi_N, \mathcal{N}\mathcal{U}_N(t;0)\,\xi_N\rangle \leq C\left[\langle \xi_N, \mathcal{N}\xi_N\rangle + N\varepsilon^{1/2-\delta}\right].$$

Thanks to Eqs. (2.14), (2.15), this bound allows to prove the estimates (3.12), (3.13). This concludes the sketch of the proof of Theorem 3.5.

References

[1] Laurent Amour, Mohamed Khodja, and Jean Nourrigat, *The semiclassical limit of the time dependent Hartree-Fock equation: the Weyl symbol of the solution*, Anal. PDE **6** (2013), no. 7, 1649–1674, DOI 10.2140/apde.2013.6.1649. MR3148063

[2] Agissilaos Athanassoulis, Thierry Paul, Federica Pezzotti, and Mario Pulvirenti, *Strong semiclassical approximation of Wigner functions for the Hartree dynamics*, Atti Accad. Naz. Lincei Rend. Lincei Mat. Appl. **22** (2011), no. 4, 525–552. MR2904998

[3] Volker Bach, *Error bound for the Hartree-Fock energy of atoms and molecules*, Comm. Math. Phys. **147** (1992), no. 3, 527–548. MR1175492

[4] Volker Bach, Sébastien Breteaux, Sören Petrat, Peter Pickl, and Tim Tzaneteas, *Kinetic energy estimates for the accuracy of the time-dependent Hartree-Fock approximation with Coulomb interaction* (English, with English and French summaries), J. Math. Pures Appl. (9) **105** (2016), no. 1, 1–30, DOI 10.1016/j.matpur.2015.09.003. MR3427937

[5] Claude Bardos, François Golse, Alex D. Gottlieb, and Norbert J. Mauser, *Mean field dynamics of fermions and the time-dependent Hartree-Fock equation* (English, with English and French summaries), J. Math. Pures Appl. (9) **82** (2003), no. 6, 665–683, DOI 10.1016/S0021-7824(03)00023-0. MR1996777

[6] Niels Benedikter, Vojkan Jakšić, Marcello Porta, Chiara Saffirio, and Benjamin Schlein, *Mean-field evolution of fermionic mixed states*, Comm. Pure Appl. Math. **69** (2016), no. 12, 2250–2303, DOI 10.1002/cpa.21598. MR3570479

[7] Niels Benedikter, Marcello Porta, Chiara Saffirio, and Benjamin Schlein, *From the Hartree dynamics to the Vlasov equation*, Arch. Ration. Mech. Anal. **221** (2016), no. 1, 273–334, DOI 10.1007/s00205-015-0961-z. MR3483896

[8] Niels Benedikter, Marcello Porta, and Benjamin Schlein, *Mean-field evolution of fermionic systems*, Comm. Math. Phys. **331** (2014), no. 3, 1087–1131, DOI 10.1007/s00220-014-2031-z. MR3248060

[9] Niels Benedikter, Marcello Porta, and Benjamin Schlein, *Mean-field dynamics of fermions with relativistic dispersion*, J. Math. Phys. **55** (2014), no. 2, 021901, 10, DOI 10.1063/1.4863349. MR3202863

[10] Jan Dereziński and Christian Gérard, *Mathematics of quantization and quantum fields*, Cambridge Monographs on Mathematical Physics, Cambridge University Press, Cambridge, 2013. MR3060648

[11] Alexander Elgart, László Erdős, Benjamin Schlein, and Horng-Tzer Yau, *Nonlinear Hartree equation as the mean field limit of weakly coupled fermions* (English, with English and French summaries), J. Math. Pures Appl. (9) **83** (2004), no. 10, 1241–1273, DOI 10.1016/j.matpur.2004.03.006. MR2092307

[12] C. Fefferman and R. de la Llave, *Relativistic stability of matter. I*, Rev. Mat. Iberoamericana **2** (1986), no. 1-2, 119–213, DOI 10.4171/RMI/30. MR864658

[13] S. Fournais, M. Lewin, and J. P. Solovej, *The semi-classical limit of large fermionic systems*, Calc. Var. Partial Differential Equations **57** (2018), no. 4, 57:105, DOI 10.1007/s00526-018-1374-2. MR3814648

[14] Jürg Fröhlich and Antti Knowles, *A microscopic derivation of the time-dependent Hartree-Fock equation with Coulomb two-body interaction*, J. Stat. Phys. **145** (2011), no. 1, 23–50, DOI 10.1007/s10955-011-0311-y. MR2841931

[15] I. Gasser, R. Illner, P. A. Markowich, and C. Schmeiser, *Semiclassical, $t \to \infty$ asymptotics and dispersive effects for Hartree-Fock systems* (English, with English and French summaries), RAIRO Modél. Math. Anal. Numér. **32** (1998), no. 6, 699–713, DOI 10.1051/m2an/1998320606991. MR1652601

[16] Gian Michele Graf and Jan Philip Solovej, *A correlation estimate with applications to quantum systems with Coulomb interactions*, Rev. Math. Phys. **6** (1994), no. 5A, 977–997, DOI 10.1142/S0129055X94000328. Special issue dedicated to Elliott H. Lieb. MR1301362

[17] Elliott H. Lieb, *Thomas-Fermi and related theories of atoms and molecules*, Rev. Modern Phys. **53** (1981), no. 4, 603–641, DOI 10.1103/RevModPhys.53.603. MR629207

[18] Elliott H. Lieb, *Variational principle for many-fermion systems*, Phys. Rev. Lett. **46** (1981), no. 7, 457–459, DOI 10.1103/PhysRevLett.46.457. MR601336

[19] Elliott H. Lieb and Barry Simon, *The Thomas-Fermi theory of atoms, molecules and solids*, Advances in Math. **23** (1977), no. 1, 22–116, DOI 10.1016/0001-8708(77)90108-6. MR0428944

[20] Pierre-Louis Lions and Thierry Paul, *Sur les mesures de Wigner* (French, with English and French summaries), Rev. Mat. Iberoamericana **9** (1993), no. 3, 553–618, DOI 10.4171/RMI/143. MR1251718

[21] Peter A. Markowich and Norbert J. Mauser, *The classical limit of a self-consistent quantum-Vlasov equation in 3D*, Math. Models Methods Appl. Sci. **3** (1993), no. 1, 109–124, DOI 10.1142/S0218202593000072. MR1203274

[22] Heide Narnhofer and Geoffrey L. Sewell, *Vlasov hydrodynamics of a quantum mechanical model*, Comm. Math. Phys. **79** (1981), no. 1, 9–24. MR609224

[23] Sören Petrat and Peter Pickl, *A new method and a new scaling for deriving fermionic mean-field dynamics*, Math. Phys. Anal. Geom. **19** (2016), no. 1, Art. 3, 51, DOI 10.1007/s11040-016-9204-2. MR3461406

[24] Sören Petrat, *Hartree corrections in a mean-field limit for fermions with Coulomb interaction*, J. Phys. A **50** (2017), no. 24, 244004, 19, DOI 10.1088/1751-8121/aa6e0b. MR3659128

[25] Federica Pezzotti and Mario Pulvirenti, *Mean-field limit and semiclassical expansion of a quantum particle system*, Ann. Henri Poincaré **10** (2009), no. 1, 145–187, DOI 10.1007/s00023-009-0404-1. MR2496306

[26] Marcello Porta, Simone Rademacher, Chiara Saffirio, and Benjamin Schlein, *Mean field evolution of fermions with Coulomb interaction*, J. Stat. Phys. **166** (2017), no. 6, 1345–1364, DOI 10.1007/s10955-017-1725-y. MR3612230

[27] J. P. Solovej. Many Body Quantum Mechanics. Lecture Notes. Summer 2007.

[28] H. Spohn, *On the Vlasov hierarchy*, Math. Methods Appl. Sci. **3** (1981), no. 4, 445–455, DOI 10.1002/mma.1670030131. MR657065

Department of Mathematics, Eberhard Karls Universität Tübingen, Auf der Morgenstelle 10, 72076 Tübingen, Germany

Email address: `marcello.porta@uni-tuebingen.de`

Contemporary Mathematics
Volume **717**, 2018
https://doi.org/10.1090/conm/717/14439

Correlations in disordered quantum harmonic oscillator systems: The effects of excitations and quantum quenches

Houssam Abdul-Rahman, Robert Sims, and Günter Stolz

ABSTRACT. We prove spatial decay estimates on disorder-averaged position-momentum correlations in a gapless class of random oscillator models. First, we prove a decay estimate on dynamic correlations for general eigenstates with a bound that depends on the magnitude of the maximally excited mode. Then, we consider the situation of a quantum quench. We prove that the full time-evolution of an initially chosen (uncorrelated) product state has disorder-averaged correlations which decay exponentially in space, uniformly in time.

1. Introduction

The mathematical investigation of disorder effects on quantum many-body systems, including, in particular, the phenomenon of many-body localization (MBL), is still in the early stages of its development. It has recently received strong attention in the physics and quantum information theory literature, see, e.g., [**4**, **5**, **15**] for recent reviews with extensive lists of references. Most mathematical results have been for models whose study can be fully reduced to the investigation of an effective one-particle Hamiltonian (i.e., without interaction) such as the Anderson model. Only few results go beyond such models. This includes [**11**], which proposes a derivation of MBL for certain disordered quantum spin chains, based on an as yet unproven assumption on level statistics for these models. Also, [**12**] establishes exponential decay of ground state correlations for the Aubry-André model (an XXZ chain in quasi-periodic field, which maps via the Jordan-Wigner transform to interacting Fermions). Recently, fully rigorous proofs of MBL in the droplet spectrum of the XXZ chain in random field (a regime extending beyond the ground state) have been given in [**6**] and [**10**]. In particular, the latter establishes exponential clustering of all eigenstates throughout the droplet spectrum with respect to arbitrary local observables.

Models which can be fully reduced to an effective one-particle Hamiltonian include the XY spin chain in random transversal field, see the review [**2**], and the Tonks-Girardeau gas subject to a random potential [**18**]. Here we present some new results on localization properties for another such model, disordered harmonic oscillator systems, as previously studied in [**1**, **13**, **14**]. In [**13**] results on the many-body dynamics in the form of zero-velocity Lieb-Robinson bounds as well as exponential decay of dynamical correlations (exponential clustering) of the ground state and of thermal states of such systems were shown. [**14**] further investigated ground and thermal states by establishing an area law for their bipartite entanglement entropy. More recently in [**1**], area laws are proven for uniform ensembles of equal-excitation energy eigenstates where the surface area bound increases linearly in the total number of excitations (modes).

There is, of course, a long history of interesting results for deterministic oscillator models. It is well-known, see for example [**7**, **17**] and references therein, that ground states of uniformly gapped oscillator models satisfy exponential decay of correlations. Moreover, area laws for both ground and thermal states of gapped oscillator models can be found e.g. in [**9**], see also the well-referenced review [**8**]. By contrast, as in [**1**, **13**, **14**], we will consider models where the gap above the ground state energy vanishes in the thermodynamic limit. For the results we will prove, estimates like the above mentioned deterministic results will not suffice, and we instead exploit disorder effects.

Our first new result here, Theorem 2.1 below, shows that exponential clustering in disordered oscillator systems also holds for the dynamic position-momentum correlations of excited states. The bound obtained will only depend on the maximal local excitation number of these states, i.e., when expressed in terms of the corresponding free Boson system, for states with positive particle number density. This is desirable to show that the model is in the many-body localized phase, as the latter, if it exists for a given model, should cover an extended part of the energy spectrum of the system. We also point out that, as opposed to ground and thermal states, the excited states of oscillator systems are not quasi-free, a property used in most of the works on exactly solvable models mentioned above. While the creation of excited states out of the ground state of an oscillator system is a simple algebraic process, our result can still be seen as a simple example of the possibility to go beyond quasi-free states in the study of disordered many-body systems.

In our second result, Theorem 2.2, we study quenched position-momentum correlations of disordered oscillator systems. Quantum quenches and their effect on the non-equilibrium dynamics of quantum many-body systems have been frequently considered in physics, see, e.g., [**16**] for a survey with many related references, as well as [**19**] for a discussion of quantum quenches in the context of many-body localization.

In its simplest form, a quantum quench refers to the investigation of a quantum state which is initially prepared with respect to one Hamiltonian and then time-evolved with respect to another. A common scenario is as follows. Consider a system defined on a Hilbert space for which there is a natural bipartite decomposition into two subsystems, i.e. $\mathcal{H} = \mathcal{H}_1 \otimes \mathcal{H}_2$. Denote by $H = H_1 \otimes \mathbb{1} + \mathbb{1} \otimes H_2 + I$ the Hamiltonian for the full system where I represents the interaction between the two subsystems. As an initial state take a product $\varphi = \varphi_1 \otimes \varphi_2$ which is uncorrelated with respect to the bipartite decomposition. The time-evolution $\varphi_t = e^{-iHt}\varphi$ of this initial state φ under the *full* Hamiltonian dynamics may exhibit interesting behavior, for example, non-trivial correlations may develop in time due to the interaction I.

In the disordered oscillator systems considered here we will assume that the two states φ_1 and φ_2 have exponentially clustered correlations with respect to the Hamiltonians H_1 and H_2 of the subsystem. We will show that the position-momentum correlations of the state will remain globally exponentially clustered, uniform in time and in the sizes of both subsystems. In fact, Theorem 2.2 will be more general in allowing for the decomposition into an arbitrary number of subsystems.

Applications of Theorem 2.2, which we discuss in Section 6, include the case where the initial product state consists of thermal states of the subsystems, so that the result of [**13**] on exponential clustering of these states applies, or where one starts with products of eigenstates of the subsystems, so that our first result, Theorem 2.1, can be applied in the subsystems. In physical terms these applications say that if each of the subsystems is localized in the sense of exponential decay of static correlations of eigenstates and thermal states within the subsystem, then this form of quantum quench yields no thermalization. In this context we include Theorem 6.1, as a technical result, proven in Appendix A, which improves results on thermal state correlations in [**13**] by quantifying the temperature dependence.

2. Model and Results

2.1. The Model. For any dimension $d \geq 1$, we consider a coupled harmonic oscillator system, on a finite rectangular box $\Lambda := ([a_1, b_1] \times \ldots \times [a_d, b_d]) \cap \mathbb{Z}^d$, given by the Hamiltonian

$$H_\Lambda = \sum_{x \in \Lambda} \left(p_x^2 + \frac{k_x}{2} q_x^2 \right) + \sum_{\substack{\{x,y\} \subset \Lambda : \\ |x-y|=1}} \lambda (q_x - q_y)^2. \tag{1}$$

This Hamiltonian acts on the Hilbert space

$$\mathcal{H}_\Lambda = \bigotimes_{x \in \Lambda} L^2(\mathbb{R}) = L^2(\mathbb{R}^\Lambda) \tag{2}$$

and q_x and p_x are, respectively, the position and momentum operators at the site $x \in \Lambda$. By standard results, these operators are self adjoint, on suitable domains, and satisfy the commutation relations

$$[q_x, q_y] = [p_x, p_y] = 0, \text{ and } [q_x, p_y] = i\delta_{x,y}\mathbb{1} \text{ for all } x, y \in \Lambda. \tag{3}$$

Here $\delta_{x,y}$ is the Kronecker delta function.

For each $x \in \Lambda$, k_x represents a variable spring constant. We introduce disorder by allowing the sequence $\{k_x\}$ to be chosen as a sequence of i.i.d. random variables. More precisely, we will assume that their common distribution μ is absolutely continuous,

$$d\mu(k) = \nu(k)dk, \quad \text{with } \|\nu\|_\infty < \infty \text{ and } \operatorname{supp} \nu = [0, k_{max}] \tag{4}$$

for some constant $k_{\max} < \infty$.

The Hamiltonian H_Λ describes a convenient family of harmonic oscillators that interact through nearest neighbor terms with a coupling strength of $\lambda > 0$. To be clear, the second sum in (1) is taken over all undirected edges $\{\{x, y\} \subset \Lambda : |x-y| = 1\}$ where $|\cdot|$ denotes the 1-norm. With methods similar to those of [**13**], the results we prove below generalize to a larger class of disordered oscillator models; the caveat being that the assumptions on the effective one-particle Hamiltonian, i.e. the analogue of (7) below, would need to be verified on a case-by-case basis. For ease of presentation, we restrict our attention to the model defined by (1) above.

As is well known, the analysis of general oscillator systems reduces, via Bogoliubov transformation, to the analysis of an effective one-particle Hamiltonian, see, e.g., [**13**] where this is reviewed for more general systems. In the specific case of (1), the corresponding one-particle Hamiltonian is the finite volume Anderson model on $\ell^2(\Lambda)$, i.e.

$$h_\Lambda = \lambda h_{0,\Lambda} + \frac{1}{2}k, \tag{5}$$

where $h_{0,\Lambda}$ is the negative discrete Laplacian on Λ and $\frac{1}{2}k$ an i.i.d. random potential. As we work in finite volume, the spectrum of h_Λ is discrete and, under our assumption of an absolutely continuous distribution for the k_x, almost surely simple (as is seen easily by an analyticity argument).

By our assumptions, it is clear that h_Λ is self-adjoint and moreover, using that $0 \le h_{0,\Lambda} \le 4d$,

$$\sigma(h_\Lambda) \subseteq \left[\frac{1}{2}\min_{x\in\Lambda} k_x, \left(4d\lambda + \frac{k_{\max}}{2}\right)\right]. \tag{6}$$

This means that h_Λ is invertible almost surely, but its inverse will not be uniformly bounded in the disorder. In fact, for large boxes Λ, the minimum of $\sigma(h_\Lambda)$ will be close to zero with high probability, due to the presence of large clusters where all k_x are close to zero.

As a localization characteristic of h_Λ, we will assume that its *singular* eigenfunction correlators decay exponentially. More precisely, we assume that there exist constants $C < \infty$, $\eta > 0$ and $0 < s \le 1$, independent of Λ, such that

$$\mathbb{E}\left(\sup_{|g|\le 1} |\langle \delta_x, h_\Lambda^{-\frac{1}{2}} g(h_\Lambda)\delta_y\rangle|^s\right) \le Ce^{-\eta|x-y|}, \tag{7}$$

for all $x, y \in \Lambda$, where $\mathbb{E}(\cdot)$ is the disorder average and $\{\delta_x\}_{x\in\Lambda}$ the canonical basis of $\ell^2(\Lambda)$. The supremum is taken over all functions $g : \mathbb{R} \to \mathbb{C}$ with $|g(x)| \le 1$ and $g(h_\Lambda)$ is defined by the functional calculus of symmetric matrices.

The non-standard feature of (7) is the term $h_\Lambda^{-1/2}$. By the discussion above, this term does not have an a-priori norm bound and can thus not be absorbed into $g(h_\Lambda)$. This term also is the reason for the inclusion of the moment s, which increases the applicability of (7).

In the absence of the factor $h_\Lambda^{-1/2}$ the bound

$$\mathbb{E}\left(\sup_{|g|\le 1} |\langle \delta_x, g(h_\Lambda)\delta_y\rangle|\right) \le Ce^{-\eta|x-y|} \tag{8}$$

is a standard result for two regimes, e.g. [**3**]: (i) in dimension $d = 1$ (where Λ is an interval) for any choice of the distribution density ν, and (ii) in the large disorder regime for $d > 1$ (meaning that $\|\nu\|_\infty$ is sufficiently small, or k_x is rescaled by a large parameter).

The singular eigenfunction correlators in (7) were first introduced in [**13**]. As is discussed there in Appendix A, (7) holds for $d \ge 1$ and large disorder with $s = 1$ (combing Propositions A.1(b) and A.3(b)), and for $d = 1$ and any ν with $s = 1/2$ (Propositions A.1(c) and A.4(a)). Note that in the latter example, the validity of (7) with $s = 1/2$ does not trivially imply that it is also valid with s replaced by 1, due to the fact that $|\langle \delta_x, h_\Lambda^{-\frac{1}{2}} g(h_\Lambda)\delta_y\rangle|$ does not satisfy a uniform a-priori bound. Applications such as this are the reason we allow for the flexibility of s in (7).

As further discussed in Section 3, the diagonalization of h_Λ transforms H_Λ into a model of free bosons

$$H_\Lambda = \sum_{k=1}^{|\Lambda|} \gamma_k (2B_k^* B_k + 1\!\!1), \tag{9}$$

where γ_k^2 are the eigenvalues of h_Λ, and the operators $\{B_k\}$ satisfy canonical commutation relations (CCR) namely

$$[B_j, B_k] = [B_j^*, B_k^*] = 0, \quad [B_j, B_k^*] = \delta_{j,k} 1\!\!1. \tag{10}$$

In this case, there is a unique normalized vacuum state $\Omega \in \mathcal{H}_\Lambda$ corresponding to these B-operators, i.e., Ω satisfies $B_k\Omega = 0$ for all k. An explicit orthonormal basis (ONB) of eigenvectors of H_Λ is then given by

$$\psi_\alpha = \prod_{j=0}^{|\Lambda|} \frac{1}{\sqrt{\alpha_j!}} (B_j^*)^{\alpha_j} \Omega \tag{11}$$

for an excitation number configuration $\alpha = (\alpha_1, \ldots, \alpha_{|\Lambda|}) \in \mathbb{N}_0^{|\Lambda|}$ (here $\mathbb{N}_0 := \{0, 1, 2, \cdots\}$). One easily checks that these excited states satisfy

$$H_\Lambda \psi_\alpha = E_\alpha \psi_\alpha \quad \text{with} \quad E_\alpha = \sum_{k=1}^{|\Lambda|} (2\alpha_k + 1)\gamma_k \tag{12}$$

and therefore, the gap above the ground state energy of H_Λ is $2 \min_k \gamma_k$.

2.2. Dynamic Correlations in Eigenstates. One goal of this work is to estimate dynamic correlations of position and momentum operators in arbitrary eigenstates. To make this more precise, let $\tau_t(A)$ denote the Heisenberg evolution of an operator A under H_Λ, i.e.,

$$\tau_t(A) = e^{itH_\Lambda} A e^{-itH_\Lambda}, \tag{13}$$

and for any trace-class operator ρ on $\mathcal{H}_\Lambda$ take

$$\rho_t = e^{-itH_\Lambda} \rho e^{itH_\Lambda} \tag{14}$$

to be the Schrödinger evolution of ρ. In this case, if $\langle A \rangle_\rho = \mathrm{Tr}[A\rho]$ denotes the ρ-expectation of the observable A, then the Heisenberg and Schrödinger evolutions are related by $\langle \tau_t(A) \rangle_\rho = \langle A \rangle_{\rho_t}$.

It is convenient to introduce a $2|\Lambda| \times 2|\Lambda|$ correlation matrix

$$\Gamma_\rho(t, t') := \left\langle \tau_t \begin{pmatrix} q \\ p \end{pmatrix} (q^T, p^T) \right\rangle_{\rho_{t'}} - \left\langle \tau_t \begin{pmatrix} q \\ p \end{pmatrix} \right\rangle_{\rho_{t'}} \langle (q^T, p^T) \rangle_{\rho_{t'}} \tag{15}$$

which collects *mixed-time* dynamic correlations of position and momentum operators corresponding to ρ. Here $\begin{pmatrix} q \\ p \end{pmatrix}$ and (q^T, p^T) are $2|\Lambda|$ column and row vectors, the time-evolution and ρ-expectation of vectors and matrices are understood component-wise, and columns are multiplied with rows in the usual sense of matrix multiplication to form matrices. Use of these mixed-time correlations $\Gamma_\rho(t, t')$ is convenient when formulating our main results below; in one we set $t' = 0$ and in the other $t = 0$.

Our first result concerns disorder-averaged correlations in eigenstates, i.e. we consider $\rho = \rho_\alpha = |\psi_\alpha\rangle\langle\psi_\alpha|$ for some $\alpha \in \mathbb{N}_0^{|\Lambda|}$. Since eigenstates are time-invariant, we set $t' = 0$ in (15) and denote by $\Gamma_\alpha(t) := \Gamma_{\rho_\alpha}(t, 0)$. Note that

$$\langle \tau_t(q_x)\rangle_{\rho_\alpha} = \langle \tau_t(p_x)\rangle_{\rho_\alpha} = 0 \tag{16}$$

for all x, t and α (in fact, we will also see this directly in the proof of Theorem 2.1 below). Thus (15) simplifies to

$$\Gamma_\alpha(t) = \left\langle \tau_t \begin{pmatrix} q \\ p \end{pmatrix} (q^T, p^T) \right\rangle_{\rho_\alpha}. \tag{17}$$

Lastly, for a $2|\Lambda| \times 2|\Lambda|$ block-matrix

$$M = \begin{pmatrix} A & B \\ C & D \end{pmatrix}, \text{ let } M_{xy} = \begin{pmatrix} A_{xy} & B_{xy} \\ C_{xy} & D_{xy} \end{pmatrix} \tag{18}$$

be 2×2-sub-matrices with matrix norms $\|M_{xy}\|$. For definiteness we choose the latter to be the Euclidean operator norm.

THEOREM 2.1. *Assume that the effective Hamiltonian h_Λ satisfies (7), with bounds uniform in Λ. Then*

$$\mathbb{E}\left(\sup_t \|(\Gamma_\alpha(t))_{xy}\|^s \right) \le CC'(1 + \|\alpha\|_\infty)^{1+s} e^{-\eta|x-y|} \tag{19}$$

for all finite rectangular boxes $\Lambda \subset \mathbb{Z}^d$, $x, y \in \Lambda$ and $\alpha \in \mathbb{N}_0^{|\Lambda|}$. Here C, η and s are as in (7) and $C' < \infty$ depends on d, λ, s and k_{max}, but is independent of Λ.

We finally note that our proof of Theorem 2.1 in Section 4 below will show, the bound in (19) can be slightly improved to

$$\le \tilde{C}\left(1 + \min\{\|\alpha\|_1, \|\alpha\|_\infty^{1+s}\}\right) e^{-\eta|x-y|}, \tag{20}$$

which is better for excitation vectors with only a few large excitations α_j (say, just one of them).

2.3. Quenched Correlations. Our second result concerns the position and momentum correlations when a quantum quench is applied. In particular, we decompose the rectangular box Λ into M disjoint rectangular sub-boxes

$$\Lambda = \biguplus_{\ell=1}^{M} \Lambda_\ell. \tag{21}$$

For $\ell = 1, 2, \ldots, M$, consider the restrictions H_{Λ_ℓ} of the harmonic system H_Λ to Λ_ℓ. Let $H_{0,\Lambda}$ denote the Hamiltonian of the non-interacting system on $\mathcal{H}_\Lambda$,

$$H_{0,\Lambda} = \sum_{\ell=1}^{M} H_{\Lambda_\ell} \otimes \mathbb{1}_{\Lambda\setminus\Lambda_\ell}. \tag{22}$$

For each $\ell = 1, \ldots, M$, let ρ_ℓ be a state acting on the Hilbert space $\mathcal{H}_{\Lambda_\ell} := L^2(\mathbb{R}^{\Lambda_\ell})$. In particular, we will consider the cases where the ρ_ℓ are either eigenstates or thermal states of H_{Λ_ℓ}. We are interested in the Schrödinger time evolution ρ_t, under the full Hamiltonian H_Λ given in (1), of the state initially given by the product state

$$\rho := \bigotimes_{\ell=1}^{M} \rho_\ell. \tag{23}$$

This *quantum quench* is understood as a sudden change in the Hamiltonian $H_{0,\Lambda}$ at $t=0$, consisting in switching on the interactions between the subsystems H_{Λ_ℓ}.

To describe the dynamic correlations in this case we set the first argument equal to zero in (15) and define

$$\tilde{\Gamma}_\rho(t) := \Gamma_\rho(0,t). \tag{24}$$

That the local systems H_{Λ_ℓ} are initially uncorrelated means that for $x \in \Lambda_j$ and $y \in \Lambda_\ell$ with $j \neq \ell$,

$$(\tilde{\Gamma}_\rho(0))_{xy} = 0. \tag{25}$$

The following result says that if each of the subsystems H_{Λ_ℓ} is localized in the sense of exponential decay of static correlations within the subsystem, then their quantum quench, described above, yields no thermalization. More precisely, the so-called quenched dynamic correlations of the product state ρ remain exponentially decaying in the fully interacting system for all times (here we also use the local $2|\Lambda_\ell| \times 2|\Lambda_\ell|$ correlation matrices Γ_{ρ_ℓ} with $t=t'=0$ in (15)).

THEOREM 2.2. *Assume that the effective Hamiltonian h_Λ satisfies (7), with bounds uniform in Λ. Let $\rho_\ell \in \mathcal{B}(\mathcal{H}_{\Lambda_\ell})$, $\ell = 1, \dots, M$, be a family of states such that, for some $C' < \infty$, and $\eta' > 0$,*

$$\mathbb{E}\left(\|(\Gamma_{\rho_\ell})_{xy}\|^s\right) \leq C' e^{-\eta'|x-y|} \tag{26}$$

for all ℓ and all $x, y \in \Lambda_\ell$, where $0 < s \leq 1$ is as in (7).

Then, for η from (7), $\tilde{\eta} := \frac{1}{6}\min\{\eta, \eta'\}$ and $\rho = \bigotimes_\ell \rho_\ell$, there exists a constant $C'' < \infty$ such that

$$\mathbb{E}\left(\sup_{t\in\mathbb{R}} \|(\tilde{\Gamma}_\rho(t))_{xy}\|^{\frac{s}{3}}\right) \leq (C')^{1/3} C'' e^{-\tilde{\eta}|x-y|} \tag{27}$$

for all $x, y \in \Lambda$. Here C' is the constant from (26) and C'' depends on d, λ, s, k_{max} and $\tilde{\eta}$, but is independent of Λ and the number of subregions M.

This will be proven in Section 5.

By results in [**13**], special cases where condition (26) is known to hold include the ground state and thermal states of the subsystems H_{Λ_ℓ}. Theorem 2.1 above extends this to excited states. In each of these cases, (26) actually follows from (7). Theorem 2.2 allows for the additional freedom to choose different temperatures and different maximal excitation numbers in each of the subsystems, or even to choose some of the factors in the initial product state as thermal states and others as excited states. It is then of some interest to understand the dependence of the constants in (27) on these additional parameters. We include a more thorough discussion of this in Section 6 at the end of this paper.

3. Reduction to the Effective Hamiltonian

In this section, we briefly review the previously mentioned reduction of the many-body Hamiltonian H_Λ to the effective one-particle h_Λ as a means to introduce some relevant notation. Once this is done, we provide a simple lemma concerning mixed-time correlations, i.e. (15).

Keeping with the vector notation established in Section 2.2, one readily sees that the oscillator Hamiltonian H_Λ in (1) can be re-written as

$$H_\Lambda = (q^T, p^T)\begin{pmatrix} h_\Lambda & 0 \\ 0 & \mathbb{1}\end{pmatrix}\begin{pmatrix} q \\ p\end{pmatrix} \tag{28}$$

with h_Λ the effective one-particle Hamiltonian described in (5). The real non-negative matrix h_Λ can be diagonalized in terms of a real orthogonal $\mathcal{O} : \mathbb{C}^{|\Lambda|} \to \ell^2(\Lambda)$ and its transpose $\mathcal{O}^T = \mathcal{O}^{-1}$, i.e.

$$\mathcal{O}^T h_\Lambda \mathcal{O} = \gamma^2 \tag{29}$$

where $\gamma^2 = \mathrm{diag}(\gamma_k^2)$ with $1 \le k \le |\Lambda|$. Here the numbers γ_k^2 are the eigenvalues of h_Λ counted according to multiplicity. By our assumptions on the spring constants, the eigenvalues of h_Λ are almost surely positive, and we will denote by $\gamma = \mathrm{diag}(\gamma_k)$ with $\gamma_k > 0$ for all $1 \le k \le |\Lambda|$. As discussed in Section 2.1, the γ_k are almost surely non-degenerate.

As is well-known, see [**13**] for more details in this specific setting, H_Λ can be reduced to a system of free Bosons. In fact, consider the mapping

$$V^{-1} = \frac{1}{\sqrt{2}}\begin{pmatrix} \mathbb{1} & i\mathbb{1} \\ \mathbb{1} & -i\mathbb{1}\end{pmatrix}\begin{pmatrix} \gamma^{1/2}\mathcal{O}^T & 0 \\ 0 & \gamma^{-1/2}\mathcal{O}^T\end{pmatrix}. \tag{30}$$

Our assumptions guarantee this map is almost surely well-defined, invertible, and one readily checks that the product

$$\begin{pmatrix} B \\ B^*\end{pmatrix} := V^{-1}\begin{pmatrix} q \\ p\end{pmatrix} \tag{31}$$

produces a collection of operators $\{B_k\}_{k=1}^{|\Lambda|}$ on $\mathcal{H}_\Lambda$ which, together with their adjoints, satisfy the CCR, i.e. (10). Moreover, in terms of these B-operators

$$H_\Lambda = \sum_{k=1}^{|\Lambda|} \gamma_k (2B_k^* B_k + \mathbb{1}), \tag{32}$$

a model of free Bosons.

Due to the simple form of (32), the dynamics of these B-operators is

$$\tau_t\begin{pmatrix} B \\ B^*\end{pmatrix} = \begin{pmatrix} e^{-2it\gamma} & 0 \\ 0 & e^{2it\gamma}\end{pmatrix}\begin{pmatrix} B \\ B^*\end{pmatrix} \tag{33}$$

from which the dynamics of position and momentum operators readily follows,

$$\tau_t\begin{pmatrix} q \\ p\end{pmatrix} = V\begin{pmatrix} e^{-2it\gamma} & 0 \\ 0 & e^{2it\gamma}\end{pmatrix}\begin{pmatrix} B \\ B^*\end{pmatrix}, \tag{34}$$

where we have used (31). It will also be convenient to note that

$$V = \frac{1}{\sqrt{2}}\begin{pmatrix} \mathcal{O}\gamma^{-1/2} & 0 \\ 0 & \mathcal{O}\gamma^{1/2}\end{pmatrix}\begin{pmatrix} \mathbb{1} & \mathbb{1} \\ -i\mathbb{1} & i\mathbb{1}\end{pmatrix}. \tag{35}$$

As indicated in Section 2.2, much of our analysis reduces to the investigation of the mixed-time correlation function $\Gamma_\rho(t,t')$ in (15) for a state ρ. Since (34) shows that the dynamics of position and momentum operators can be expressed in terms of the B-operators, up to scalar-valued coefficients, one immediately has the following.

LEMMA 3.1. *Let ρ be a state on $\mathcal{H}_\Lambda$. Suppose that the matrix*

$$\Gamma_\rho^B = \left\langle \begin{pmatrix} B \\ B^* \end{pmatrix} (B^T, (B^*)^T) \right\rangle_\rho - \left\langle \begin{pmatrix} B \\ B^* \end{pmatrix} \right\rangle_\rho \langle (B^T, (B^*)^T) \rangle_\rho \tag{36}$$

is well-defined. Then, for all $t, t' \in \mathbb{R}$,

$$\Gamma_\rho(t,t') = V \begin{pmatrix} e^{-2i(t+t')\gamma} & 0 \\ 0 & e^{2i(t+t')\gamma} \end{pmatrix} \Gamma_\rho^B \begin{pmatrix} e^{-2it'\gamma} & 0 \\ 0 & e^{2it'\gamma} \end{pmatrix} V^T \tag{37}$$

where V is as in (35).

4. Proof of Theorem 2.1

We begin with a calculation which evaluates the eigenstate correlation matrix $\Gamma_\alpha(t)$ given by (17) in terms of the effective Hamiltonian h_Λ, using Lemma 3.1.

LEMMA 4.1. *We have the identity*

$$\begin{aligned} \Gamma_\alpha(t) \quad = \quad & \begin{pmatrix} \mathcal{O}\alpha\mathcal{O}^T & 0 \\ 0 & \mathcal{O}\alpha\mathcal{O}^T \end{pmatrix} \begin{pmatrix} h_\Lambda^{-1/2}\cos(2th_\Lambda^{1/2}) & \sin(2th_\Lambda^{1/2}) \\ -\sin(2th_\Lambda^{1/2}) & h_\Lambda^{1/2}\cos(2th_\Lambda^{1/2}) \end{pmatrix} + \\ & + \frac{1}{2} \begin{pmatrix} h_\Lambda^{-1/2} e^{-2ith_\Lambda^{1/2}} & ie^{-2ith_\Lambda^{1/2}} \\ -ie^{-2ith_\Lambda^{1/2}} & h_\Lambda^{1/2} e^{-2ith_\Lambda^{1/2}} \end{pmatrix}. \end{aligned} \tag{38}$$

Here, in a slight abuse of notation, we use α also to denote the diagonal matrix with entries α_k, $1 \le k \le |\Lambda|$.

PROOF. By orthogonality of the eigenvectors ψ_α, it is clear that each of

$$\left\langle \begin{pmatrix} B \\ B^* \end{pmatrix} \right\rangle_{\rho_\alpha}, \langle BB^T \rangle_{\rho_\alpha}, \text{ and } \langle B^*(B^*)^T \rangle_{\rho_\alpha} \tag{39}$$

vanish identically. We note that this and (34) implies (16). Moreover, using also the commutation relations (10), we find that for all $1 \le j, k \le |\Lambda|$,

$$\langle B_k^* B_j \rangle_{\rho_\alpha} + \delta_{j,k} = \langle B_j B_k^* \rangle_{\rho_\alpha} = \langle B_j^* \psi_\alpha, B_k^* \psi_\alpha \rangle = (\alpha_j + 1)\delta_{j,k}\,. \tag{40}$$

and therefore, for $\Gamma_{\rho_\alpha}^B$ as in (36), we have that

$$\Gamma_{\rho_\alpha}^B = \begin{pmatrix} 0 & (\alpha + \mathbb{1}) \\ \alpha & 0 \end{pmatrix} \tag{41}$$

An application of Lemma 3.1 yields

$$\Gamma_\alpha(t) = V \begin{pmatrix} 0 & (\alpha+\mathbb{1})e^{-2i\gamma t} \\ \alpha e^{2i\gamma t} & 0 \end{pmatrix} V^T. \tag{42}$$

A short calculation shows that

$$\frac{1}{2} \begin{pmatrix} \mathbb{1} & \mathbb{1} \\ -i\mathbb{1} & i\mathbb{1} \end{pmatrix} \begin{pmatrix} 0 & (\alpha+\mathbb{1})e^{-2i\gamma t} \\ \alpha e^{2i\gamma t} & 0 \end{pmatrix} \begin{pmatrix} \mathbb{1} & -i\mathbb{1} \\ \mathbb{1} & i\mathbb{1} \end{pmatrix} \tag{43}$$

can be rewritten as

$$\begin{pmatrix} \alpha & 0 \\ 0 & \alpha \end{pmatrix} \begin{pmatrix} \cos(2\gamma t) & \sin(2\gamma t) \\ -\sin(2\gamma t) & \cos(2\gamma t) \end{pmatrix} + \frac{1}{2} \begin{pmatrix} e^{-2i\gamma t} & 0 \\ 0 & e^{-2i\gamma t} \end{pmatrix} \begin{pmatrix} \mathbb{1} & i\mathbb{1} \\ -i\mathbb{1} & \mathbb{1} \end{pmatrix}. \tag{44}$$

Using the form of V and V^{-1} in (35) and (30) this gives

$$\begin{aligned}\Gamma_\alpha(t) &= \begin{pmatrix} \mathcal{O}\gamma^{-1/2} & 0 \\ 0 & \mathcal{O}\gamma^{1/2}\end{pmatrix}\begin{pmatrix} \alpha\cos(2\gamma t) & \alpha\sin(2\gamma t) \\ -\alpha\sin(2\gamma t) & \alpha\cos(2\gamma t)\end{pmatrix}\begin{pmatrix} \gamma^{-1/2}\mathcal{O}^T & 0 \\ 0 & \gamma^{1/2}\mathcal{O}^T\end{pmatrix} \\ &+\frac{1}{2}\begin{pmatrix} \mathcal{O}\gamma^{-1/2} & 0 \\ 0 & \mathcal{O}\gamma^{1/2}\end{pmatrix}\begin{pmatrix} e^{-2i\gamma t} & ie^{-2i\gamma t} \\ -ie^{-2i\gamma t} & e^{-2i\gamma t}\end{pmatrix}\begin{pmatrix} \gamma^{-1/2}\mathcal{O}^T & 0 \\ 0 & \gamma^{1/2}\mathcal{O}^T\end{pmatrix}. \end{aligned} \tag{45}$$

This is the same as (38), due to

$$\mathcal{O}\gamma^{-1/2}\alpha\cos(2\gamma t)\gamma^{-1/2}\mathcal{O}^T = \mathcal{O}\alpha\mathcal{O}^T h_\Lambda^{-1/2}\cos(2th_\Lambda^{-1/2}) \tag{46}$$

and similar consequences of the functional calculus. □

We can now present the proof of Theorem 2.1.

PROOF. (of Theorem 2.1) We first consider the most singular case, and then comment on how the remaining cases follow similarly.

Lemma 4.1 demonstrates that

$$\langle\tau_t(q_x)q_y\rangle_\alpha = \langle\delta_x, \mathcal{O}\alpha\mathcal{O}^T h_\Lambda^{-1/2}\cos(2th_\Lambda^{1/2})\delta_y\rangle + \frac{1}{2}\langle\delta_x, h_\Lambda^{-1/2}e^{-2ith_\Lambda^{1/2}}\delta_y\rangle. \tag{47}$$

Using (7), it is clear that

$$\mathbb{E}\left(\sup_{t\in\mathbb{R}}\left|\langle\delta_x, h_\Lambda^{-1/2}e^{-2ith_\Lambda^{1/2}}\delta_y\rangle\right|^s\right) \le Ce^{-\eta|x-y|}, \tag{48}$$

and so we need only estimate the first term in (47) above. If the eigenvalues of h_Λ are non-degenerate, which holds almost surely, we can write

$$\mathcal{O}\alpha\mathcal{O}^T = \sum_{j=1}^{|\Lambda|}\alpha_j\chi_{\{\gamma_j^2\}}(h_\Lambda) = \sum_{a=0}^{\|\alpha\|_\infty} a\chi_{J(a)}(h_\Lambda). \tag{49}$$

Here $\chi_{\{\gamma_j^2\}}(h_\Lambda)$ is the projection onto the eigenvector of h_Λ to γ_j^2, and $\chi_{J(a)}(h_\Lambda)$ is the spectral projection for h_Λ onto $J(a) := \{\gamma_j^2 : \alpha_j = a\}$. Given this, one immediately sees that

$$\left|\langle\delta_x, \mathcal{O}\alpha\mathcal{O}^T h_\Lambda^{-1/2}\cos(2th_\Lambda^{1/2})\delta_y\rangle\right|^s \le \sum_{a=0}^{\|\alpha\|_\infty} a^s\left|\langle\delta_x, h_\Lambda^{-1/2}\chi_{J(a)}(h_\Lambda)\cos(2th_\Lambda^{1/2})\delta_y\rangle\right|^s \tag{50}$$

and therefore an application of (7) again implies

$$\mathbb{E}\left(\sup_{t\in\mathbb{R}}\left|\langle\delta_x, \mathcal{O}\alpha\mathcal{O}^T h_\Lambda^{-1/2}\cos(2th_\Lambda^{1/2})\delta_y\rangle\right|^s\right) \le C\|\alpha\|_\infty^s(\|\alpha\|_\infty+1)e^{-\eta|x-y|}. \tag{51}$$

This completes the argument for the most singular correlations.

As is clear from Lemma 4.1, the other correlations in the 2×2-matrix $(\Gamma_\alpha(t))_{xy}$ produce similar terms. These terms require bounds on eigenfunction correlators less singular than (7), in the sense that the term $h_\Lambda^{-1/2}$ is replaced by $\mathbb{1}$ or $h_\Lambda^{1/2}$. They can be bounded by (7) due to the uniform spectral bound (6). For example, associating $\tilde{g}(x) := x^{1/2}g(x)$ with each g such that $|g|\le 1$, one gets

$$\mathbb{E}\left(\sup_{|g|\le 1}|\langle\delta_x, g(h_\Lambda)\delta_y\rangle|^s\right) \le (4d\lambda + \frac{k_{max}}{2})^{s/2}\mathbb{E}\left(\sup_{|g|\le 1}|\langle\delta_x, h_\Lambda^{-1/2}g(h_\Lambda)\delta_y\rangle|^s\right), \tag{52}$$

and similar for $\mathbb{E}\left(\sup_{|g|\le 1}|\langle\delta_x, h_\Lambda^{1/2}g(h_\Lambda)\delta_y\rangle|^s\right)$.

Finally, the bound in terms of $\|\alpha\|_1$ in (20) follows by directly considering the middle term in (49).

This completes the proof of Theorem 2.1. □

5. Proof of Theorem 2.2

By Lemma 3.1, the *qp*-correlations (24) corresponding to the time-evolution of any initially chosen density matrix ρ can be evaluated as:

$$\tilde{\Gamma}_\rho(t) = \left\langle \begin{pmatrix} q \\ p \end{pmatrix} (q^T, p^T) \right\rangle_{\rho_t} - \left\langle \begin{pmatrix} q \\ p \end{pmatrix} \right\rangle_{\rho_t} \langle (q^T, p^T) \rangle_{\rho_t} \tag{53}$$

$$= V \begin{pmatrix} e^{-2it\gamma} & 0 \\ 0 & e^{2it\gamma} \end{pmatrix} \Gamma_\rho^B \begin{pmatrix} e^{-2it\gamma} & 0 \\ 0 & e^{2it\gamma} \end{pmatrix} V^T$$

where V and Γ_ρ^B are as in (35) and (36), respectively. For our arguments here, we prefer to re-express this in terms of the time-zero *qp*-correlations, i.e., we write

$$\tilde{\Gamma}_\rho(t) = V_t \Gamma_\rho V_t^T, \tag{54}$$

where we have set

$$V_t = V \begin{pmatrix} e^{-2it\gamma} & 0 \\ 0 & e^{2it\gamma} \end{pmatrix} V^{-1} = \begin{pmatrix} \cos(2th_\Lambda^{1/2}) & h_\Lambda^{-1/2}\sin(2th_\Lambda^{1/2}) \\ -h_\Lambda^{1/2}\sin(2th_\Lambda^{1/2}) & \cos(2th_\Lambda^{1/2}) \end{pmatrix}. \tag{55}$$

The final equality is a direct calculation.

By (54) one has, for any $x, y \in \Lambda$,

$$(\tilde{\Gamma}_\rho(t))_{xy} = \sum_{z,z' \in \Lambda} (V_t)_{xz} (\Gamma_\rho)_{zz'} (V_t^T)_{z'y} \tag{56}$$

with 2×2-matrices $(\tilde{\Gamma}_\rho(t))_{xy}$, $(V_t)_{xz}$, $(\Gamma_\rho)_{zz'}$ and $(V_t^T)_{z'y}$ defined according to (18).

Similar to the arguments in Section 4, our basic assumption (7), guarantees the existence of $\tilde{C} < \infty$, depending on d, λ, s and k_{max}, such that

$$\mathbb{E}\left(\sup_{t\in\mathbb{R}} \|(V_t)_{xy}\|^s \right) \le \tilde{C} e^{-\eta|x-y|} \tag{57}$$

for every rectangular box Λ and all $x, y \in \Lambda$. It is clear that the same bound also holds for V_t^T.

For the product state $\rho = \otimes_{\ell=1}^M \rho_\ell$ in (23), the *qp*-correlation matrix is the direct sum of the correlation matrices of the factors ρ_ℓ. More precisely, for $x, y \in \Lambda$,

$$(\Gamma_\rho)_{xy} = \begin{cases} (\Gamma_{\rho_\ell})_{xy} & \text{if } x, y \in \Lambda_\ell \text{ for some } \ell, \\ 0 & \text{otherwise.} \end{cases} \tag{58}$$

Thus, by condition (26)

$$\mathbb{E}\left(\|(\Gamma_\rho)_{xy}\|^s \right) \le C' e^{-\eta'|x-y|} \tag{59}$$

for all $x, y \in \Lambda$.

For all $x, y \in \Lambda$ we have that

$$\mathbb{E}\left(\sup_{t\in\mathbb{R}} \|(\tilde{\Gamma}_\rho(t))_{xy}\|^{s/3} \right) \le \sum_{z,z'\in\Lambda} \mathbb{E}\left(\sup_{t\in\mathbb{R}} \|(V_t)_{xz}\|^s \right)^{1/3} \times$$

$$\times \mathbb{E}\left(\|(\Gamma_\rho)_{zz'}\|^s \right)^{1/3} \mathbb{E}\left(\sup_{t\in\mathbb{R}} \|(V_t^T)_{z'y}\|^s \right)^{1/3}, \tag{60}$$

where we have used (56) and Hölder's inequality. Thus (57) and (59) yield

$$\begin{aligned} \mathbb{E}\left(\sup_{t\in\mathbb{R}} \|(\tilde{\Gamma}_\rho(t))_{xy}\|^{s/3}\right) &\leq \tilde{C}^{2/3} C'^{\frac{1}{3}} \sum_{z,z'\in\Lambda} e^{-\eta|x-z|/3} e^{-\eta'|z-z'|/3} e^{-\eta|z'-y|/3} \\ &\leq C'' e^{-\tilde{\eta}|x-y|}. \end{aligned} \tag{61}$$

Here one may take $\tilde{\eta} = \frac{1}{6}\min\{\eta,\eta'\}$ and

$$C'' = \tilde{C}^{2/3}(C')^{1/3}\left(\frac{2}{1-e^{-\tilde{\eta}}}\right)^{2d}. \tag{62}$$

6. Applications of Theorem 2.2

(i) As a first application of Theorem 2.2 we consider the case where the factors in the product state (23) are thermal states of the subsystems. Assumption (26) in Theorem 2.2 is then a consequence of Theorem 6.1 in [**13**] on the position-momentum correlations of thermal states of oscillator systems. We start by stating an improved version of this result, which makes the temperature dependence of the bound explicit, a fact of some interest by itself which was not addressed in [**13**]. In Appendix A we sketch the modifications of the argument in [**13**] needed to get this improvement.

While more general systems are considered in [**13**], we will continue to focus on the model (1). Here we only require the general assumption (4) on the distribution of the k_x, and, in particular, we do not require to be in a fully localized regime as needed for (7).

THEOREM 6.1. *For a rectangular box $\Lambda \subset \mathbb{Z}^d$ and $\beta \in (0,\infty)$, let H_Λ be given by (1), $\rho_\beta = e^{-\beta H_\Lambda}/\operatorname{Tr}[e^{-\beta H_\Lambda}]$ its thermal states, and $\Gamma_{\rho_\beta} = \Gamma_{\rho_\beta}(0,0)$ their static position-momentum correlation matrices.*

There exist $C<\infty$ and $\mu>0$, dependent on d, λ and the distribution of the random variables k_x, but independent of Λ and β, such that

$$\mathbb{E}\left(\|(\Gamma_{\rho_\beta})_{xy}\|^{\frac{1}{2}}\right) \leq C \max\left\{1, \frac{1}{\beta}\right\} e^{-\mu|x-y|} \tag{63}$$

for all $x, y \in \Lambda$.

In Section A below we will briefly discuss how the β dependence in (63) can be extracted from the bounds provided in [**13**].

Let us consider the quantum quench with respect to the decomposition $\Lambda = \biguplus_{\ell=1}^M \Lambda_\ell$, and assume that the local states are the thermal states of H_{Λ_ℓ} with inverse temperatures β_ℓ, $\ell = 1,\dots,M$, i.e.,

$$\rho_{\ell,\beta_\ell} = \frac{e^{-\beta_\ell H_{\Lambda_\ell}}}{\operatorname{Tr}[e^{-\beta_\ell H_{\Lambda_\ell}}]}. \tag{64}$$

Then condition (26) is satisfied by Theorem 6.1 when applied to each of the local Hamiltonians. In this case, and for the remainder of this section, we will further assume that (7) holds with $s = 1/2$. As is discussed in Section 2.1, this will be the case for the model we are considering when either $d = 1$ or $d \geq 1$ and the disorder is large. Now let

$$\rho_{\beta_1,\dots,\beta_M} := \bigotimes_{\ell=1}^M \rho_{\ell,\beta_\ell}. \tag{65}$$

Theorem 2.2 implies that

$$\mathbb{E}\left(\sup_t \|(\tilde{\Gamma}_{\rho_{\beta_1,\ldots,\beta_M}}(t))_{xy}\|^{\frac{1}{6}}\right) \leq C' \max\left\{1, \beta^{-1/3}\right\} e^{-\tilde{\eta}|x-y|} \tag{66}$$

for all $x, y \in \Lambda$. Here $\beta = \min_\ell \beta_\ell$ and $\tilde{\eta} = \frac{1}{6}\min\{\eta, \mu\}$ where η and μ are as in (7) and (63), respectively, and C' is independent of Λ, β and the number of subsystems M.

(ii) Next we discuss the case where the initial state is a product of eigenstates of the subsystems. Fix a nonnegative integer $N < \infty$, and let $\alpha_\ell \in \mathbb{N}_0^{|\Lambda_\ell|}$ with $\|\alpha_\ell\|_\infty \leq N$ for all $\ell = 1, \ldots, M$. Consider any family $\rho_{\alpha_\ell} = |\psi_{\alpha_\ell}\rangle\langle\psi_{\alpha_\ell}|$, $\ell = 1, \ldots, M$, of eigenstates of H_{Λ_ℓ} corresponding to the excitation vectors α_ℓ, with ψ_{α_ℓ} given by (11) when used for the subsystem Λ_ℓ . Theorem 2.1 implies that condition (26) is satisfied for all $\ell = 1, \ldots, M$, in particular, there exist constants $C' > 0$ and $\eta < \infty$ such that

$$\mathbb{E}\left(\|(\Gamma_{\rho_{\alpha_\ell}})_{xy}\|^{\frac{1}{2}}\right) \leq C'(1 + \|\alpha_\ell\|_\infty)^{\frac{3}{2}} e^{-\eta|x-y|}, \tag{67}$$

for all ℓ and all $x, y \in \Lambda_\ell$. Here η is as in (7), which we have again taken to hold with $s = 1/2$, and C' is independent of Λ, N, and of M. With ρ_α the corresponding product, i.e. $\rho_\alpha = \otimes_{\ell=1}^M \rho_{\alpha_\ell}$, an application of Theorem 2.2 shows that

$$\mathbb{E}\left(\sup_t \|(\tilde{\Gamma}_{\rho_\alpha}(t))_{xy}\|^{\frac{1}{6}}\right) \leq \tilde{C}(1 + N)^{\frac{1}{2}} e^{-\frac{\eta}{6}|x-y|} \tag{68}$$

for all $x, y \in \Lambda$. Here again $\tilde{C} < \infty$ is independent of Λ, the number M of decompositions, and of the highest excitation N, and moreover, η is as above.

(iii) One can combine the cases (i) and (ii) and consider a product state ρ as in (23) where each local state is either a thermal state or an eigenstate of the Hamiltonian H_{Λ_ℓ}. The arguments in cases (i) and(ii) above provide a proof of the following result, which summarizes all the cases considered so far.

Corollary 6.2. *Fix $\beta > 0$ and $N < \infty$. Let $\rho = \otimes_{\ell=1}^M \rho_\ell$ where each of the local states ρ_ℓ is either a thermal state of H_{Λ_ℓ} with inverse temperature $\beta_\ell \in [\beta, \infty)$, or an eigenstate associated with an excitation vector α_ℓ such that $\|\alpha_\ell\|_\infty \leq N$. If $\tilde{\eta} = \frac{1}{6}\min\{\eta, \mu\}$, where η is as in (7) with $s = 1/2$ and μ is as in (63), then there exists $C < \infty$ such that*

$$\mathbb{E}\left(\sup_t \|(\tilde{\Gamma}_\rho(t))_{xy}\|^{\frac{1}{6}}\right) \leq C \max\left\{(1 + N)^{\frac{3}{2}}, \frac{1}{\beta}\right\}^{\frac{1}{3}} e^{-\tilde{\eta}|x-y|} \tag{69}$$

for all $x, y \in \Lambda$. Here C is independent of Λ, N, M and of β.

(iv) In the extreme case where each subsystem consists of only one site, i.e., $M = |\Lambda|$, the initial Hamiltonian $H_{0,\Lambda}$ is a system of non-interacting harmonic oscillators over the d dimensional lattice Λ,

$$H_{0,\Lambda} = \sum_{x \in \Lambda} H_{\{x\}} \otimes \mathbb{1}_{\Lambda \setminus \{x\}}, \tag{70}$$

where $H_{\{x\}}$ is the one dimensional harmonic oscillator

$$H_{\{x\}} = p_x^2 + \frac{k_x}{2} q_x^2. \tag{71}$$

The eigenstates of $H_{\{x\}}$ are known to be the Hermite functions

$$\phi_{n_x}(q_x) = \frac{1}{\sqrt{2^{n_x} n_x!}} \left(\frac{\sqrt{2k_x}}{2\pi}\right)^{\frac{1}{4}} H_{n_x}\left(\sqrt[4]{\frac{k_x}{2}} q_x\right) e^{-\frac{\sqrt{2k_x}}{4} q_x^2}, \tag{72}$$

where $n_x \in \mathbb{N}_0$ the excitation number at vertex $x \in \Lambda$, $H_{n_x}(\cdot)$ is the Hermite polynomial of degree n_x. In this special case, the following corollary improves on the bound in (68) for the correlations of the dynamics of the product state

$$\rho = \bigotimes_{x \in \Lambda} \rho_{n_x}, \text{ where } \rho_{n_x} = |\phi_{n_x}\rangle\langle\phi_{n_x}|. \tag{73}$$

COROLLARY 6.3. *Let ρ be as in (73) and let $N = \max_x n_x$. If η is as in (7) with $s = 1/2$, then there exists $C < \infty$, independent of Λ and N, such that*

$$\mathbb{E}\left(\sup_t \|(\tilde{\Gamma}_\rho(t))_{xy}\|^{\frac{1}{6}}\right) \le C(1+2N)^{\frac{1}{6}} e^{-\frac{\eta}{6}|x-y|} \tag{74}$$

for all $x, y \in \Lambda$.

PROOF. In the current case the local Hamiltonians $h_{\{x\}}$ reduce to the single numbers $k_x/2$. This means that in the case of an n_x excitation at site x, the correlation matrix from (38) reduces to the 2×2-matrix

$$\Gamma_{n_x}(0) = \begin{pmatrix} \frac{1}{\sqrt{2}} k_x^{-\frac{1}{2}}(1+2n_x) & \frac{1}{2}i \\ -\frac{1}{2}i & \frac{1}{2\sqrt{2}} k_x^{\frac{1}{2}}(1+2n_x) \end{pmatrix}. \tag{75}$$

Since k_x is a random variable with a bounded density ν and supported on the compact set $[0, k_{\max}]$, one gets

$$\mathbb{E}(\max\{k_x^{\frac{1}{2}}, k_x^{-\frac{1}{2}}\}) \le 2\|\nu\|_\infty \max\left\{(k_{\max})^{\frac{1}{2}}, (k_{\max})^{\frac{3}{2}}/3\right\}. \tag{76}$$

Hence, there exists $C' < \infty$ such that

$$\mathbb{E}\left(\|\Gamma_{n_x}(0)\|\right) \le C'(1+2n_x). \tag{77}$$

Given (77), the full correlation matrix for the product state $\rho = \otimes_x \rho_{n_x}$ satisfies

$$\mathbb{E}\left(\|(\Gamma_\rho)_{xy}\|\right) \le C'(1+2N)\delta_{x,y}. \tag{78}$$

Arguing as in (60), see also (61), we conclude that

$$\mathbb{E}\left(\sup_{t\in\mathbb{R}} \|(\tilde{\Gamma}_\rho(t))_{xy}\|^{\frac{1}{6}}\right) \le \tilde{C}^{\frac{2}{3}} \sum_{z,z' \in \Lambda} e^{-\eta|x-z|/3} \mathbb{E}\left(\|(\Gamma_\rho)_{zz'}\|^{\frac{1}{2}}\right)^{\frac{1}{3}} e^{-\eta|z'-y|/3} \tag{79}$$

where $\tilde{C}$ is as in (57) with $s = 1/2$. Since Holder and (78) imply that

$$\mathbb{E}\left(\|(\Gamma_\rho)_{zz'}\|^{\frac{1}{2}}\right) \le \mathbb{E}\left(\|(\Gamma_\rho)_{zz'}\|\right)^{\frac{1}{2}} \le \sqrt{C'(1+2N)}\delta_{z,z'} \tag{80}$$

the claim in (74) now follows as in the end of the proof of Theorem 2.2. □

Appendix A. Proof of Theorem 6.1

In the following we use and refine several results from [**13**] to prove Theorem 6.1. We start by noting that these results are only formulated for cubes in [**13**], but that they extend to the rectangular boxes considered here.

By Lemma 5.4 of [**13**],

$$(\Gamma_{\rho_\beta})_{xy} = \frac{1}{2}\begin{pmatrix} \langle \delta_x, h_\Lambda^{-1/2}\varphi(h_\Lambda)\delta_y\rangle & i\delta_{xy} \\ -i\delta_{xy} & \langle \delta_x, h_\Lambda^{1/2}\varphi(h_\Lambda)\delta_y\rangle \end{pmatrix}, \tag{81}$$

where we have set $\varphi(t) = \coth(\beta t^{1/2})$. Thus for (63) it suffices to show that

$$\mathbb{E}\left(|\langle \delta_x, h_\Lambda^{\pm 1/2}\varphi(h_\Lambda)\delta_y\rangle|^{1/2}\right) \le C\max\left\{1, \frac{1}{\beta}\right\} e^{-\mu|x-y|}. \tag{82}$$

Expanding $\mathbb{1} = \sum_z |\delta_z\rangle\langle\delta_z|$ and an application of Hölder's inequality show that the left hand side of (82) can be bounded by

$$\sum_z \left(\mathbb{E}(|\langle \delta_x, h_\Lambda^{\pm 1/2}\delta_z\rangle|)\right)^{1/2} \left(\mathbb{E}(|\langle \delta_z, \varphi(h_\Lambda)\delta_y\rangle|)\right)^{1/2}. \tag{83}$$

The two factors in the sum can both be bounded using Proposition A.3(c) of [**13**] and the method of its proof, respectively. For the first factor we can cite Proposition A.3(c) directly to conclude the existence of $C_1 < \infty$ and $\mu_1 > 0$ such that

$$\mathbb{E}(|\langle \delta_x, h_\Lambda^{\pm 1/2}\delta_z\rangle|) \le C_1 e^{-\mu_1|x-z|} \tag{84}$$

for all x and z.

To understand the β-dependence of the second factor, we need to analyze the proof of Proposition A.3(c) of [**13**]. It requires splitting low and high energies of h_Λ. At low energies, we can use localization of h_Λ: Our assumptions yield that there exists $E_0 > 0$ such that the Green function of h_Λ has localized s-fractional moments in $[0, E_0]$ for all $s \in (0,1)$, e.g. [**3**]. By Proposition A.3(b) of [**13**] this implies the existence of $C_2 < \infty$ and $\mu_2 > 0$ such that

$$\mathbb{E}\left(\sup_{|g|\le 1} |\langle \delta_z, h_\Lambda^{-1/2} g(h_\Lambda)\chi_{[0,E_0]}(h_\Lambda)\delta_y\rangle|\right) \le C_2 e^{-\mu_2|z-y|}. \tag{85}$$

Using the elementary bound

$$|\varphi(t)| \le \frac{\beta\sqrt{E_0}+1}{\beta} t^{-\frac{1}{2}} \tag{86}$$

for all $t \in [0, E_0]$, (85) gives

$$\mathbb{E}\left(|\langle \delta_z, \varphi(h_\Lambda)\chi_{[0,E_0]}(h_\Lambda)\delta_y\rangle|\right) \le C_2\frac{\beta\sqrt{E_0}+1}{\beta} e^{-\mu_2|z-y|}. \tag{87}$$

We will further prove that there are $C_3 < \infty$ and $\mu_3 > 0$ such that

$$\begin{aligned} \mathbb{E}\left(|\langle \delta_z, \varphi(h_\Lambda)\chi_{(E_0,\infty)}(h_\Lambda)\delta_y\rangle|\right) &\le C_3 \coth^2(\beta\sqrt{E_0})\, e^{-\mu_3|z-y|} \\ &\le C_3\left(1 + \frac{1}{\beta\sqrt{E_0}}\right)^2 e^{-\mu_3|z-y|} \end{aligned} \tag{88}$$

for all z and y.

Inserting all of (84), (87) and (88) into (83), ultimately gives the bound (82).

We still owe the proof of the first claim in (88). This is done by an analysis of the proof of Proposition A.3(c) in [**13**]. This proof, see (A.15) in [**13**], uses that

$$|\langle \delta_z, \varphi(h_\Lambda)\chi_{(E_0,\infty)}(h_\Lambda)\delta_y\rangle| \le C' \int_\Gamma |\langle \delta_z, (h_\Lambda - \zeta)^{-1}\chi_{(E_0,M]}(h_\Lambda)\delta_y\rangle|\,|d\zeta|, \tag{89}$$

where Γ is the rectangular contour with vertices $E_0 \pm i$ and $(M+1) \pm i$, M the a-priori upper bound for $\sigma(h_\Lambda)$ from (6), and $C' = \max\{|\varphi(\zeta)| : \zeta \in \Gamma\}/(2\pi)$.

Using the elementary bound $|\coth(\zeta)| \le \coth^2(\operatorname{Re}\zeta)$ for $\zeta \in \mathbb{C} \setminus \{0\}$ one has

$$|\varphi(\zeta)| \le \coth^2(\beta \operatorname{Re}\zeta^{\frac{1}{2}}) = \coth^2\left(\beta\sqrt{\frac{\operatorname{Re}\zeta + |\zeta|}{2}}\right). \tag{90}$$

This means that

$$C' \le \frac{1}{2\pi}\coth^2\left(\beta \min_{\zeta\in\Gamma}\sqrt{\frac{\operatorname{Re}\zeta + |\zeta|}{2}}\right) = \frac{1}{2\pi}\coth^2\left(\beta\sqrt{E_0}\right). \tag{91}$$

The argument in [**13**] shows that the (β-independent) integral in (89) is bounded by $C''e^{-\mu_3|z-y|}$ for some $C'' < \infty$, $\mu_3 > 0$. Combined with (91) this yields (88).

References

[1] Houssam Abdul-Rahman, *Entanglement of a class of non-Gaussian states in disordered harmonic oscillator systems*, J. Math. Phys. **59** (2018), no. 3, 031904, 17, DOI 10.1063/1.5000708. MR3777285

[2] Houssam Abdul-Rahman, Bruno Nachtergaele, Robert Sims, and Günter Stolz, *Localization properties of the disordered XY spin chain: a review of mathematical results with an eye toward many-body localization*, Ann. Phys. **529** (2017), no. 7, 201600280, 17, DOI 10.1002/andp.201600280. MR3671049

[3] Michael Aizenman and Simone Warzel, *Random operators*, Graduate Studies in Mathematics, vol. 168, American Mathematical Society, Providence, RI, 2015. Disorder effects on quantum spectra and dynamics. MR3364516

[4] Kartiek Agarwal, Ehud Altman, Eugene Demler, Sarang Gopalakrishnan, David A. Huse, and Michael Knap, *Rare-region effects and dynamics near the many-body localization transition*, Ann. Phys. **529** (2017), no. 7, 201600326, 18, DOI 10.1002/andp.201600326. MR3671051

[5] E. Altman and R. Vosk, *Universal dynamics and renormalization in many body localized systems*, Annual Review of Condensed Matter Physics **6** (2015), 383–409

[6] Vincent Beaud and Simone Warzel, *Low-energy Fock-space localization for attractive hard-core particles in disorder*, Ann. Henri Poincaré **18** (2017), no. 10, 3143–3166, DOI 10.1007/s00023-017-0591-0. MR3697191

[7] M. Cramer and J. Eisert, *Correlations, spectral gap and entanglement in harmonic quantum systems on generic lattices*, New J. Phys. **8** (2006), no. May, 71.1–71.24, DOI 10.1088/1367-2630/8/5/071. MR2245993

[8] J. Eisert, M. Cramer, and M. B. Plenio, *Colloquium: Area laws for the entanglement entropy*, Rev. Modern Phys. **82** (2010), no. 1, 277–306, DOI 10.1103/RevModPhys.82.277. MR2629609

[9] M. B. Plenio, J. Eisert, J. Dreißig, and M. Cramer, *Entropy, entanglement, and area: analytical results for harmonic lattice systems*, Phys. Rev. Lett. **94** (2005), no. 6, 060503, 4, DOI 10.1103/PhysRevLett.94.060503. MR2135112

[10] A. Elgart, A. Klein and G. Stolz, *Many-body localization in the droplet spectrum of the random XXZ quantum spin chain*, J. Funct. Anal., in press, online at https://doi.org/10.1016/j.jfa.2017.11.001

[11] John Z. Imbrie, *On many-body localization for quantum spin chains*, J. Stat. Phys. **163** (2016), no. 5, 998–1048, DOI 10.1007/s10955-016-1508-x. MR3493184

[12] Vieri Mastropietro, *Localization in interacting fermionic chains with quasi-random disorder*, Comm. Math. Phys. **351** (2017), no. 1, 283–309, DOI 10.1007/s00220-016-2773-x. MR3613505

[13] Bruno Nachtergaele, Robert Sims, and Günter Stolz, *Quantum harmonic oscillator systems with disorder*, J. Stat. Phys. **149** (2012), no. 6, 969–1012, DOI 10.1007/s10955-012-0652-1. MR3006316

[14] Bruno Nachtergaele, Robert Sims, and Günter Stolz, *An area law for the bipartite entanglement of disordered oscillator systems*, J. Math. Phys. **54** (2013), no. 4, 042110, 24, DOI 10.1063/1.4802029. MR3088230

[15] R. Nandkishore and D. A. Huse, *Many body localization and thermalization in quantum statistical mechanics*, Annual Review of Condensed Matter Physics **6** (2015), 15–38

[16] A. Polkovnikov and K. Sengupta and A. Silva and M. Vengalattore, *Colloquium: Nonequilibrium dynamics of closed interacting quantum systems*, Rev. Mod. Phys. **83** (2011), 863.

[17] Norbert Schuch, J. Ignacio Cirac, and Michael M. Wolf, *Quantum states on harmonic lattices*, Comm. Math. Phys. **267** (2006), no. 1, 65–92, DOI 10.1007/s00220-006-0049-6. MR2238904

[18] Robert Seiringer and Simone Warzel, *Decay of correlations and absence of superfluidity in the disordered Tonks-Girardeau gas*, New J. Phys. **18** (2016), no. March, 035002, 14, DOI 10.1088/1367-2630/18/3/035002. MR3484875

[19] M. Serbyn and Z. Papić and D. Abanin, *Quantum quenches in the many-body localized phase*, Physical Review B **90** (2014), 174302

Department of Mathematics, University of Arizona, Tucson, Arizona 85721
Email address: `houssam@math.arizona.edu`

Department of Mathematics, University of Arizona, Tucson, Arizona 85721
Email address: `rsims@math.arizona.edu`

Department of Mathematics, University of Alabama at Birmingham, Birmingham, Alabama 35294
Email address: `stolz@uab.edu`

Contemporary Mathematics
Volume **717**, 2018
https://doi.org/10.1090/conm/717/14440

The lattice Anderson model with discrete disorder

John Z. Imbrie

ABSTRACT. Consider the Anderson tight-binding model on the lattice in any dimension, with single-site potential having a discrete distribution taking N values, with N large. We discuss recent work elucidating mechanisms by which randomness localizes eigenfunctions, smooths out eigenvalue distributions, and produces eigenvalue separation.

1. Introduction

Quantum systems with disorder present particular mathematical challenges. In order to understand the phenomenon of localization in such systems, it is important to get control over eigenvalue statistics, especially the density of states. The Anderson model [**And58**] describes a particle on a lattice with a random potential. If the distribution of the potential has some regularity, then one may control the density of states using the Wegner argument [**Weg81**]. But this breaks down if the potential has a discrete distribution.

The case of a binary distribution has been called the Anderson-Bernoulli model. Localization in one dimension has been proven on the lattice [**CKM87**, **SVW98**] and in the continuum [**DSS02**]. Further results on the lattice in one dimension include improved regularity of the density of states for weak disorder [**Bou12**, **Bou14**]. In higher dimensions, results on localization have been confined to the continuum case [**BK05**], where localization was established near the bottom of the spectrum, using a quantitative form of the unique continuation principle. This led to several generalizations in other cases involving singular potentials [**GK07**, **GHK07**, **GK13**, **KT16**]. Recently, the author has obtained results on localization and eigenvalue statistics for all energies for the lattice Anderson model with discrete disorder distribution taking N values, with N large [**Imb17**]. Here we present some of the key ideas in this work.

In the course of the analysis, it was essential to obtain a degree of energy level separation. This is a particular challenge in the case of a discrete distribution, because degeneracies appear with positive probability in small subsystems. Thus it is essential to demonstrate that as larger neighborhoods of a subsystem are taken into account, the disorder has a tendency to break these degeneracies. However, due to the exponential decay of influence of distant random variables, the gaps between eigenvalues may be extremely small.

2010 *Mathematics Subject Classification.* Primary 82D30, 82B44.

It should be evident that controlling the statistical separation of eigenvalues is important tool for the mathematical study of disordered quantum systems. The author's proof of many-body localization [**Imb16b**] was dependent on an assumption on level-separation for that system. The usual Minami estimate [**Min96**] is not available for discrete disorder distributions. So in order to handle the level-separation problem, we use a method of iterated Schur complementation. A sequence of reductions to lower-dimensional problems with associated effective Hamiltonians allows a precise examination of the dependence on the disorder. The reductions are accomplished by eliminating sites or regions where the spectrum of the effective Hamiltonian is disjoint from the energy interval of interest. The regions that remain may be called resonant regions or resonant blocks. One may then examine the way the spectrum of the effective Hamiltonian for a resonant block depends on the disorder. We use the disorder in the vicinity of a resonant block to produce eigenvalue movement and separation.

2. Model and Results

We work on a rectangular sublattice $\Lambda \subset \mathbb{Z}^d$. The Anderson model Hamiltonian is

$$H = -\gamma\Delta + v, \tag{2.1}$$

where Δ is the lattice Laplacian and v is multiplication by the lattice potential v_x, $x \in \Lambda$. We assume that $\{v_x\}_{x\in\Lambda}$ is a collection of iid random variables, each with a uniform distribution on $\{0, \frac{1}{N-1}, \frac{2}{N-1}, \dots, 1\}$, with N an integer much greater than 1. This generalizes the Anderson-Bernouilli model, which corresponds to $N = 2$. The diagonal entries of H are $2d\gamma + v_x$, and the off-diagonal part of H is γJ, where

$$J_{xy} = \begin{cases} 1, & \text{if } |x-y| = 1; \\ 0, & \text{otherwise,} \end{cases} \tag{2.2}$$

and $|x| = \sum_{i=1}^d |x_i|$.

In stating our results, we use a parameter $p > d$, a sufficiently large constant that serves as the exponent for log-Hölder continuity of the density of states. We take N sufficiently large, depending on p. Then we take $\gamma \le \varepsilon^{20}$, where $\varepsilon \equiv \frac{1}{N-1}$. Let $I_\delta(E) = [E-\delta, E+\delta]$, and let $\mathcal{N}(I)$ denote the number of eigenvalues of H in I. Let $\{E_\beta, \varphi_\beta\}_{\beta=1,\dots,|\Lambda|}$ denote the eigenvalues and associated normalized eigenvectors of H.

THEOREM 2.1. *Choose a sufficiently large p. Then for N sufficiently large (depending on p) and γ sufficiently small (depending on N),*

$$\mathbb{E}\mathcal{N}\big(I_\delta(E)\big) \le |\Lambda|(\log_\gamma \delta)^{-p}. \tag{2.3}$$

for any rectangle Λ and any $\delta \in [\gamma^{\operatorname{diam}(\Lambda)/2}, 1]$. Furthermore, the eigenfunction correlator satisfies

$$\mathbb{E}\sum_\beta |\varphi_\beta(x)\varphi_\beta(y)| \le \big(|x-y| \vee 1\big)^{-(p/2-d-1)}. \tag{2.4}$$

Finally, the minimum eigenvalue spacing satisfies

$$P\Big(\min_{\beta\neq\tilde\beta} |E_\beta - E_{\tilde\beta}| < \delta\Big) \le |\Lambda|^2(\log_\gamma \delta)^{-(p/2-1)}, \tag{2.5}$$

for any rectangle Λ *and any* $\delta \in [\gamma^{\mathrm{diam}(\Lambda)/2}, 1]$.

Although we only establish power-law decay of the eigenfunction correlator, we do obtain exponential decay of the eigenfunctions, in the sense that

$$P\Big(\max_{y:\,|y-x|\geq R}\sum_{\beta}|\varphi_\beta(x)\varphi_\beta(y)|\gamma^{-|x-y|/5} > 1\Big) \leq R^{-(p/2-4d-1)}. \tag{2.6}$$

Note that (2.5) implies that the probability of an exact degeneracy in the spectrum of H decreases as a large power of diam(Λ).

3. Resonant Blocks and Random-Walk Expansions

3.1. Schur Complement. The following lemma from [**IM16**] shows how to reduce to an effective Hamiltonian on a lower-dimensional subspace. The Schur complement gives an accurate picture of the spectrum in some neighborhood about a target energy E.

LEMMA 3.1. *Let* K *be a* $(p+q)\times(p+q)$ *symmetric matrix in block form,* $K = \left(\begin{smallmatrix} A & B \\ C & D\end{smallmatrix}\right)$, *with* A *a* $p\times p$ *matrix,* D *a* $q\times q$ *matrix, and* $C = B^T$. *Assume that* $\|(D-E)^{-1}\| \leq \tilde{\varepsilon}^{-1}$, $\|B\| \leq \tilde{\gamma}$, $\|C\| \leq \tilde{\gamma}$. *Define the Schur complement with respect to* λ:

$$F_\lambda \equiv A - B(D-\lambda)^{-1}C. \tag{3.1}$$

Let $\tilde{\varepsilon}$ *and* $\tilde{\gamma}/\tilde{\varepsilon}$ *be small, and* $|\lambda - E| \leq \tilde{\varepsilon}/2$. *Then*

(1) *If* φ *is an eigenvector for* F_λ *with eigenvalue* λ, *then* $\big(\varphi, -(D-\lambda)^{-1}C\varphi\big)$ *is an eigenvector for* K *with eigenvalue* λ, *and all eigenvectors of* K *with eigenvalue* λ *are of this form.*

(2)

$$\|F_\lambda - F_E\| \leq 2\Big(\frac{\tilde{\gamma}}{\tilde{\varepsilon}}\Big)^2|\lambda - E|. \tag{3.2}$$

(3) *The spectrum of* K *in* $[E-\tilde{\varepsilon}/2, E+\tilde{\varepsilon}/2]$ *is in close agreement with that of* F_E *in the following sense. If* $\lambda_1 \leq \lambda_2 \leq \ldots \leq \lambda_m$ *are the eigenvalues of* K *in* $[E-\tilde{\varepsilon}/2, E+\tilde{\varepsilon}/2]$, *then there are corresponding eigenvalues* $\tilde{\lambda}_1 \leq \tilde{\lambda}_2 \leq \ldots \leq \tilde{\lambda}_m$ *of* F_E, *and* $|\lambda_i - \tilde{\lambda}_i| \leq 2(\tilde{\gamma}/\tilde{\varepsilon})^2|\lambda_i - E|$.

PROOF. The first statement follows from the fact that $\big(\left(\begin{smallmatrix} A & B \\ C & D\end{smallmatrix}\right) - \lambda\big)\left(\begin{smallmatrix} \varphi \\ \tilde{\varphi}\end{smallmatrix}\right) = 0$ if and only if $C\varphi + (D-\lambda)\tilde{\varphi} = 0$ and $(F_\lambda - \lambda)\varphi = 0$. The second statement follows by writing

$$F_E - F_\lambda = B(D-E)^{-1}(\lambda - E)(D-\lambda)^{-1}C. \tag{3.3}$$

and using the assumed bounds on each factor. The last statement follows from Weyl's inequality and (3.2). □

The lemma provides an algorithm for finding the eigenvalues of K near E. Again using Weyl's inequality and (3.2), we see that the eigenvalues of F_λ are Lipschitz continuous functions of λ, with a small Lipschitz constant. Therefore, the eigenvalues of K near E may be found using a fixed point argument, to solve the condition $\lambda \in \operatorname{spec} F_\lambda$.

3.2. First Step. The spectrum of $H = -\gamma\Delta + v$ is contained in the interval $[0, 1+4d\gamma]$. Let us work in the vicinity of some energy E in this interval. Each v_x is a multiple of $\frac{1}{N-1}$ in $[0,1]$ in $[0,1]$, so the spacing between these values is $\varepsilon \equiv \frac{1}{N-1}$. Let us define $\varepsilon_1 \equiv \varepsilon/3$. We say that a site x is resonant to E if $v_x + 2d\gamma \in I_{\varepsilon_1}(E)$. It should be clear that at most one potential value v_x will permit x to be resonant to E. Thus the probability that x is resonant to E is bounded by $\frac{1}{N} < \varepsilon$. Since N is large, the set of resonant sites will typically be a dilute subset of Λ. We denote this subset by $R^{(1)}$.

It is natural to use the set of resonant sites as the basis for a block decomposition of the Hamiltonian:

$$H = \begin{pmatrix} A^{(1)} & B^{(1)} \\ C^{(1)} & D^{(1)} \end{pmatrix}, \tag{3.4}$$

with $A^{(1)}$ denoting the restriction of H to the subspace with indices in $R^{(1)}$, and $D^{(1)}$ denoting the restriction to the space with indices in $\Lambda \setminus R^{(1)}$. This leads to the Schur complement

$$F_\lambda^{(1)} \equiv A^{(1)} - B^{(1)}(D^{(1)} - \lambda)^{-1}C^{(1)}. \tag{3.5}$$

We may decompose $D^{(1)}$ into diagonal and off-diagonal parts:

$$D^{(1)} = W^{(1)} - V^{(1)}, \tag{3.6}$$

with $W_{xy}^{(1)} = (2d\gamma + v_x)\delta_{xy}$ and $V_{xy}^{(1)} = \gamma J_{xy}$. Let us take $\lambda - E \le \varepsilon_1/2$, so that $(2d\gamma + v_x - \lambda)^{-1} \le 2/\varepsilon_1$ for $x \in \Lambda \setminus R^{(1)}$. Then the Neumann series

$$(D^{(1)} - \lambda)^{-1} = (W^{(1)} - \lambda)^{-1} + (W^{(1)} - \lambda)^{-1}V^{(1)}(W^{(1)} - \lambda)^{-1} + \dots \tag{3.7}$$

converges for γ small. Since $V^{(1)}$ induces nearest-neighbor steps, we may expand the matrix products to obtain a random-walk expansion for $\left[B^{(1)}(D^{(1)} - \lambda)^{-1}C^{(1)}\right]_{xy}$ as a sum of graphs from x to y, with bounds that decay exponentially in the length of the walk. Thus the Schur complement matrix $F_\lambda^{(1)}$ has long-range (but exponentially decaying) interactions.

To set the stage for the next step, we introduce a sequence of length scales,

$$L_k = L_0 2^k, k = 1, 2, \dots, \tag{3.8}$$

where L_0 to be a large integer, whose choice will depend on the value of p. We form connected components of $R^{(1)}$ by declaring that x and y are connected if $|x - y| \le L_1^\alpha$, with $\alpha \equiv \frac{3}{2}$. Let B_1 denote one of these components. Then let $\bar{B}_1$ denote the set of lattice points within a distance $2L_1$ of B_1. Each component or block B_1 will be considered an indivisible unit (the analog of a site) in subsequent steps. We need to distinguish between small blocks satisfying $\operatorname{diam}(B_1) \le L_1$ and the rest. The small blocks are called *isolated* because the distance to other blocks is much greater than their diameter. For each isolated component, we define a localized version of $F_\lambda^{(1)}$:

$$[\tilde{F}_\lambda^{(1)}(B_1)]_{xy} \equiv A_{xy}^{(1)} - \sum_{g_1 : x \to y,\ g_1 \subseteq \bar{B}_1} \prod_{i=1}^{m} \frac{1}{2d\gamma + v_{x_i} - \lambda} \prod_{j=0}^{m} V_{x_j x_{j+1}}^{(1)}. \tag{3.9}$$

We have written out the random-walk expansion explicitly as a sum of walks g_1 from x to y, but here the walks are required to visit only sites within $\bar{B}_1$. The point is to specify for the next step what constitutes the "diagonal" part of the effective Hamiltonian $F_\lambda^{(1)}$.

3.3. k^{th} Step. As in the first step, we need to define a notion of resonance. But this will only apply to the isolated blocks, because the others are still too singular to participate in the next random-walk expansion (they are too large). We define the energy window width for the kth step:

$$\varepsilon_k \equiv \gamma^{1.6L_k}, \text{ for } k > 1. \tag{3.10}$$

This is designed to be a bit larger than anticipated bounds on the terms discarded in (3.9). In the kth step, we are working with blocks B_{k-1} that are the connected components of the $(k-1)^{\text{st}}$ resonant set $R^{(k-1)}$. Here connectivity is defined using connections of length L_{k-1}^{α}. For each block we have a matrix $\tilde{F}_\lambda^{(k-1)}(B_{k-1})$ that is obtained by restricting the set of graphs that define $F_\lambda^{(k-1)}$ to those that remain within $\bar{B}_{k-1}$. Here the set $\bar{B}_{k-1}$ is essentially the neighborhood of width $2L_{k-1}$ about B_{k-1}, but its boundary is constructed so as to maintain a distance $L_j^{\sqrt{\alpha}}$ from the sets $\bar{B}_j$ from prior scales. These constructions are modeled on those in [**FS83**].

DEFINITION 3.2. Let B_{k-1} be a component of $R^{(k-1)}$ with $k \geq 2$. We say that B_{k-1} is **resonant in step** k if it is isolated in step k and if

$$\operatorname{dist}\big(\operatorname{spec} \tilde{F}_E^{(k-1)}(B_{k-1}), E_k\big) \leq \varepsilon_k. \tag{3.11}$$

The new resonant set $R^{(k)}$ is obtained by deleting from $R^{(k-1)}$ all of its components that are isolated but not resonant in step k. As in the first step, $R^{(k)}$ determines the block decomposition

$$F_\lambda^{(k-1)} = \begin{pmatrix} A^{(k)} & B^{(k)} \\ C^{(k)} & D^{(k)} \end{pmatrix}. \tag{3.12}$$

Then we may define the next Schur complement matrix

$$F_\lambda^{(k)} = A^{(k)} - B^{(k)}(D^{(k)} - \lambda)^{-1}C^{(k)}. \tag{3.13}$$

In order to generate the random-walk expansion in the kth step, we restrict to the neighborhood $|\lambda - E| \leq \varepsilon_k/2$ and write

$$D^{(k)} = W^{(k)} - V^{(k)}, \tag{3.14}$$

where $W^{(k)}$ is a block diagonal matrix formed out of blocks $\tilde{F}_\lambda^{(k-1)}(B_{k-1})$. The "off-diagonal" matrix $V^{(k)}$ contains the long graphs not included in $\tilde{F}_\lambda^{(k-1)}(B_{k-1})$. Then we expand $(D^{(k)} - \lambda)^{-1}$ in a Neumann series:

$$(D^{(k)} - \lambda)^{-1} = (W^{(k)} - \lambda)^{-1} + (W^{(k)} - \lambda)^{-1}V^{(k)}(W^{(k)} - \lambda)^{-1} + \dots. \tag{3.15}$$

Working inductively, we can use bounds on the random-walk expansion to prove a Lipschitz bound

$$\|\tilde{F}_\lambda^{(k-1)}(B_{k-1}) - \tilde{F}_E^{(k-1)}(B_{k-1})\| \leq \gamma|\lambda - E_k|. \tag{3.16}$$

This is less than $\varepsilon_k/6$, because $|\lambda - E| \leq \varepsilon_k/2$. This series involves only nonresonant blocks satisfying $\operatorname{dist}\big(\operatorname{spec} \tilde{F}_E^{(k-1)}(B_{k-1}), E\big) \geq \varepsilon_k$, and so $\|(W^{(k)} - \lambda)^{-1}\| \leq 3\varepsilon_k^{-1}$. Although this is a large factor, the graphs contributing to $V^{(k)}$ extend at least a distance $4L_{k-1} = 2L_k$ (from B_{k-1} to $\bar{B}_{k-1}^{\mathrm{c}}$ and back). Our inductive bounds on the random-walk expansion imply decay like $\gamma^{.85R}$ for graphs of length R.

Thus $\|V^{(k)}\| \leq \gamma^{1.7L_k}$, which is sufficient to control a factor of $\varepsilon_k^{-1} = \gamma^{-1.6L_k}$ (compare with the condition $\tilde{\gamma}/\tilde{\varepsilon} \ll 1$ in Lemma 3.1). See [**Imb17**] for further details on the random-walk expansion bounds.

4. Density of States Smoothing from Environmental Disorder

The basic mechanism for demonstrating some degree of smoothness in the density of states is the reduction of the size of $R^{(k)}$ as k grows and ε_k shrinks. This goes hand in hand with a decrease in $\mathcal{N}(I_\delta(E))$ as δ decreases. Consider the fate of a block B_{k-1} in the kth step. There are four possibilities. The block may fail to be resonant in step k, in which case it is not part of $R^{(k)}$. It may join up with other blocks in forming a component B_k of $R^{(k)}$. It may survive to $R^{(k)}$ on account of its not being isolated. Finally, it may be isolated and resonant in step $k-1$ and remain resonant in step k. In the first case, the requisite reduction in the size of $R^{(k)}$ does occur. In the second and third cases, there is no reduction, but the probability is small, due to the large size of the blocks involved. The fourth case requires careful analysis, because we need to argue for the rarity of the set of potentials in $\bar{B}_k \setminus \bar{B}_{k-1}$ that permit the continued resonance of B_{k-1} in step k.

4.1. Influential Sites. We begin the analysis by establishing the existence of sites with significant influence on the relevant eigenvalues. Let H_X denote the matrix obtained by restricting H to $X \subseteq \Lambda$. It is useful to note that $\tilde{F}_\lambda^{(k-1)}(B_{k-1})$ is the same as what we would have obtained for $F_\lambda^{(k-1)}(B_{k-1})$, if we had begun with $H_{\bar{B}_{k-1}}$. (This is the effect of restricting graphs to $\bar{B}_{k-1}$.) Thus, any solution to $\lambda \in \operatorname{spec} \tilde{F}_\lambda^{(k-1)}(B_{k-1})$ in $I_{\varepsilon_k/2}(E_k)$ corresponds to an eigenvalue of $H_{\bar{B}_{k-1}}$, via repeated application of Lemma 3.1. Likewise, the eigenfunctions φ_λ corresponding to each such solution map to an eigenfunction ψ_λ of $H_{\bar{B}_{k-1}}$. The lemma shows that we may write this map as $\psi_\lambda = G_\lambda^{(k-1)}\varphi_\lambda$, where $G_\lambda^{(k-1)}$ is given by a product of operators $\left(\begin{smallmatrix} I \\ -(D^{(j)}-\lambda)^{-1}C^{(j)}\end{smallmatrix}\right)$ mapping functions on $R^{(j)}$ to functions on $R^{(j-1)}$ (working all the time in $\bar{B}_{k-1}$).

LEMMA 4.1. *Let B_{k-1} be isolated in step k, i.e.* $\operatorname{diam}(B_{k-1}) \leq L_{k-1}$. *Let $\psi = G_\lambda^{(k-1)}\varphi$, with $(\tilde{F}_\lambda^{(k-1)}(B_{k-1}) - \lambda)\varphi = 0$, $\|\varphi\| = 1$, and $\lambda \in I_{\varepsilon_k/2}(E)$. Then $(H_{\bar{B}_{k-1}} - \lambda)\psi = 0$. For any y with* $\operatorname{dist}(y, \bar{B}_{k-1}) = 1$, *define the influence of y as*

$$\mathcal{I}_\psi(y) = \left| \sum_{x \in \bar{B}_{k-1},\, |x-y|=1} \psi(x) \right|. \tag{4.1}$$

If $\operatorname{diam}(\bar{B}_{k-1}) < \operatorname{diam}(\Lambda)$, *then for γ small, there exists at least one $y \in \Lambda \setminus \bar{B}_{k-1}$ with $\mathcal{I}_\psi(y) \geq \gamma^{3.1L_{k-1}}$.*

PROOF. Since φ is normalized, there must be a point $\bar{x}$ of B_{k-1} such that $|\varphi(\bar{x})| \geq |B_{k-1}|^{-1/2}$. Let us use $\bar{x}$ as the origin of coordinates in Λ. Then let us assume that the z-coordinate runs toward a boundary face of Λ that contains no points of $\bar{B}_{k-1}$ – this is possible because $\operatorname{diam}(\bar{B}_{k-1}) < \operatorname{diam}(\Lambda)$. The construction of $\bar{B}_{k-1}$ ensures that it extends no further than a distance $2.05L_{k-1}$ from B_{k-1}. Let $z_{\max}$ denote the maximal z-coordinate for points in $\bar{B}_{k-1}$. Then $z_{\max} \leq 3.05L_{k-1}$.

Working toward a proof by contradiction, let us suppose that there is no site $y \in \Lambda \setminus \bar{B}_{k-1}$ with $\mathcal{I}_\psi(y) \geq \gamma^{3.1L_{k-1}}$. Then for each x_0 in the top layer at $z = z_{\max}$

we have $|\psi(x_0)| < \gamma^{3.1L_{k-1}}$. (Each site y with d^{th} coordinate $z_{\max}+1$ is in Λ and is adjacent to no more than one site of $\bar{B}_{k-1}$, so the sum in (4.1) reduces to a single term.)

Let x be a site of $\bar{B}_{k-1}$ that is immediately below a top-layer site x_0 of $\bar{B}_{k-1}$. Let $y_1, \ldots, y_{2d-1}$ denote the other neighbors of x_0. Then

$$-\gamma\left(\psi(x) + \sum_{i-1}^{2d-1} \psi(y_i)\right) + (2d\gamma + v_{x_0} - \lambda)\psi(x_0) = 0, \tag{4.2}$$

where we put $\psi(x) = 0$ for $x \notin \bar{B}_{k-1}$. Observing that $\lambda \in [0, 1+4d\gamma]$, $v_{x_0} \in [0,1]$, we have that

$$|2d\gamma + v_{x_0} - \lambda| \leq 1 + 2d\gamma. \tag{4.3}$$

Hence

$$|\psi(x)| \leq [\tfrac{1}{\gamma}(1+2d\gamma) + 2d - 1]\gamma^{3.1L_{k-1}} = (\tfrac{1}{\gamma} + 4d - 1)\gamma^{3.1L_{k-1}} \leq \tfrac{2}{\gamma}\gamma^{3.1L_{k-1}}. \tag{4.4}$$

The remaining sites x in the second layer lie below a site $y \notin \bar{B}_{k-1}$, which then must satisfy $\mathcal{I}_\psi(y) < \gamma^{3.1L_{k-1}}$. We have already established that the other neighbors of y satisfy $|\psi(y_i)| \leq \gamma^{3.1L_{k-1}}$. Therefore, $|\psi(x)| \leq 2d\gamma^{3.1L_{k-1}}$ (otherwise, even after a cancellation with the other neighbors, $\mathcal{I}_\psi(y)$ would be too large). Thus (4.4) holds for all sites in the second layer (for γ small).

We continue this argument on successive layers, obtaining a bound

$$|\psi(x)| \leq \left(\tfrac{2}{\gamma}\right)^{z_{\max}-z}\gamma^{3.1L_{k-1}} \leq (2\gamma)^{.05L_{k-1}} \tag{4.5}$$

for the layer with d-coordinate $z \geq 0$. Thus we learn that $|\psi(\bar{x})| = |\varphi(\bar{x})| \leq (2\gamma)^{.05L_{k-1}}$. This contradicts our assumption that $|\varphi(\bar{x})| \geq |B_{k-1}|^{-1/2} \geq (2L_{k-1}+1)^{-d/2}$. Hence there must be at least one influential site $y \in \Lambda \setminus \bar{B}_{k-1}$ satisfying $\mathcal{I}_\psi(y) \geq \gamma^{3.1L_{k-1}}$. □

Note that this argument does not work for every lattice. For example, on the Kagome lattice, some points are connected to more than one point at the next level down. See Fig. 1. Indeed, on the Kagome lattice there may be compactly supported eigenfunctions, which would be immune to the influence of disorder on neighboring sites. In contrast, on the rectangular lattice, eigenfunctions cannot exhibit uniform decay that is faster than an exponential [**CS83**].

4.2. Local Degeneracy of the Spectrum. We will need to follow the behavior of the number of eigenvalues in small windows around various energies. Define for each k and each component of $R^{(k)}$

$$\hat{n}_k(B_k) = \text{the number of eigenvalues of } \tilde{F}_E^{(k)}(B_k) \text{ in } I_{\varepsilon_{k+1}}(E). \tag{4.6}$$

Here we count eigenvalues with multiplicity.

PROPOSITION 4.2. *Let L_0 be sufficiently large. Take $\varepsilon = \frac{1}{N-1}$ to be sufficiently small, depending on L_0, and take $\gamma \leq \varepsilon^{20}$. Assume that B_{k-1} is isolated and resonant in step k with respect to energy E, and that diam$(\bar{B}_{k-1}) <$ diam(Λ). Assume that B_{k-1} remains isolated in step $k+1$, so that $B_k = B_{k-1}$. Fix all v_y for $y \in \bar{B}_{k-1}$. These determine a particular $\bar{y}$ such that dist$(\bar{y}, \bar{B}_{k-1}) = 1$. Fix all remaining $v_y \in \bar{B}_k$, $y \neq \bar{y}$. Then*

$$\hat{n}_k(B_k) \leq \hat{n}_{k-1}(B_{k-1}). \tag{4.7}$$

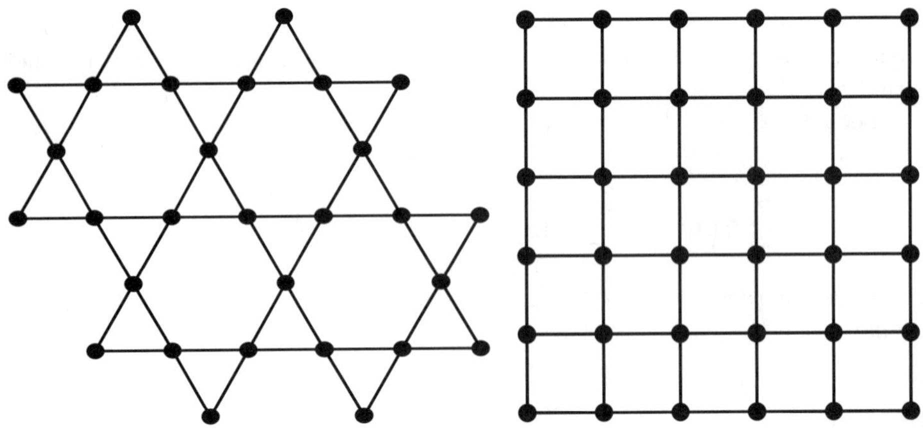

FIGURE 1. On the Kagome lattice (left), some points are connected to more than one point with smaller z-coordinate. Each point on the rectangular lattice (right) is singly connected to the layer below.

Furthermore,

$$\hat{n}_k(B_k) < \hat{n}_{k-1}(B_{k-1}) \tag{4.8}$$

for all but one value of $v_{\bar{y}}$.

We outline the main ideas behind the proof. To begin, we need to produce a useful representation for

$$\Delta F_\lambda \equiv \tilde{F}_\lambda^{(k-1)}(B_{k-1}) - \tilde{F}_\lambda^{(k)}(B_k), \tag{4.9}$$

which represents the change in the effective Hamiltonian as we expand the domain from $\bar{B}_{k-1}$ to $\bar{B}_k$. This will lead to a choice of $\bar{y}$ as a site adjacent to $\bar{B}_{k-1}$ with maximum influence on the set of eigenvalues in $I_{\varepsilon_k/2}(E)$, based on the leading term in this representation. Lemma 4.1 provides a lower bound on this influence. We show that higher-order terms are either smaller than the leading term, or independent of $v_{\bar{y}}$. Then it will be possible to demonstrate sufficient eigenvalue movement to obtain (4.8).

Let λ_0 be the closest eigenvalue of $\tilde{F}_E^{(k-1)}(B_{k-1})$ to E. We can assume that $\lambda_0 \in I_{\varepsilon_k/9}(E)$, because otherwise we would have that $\hat{n}(B_k) = 0$, due to the rapid contraction of energy windows. Let λ be the solution to $\lambda \in \operatorname{spec} \tilde{F}_\lambda^{(k-1)}(B_{k-1})$ that is closest to λ_0. Using the Lipschitz continuity of $\tilde{F}_\lambda^{(k-1)}(B_{k-1})$, there is a solution satisfying $|\lambda - E| \le \varepsilon_k/6$. Define

$$\hat{n} = \text{the number of eigenvalues of } \tilde{F}_\lambda^{(k-1)}(B_{k-1}) \text{ in } I_{\varepsilon_k/2}(E_{k+1}). \tag{4.10}$$

Then we have that

$$\hat{n} \le \hat{n}_{k-1}(B_{k-1}), \tag{4.11}$$

because eigenvalues of $\tilde{F}_E^{(k-1)}(B_{k-1})$ outside of $I_{\varepsilon_k}(E)$ cannot migrate into $I_{\varepsilon_k/2}(E)$ with the change $E \to \lambda$.

Using resolvent identities, one can show that

$$\Delta F_\lambda = \tilde{G}_\lambda^{(k-1)\mathrm{tr}} \Gamma \big(H_{\bar{B}_k \setminus B_k} - \lambda\big)^{-1} \Gamma \tilde{G}_\lambda^{(k-1)}, \tag{4.12}$$

where

$$\Gamma_{xy} = \begin{cases} \gamma, & \text{if } |x-y| = 1 \text{ with one in } \bar{B}_{k-1} \text{ and the other in } \bar{B}_k; \\ 0, & \text{otherwise.} \end{cases} \tag{4.13}$$

Here $\tilde{G}_\lambda^{(k-1)} : L^2(B_{k-1}) \to L^2(\bar{B}_{k-1} \backslash B_{k-1})$ is the operator providing the extension of the eigenfunction from B_{k-1} to $\bar{B}_{k-1} \setminus B_{k-1}$. Thus $G_\lambda^{(k-1)} = \big(\begin{smallmatrix} I \\ \tilde{G}_\lambda^{(k-1)} \end{smallmatrix}\big)$. We may interpret (4.12) graphically: the operators $\tilde{G}_\lambda^{(k-1)}$ involve graphs extending from B_{k-1} to its boundary, then Γ produces a step into $\bar{B}_k \setminus \bar{B}_{k-1}$, and $\big(H_{\bar{B}_k \backslash B_k} - \lambda\big)^{-1}$ connects the two ends of the graph with walks in $\bar{B}_k \setminus B_k$. All these graphical expansions are under control, with bounds $\sim \gamma^{.85R}$ for graphs of length R.

Let us use a basis of normalized eigenvectors $\{\varphi_1, \ldots, \varphi_{\hat{n}}, \varphi_{\hat{n}+1}, \ldots, \varphi_n\}$ corresponding to eigenvalues $\lambda_1, \ldots, \lambda_{\hat{n}}, \lambda_{\hat{n}+1}, \ldots, \lambda_n$. of $\tilde{F}_\lambda^{(k-1)}(B_{k-1})$. Here $\lambda = \lambda_1, \lambda_2, \ldots, \lambda_{\hat{n}}$ are the eigenvalues in $I_{\varepsilon_k/2}(E_{k+1})$, and $n = |B_{k-1}|$ is the number of sites in B_{k-1}. In this basis,

$$\Delta F_{\lambda,\beta\tilde{\beta}} = \big\langle \varphi_\beta, \tilde{G}_\lambda^{(k-1)\mathrm{tr}} \Gamma \big(H_{\bar{B}_k \backslash B_k} - \lambda\big)^{-1} \Gamma \tilde{G}_\lambda^{(k-1)} \varphi_{\tilde{\beta}} \big\rangle. \tag{4.14}$$

Define for any $y \in \bar{B}_k \setminus \bar{B}_{k-1}$ with $\operatorname{dist}(y, \bar{B}_{k-1}) = 1$

$$\chi_y(x) = \begin{cases} 1, & \text{if } |x-y| = 1 \text{ and } x \in \bar{B}_{k-1}; \\ 0, & \text{otherwise.} \end{cases} \tag{4.15}$$

This is the indicator function for the set of sites in $\bar{B}_{k-1}$ that are adjacent to y. Then put

$$a_\beta(y) \equiv \langle \chi_y, \tilde{G}_\lambda^{(k-1)} \varphi_\beta \rangle. \tag{4.16}$$

The vector $\mathbf{a}^{(\mathrm{r})}(y) = (a_1(y), \ldots, a_{\hat{n}}(y))$ – in particular its length-squared $|\mathbf{a}^{(\mathrm{r})}(y)|^2 = \sum_{\beta=1}^{\hat{n}} a_\beta(y)^2$ – is a measure of the influence of v_y on the family of eigenvalues $\{\lambda_1, \ldots, \lambda_{\hat{n}}\}$. Choose $\bar{y} \in \Lambda$ with $\operatorname{dist}(\bar{y}, \bar{B}_{k-1}) = 1$ to be a site that maximizes $|\mathbf{a}^{(\mathrm{r})}(y)|$ from amongst all neighbors of $\bar{B}_{k-1}$. Lemma 4.1 implies that $|a_1(y)| \geq \gamma^{3.1L_{k-1}}$ for at least one y adjacent to $\bar{B}_{k-1}$. Hence $|\mathbf{a}^{(\mathrm{r})}(\bar{y})| \geq \gamma^{3.1L_{k-1}}$.

Let us write

$$\Delta F_{\lambda,\beta\tilde{\beta}} = \sum_{xy} a_\beta(x) K(x,y) a_{\tilde{\beta}}(y), \tag{4.17}$$

where

$$K(x,y) \equiv \gamma^2 \big(H_{\bar{B}_k \backslash B_k} - \lambda\big)^{-1}_{xy} = K_0(x,y) + K_1(x,y) + K_2(x,y). \tag{4.18}$$

for $x, y \in \bar{B}_k \setminus \bar{B}_{k-1}$ and adjacent to $\bar{B}_{k-1}$. Here K_0 is the sum of graphs for $\big(H_{\bar{B}_k \backslash \bar{B}_{k-1}} - \lambda\big)^{-1}$ that do not include the site $\bar{y}$ (which means that K_0 is independent of $v_{\bar{y}}$). For K_1, only the trivial graph of at $\bar{y}$ is included; thus

$$K_1(x,y) = \frac{\gamma^2 \delta_{x\bar{y}} \delta_{\bar{y}y}}{v_{\bar{y}} + 2d\gamma - \lambda}. \tag{4.19}$$

The remaining graphs make up K_2; they must contain $\bar{y}$ and have at least three steps.

To proceed further, we would need to make another Schur complement, to focus on the modes $\{\varphi_1, \ldots, \varphi_{\hat{n}}\}$. But let us avoid this complication by assuming that $n = \hat{n}$. The following proposition gives the estimates needed for the proof of Proposition 4.2.

PROPOSITION 4.3. *Let L_0 be sufficiently large. Take $\varepsilon = \frac{1}{N-1}$ to be sufficiently small, depending on L_0, and take $\gamma \leq \varepsilon^{20}$. Assume that B_{k-1} is isolated and resonant in step k with respect to energy E_k, and that B_{k-1} remains isolated in step $k+1$, so that $B_{k-1} = B_k$. Given E, there is a decomposition*
(4.20)
$$\left(\tilde{F}_\lambda^{(k-1)}(B_{k-1}) - \tilde{F}_E^{(k)}(B_k)\right)_{\beta\tilde{\beta}} = \frac{\gamma^2}{v_{\bar{y}} + 2d\gamma - \lambda} a_\beta^{(r)}(\bar{y}) a_{\tilde{\beta}}^{(r)}(\bar{y}) + \mathcal{C}_{\beta\tilde{\beta}} + \mathcal{R}(v_{\bar{y}})_{\beta\tilde{\beta}},$$
where $\mathcal{C}$ is independent of $v_{\bar{y}}$, and

$$\|\mathcal{C}\| \leq \gamma\varepsilon_k \tag{4.21}$$
$$\|\mathcal{R}(v_{\bar{y}})\| \leq \gamma^{2.5}|\mathbf{a}^{(r)}(\bar{y})|^2. \tag{4.22}$$

PROOF. We put
(4.23)
$$\tilde{F}_\lambda^{(k-1)}(B_{k-1}) - \tilde{F}_E^{(k)}(B_k) = \left(\tilde{F}_\lambda^{(k-1)}(B_{k-1}) - \tilde{F}_\lambda^{(k)}(B_k)\right) - \left(\tilde{F}_\lambda^{(k)}(B_k) - \tilde{F}_E^{(k)}(B_k)\right).$$

Consider the first term in (4.23), which corresponds to (4.17). The K_1 term appears explicitly in (4.20). The K_0 term may be denoted $\mathcal{C}^{(1)}$; it is independent of $v_{\bar{y}}$, and has a minimum decay distance $4L_{k-1} = 2L_k$ from the trip from B_{k-1} to $\bar{B}_k \setminus \bar{B}_{k-1}$ and back. Thus $\|\mathcal{C}^{(1)}\| \leq \gamma^{.85 \cdot 2L_k} = \gamma^{1.7L_k} \leq \frac{1}{2}\gamma\varepsilon_k$, since $\varepsilon_k = \gamma^{1.6L_k}$. Denoting the K_2 term by $\mathcal{R}^{(1)}(v_{\bar{y}})$, recall that $\bar{y}$ is defined as the site that maximizes $|\mathbf{a}^{(r)}(y)|$, so $|\mathbf{a}^{(r)}(y)| \leq |\mathbf{a}^{(r)}(\bar{y})|$ for all y adjacent to $\bar{B}_{k-1}$. Therefore,

$$\|\mathcal{R}^{(1)}(v_{\bar{y}})\| \leq \sum_{xy} \|\mathbf{a}^{(r)}(x)\mathbf{a}^{(r)}(y)^{\mathrm{tr}}\||K_2(x,y)| \leq \tfrac{1}{2}\gamma^{2.5}|\mathbf{a}^{(r)}(\bar{y})|^2. \tag{4.24}$$

We have used the fact that the norm of an outer product matrix $\mathbf{u}\mathbf{w}^{\mathrm{tr}}$ is bounded by $|\mathbf{u}||\mathbf{w}|$. Also, we have used $\sum_{xy} |K_2(x,y)| \leq \frac{1}{2}\gamma^{2.5}$, because the relevant graphs are tied to a single point, $\bar{y}$, and have minimum length 3 (the leading term with length 2 is in K_0). This is a critical bound that controls remainders in terms of $|\mathbf{a}^{(r)}(\bar{y})|^2$, which governs the lower bound on the leading term.

The second term of (4.23) may be written as a sum of $\mathcal{C}^{(2)} + \mathcal{R}^{(2)}(v_{\bar{y}})$, with $\mathcal{C}^{(2)} = \tilde{F}_\lambda^{(k-1)}(B_{k-1}) - \tilde{F}_E^{(k-1)}(B_{k-1})$ and $\mathcal{R}^{(2)}(v_{\bar{y}}) = \Delta F_E - \Delta F_\lambda$. Note that $\mathcal{C}^{(2)}$ is independent of the potentials in $\bar{B}_k \setminus \bar{B}_{k-1}$. It is bounded by $\frac{1}{2}\gamma\varepsilon_k$, by the Lipschitz continuity estimate for $\tilde{F}_\lambda^{(k-1)}(B_{k-1})$, since $|\lambda - E| \leq \varepsilon_k/2$. The term $\mathcal{R}^{(2)}(v_{\bar{y}})$ also involves a difference $|\lambda - E|$, and as above the graphs involved span a distance $4L_{k-1}$. Hence it is bounded by $\gamma^{1.7L_k}\varepsilon_k \leq \gamma^{3.3L_k} \leq \frac{1}{2}\gamma^{2.5}|\mathbf{a}^{(r)}(\bar{y})|^2$, because $|\mathbf{a}^{(r)}(\bar{y})|^2 \geq (\gamma^{3.1L_{k-1}})^2 = \gamma^{3.1L_k}$. If we write $\mathcal{C} = \mathcal{C}^{(1)} + \mathcal{C}^{(2)}$, $\mathcal{R}(v_{\bar{y}}) = \mathcal{R}^{(1)}(v_{\bar{y}}) + \mathcal{R}^{(2)}(v_{\bar{y}})$, and combine the bounds proven above, we obtain (4.21), (4.22). □

PROOF OF PROPOSITION 4.2. We will again simplify the analysis by assuming that $n = \hat{n}$. Then the inequality $\hat{n}_k(B_k) \leq \hat{n}$ is true by construction, as $\hat{n}$ is the dimension of the matrix $\tilde{F}_E^{(k)}(B_k)$. Combining this fact with (4.11), we obtain (4.7).

With $\lambda \in [0, 1 + 4d\gamma]$, take any v_1, v_2 in the set of allowed potential values $\{0, \frac{1}{N-1}, \frac{2}{N-1}, \ldots, 1\}$. A straightforward calculation shows that for small γ,

$$\left| \frac{1}{v_1 + 2d\gamma - \lambda} - \frac{1}{v_2 + 2d\gamma - \lambda} \right| \geq \frac{1}{N}. \tag{4.25}$$

Consequently, there are N distinct values of $(v + 2d\gamma - \lambda)^{-1}$ as v varies, and the minimum gap between these values is $\frac{1}{N}$.

Consider the case $\hat{n} = 1$. Then the matrix $a_\beta^{(\mathrm{r})}(\bar{y}) a_{\tilde{\beta}}^{(\mathrm{r})}(\bar{y})$ reduces to a number $|a_1^{(\mathrm{r})}(\bar{y})|^2 \geq \gamma^{3.1L_k}$. Likewise $\tilde{F}_E^{(k)}(B_k)$ and $\tilde{F}_\lambda^{(k-1)}(B_{k-1})$ are numbers, with the latter independent of $v_{\bar{y}}$. It is evident from Proposition 4.3 and (4.25) that the set of values that $\tilde{F}_E^{(k)}(B_k)$ takes as $v_{\bar{y}}$ varies is spaced apart by at least $(\gamma^2/N)|a_1^{(\mathrm{r})}(\bar{y})|^2 \gg \varepsilon_{k+1}$. Hence there is at most one value of $v_{\bar{y}}$ such that $\tilde{F}_E^{(k)}(B_k)$ lies in $I_{\varepsilon_{k+1}}(E)$.

Now consider the case $\hat{n} > 1$. The spread of a Hermitian matrix is defined as the difference between its largest and smallest eigenvalues. Weyl's inequality implies that

$$\mathrm{spread}(M_1 + M_2) \geq |\mathrm{spread}(M_1) - \mathrm{spread}(M_2)|. \tag{4.26}$$

Put $M_1 = \mathbf{a}^{(\mathrm{r})}(\bar{y}) \mathbf{a}^{(\mathrm{r})}(\bar{y})^{\mathrm{tr}}$. This is a rank-one matrix, so its spread is equal to its nonzero eigenvalue $|\mathbf{a}^{(\mathrm{r})}(\bar{y})|^2 \geq \gamma^{3.1L_k} \gg \varepsilon_{k+1}$.

From (4.20) we have that

$$\tilde{F}_E^{(k)}(B_k) = \tilde{F}_\lambda^{(k-1)}(B_{k-1}) - \frac{\gamma^2}{v_{\bar{y}} + 2d\gamma - \lambda} M_1 - \mathcal{C} - \mathcal{R}(v_{\bar{y}}). \tag{4.27}$$

Let $M_2 = f_\lambda^{(k-1)} - \mathcal{C}$. It should be clear that

$$\left| \mathrm{spread}(M_2) - \mathrm{spread}\left(\frac{\gamma^2}{v_{\bar{y}} + 2d\gamma - \lambda} M_1 \right) \right| \leq \frac{\gamma^2}{3N} |\mathbf{a}^{(\mathrm{r})}(\bar{y})|^2 \tag{4.28}$$

for at most one value of $v_{\bar{y}}$. (The values of the second spread are spaced out by at least $(\gamma^2/N)|a_1^{(\mathrm{r})}(\bar{y})|^2$, so cancellation as in (4.2) can happen at most once.) The bound on $\|\mathcal{R}(v_{\bar{y}})\|$ is much smaller than this, so

$$\mathrm{spread}\big(\tilde{F}_E^{(k)}(B_k)\big) \geq \Big(\frac{\gamma^2}{3N} - 2\gamma^{2.5}\Big) |\mathbf{a}^{(\mathrm{r})}(\bar{y})|^2 \geq \frac{\gamma^2}{4N} \gamma^{3.1L_k} \gg \varepsilon_{k+1} = \gamma^{3.2L_k}, \tag{4.29}$$

for all but one value of $v_{\bar{y}}$. Hence for all but one value of $v_{\bar{y}}$, at least one eigenvalue of $\tilde{F}_E^{(k)}(B_k)$ must fall outside of $I_{\varepsilon_{k+1}}(E)$. Thus we have demonstrated that $\hat{n}_k(B_k) < \hat{n}$ for all but one value of $v_{\bar{y}}$. Combining this with (4.11), we obtain (4.8). □

REMARK 4.4. From (4.29) we see the importance of the fact that ε_{k+1}, the next energy window, fits inside the proven spread for $\tilde{F}_E^{(k)}(B_k)$, originating from the lower bound on the influence of $\bar{y}$. We have already seen that ε_{k+1} must be larger than any of the terms not included in $\tilde{F}_E^{(k)}(B_k)$, specifically graphs extending farther than $2L_k$ from B_k. It is inevitable that there is a gap between the decay rates that can be established for upper bounds and for lower bounds. Since the upper bound for step $k+1$ must fit inside the lower bound for step k, we are forced to use a sequence of length scales L_k that increase geometrically. Consequently, we are only able to make use of a single random potential per annular region on each scale. We obtain a Cantor-like spreading of the eigenvalue distribution that develops with the logarithm of the length scale; this leads to log-Hölder continuity of the density of states as in (2.3).

4.3. Percolation Estimates. Proposition 4.2 gives the key result on the manner in which $\hat{n}_k(B_k)$, the local degeneracy of the spectrum, has a strong tendency to decrease with increasing k. The fact that $\hat{n}_k(B_k)$ fails to decrease for at most one value of $v_{\bar{y}}$ can be translated into probabilistic terms to demonstrate that the resonant regions $R^{(k)}$ get more and more dilute as k increases. The resonant blocks B_k can be considered components of a multi-scale percolation problem. Accurate control of this problem is a prerequisite for obtaining our main results. Here we discuss the problem in broad terms; the details may be found in [**Imb17**].

The first step blocks B_1 are actually the connected components of a site percolation problem, where the probability that a site is occupied is bounded by $\frac{1}{N} < \varepsilon$. Unlike the usual site percolation problem on $\mathbb{Z}^d$, connectedness is defined based on linking sites up to a distance L_1^α. Nevertheless, one can estimate the connectivity function (the probability that x and y are in the same percolation cluster) by summing over sequences of occupied sites, each separated from the previous one by a distance no greater than L_1^α, and including a factor of ε for each site. In the kth step, components from the $(k-1)^{\text{st}}$ step are either resonant or not. The resonant ones are joined using connections up to a distance L_k^α. Thus we have a competition between the increasing distance over which connections are made and the decreasing probability that components remain resonant.

The simplest situation occurs when a component satisfies $\hat{n}_{k-1}(B_{k-1}) = 1$, *i.e.* the associated local Hamiltonian $\tilde{F}_E^{(k-1)}(B_{k-1})$ has only a single eigenvalue in $I_{\varepsilon_k}(E)$. Then if no new connection is made, then Proposition 4.2 implies that the probability that the component remains resonant is bounded by $\frac{1}{N-1} = \varepsilon$. (The probability in question is actually a conditional probability. Since we condition on the B_{k-1}, $\bar{B}_{k-1}$, and the set of potentials in $\bar{B}_{k-1}$, this eliminates one of the possible values for $v_{\bar{y}}$, as it would introduce a component B_1 adjacent to $\bar{B}_{k-1}$, which is incompatible with the definition of $\bar{B}_{k-1}$. Hence the denominator is $N-1$, not N. A similar situation occurs for ordinary site percolation: if B is a component, then neighbors of B are necessarily unoccupied.) If a single-site component B_1 remains resonant through step k, then the probability is bounded by $\varepsilon^k \le L_k^{-q}$, provided ε is taken sufficiently small, depending on L_0 and q. (Recall that $L_k \equiv L_0 2^k$.) Thus we see that the procedure naturally leads to a decrease in probability as a large inverse power of the length scale. This is sufficient also to handle sums over $\sim L_k^{\alpha d}$ sites when making connections, provided $q > \alpha d$. Thus we should expect the connectivity function to decay as a large power $p = q - \alpha d$, when q is large enough. However, there is a complication in the multi-scale situation when two components, each with probability L_{k-1}^{-p}, are connected at a distance L_k^α. Then we require $(L_{k-1}^{-p})^2 \le L_k^{-\alpha p}$, which means that α must be less than 2. On the other hand, to preserve exponential decay of the random-walk expansion, we require that $\alpha > 1$, so that the fraction of decay lost due to gaps at resonant blocks is summable. These limitations on α go back to [**FS83**].

Finally, let us consider the complication that arises when a component has multiple eigenvalues in the window, *i.e.* $\hat{n}_{k-1}(B_{k-1}) > 1$. (This is only an issue if no new connections are made, since new connections entail sufficient smallness, as discussed above.) Then we may fail to get the anticipated factor of ε when going to the next scale. However, Proposition 4.2 shows that if this happens, $\hat{n}_k(B_k) < \hat{n}_{k-1}(B_{k-1}) = 1$. Thus there can be a delay in the accumulation of factors of ε while the degeneracy works its way down to 1. However, an n-fold degeneracy can only occur for a component of size n or more. Thus we have n factors of ε available at the start, and these are sufficient to make up for what is missing due to degeneracies. As mentioned earlier, the density of states problem is intertwined with the problem of degeneracy breaking or level separation.

Let us conclude this section by stating a bound on the connectivity function for the percolation problem in step k. It is a straightforward application of the percolation estimates of [**Imb17**]. Let $\mathcal{Q}_{xy}^{(k)}$ denote the probability that x and y are in the same component B_k of $R^{(k)}$. Then for any sufficiently large p, we may choose N sufficiently large (and γ sufficiently small) so that

$$\mathcal{Q}_{xy}^{(k)} \le \varepsilon^{1/4}\big(|x-y| \vee L_{k-1}\big)^{-p}. \tag{4.30}$$

5. Proof of Main Results

5.1. Density of States. We are ready to prove statement (2.3) of Theorem 2.1, which is that $\mathbb{E}\mathcal{N}(I_\delta(E)) \le |\Lambda|(\log_\gamma \delta)^{-p}$. As we are primarily interested in small δ, let us consider the case where

$$\varepsilon_{k+1}/3 < \delta \le \varepsilon_k/3 \tag{5.1}$$

for some $k \ge 2$. In Theorem 2.1 we restrict to $\delta > \gamma^{\operatorname{diam}(\Lambda)/2}$; this is to ensure that value of k determined by (5.1) is such that that $\operatorname{diam}(\bar{B}_{k-1}) < \operatorname{diam}(\Lambda)$, as required by Proposition 4.2. (Once this is no longer true, we run out of random variables to continue the smoothing process.)

By repeated application of Lemma 3.1, we can see that the number of eigenvalues in $I_{\varepsilon_k/3}(E)$ is bounded by $|R^{(k)}|$, the dimension of the space on which $F_E^{(k)}$ acts. The probability that $x \in R^{(k)}$ may be bounded by L_{k-1}^{-p} by taking $y = x$ in the connectivity function (4.30). Then using (5.1) and $\varepsilon_{k+1} = \gamma^{6.4L_{k-1}}$, we obtain

$$\begin{aligned}\mathbb{E}\mathcal{N}\big(I_\delta(E)\big) &\le \mathbb{E}\mathcal{N}\big(I_{\varepsilon_k/3}(E)\big) \le \sum_x P\big(x \in R^{(k)}\big)\\ &\le |\Lambda|\varepsilon^{1/4}L_{k-1}^{-p} = |\Lambda|\varepsilon^{1/4}\big(\tfrac{1}{6.4}\log_\gamma \varepsilon_{k+1}\big)^{-p} \le |\Lambda|(\log_\gamma \delta)^{-p}.\end{aligned} \tag{5.2}$$

5.2. Energy-Following Procedure. Before proceeding to the other statements in Theorem 2.1, we need to discuss a new tool, which we call the energy-following procedure (EFP). There is a problem that arises from the fact that the eigenfunction correlator involves a sum over all eigenvalues, not just those that lie in a predetermined interval. Likewise, the statement on the minimum eigenvalue spacing applies to all eigenvalues of H. One cannot simply choose eigenvalues as the center points for energy windows, because they depend on the potentials throughout Λ, and we would lose the independence that is a crucial aspect of our percolation estimates. Nor can one consider a complete system of overlapping energy windows, because our windows are too narrow (and the probabilities are not small enough) to control the sum over all cases. (This strategy only works if the density of states is Hölder continuous with exponent greater than $\frac{1}{2}$ – see [**CGK09**].) The energy-following procedure was first introduced in [**IM16**] to deal with these issues in the context of the block Anderson model.

To deal with this issue, we construct all the eigenvalues through a sequence of successive approximations that are generated as the length-scale induction proceeds. Thus we allow the center point for energy windows to depend on k. To initiate the procedure, we choose a site $x \in \Lambda$ and let $E_1 = 2d\gamma + v_x$, the diagonal entry of H at x. Then we choose E_2 to be one of the solutions to $\lambda \in \operatorname{spec} \tilde{F}_\lambda^{(1)}(B_{x,1})$ in $I_{\varepsilon_1/3}$, where $B_{x,1}$ is the component of $R^{(1)}$ containing x. Continuing in this fashion, we obtain in each step a better approximate eigenvalue. Once L_k exceeds the diameter of Λ, the approximate eigenvalue becomes an exact eigenvalue of H. One can show that the EFP is complete, in the sense that every eigenvalue of H can be obtained by some sequence of choices $E_1, E_2, \ldots$. The key feature of the procedure is its *quasilocality*. This is the property that each E_k depends only on the potential in a neighborhood of x whose radius is some multiple of L_k (and subsequent corrections to the energy are exponentially small in L_k, with probability $\sim 1 - L_k^{-p}$). The way the energies are chosen, each $B_{x,k}$ is automatically resonant, but we preserve the conditional independence of events defined in terms of the potentials outside $\bar{B}_{x,k}$. Thus with minor modifications (in particular p becomes $p/2 - 1$), we obtain percolation estimates such as (4.30). The fact that the initial basis can be deformed into the eigenvector basis by means of quasilocal rotations has played a major role in the theory of many-body localization, see [**IRS17**] for a review. This was first established in the Anderson model in [**Imb16a**], and in the many-body context in [**Imb16b**], under an assumption on level statistics.

5.3. Eigenfunction Correlator and Level Spacing. Here we give a brief discussion of the proofs of the remaining statements in Theorem 2.1. We run the EFP up to the final step $\bar{k}$, at which point $\bar{B}_{x,\bar{k}} = \Lambda$. In the EFP each eigenfunction can be written as $G_{E_{\bar{k}}}^{(\bar{k}-1)} \varphi^{(\bar{k}-1)}$, for some eigenvector $\varphi^{(\bar{k}-1)}$ of $F_{E_{\bar{k}}}^{(\bar{k}-1)}$ (here $E_{\bar{k}}$ is the corresponding eigenvalue). Then we have

(5.3)
$$\sum_\beta |\varphi_\beta(y_1)\varphi_\beta(y_2)| \le \sum_x \sum_{\beta \text{ reachable from } x} \left|\big(G_{E_{\bar{k}}}^{(\bar{k}-1)} \varphi_\beta^{(\bar{k}-1)}\big)(y_1)\big(G_{E_{\bar{k}}}^{(\bar{k}-1)} \varphi_\beta^{(\bar{k}-1)}\big)(y_2)\right|.$$

In the second sum on the right-hand side, we are summing over all possible choices in the energy-following procedure, starting at x. (Each β may be reachable from more than one x.) Since $G_{E_{\bar{k}}}^{(\bar{k}-1)}$ decays exponentially, we may rewrite this as

$$\sum_\beta |\varphi_\beta(y_1)\varphi_\beta(y_2)| \le \sum_{x,z_1,z_2} \gamma^{.85|y_1-z_1|}\gamma^{.85|y_2-z_2|} N_{x,z_1,z_2}, \tag{5.4}$$

where $N_{x,y,z}$ denotes the number of eigenvalues of H that can be reached via the EFP with starting point x, and with a resonant region $B_{x,\bar{k}-1}$ that includes y and z. In order to bound $\mathbb{E}N_{x,y,z}$, we need to control the sum over choices in the EFP. These are never more numerous than the size n of the blocks involved, and a factor ε^n is available from the percolation estimates. Thus $\mathbb{E}N_{x,y,z}$ can be controlled as if it were a 3-point connectivity function for the percolation problem, and as in (4.30) we obtain decay as a large inverse power of $\operatorname{diam}(\{x, y, z\})$. As a result, we can take the expectation of (5.4) and perform the sums over x, y, z to obtain

$$\mathbb{E} \sum_\beta |\varphi_\beta(x)\varphi_\beta(y)| \le \big(|x-y| \vee 1\big)^{-(p/2-d-1)}, \tag{5.5}$$

which is (2.4).

We proceed to the level-spacing statement in Theorem 2.1. We may write

$$P\Big(\min_{\beta\neq\tilde{\beta}} |E_\beta - E_{\tilde{\beta}}| < \delta\Big) \leq \tfrac{1}{2}\sum_x \mathbb{E}N_x(\delta), \tag{5.6}$$

where $N_x(\delta)$ is the number of eigenvalues λ_0 of H that can be reached via the EFP starting at x, and which have another eigenvalue in $I_\delta(\lambda_0)$. Then (2.5) follows from the estimate

$$\mathbb{E}N_x(\delta) \leq 2|\Lambda|(\log_\gamma \delta)^{-(p/2-1)}, \tag{5.7}$$

where we take $\delta \in [\gamma^{\mathrm{diam}(\Lambda)}, \gamma]$. Let us define k by the inequality

$$\varepsilon_{k+1}/4 < \delta \leq \varepsilon_k/4, \tag{5.8}$$

and again we consider only the case $k \geq 2$. When we run the EFP starting at x, we have the condition that another eigenvalue is within δ of the final eigenvalue $E_{\bar{k}}$. This implies that one of two things happen. In case 1, the multiplicity $\hat{n}_{k-1}(B_{x,k-1})$ is greater than 1. Then the percolation estimates produce a bound $L_{k-1}^{-(p/2-1)}$ due to the continuous failure of the multiplicity to drop to 1. (In the EFP, we follow an eigenvalue, so the multiplicity cannot drop below 1, but Proposition 4.2 still leads to a factor of ε if it fails to drop below 2.) In case 2, there is a block B_{k-1} other than $B_{x,k-1}$ satisfying

$$\mathrm{dist}\big(\mathrm{spec}\, \tilde{F}_{E_k}^{(k-1)}(B_{k-1}), E_k\big) \leq \varepsilon_k, \tag{5.9}$$

so it is resonant in step k and survives to $R^{(k)}$. Thus as in the proof of the density of states bound (2.3), we obtain a bound of $|\Lambda| L_{k-1}^{-p}$ on the expectation for case 2. Combining the two cases, we obtain (5.7) to complete the proof.

References

[And58] P. Anderson, *Absence of diffusion in certain random lattices*, Phys. Rev. **109** (1958), 1492–1505.

[BK05] Jean Bourgain and Carlos E. Kenig, *On localization in the continuous Anderson-Bernoulli model in higher dimension*, Invent. Math. **161** (2005), no. 2, 389–426, DOI 10.1007/s00222-004-0435-7. MR2180453

[Bou12] J. Bourgain, *On the Furstenberg measure and density of states for the Anderson-Bernoulli model at small disorder*, J. Anal. Math. **117** (2012), 273–295, DOI 10.1007/s11854-012-0022-6. MR2944098

[Bou14] J. Bourgain, *An application of group expansion to the Anderson-Bernoulli model*, Geom. Funct. Anal. **24** (2014), no. 1, 49–62, DOI 10.1007/s00039-014-0260-0. MR3177377

[CGK09] Jean-Michel Combes, François Germinet, and Abel Klein, *Generalized eigenvalue-counting estimates for the Anderson model*, J. Stat. Phys. **135** (2009), no. 2, 201–216, DOI 10.1007/s10955-009-9731-3. MR2505733

[CKM87] René Carmona, Abel Klein, and Fabio Martinelli, *Anderson localization for Bernoulli and other singular potentials*, Comm. Math. Phys. **108** (1987), no. 1, 41–66. MR872140

[CS83] Walter Craig and Barry Simon, *Log Hölder continuity of the integrated density of states for stochastic Jacobi matrices*, Comm. Math. Phys. **90** (1983), no. 2, 207–218. MR714434

[DSS02] David Damanik, Robert Sims, and Günter Stolz, *Localization for one-dimensional, continuum, Bernoulli-Anderson models*, Duke Math. J. **114** (2002), no. 1, 59–100, DOI 10.1215/S0012-7094-02-11414-8. MR1915036

[FS83] Jürg Fröhlich and Thomas Spencer, *Absence of diffusion in the Anderson tight binding model for large disorder or low energy*, Comm. Math. Phys. **88** (1983), no. 2, 151–184. MR696803

[GHK07] François Germinet, Peter D. Hislop, and Abel Klein, *Localization for Schrödinger operators with Poisson random potential*, J. Eur. Math. Soc. (JEMS) **9** (2007), no. 3, 577–607, DOI 10.1016/j.crma.2005.09.001. MR2314108

[GK07] François Germinet and Abel Klein, *Localization for a continuum Cantor-Anderson Hamiltonian*, Adventures in mathematical physics, Contemp. Math., vol. 447, Amer. Math. Soc., Providence, RI, 2007, pp. 103–112, DOI 10.1090/conm/447/08685. MR2423574

[GK13] François Germinet and Abel Klein, *A comprehensive proof of localization for continuous Anderson models with singular random potentials*, J. Eur. Math. Soc. (JEMS) **15** (2013), no. 1, 53–143, DOI 10.4171/JEMS/356. MR2998830

[IM16] John Z. Imbrie and Rajinder Mavi, *Level spacing for non-monotone Anderson models*, J. Stat. Phys. **162** (2016), no. 6, 1451–1484, DOI 10.1007/s10955-016-1461-8. MR3463783

[Imb16a] John Z. Imbrie, *Multi-scale Jacobi method for Anderson localization*, Comm. Math. Phys. **341** (2016), no. 2, 491–521, DOI 10.1007/s00220-015-2522-6. MR3440194

[Imb16b] John Z. Imbrie, *On many-body localization for quantum spin chains*, J. Stat. Phys. **163** (2016), no. 5, 998–1048, DOI 10.1007/s10955-016-1508-x. MR3493184

[Imb17] ———, *Localization and eigenvalue statistics for the lattice anderson model with discrete disorder*, arXiv:1705.01916.

[IRS17] John Z. Imbrie, Valentina Ros, and Antonello Scardicchio, *Local integrals of motion in many-body localized systems*, Ann. Phys. **529** (2017), no. 7, 201600278, 23, DOI 10.1002/andp.201600278. MR3671048

[KT16] Abel Klein and C. S. Sidney Tsang, *Quantitative unique continuation principle for Schrödinger operators with singular potentials*, Proc. Amer. Math. Soc. **144** (2016), no. 2, 665–679, DOI 10.1090/proc12734. MR3430843

[Min96] Nariyuki Minami, *Local fluctuation of the spectrum of a multidimensional Anderson tight binding model*, Comm. Math. Phys. **177** (1996), no. 3, 709–725. MR1385082

[SVW98] C. Shubin, R. Vakilian, and T. Wolff, *Some harmonic analysis questions suggested by Anderson-Bernoulli models*, Geom. Funct. Anal. **8** (1998), no. 5, 932–964, DOI 10.1007/s000390050078. MR1650106

[Weg81] Franz Wegner, *Bounds on the density of states in disordered systems*, Z. Phys. B **44** (1981), no. 1-2, 9–15, DOI 10.1007/BF01292646. MR639135

Department of Mathematics, University of Virginia, Charlottesville, Virginia, 22904

Email address: imbrie@virginia.edu

Contemporary Mathematics
Volume **717**, 2018
https://doi.org/10.1090/conm/717/14441

Interacting fermions with quasi-random disorder

Vieri Mastropietro

ABSTRACT. We discuss the effect of many-body interactions in fermionic systems with strong quasi-random disorder (quasi-periodic potential). By assuming Diophantine conditions on the frequency and density and using renormalization-group methods, we prove the convergence of the expansion of the many-body correlations up to zero temperature and their exponential decay. This is true, in particular, for the interacting spinless fermionic Aubry-André model, for two coupled chains of spinless interacting disordered fermions, and for an array of noninteracting chains with the same disorder in all the chains. Zero-temperature exponential decay is a signal of persistence of localization in the many-body ground state in presence of an interaction. In other systems, however, as in the presence of spin or when more than two interacting chains are coupled, additional relevant processes prevent convergence up to zero temperature, suggesting a lack of localization at least for certain quantities. This is in agreement with recent experiments performed on cold atoms and with numerical simulations.

1. Introduction

Strong disorder in non-interacting many body quantum systems produces a phenomenon called *Anderson localization* [**1**]: the eigenfunctions of the single-particle Hamiltonian are localized and decay exponentially with the distance from some center. A basic question is what happens when an interaction is present and the single-particle description is no longer valid; if localization persists in the presence of an interaction, the system is said to be in a many-body localized (MBL) phase. The existence of such a phase would have interesting consequences for the statistical mechanics of isolated quantum systems; while normal systems are expected to approach asymptotically a thermal state (due to interaction, "the system acts as its own bath"), this does not happen in an MBL phase [**2**],[**3**]. Establishing the very existence of such a phase is however quite a nontrivial problem; early theoretical arguments support persistence of localization [**4**], but they are based on formal series expansions and non-perturbative effects could produce delocalization.

Let us introduce a class of systems in which the interplay of disorder and interaction can be conveniently studied. The basic model has Hamiltonian H_N, a linear operator on the Hilbert space $\mathcal{H}_N$ which is a suitable subset of $\ell_2(\mathbb{Z}^N)$ containing only antisymmetric functions, that is

$$\psi(x_1, .., x_i, .., x_j, ...) = -\psi(x_1, .., x_j, .., x_i, ...), \tag{1.1}$$

$x \in \mathbb{Z}$; the Hamiltonian is given by

$$H_N = J \sum_{i=1}^{N} \nabla_x + \sum_{i=1}^{N} \phi_{x_i} + U \sum_{i,j=1}^{N} v(x_i - x_j), \tag{1.2}$$

where $v(x-y) = \delta_{x,y+1}$ is a short range potential and and $\nabla_z f(z) = f(z-1) + f(z+1) - 2f(z)$. The above model describes N fermionic particles hopping on a one-dimensional lattice with rate J, interacting with coupling U and subject to an on-site potential ϕ_x which takes into account the effect of disorder. With a suitable choice of boundary conditions the above model is equivalent to the XXZ spin chain with coordinate dependent magnetic field.

An immediate generalization of the above model consists in considering spinful particles; if $v(x-y) = \delta_{x,y}$ the model is known as Hubbard model with disorder. Note that with spinless fermions one cannot consider local interactions. Another generalization consists in coupling several chains with the same disorder along the chains. The Hamiltonian is

$$H_N = J \sum_{i=1}^{N} \nabla_x + \sum_{i=1}^{N} \phi_{x_i} + J_\perp \sum_{i=1}^{N} \nabla_{y_i} + U \sum_{i,j=1}^{N} v(x_i - x_j) \tag{1.3}$$

with $y = 1, 2, ..L$ and periodic boundary conditions are imposed in the y variable. The above Hamiltonian describes L chains, with the same disorder in the chains, intra-chain hopping J, intra-chain interaction U, and inter-chain hopping $J_\perp$.

The Hamiltonian (1.2) and (1.3) becomes the sum of single-particle Hamiltonians when the interaction is switched off, $U = 0$; the N-particle eigenfunctions are simply obtained by antisymmetric combinations of single-particle eigenfunctions. In the presence of interaction the single-particle description breaks down and the interplay between disorder and interaction can produce intricate effects.

We have yet to describe how the disorder ϕ_x is realized: there are two main possibilities, one in which the disorder is represented as a random variable (*random disorder*) and another in which is an external quasi-periodic potential (*quasi-random disorder*). In the first kind of disorder, the one originally introduced by Anderson [**1**], the ϕ_x are independent identically distributed random variables. Localization in the *absence* of interaction at a single-particle level with strong disorder in any dimension was proved in [**5**], [**6**] (see also [**7**]). As a consequence the correlations in the many-body noninteracting problem decay exponentially [**8**]. The effect of many-body interaction with random disorder have been considered in [**9**], [**10**] and in the recent [**11**], [**12**]. Full MBL has been established in a spin system with random disorder, equivalent to an interacting fermionic system, in [**13**], [**14**], assuming a property of limited level attraction (proven up to now only in the singe particle case).

Let us consider now the other kind of disorder, the quasi-random one. A typical choice consists in describing ϕ_x by an incommensurate Aubry-André potential [**15**]

$$\phi_x = \Delta \cos(2\pi(\omega x + \theta)). \tag{1.4}$$

Even if this kind of disorder may look more artificial than the random one, it is the one effectively realized in experiments with cold atoms [**16**], [**17**]. A detailed understanding has been reached in the absence of interaction and hopping, $U = J_\perp = 0$, when the Hamiltonian is sum of almost-Mathieu Hamiltonians,

$$H_A(x) = J\nabla_x + \Delta \cos(2\pi(\omega x + \theta)), \tag{1.5}$$

with ω irrational. The behavior of the eigenfunctions of H_A (1.5) depends crucially on the ratio $\frac{\Delta}{J}$ between the disorder and the hopping; for diophantine ω and almost all θ:

- if $\frac{\Delta}{J} < \frac{1}{2}$, that is, for weak disorder, H_A has purely absolutely continuous spectrum and there are Floquet solutions [**18**], [**19**], [**20**],[**21**], [**22**];
- if $\frac{\Delta}{J} > \frac{1}{2}$, that is, for large disorder, H_A has only pure point spectrum with exponentially decaying eigenfunctions. That is, Anderson localization occurs (see [**23**], [**24**], [**25**],[**26**]).

A metal-insulator transition therefore takes place as the strength of the disorder is varied, a feature making quasi-random disorder somewhat similar to random disorder in three dimensions. The ground-state fermionic correlations in the non-interacting case were studied in [**27**] and [**28**], and in the presence of interaction in the weak disorder delocalized regime in [**29**] and [**30**].

While the single-particle behavior of fermions with Aubry-André disorder is mathematically quite well understood, much less is known when the interaction U is considered. We recall that the basic difficulty in the analysis of the non-interacting single-particle Hamiltonian H_A resides in the fact that one uses a perturbative approach around the limit $J = 0$, which has a dense point spectrum, so that the eigenvalues are, to any order of approximation, of infinite multiplicity. This makes perturbation theory quite delicate and is reflected by the presence of small divisors, similar to those appearing in KAM tori, of the form, if ϕ_x is given by (1.4),

$$\phi_x - \phi_y, \tag{1.6}$$

which can be arbitrarily small for large $x - y$. In order to control such divisors one can assume that the frequency ω is a Diophantine number, that is, verifying

$$||\omega x|| \geq C_0 |x|^{-\tau} \quad \forall x \in \mathbb{Z}/\{0\} \tag{1.7}$$

where $||.||$ is the norm on the one-dimensional torus of period 1, and C_0, τ are suitable positive constants. With these conditions KAM methods, for instance, allow small divisors of such form to be dealt with. In the presence of interactions the small divisors in the expansion for the N-particle eigenfunctions are much more complex: they are of the form

$$\sum_{i=1}^{N} \phi_{x_i} - \sum_{i=1}^{N} \phi_{y_i}, \tag{1.8}$$

and even in the case $N = 2$ it is unknown how to control them; e.g., see [**31**]. In the case of several non-interacting chains, that is, $U = 0, J_\perp \neq 0$, the small divisors are instead still of the form (1.6) and localization at finite L should be provable with an extension of a lemma in [**32**].

On the physical side, the interplay of quasi-periodic disorder, interaction, and hopping between chains has been widely investigated, by numerical simulations and experiments on cold atoms; the latter can be considered essentially quantum simulators, as the interactions are tunable so that parameters can be changed. Modulo finite-size effects, there is numerical evidence that a system of fermions with Hamiltonian H_N (1.2) with quasi-random disorder (1.4) has an MBL phase, that is localization persists in the presence of interaction, at least if the interaction is not too strong and the fermions are spinless (see [**33**]). Observation of MBL has been claimed in experiments on cold atoms [**16**], in which a physical system of quantum fermionic particles with evolution driven by (1.2) with quasi-random disorder (1.4) has been realized; the lattice and the quasi-random disorder has been realized by superimposing two laser beams with incommensurate frequency and information on the localization properties is obtained by observing the time evolution of a suitable state. Subsequent numerical simulations [**34**] with spinful particles have shown a difference regarding the localization properties with the spinless case; certain physical observables show localized behavior (in particular the ones measured in the cold-atom experiments), while others have a delocalized behavior. Finally an array of chains described by the Hamiltonian (1.2) with quasi-random disorder (1.4) has been realized with cold atoms in [**17**]; in that case localized behavior is found in the absence of interaction U, while with interactions a delocalized behavior is found. In conclusion, numerics and physical experiments seem to suggest a rather complex interplay between the interaction, the hopping, and the disorder.

2. A statistical mechanics approach

We analyze the problem using a statistical approach: instead of considering the expansion for the eigenfunctions of the N-particle problem, we compute some thermodynamical average in the limit $N \to \infty$, like the particle correlations in the grand canonical ensemble. If the zero-temperature limit can be taken (as we will see this is not always possible), such correlations give information on the many-body ground state.

In the grand canonical ensemble, one has as Hilbert space the Fock space defined as $\mathcal{F} = \bigoplus_{N=0}^{\infty} \mathcal{H}_N$ and as Hamiltonian on the Fock space, $H = \bigoplus_{N=0}^{\infty} H_N$. The grand canonical partition partition function is

$$Z = \mathrm{Tr}_{\mathcal{F}} e^{-\beta(H-\mu N)} = \sum_{N=0}^{\infty} \mathrm{Tr}_{\mathcal{H}_N} e^{-\beta(H_N-\mu N)}, \tag{2.1}$$

where μ is the chemical potential; the average of an operator O on the Fock space is denoted by

$$< O >= \frac{\mathrm{Tr}_{\mathcal{F}} e^{-\beta(H-\mu N)} O}{Z}. \tag{2.2}$$

On the Fock space one can define creation and annihilation operations $a_x^{\pm}$ verifying anticommutation relations. The Fock-space Hamiltonian corresponding to (1.2) is given by

$$H = -J\sum_x (a^+_{x+1} a^-_{x,y} + a^+_{x-1} a^-_x) + \sum_x \phi_x a^+_x a^-_x + U \sum_x a^+_x a^-_x a^+_{x+1} a^-_{x+1}. \tag{2.3}$$

We fix the phase $\theta = 0$ for definiteness and we will focus on the 2-point function

$$S_{U,J,\Delta}(\mathbf{x}, \mathbf{y}) =< T a^-_{\mathbf{x}} a^+_{\mathbf{y}} > l, \tag{2.4}$$

where

$$a^{\pm}_{\mathbf{x}} = e^{(H-\mu N)x_0} a^{\pm}_x e^{-(H-\mu N)x_0} \tag{2.5}$$

and $\mathbf{x} = (x, x_0)$. It is natural to compute $S_{U,J,\Delta}(\mathbf{x}, \mathbf{y})$ as a series expansion around some solvable limit. There are two main solvable limits; one is the *molecular limit* $J = U = 0$ in which the particles are linked to a site (no hopping), while the other is the *metallic limit* $J = \Delta = 0$ where the particles are delocalized.

In the molecular limit the 2-point function can be easily computed and one gets ($\Delta = 1$ for definiteness), if $S_{0,0,1}(\mathbf{x}, \mathbf{y}) \equiv g(\mathbf{x}, \mathbf{y})$ (called the propagator), then

$$g(\mathbf{x}, \mathbf{y}) = \delta_{x,y}\{\frac{e^{-(x_0-y_0)(\phi_x-\mu)}}{1+e^{-\beta(\phi_x-\mu)}}\mathbf{I}(x_0-y_0>0) - \frac{e^{-(\beta+x_0-y_0)(\phi_x-\mu)}}{1+e^{-\beta(\phi_x-\mu)}}\mathbf{I}(x_0-y_0\le 0)\},$$

which can be rewritten as

$$g(\mathbf{x}, \mathbf{y}) = \delta_{x,y}\frac{1}{\beta}\sum_{k_0=\frac{2\pi}{\beta}(n_0+\frac{1}{2})} \widehat{g}(x, k_0)e^{-ik_0(x_0-y_0)} \tag{2.6}$$

with

$$\widehat{g}(x, k_0) = \frac{1}{-ik_0 + \cos 2\pi(\omega x) - \mu}. \tag{2.7}$$

Note that the correlations are obviously ultra-localized: they vanish when $x \neq y$. The occupation number of the many-body system at a fixed chemical potential $\mu \neq \phi_x$ is given by

$$\lim_{\beta\to\infty} g(\mathbf{x}, \mathbf{y})|_{x=y, x_0-y_0=0^-} = -\chi(\phi_x \le \mu). \tag{2.8}$$

This is the usual Fermi distribution, saying that that the lowest energy state is obtained by filling all the states with energy smaller than μ.

The 2-point function $S_{U,J,\Delta}(\mathbf{x}, \mathbf{y})$ in the strong-disorder regime can be written as a power series expansion in J and U around the molecular limit, with coefficients expressed in terms of Feynman diagrams whose value is given by sums of products of $g(\mathbf{x}, \mathbf{y})$ (2.7). In the noninteracting case such diagrams are simply trivial trees while in the presence of interaction $U \neq 0$ the diagrams have *loops*. It is apparent by (2.7) that a small-divisor problem appears, as, due to the irrationality of ω, $\phi_x - \mu$ can be arbitrarily small. We introduce $\bar{x}$ the *Fermi coordinate* defined by

$$\mu = \cos 2\pi\omega\bar{x} \tag{2.9}$$

with $\bar{x} \notin \mathbb{N}$. The propagators are large if ωx is close to $\pm\omega\bar{x}$ mod 1; setting $x = x' + \rho\bar{x}$ with $\rho = \pm$, we have $\phi_x - \phi_{\bar{x}} = \rho v_0(\omega x') \bmod 1 + O((\omega x') \bmod 1)^2)$, v_0 a constant. In order to bound the propagator we can assume, in addition to the Diophantine condition for ω, a condition for $\bar{x}$,

$$||\omega x \pm 2\omega\bar{x}|| \ge C_0|x|^{-\tau} \quad \forall x \in \mathbb{Z}/\{0\}. \tag{2.10}$$

Therefore by (2.10)

$$||(\omega x')|| = ||\omega(x - \rho\bar{x})|| \ge C|x|^{-\tau}. \tag{2.11}$$

The propagator is therefore bounded by $O(|x|^\tau)$ and this implies, in the noninteracting case, a naive bound for a graph at order n of the form $O(J^n n!^\tau)$. In the interacting case the graphs are no longer trees but they are loop graphs whose number is not $O(n!)$ but $O(n!^2)$; therefore *even if* no small divisors had been present (but they are), so that each term of the expansion would be $O(C^n)$, there would be a convergence problem in the series (even taking into account the $n!^{-1}$ from the Taylor expansion of the exponential), which would be absent in the noninteracting case. On the other hand, the structure of the small divisors would be similar to the single-particle one, that is, no divisors of the form (1.8). In conclusion the expansion for the eigenfunctions of H_N is in terms of tree graphs with divisors of the form (1.8); the expansion for the correlations contains more complex graphs (loop graphs instead of trees), but the divisors are similar to the ones in the noninteracting case.

Finally, it is useful to compare with the weak-disorder regime, in which the correlations can be written as series in U, Δ expressed by loop graphs whose value is the integral over the momenta k_0, k of propagators (that is, the 2-point function in the $U = \Delta = 0$ case, $J = 1$), of the form

$$\frac{1}{-ik_0 + \cos k - \mu}. \tag{2.12}$$

The zero-temperature propagator is singular at $k_0 = 0, k = \pm p_F$, if $\mu = \cos p_F$; p_F are called *Fermi momenta.* Note the remarkable similarity between between the propagator (2.7) and (2.12); one becomes the other by changing the momentum k to ωx. This is a manifestation of Aubry duality [**15**].

3. The interacting Aubry-André model

In [**35**] and [**36**] the following theorem, announced in [**37**], was proved.

THEOREM 3.1. *Let us consider the 2-point function $S_{U,J,1}(\mathbf{x}, \mathbf{y})$ (2.4) with Hamiltonian given by (2.3) in the spinless case, with $\mu = \cos 2\pi\omega\bar{x} + \nu$. Assume Diophantine conditions (1.7) and (2.10). There exists an ε_0 such that, for $|J|, |U| \leq \varepsilon_0$ it is possible to choose $\nu = \nu(J, U)$ so that, for any $N \in \mathbb{N}$, the limit $\beta \to \infty$ exists and*

$$|S_{U,J,1}(\mathbf{x}, \mathbf{y})| \leq Ce^{-\xi|x-y|} \log(1 + \min(|x||y|))^{\tau} \frac{1}{1 + (\Delta|x_0 - y_0|)|)^N} \tag{3.1}$$

with $\Delta = (1+\min(|x|, |y|))^{-\tau}$, $\xi = |\log(\max(|\varepsilon|, |U|))|$ and C is a suitable constant.

The exponential decay in the coordinate difference of the zero-temperature correlations and the much slower decay in the temporal direction are a signature of localization of the many-body ground state. The proof is based on the convergence of the expansion in the hopping J and the interaction U, which is valid uniformly in the temperature. The hopping and interaction change the ultralocalized behavior in exponential decay. Note the role of the counterterm ν, necessary to fix the value of the interacting chemical potential.

A similar result is valid (see [**35**]), when $\bar{x}$ is half integer

$$\bar{x} = n + \frac{1}{2}, \quad n \in \mathbb{N}, \tag{3.2}$$

corresponding to a choice of the chemical potential in a gap of the noninteracting spectrum. In such a case the bound (3.1) is still true with Δ given by

$$\Delta = O(J^{2\bar{x}}) \tag{3.3}$$

The proof of Theorem 3.1 is based on renormalization-group (RG) methods. This technique has been applied both to Lindtsedt series for KAM tori (see [**40**], [**41**]) and to interacting fermionic models (see for instance [**42**] for a review). It is therefore a natural approach for this problem.

The starting point is to write the propagators as sums of propagators using a smooth decomposition of the unity. We introduce a function $\chi_h(t, k_0) \in C^\infty(\mathbb{T}\times\mathbb{R})$, such that $\chi_h(t, k_0) = \chi_h(-t, -k_0)$ and $\chi_h(t, k_0) = 1$, if $\sqrt{k_0^2 + v_0^2||t||^2} \leq a\gamma^{h-1}$ and $\chi_h(t, k_0) = 0$ if $\sqrt{k_0^2 + v_0^2||t||^2} \geq a\gamma^h$ with a and $\gamma > 1$ suitable constants. We choose a so that the supports of $\chi_0(\omega(x - \bar{x}), k_0)$ and $\chi_0(\omega(x + \bar{x}), k_0)$ are disjoint.

We also define $\chi^{(1)}(\omega x, k_0) = 1 - \chi_0(\omega(x-\bar{x}), k_0) - \chi_0(\omega(x+\bar{x}), k_0)$. We can then write

$$g(\mathbf{x}, \mathbf{y}) = g^{(1)}(\mathbf{x}, \mathbf{y}) + g^{(\le 0)}(\mathbf{x}, \mathbf{y}) \tag{3.4}$$

and

$$g^{(\le 0)}(\mathbf{x}, \mathbf{y}) = \sum_{\rho=\pm} g_\rho^{(\le 0)}(\mathbf{x}, \mathbf{y}), \tag{3.5}$$

where

$$g^{(1)}(\mathbf{x}, \mathbf{y}) = \frac{\delta_{x,y}}{\beta} \sum_{k_0} \chi^{(1)}(\omega x, k_0) \frac{e^{-ik_0(x_0-y_0)}}{-ik_0 + \cos 2\pi(\omega x) - \cos 2\pi(\omega \bar{x})}$$

$$g_\rho^{(\le 0)}(\mathbf{x}, \mathbf{y}) = \frac{\delta_{x,y}}{\beta} \sum_{k_0} \chi_0(\omega(x-\rho\bar{x}), k_0) \frac{e^{-ik_0(x_0-y_0)}}{-ik_0 + \cos 2\pi(\omega x) - \cos 2\pi(\omega \bar{x})}. \tag{3.6}$$

The propagator (3.5) is naturally written in terms of two propagators labeled by the index $\rho = \pm$; this produces an effective description of the system in terms of two kind of fermions with ωx close to $(\omega \bar{x})$ or $-(\omega \bar{x})$.

We can also write $\chi_0(\omega(x-\rho\bar{x}), k_0) = \sum_{k=-\infty}^{0} f_k(\omega x', k_0)$ wilth $f_k(\omega x', k_0) = \chi_k(\omega(x-\rho\bar{x}), k_0) - \chi_{k-1}(\omega(x-\rho\bar{x}), k_0)$, and

$$\widehat{g}_\rho^{(\le 0)}(k_0, \omega x) = \sum_{k=-\infty}^{0} \widehat{g}_\rho^{(k)}(k_0, \omega x) \tag{3.7}$$

with $\widehat{g}^{(k)}(k_0, \omega x) = O(2^{-k})$. The introduction of scales naturally allows the definition of an unambiguous class of subdiagrams with a certain scale h, so that all the propagators in the subdiagrams have scale at least as large as h (and at least one has scale h), while the external ones have scale smaller than h. We can naturally distinguish between two class of subdiagrams, called *resonant* and *nonresonant*; the first are such that the external lines have the same x' while the second are such that at least two have different x'. Both classes of terms have a positive scaling dimension, independently of the number of external lines, a fact that could prevent summing over all the scales unless some improvement is found. In the RG terminology, a theory with positive scaling dimensions is said to be *nonrenormalizable*.

In the case of nonresonant terms, the external propagators have scales smaller than $\bar{h} < h$, so their size is $O(\gamma^{-\bar{h}})$. Therefore if two external propagators have similar (but not equal) small size, then the difference of their coordinates x_1', x_2' is large if $\gamma^{\bar{h}}$ is small and this produces a "gain," as passing from x_1 to x_2 one needs a huge number of vertices contained in the subdiagram. That is, if $(\omega x_1')_{\text{mod}1} \sim (\omega x_2')_{\text{mod}1} \sim \gamma^{\bar{h}}$ then by the Diophantine condition,

$$2\gamma^{\bar{h}} \ge ||\omega(x_1' - x_2')|| \ge C_0 |x_1' - x_2'|^{-\tau}, \tag{3.8}$$

so that $|x_1' - x_2'| \ge \bar{C} \gamma^{-\bar{h}\tau^{-1}}$. This has the effect that one can use a small factor coming from the high power of ε, U necessary to produce an high difference of coordinates to control the size of the nonresonant terms. To implement this idea one has to face the fact that several subdiagrams can be encapsulated one within the other, with the same external lines, so that the gain factor has to be distributed among them (see [**35**], [**36**]). The resulting gain is sufficient to make the scaling dimension of such terms negative, so that the sum over the scales can be done.

Regarding the resonant terms, the crucial observation is that the external lines have the same ρ-index. This is due to the fact that any couple x_i, x_j must be integers; if $\rho_i = -\rho_j$ than $x_i - x_j = \pm 2\bar{x}$ and this is impossible as by (2.10) $2\bar{x}$ cannot be integer. Therefore, if the external lines in the resonant terms have also the same $\mathbf{x}'$ the value is zero due to the anticommutativity of fermions. This produces a gain factor in all the resonant terms with more than four fields making them irrelevant. The only relevant terms are the quadratic ones producing a finite renormalization of the chemical potential; their size is controlled by choosing properly the counterterm ν. The fermionic nature of the particles play also a crucial role in taking care of the combinatorial growth due to the presence of loops in the graphs; as we mentioned the number of diagrams is $O(n!^2)$, a factor apparently preventing convergence. However, due to cancellations in the signs following from anticommutativity, one can prove that the sum of diagrams verifies a summable bound.

Note the crucial role of the assumption of *spinless* particles in the above analysis; if the particle are *spinful* no gain is present controlling the resonant quartic terms. According to renormalization-group methods, such terms correspond to new running coupling constants which increase when iterating the RG and could destroy convergence. Therefore one can repeat the proof of Theorem 3.1 in the case of spinful fermions, but in that case the different behavior of the resonant terms prevent of taking the $T \to 0$ limit, see [**38**]. This could signal a possible divergence of the expansion, at least for some observables, and consequently its delocalization. Recent numerical simulations [**34**] have indeed found a different behavior in the spinful version of the model (1.2) with respect to the spinless case. In particular, certain spin-dependent quantities show a delocalized behavior, in contrast with others, the one measured in experiments with cold atoms [**16**]. This is in agreement with our analysis, in which convergence of the expansion for the 2-point function up to zero temperature can be proved only in the spinless case, while in presence of spin an obstruction to convergence is provided by the presence of an extra relevant process.

4. Coupled interacting chains

We discuss now the case of several coupled interacting chains with quasi-random disorder, the same in all the chains. The Hamiltonian on the Fock space is

$$H = J\sum_{x,y}(a^+_{x+1,y}a^-_{x,y} + a^+_{x-1,y}a^-_{x,y}) + \Delta\sum_{x,y}\cos(2\pi\omega x)a^+_{x,y}a^-_{x,y} +$$

$$U\sum_{x,y}a^+_{x,y}a^-_{x,y}a^+_{x+1,y}a^-_{x+1,y} + J_\perp\sum_{x,y}(a^+_{x,y+1}a^-_{x,y} + a^+_{x,y}a^-_{x,y+1}). \tag{4.1}$$

It is convenient to write $a^\pm_{x,y} = \frac{1}{L}\sum_l e^{\pm ily}\widehat{a}^\pm_{x,l}$, where $l = 2\pi\frac{n}{L}$ with $n = 0, .., L-1$, so that the Hamiltonian can be rewritten in the following way:

$$H = J\frac{1}{L}\sum_{x,l}(\widehat{a}^+_{x+1,l}\widehat{a}^-_{x,l} + \widehat{a}^+_{x-1,l}\widehat{a}^-_{x,l}) + \frac{\Delta}{L}\sum_{x,l}\cos(2\pi\omega x)\widehat{a}^+_{x,l}\widehat{a}^-_{x,l} +$$

$$\frac{J_\perp}{L}\sum_{x,l}\cos l(\widehat{a}^+_{x,l}\widehat{a}^-_{x,l} + \widehat{a}^+_{x,l}\widehat{a}^-_{x,l}) + U\sum_x \frac{1}{L^4}\sum_{l_1,l_2,l_3,l_4}\widehat{a}^+_{x,l_1}\widehat{a}^-_{x,l_2}\widehat{a}^+_{x+1,l_3}\widehat{a}^-_{x+1,l_4}\delta(l_1 - l_2 + l_3$$

In the *molecular limit* $U = J = 0$ one has (setting $\Delta = 1$ for definiteness)

$$H_0 - \mu N = \frac{1}{L}\sum_{x,l}(\cos(2\pi\omega x) - \mu_l)\widehat{a}^+_{x,l}\widehat{a}^-_{x,l}, \tag{4.2}$$

with

$$\mu_l = \mu - J_\perp \cos l \equiv \cos 2\pi\bar{x}_l, \tag{4.3}$$

and the 2-point correlation is given by

$$g_l(k_0, x) = \frac{1}{-ik_0 + \cos 2\pi\omega x - \cos 2\pi\omega\bar{x}_l}. \tag{4.4}$$

With respect to the single chain, there are now several kinds of particles with different chemical potentials; if there is no interaction they are independent. We can repeat a renormalization-group analysis similar to the previous one, defining in a similar way the resonant and nonresonant terms; the first are such that the external lines have the same x' while the second are such that at least two have different x'. Note however that the propagators carry a label l in addition to the label ρ. In order to bound the divisors one has to assume extra conditions, which are called *Melnikov condtions* in the theory of dynamical systems. In a large relative measure set of $J_\perp$ we can assume that

$$||\omega x \pm 2\omega\bar{x}_l|| \geq C_0|x|^{-\tau'} \qquad \forall x \in \mathbb{Z}/\{0\} \tag{4.5}$$

and

$$||\omega x \pm \omega\bar{x}_l \pm \omega\bar{x}_{l'}|| \geq C_0|x|^{-\tau'} \qquad \forall x \in \mathbb{Z}/\{0\}. \tag{4.6}$$

The first condition (4.5) is used to bound the propagators $|g_l(k_0, x)| \leq C|x|^\tau$. The second condition (4.6) is used to control the nonresonant terms; if x_i and x_j are coordinates of the external lines with scale h', external to a cluster with scale h

$$\begin{aligned} 2\gamma^{h'} &\geq ||(\omega x_i')|| + ||(\omega x_j')|| \geq ||\omega(x_i' - x_j')|| = \\ &||\omega(\rho_i\bar{x}_{l_i} - \rho_j\bar{x}_{l_j}) + \omega M|| \geq \frac{C_0}{|M|^{\tau'}} \end{aligned} \tag{4.7}$$

where $M = x_i - x_j$ is an integer smaller than the number of vertices in the subdiagram. This says that the nonresonant terms have a number of vertices $m \geq \bar{C}\gamma^{-\bar{h}/\tau'}$, and by the factor ε^m associated to the subgraph one can extract a small factor making them irrelevant.

Regarding the resonant terms, let us consider two external lines with coordinates x_i, x_j; since $x_i - x_j \in \mathbb{Z}$ and $x_i' = x_j'$ it follow that $\rho_j = \rho_j$ as if $\rho_i = -\rho_j$ then $2\bar{x} + (\cos l_i + \cos l_j)aJ_\perp$ must be integer or vanishing, which is impossible except in a zero-measure set. Moreover $\cos l_i = \cos l_j$ as in the resonances $\bar{x}_i - \bar{x}_j = (\cos l_i - \cos l_j)aJ_\perp = 0$. If $L = 2$, that is, the array is composed of only two chains then $l = (0, \pi)$, so that the external lines of a resonant subgraph have the same ρ and l; this has the effect that the monomials with ≥ 4 are irrelevant. This is not the case with a higher number of chains, where resonant terms with external lines with different l are possible. For instance in the problem of three chains $L = 3$ one has $l = 2\pi/3, 4\pi/3, 6\pi/3$ and $\bar{x}_1 = \bar{x} - J_\perp/2, \bar{x}_1 = \bar{x} - J_\perp/2, \bar{x}_3 = \bar{x}$. As in the case of spinful fermions in a chain, this generates a relevant process which can ruin convergence. In [**39**] the following theorem can be found.

THEOREM 4.1. *Let us consider the 2-point function* $< \hat{a}^-_{\mathbf{x},l}\hat{a}^+_{\mathbf{y},l} >$ *with Hamiltonian given by (4.2) in the spinless case, with chemical potential* $\mu + \nu_l$*. If* $U, J_\perp,$

and J are small and assuming (4.5) and (4.6), for suitable ν_l, then:

(1) *if $L=2$ for $\beta\to\infty$ then for any integer N and a suitable constant C_N*

$$|<\widehat{a}^-_{\mathbf{x},l}\widehat{a}^+_{\mathbf{y},l}>|\le e^{-\xi|x-y|}\frac{C_N|\log\Delta|}{1+(\Delta|x_0-y_0|)^N},\tag{4.8}$$

with $\xi=|\log\varepsilon|$, $\varepsilon=\max(|J|,|U|)$ and $\Delta=(1+\min(|x|,|y|))^{-\tau}$.

(2) *If $L\ge 3$, $U=0$ then for $\beta\to\infty$ (4.8) holds.*

(3) *If $L\ge 3$, then (4.8) holds for $\beta|U|\le 1$, with $\xi=\max(|\log\varepsilon|,\beta^{-1})$, $\Delta=\max((1+\min(|x|,|y|))^{-\tau},\beta^{-1})$.*

In the case of two chains (case 1) the 2-point function decays at zero-temperature exponentially in the direction of the chains, and a very weak decay is present in the imaginary time direction (faster than any power but with rate decreasing as x,y increase); this is very similar to what happens in the single chain case and indicates localization of the ground state with or without interaction. In contrast, for a greater number of chains the interaction produces a qualitative difference; in absence of many-body interaction, zero-temperature exponential decay is found for any number of chains (case 2), while in the presence of an interaction convergence of the expansion holds only up to a finite temperature (case 3). The reason is that when $L\ge 3$ there are extra relevant terms which ncrease as the RG is iterated. As the presence of diverging directions in the RG flow signals an instability which can imply delocalization, this is in agreement with the behavior observed in cold-atom experiments [**17**], in which absence of localization is found in an array of chains (except when there is no interaction, when localization is found), and localization in the single chain case; moreover, we find localization with two chains in the spinless case, a prediction accessible in principle to future experiments.

We briefly discuss finally why the Melnikov conditions (4.5) and (4.6) are true in a large relative measure set of values, that is, if $|J_\perp|\le\varepsilon_0$ in a set of whose complement has measure $O(C_L\varepsilon_0^{1+\alpha})$, $\alpha\ge 0$, C_L an L-dependent constant. Indeed if (4.5) is true, then if $\cos l\ne 0$,

$$\begin{aligned}&C_0|x|^{-\tau}\le||\omega x\pm 2\bar{x}||\le\\&||\omega x\pm 2(\bar{x}+aJ_\perp\cos l)||+2|aJ_\perp\cos l|\le C_0|x|^{-\tau'}+C|\varepsilon_0\cos la,\end{aligned}\tag{4.9}$$

so that if $\tau'>\tau+1$, $C_0/2|x|^{-\tau}\le C_0|x|^{-\tau}(1-|x|^{\tau-\tau'})\le C|a\varepsilon_0\cos l|$ for $|x|\ge 2$ and $|x|\ge(2C\varepsilon_0\cos la|/C_0)^{\frac{-1}{\tau}}=\mathrm{N}_0$. The subset I of $J_\perp$ not verifying (4.5) is defined by the condition, for $-1\le s\le 1$

$$f(s)=\omega x\pm 2(\bar{x}+J_\perp(s)\cos la)=sC_0|x|^{-\tau'},\tag{4.10}$$

and $\frac{\partial f}{\partial s}=\frac{\partial f}{\partial J_\perp}\frac{\partial J_\perp}{\partial s}=C_0|x|^{-\tau'}$, so the measure of the region in which (4.5) is not true is

$$\int_I dJ_\perp=\sum_l\sum_{n\ge\mathrm{N}_0}\int_{-1}^1|\frac{dJ_\perp}{ds}|ds\le\sum_{l,\cos l\ne 0}\frac{C}{\cos l}\sum_{x\ge\mathrm{N}_0}|x|^{-\tau'}\le C_L|\varepsilon_0|^{\frac{\tau'-1}{\tau}}.\tag{4.11}$$

Choosing $\frac{\tau'-1}{\tau}>1$, that is, $\tau'>\tau+1$ we have that for $|J_\perp|\le\varepsilon_0$ the relative measure of the excluded $J_\perp$ is $O(C_L\varepsilon_0^{\tau'-\tau-1})$, hence vanishing if $\varepsilon_0\to 0$. A similar procedure can be repeated for the second Melnikov condition; if $\cos l_i\pm\cos l_j\ne 0$, then

$$C_0|x|^{-\tau} \le ||\omega x \pm (\bar{x} + a\cos l_i J_\perp) \pm (\bar{x} + a\cos l_j J_\perp)|| +$$
$$|J_\perp a(\cos l_i \pm \cos l_j)| \le C_0|x|^{-\tau'} + C|\varepsilon_0(\cos l_i \pm \cos l_j)a|,$$

from which $|x| \ge (2C\varepsilon_0|(\cos l_i \pm \cos l_j)a|/C_0)^{\frac{-1}{\tau}}$; one then proceeds as above with $|\cos l_i \pm \cos l_j|$ replacing $|\cos l_i|$.

References

[1] P. W. Anderson:*Absence of Diffusion in Certain Random Lattices.* Phys. Rev. 109, 1492 (1958)

[2] R. Nandkshore, D. Huse: *Many body localization and thermalization in quantum statistical mechanics* Annual Review of Condensed Matter Physics, Vol. 6: 15-38 (2015)

[3] Sheldon Goldstein, David A. Huse, Joel L. Lebowitz, and Roderich Tumulka, *Macroscopic and microscopic thermal equilibrium*, Ann. Phys. **529** (2017), no. 7, 201600301, 20, DOI 10.1002/andp.201600301. MR3671055

[4] D.M. Basko, I. Alteiner, B. L. Altshuler: *Metal-insulator transition in a weakly interacting many-electron system with localized single-particle states.* Ann. Phys. 321, 1126 (2006)

[5] Jürg Fröhlich and Thomas Spencer, *Absence of diffusion in the Anderson tight binding model for large disorder or low energy*, Comm. Math. Phys. **88** (1983), no. 2, 151–184. MR696803

[6] Michael Aizenman and Stanislav Molchanov, *Localization at large disorder and at extreme energies: an elementary derivation*, Comm. Math. Phys. **157** (1993), no. 2, 245–278. MR1244867

[7] Michael Aizenman and Simone Warzel, *Random operators*, Graduate Studies in Mathematics, vol. 168, American Mathematical Society, Providence, RI, 2015. Disorder effects on quantum spectra and dynamics. MR3364516

[8] M. Aizenman and G. M. Graf, *Localization bounds for an electron gas*, J. Phys. A **31** (1998), no. 32, 6783–6806, DOI 10.1088/0305-4470/31/32/004. MR1715186

[9] Abel Klein and J. Fernando Perez, *Localization in the ground state of a disordered array of quantum rotators*, Comm. Math. Phys. **147** (1992), no. 2, 241–252. MR1174412

[10] Michael Aizenman and Simone Warzel, *Localization bounds for multiparticle systems*, Comm. Math. Phys. **290** (2009), no. 3, 903–934, DOI 10.1007/s00220-009-0792-6. MR2525644

[11] Vincent Beaud and Simone Warzel, *Low-energy Fock-space localization for attractive hard-core particles in disorder*, Ann. Henri Poincaré **18** (2017), no. 10, 3143–3166, DOI 10.1007/s00023-017-0591-0. MR3697191

[12] A. Elgart, A. Klein, G. Stolz *Many Body localization in the droplet spectrum of the random XXZ quantum spin chain* arxiv:1703.07483

[13] John Z. Imbrie, *On many-body localization for quantum spin chains*, J. Stat. Phys. **163** (2016), no. 5, 998–1048, DOI 10.1007/s10955-016-1508-x. MR3493184

[14] J. Imbrie: *Diagonalization and Many-Body Localization for a Disordered Quantum Spin Chain.* Phys. Rev. Lett. 117, 027201 (2016)

[15] Serge Aubry and Gilles André, *Analyticity breaking and Anderson localization in incommensurate lattices*, Group theoretical methods in physics (Proc. Eighth Internat. Colloq., Kiryat Anavim, 1979), Ann. Israel Phys. Soc., vol. 3, Hilger, Bristol, 1980, pp. 133–164. MR626837

[16] Michael Schreiber, Sean S. Hodgman, Pranjal Bordia, Henrik P. Lüschen, Mark H. Fischer, Ronen Vosk, Ehud Altman, Ulrich Schneider, and Immanuel Bloch, *Observation of many-body localization of interacting fermions in a quasirandom optical lattice*, Science **349** (2015), no. 6250, 842–845, DOI 10.1126/science.aaa7432. MR3409723

[17] P. Bordia, H Lüschen, S. Hodgman, M. Schreiber, I. Bloch, U. Schneider: *Coupling Identical 1D Many-Body Localized Systems* Phys. Rev. Lett. 116, 140401 (2016)

[18] E. I. Dinaburg and Ja. G. Sinaĭ, *The one-dimensional Schrödinger equation with quasiperiodic potential* (Russian), Funkcional. Anal. i Priložen. **9** (1975), no. 4, 8–21. MR0470318

[19] L. A. Pastur, *Spectra of random selfadjoint operators* (Russian), Uspehi Mat. Nauk **28** (1973), no. 1(169), 3–64. MR0406251

[20] J. Bellissard, R. Lima, and D. Testard, *A metal-insulator transition for the almost Mathieu model*, Comm. Math. Phys. **88** (1983), no. 2, 207–234. MR696805

[21] L. H. Eliasson, *Floquet solutions for the 1-dimensional quasi-periodic Schrödinger equation*, Comm. Math. Phys. **146** (1992), no. 3, 447–482. MR1167299

[22] Jürgen Moser and Jürgen Pöschel, *An extension of a result by Dinaburg and Sinaĭ on quasiperiodic potentials*, Comment. Math. Helv. **59** (1984), no. 1, 39–85, DOI 10.1007/BF02566337. MR743943

[23] Ya. G. Sinaĭ, *Anderson localization for one-dimensional difference Schrödinger operator with quasiperiodic potential*, J. Statist. Phys. **46** (1987), no. 5-6, 861–909, DOI 10.1007/BF01011146. MR893122

[24] J. Fröhlich, T. Spencer, and P. Wittwer, *Localization for a class of one-dimensional quasi-periodic Schrödinger operators*, Comm. Math. Phys. **132** (1990), no. 1, 5–25. MR1069198

[25] Svetlana Ya. Jitomirskaya, *Metal-insulator transition for the almost Mathieu operator*, Ann. of Math. (2) **150** (1999), no. 3, 1159–1175, DOI 10.2307/121066. MR1740982

[26] Artur Avila and Svetlana Jitomirskaya, *The Ten Martini Problem*, Ann. of Math. (2) **170** (2009), no. 1, 303–342, DOI 10.4007/annals.2009.170.303. MR2521117

[27] G. Benfatto, G. Gentile, and V. Mastropietro, *Electrons in a lattice with an incommensurate potential*, J. Statist. Phys. **89** (1997), no. 3-4, 655–708, DOI 10.1007/BF02765540. MR1658711

[28] G. Gentile and V. Mastropietro, *Anderson localization for the Holstein model*, Comm. Math. Phys. **215** (2000), no. 1, 69–103, DOI 10.1007/s002200215069. MR1800918

[29] Vieri Mastropietro, *Small denominators and anomalous behaviour in the incommensurate Hubbard-Holstein model*, Comm. Math. Phys. **201** (1999), no. 1, 81–115, DOI 10.1007/s002200050550. MR1669429

[30] V.Mastropietro: *Dense gaps in the interacting Aubry-André model* Phys. Rev. B 93, 245154 (2016)

[31] D.L. Shepelyansky Two interacting particles in the Harper model. Phys Rev B 54 (21):14896 (1996)

[32] L. H. Eliasson, *Perturbations of linear quasi-periodic system*, Dynamical systems and small divisors (Cetraro, 1998), Lecture Notes in Math., vol. 1784, Springer, Berlin, 2002, pp. 1–60, DOI 10.1007/978-3-540-47928-4_1. MR1924910

[33] S. Iyer, V. Oganesyan, G. Refael, D. A. Huse: *Many-body localization in a quasiperiodic system.* Phys. Rev. B 87, 134202 (2013)

[34] P. Prelovsek, O.S. Barisic, M. Znidaric *Absence of full many body localization in disordered Hubbard chain* Phys. Rev. B 94, 241104 (2016)

[35] Vieri Mastropietro, *Localization in the ground state of an interacting quasi-periodic fermionic chain*, Comm. Math. Phys. **342** (2016), no. 1, 217–250, DOI 10.1007/s00220-015-2498-2. MR3455150

[36] Vieri Mastropietro, *Localization in interacting fermionic chains with quasi-random disorder*, Comm. Math. Phys. **351** (2017), no. 1, 283–309, DOI 10.1007/s00220-016-2773-x. MR3613505

[37] V. Mastropietro *Localization of interacting fermions in the Aubry-André model* Phys. Rev. Lett. 115, 180401 (2015)

[38] V. Mastropietro Interacting spinning fermions with quasi-random disorder. Ann. der Physik (2017).

[39] V.Mastropietro Coupled identical localized fermionic chains with quasirandom disorder. Phys. Rev. B 95, 075155 (2017)

[40] Giovanni Gallavotti, *Twistless KAM tori*, Comm. Math. Phys. **164** (1994), no. 1, 145–156. MR1288156

[41] G. Gentile and V. Mastropietro, *Methods for the analysis of the Lindstedt series for KAM tori and renormalizability in classical mechanics. A review with some applications*, Rev. Math. Phys. **8** (1996), no. 3, 393–444, DOI 10.1142/S0129055X96000135. MR1388257

[42] V.Mastropietro *Constructive Renormalization* World Scientific 2006

Dipartimento di Matematica, Universitá di Milano, Via C. Saldini 50, 20133, Milano, Italy

Email address: `vieri.mastropietro@unimi.it`

Contemporary Mathematics
Volume **717**, 2018
https://doi.org/10.1090/conm/717/14442

Local density approximation for almost-bosonic anyons

M. Correggi, D. Lundholm, and N. Rougerie

Abstract. We discuss the average-field approximation for a trapped gas of non-interacting anyons in the quasi-bosonic regime. In the homogeneous case, i.e., for a confinement to a bounded region, we prove that the energy in the regime of large statistics parameter, i.e., for "less-bosonic" anyons, is independent of boundary conditions and of the shape of the domain. When a non-trivial trapping potential is present, we derive a local density approximation in terms of a Thomas-Fermi-like model. The results presented here mainly summarize [Anal. PDE 10 (2017), 1169-1200] with additional remarks and strengthening of some statements.

1. Introduction

The physics of identical quantum particles in two-dimensions (2D) is much richer than in the three-dimensional world, where the symmetry under exchange admits only two opposite representations leading to *bosonic* (symmetric) and *fermionic* (antisymmetric) particles. However, in 2D, the way two or more particles are exchanged plays an important role in classifying the species of identical particles. There is indeed room for other statistics different from the usual Bose-Einstein and Fermi-Dirac ones, which go under the name of *intermediate* or *fractional* statistics. The corresponding particles are named *anyons* (a name originally introduced by F. Wilczek in [**Wil82**]) and include bosons and fermions as special cases. Mathematically speaking, the reason behind these unconventional features is the exchange symmetry: since the particles are indistinguishable the Hilbert space containing the states of the system must be the space of a one-dimensional[1] representation of the symmetry group. In three or more dimensions this group is the permutation group, which has only two inequivalent irreducible representations, given by the space of symmetric and antisymmetric functions. On the opposite, in 2D, there are many inequivalent ways of exchanging the particles and the symmetry group to consider is the braid group which has infinitely many irreducible representations. Topologically, this difference is due to the fact that a circle is always contractible to a point in three or more dimensions, while it is not in 2D.

Another way of understanding the emergence of fractional statistics is simply by observing that for identical particles all the physics must be invariant under exchange. This has to be true at least for the probability density and therefore the

[1] It is also possible to consider higher dimensional representations of the symmetry group, which lead to generalizations of anyons going under the name of *plektons* or *non-abelian anyons.*

modulus of the wave function $|\Psi(\mathbf{r}_1,\ldots,\mathbf{r}_N)|$ can not change when two particles are exchanged:

$$|\Psi(\ldots,\mathbf{r}_i,\ldots,\mathbf{r}_j,\ldots)| = |\Psi(\ldots,\mathbf{r}_j,\ldots,\mathbf{r}_i,\ldots)|.$$

In 3D this implies that only the sign of Ψ can flip under exchange, while in two dimensions the wave function can acquire a generic phase factor, i.e.,

$$\Psi(\ldots,\mathbf{r}_i,\ldots,\mathbf{r}_j,\ldots) = e^{i\pi\alpha}\Psi(\ldots,\mathbf{r}_j,\ldots,\mathbf{r}_i,\ldots), \tag{1.1}$$

where $\alpha \in \mathbb{R}$ can be *any* real number, which is called the *statistics parameter*. Without loss of generality one can however assume that

$$\alpha \in (-1,1], \tag{1.2}$$

where the points 0 and 1 (and -1 by periodicity) describe the usual bosonic and fermionic particles.

Whether such anyonic particles do really exist is still a debated question within the physics community. Although elementary particles live in a three-dimensional world and therefore can never be anyonic, the possibility to observe some quasi-particles obeying to fractional statistics is more than plausible. In particular physical models involving anyonic particles have been proposed in relation to the fractional quantum Hall effect. We refer to [**Kha05, LR16, Myr99, Wil90**] and references therein for a more detailed discussion of this point.

In order to describe non-interacting anyonic particles in a trap, one has thus to face the problem of defining suitable Schrödinger operators of the form $\sum(-\Delta_i + V(\mathbf{r}_i))$ on a Hilbert space of functions satisfying (1.1) [**DFT97**]. This poses hard technical questions since (1.1) implies that anyonic wave functions are in general multi-valued. However, there is an alternative but equivalent approach which we are going to adopt in this note: instead of considering multi-valued functions satisfying (1.1), which goes under the name of *anyonic gauge*, one can rewrite the Hamiltonian in the *magnetic gauge*, i.e., set, at least formally,

$$\Psi(\mathbf{r}_1,\ldots,\mathbf{r}_N) =: \prod_{j<k} e^{i\alpha\phi_{jk}}\Phi(\mathbf{r}_1,\ldots,\mathbf{r}_N), \tag{1.3}$$

where $\phi_{jk} = \arg(\mathbf{r}_j - \mathbf{r}_k)$ is the angle between the vectors $\mathbf{r}_j$ and $\mathbf{r}_k$. With such a choice it is easy to see that (1.1) holds true if and only if $\Phi \in L^2_{\mathrm{sym}}(\mathbb{R}^{2N})$, i.e., Φ must be *symmetric* under exchange, i.e., bosonic. There is however a price to pay for this simplification, i.e., the Schrödinger operator acting on Φ becomes

$$H = \sum_{j=1}^{N} \left\{(-i\nabla_j + \alpha\mathbf{A}_j)^2 + V(\mathbf{r}_j)\right\}, \tag{1.4}$$

i.e., each particle carries a magnetic flux generated by the Aharonov-Bohm magnetic potential

$$\mathbf{A}_j(\mathbf{r}_1,\ldots,\mathbf{r}_N) = \sum_{k\neq j} \frac{(\mathbf{r}_j - \mathbf{r}_k)^{\perp}}{|\mathbf{r}_j - \mathbf{r}_k|^2}, \tag{1.5}$$

where $\mathbf{r}^{\perp} := (-y, x)$. To discuss self-adjointness of H, a typical approach is to consider the symmetric operator defined by H on the domain of functions with support away from the diagonals $\mathbf{r}_j = \mathbf{r}_k$. Such an operator admits several self-adjoint extensions (see, e.g., [**AT98, BS92**] for the discussion of the two-particle case), but one can at least identify the Hamiltonian with its Friedrichs extension [**LS13**, Sect. 2.2].

An alternative approach to the rigorous definition of anyonic Schrödinger operators is through a sort of regularization of (1.4): since the singularities of H live on the diagonals $\mathbf{r}_j = \mathbf{r}_k$, it seems reasonable to replace the singular magnetic potential (1.5) with its cut-off version obtained as

$$\mathbf{A}_j^R(\mathbf{r}_1, \ldots, \mathbf{r}_N) = \sum_{k \neq j} (\mathbf{r}_j - \mathbf{r}_k)^\perp \left[\frac{\mathbb{1}_{\{|\mathbf{r}_j - \mathbf{r}_k| > R\}}}{|\mathbf{r}_j - \mathbf{r}_k|^2} + \frac{\mathbb{1}_{\{|\mathbf{r}_j - \mathbf{r}_k| \leqslant R\}}}{R^2} \right], \tag{1.6}$$

for some $R > 0$. In other words the point-like anyons are replaced with extended particles exactly in the same way in classical electrodynamics giving a spatial dimension to point charges makes the theory more regular. It is thus clear why the so-obtained particles are named *extended anyons*. The corresponding Schrödinger operator H_R is essentially self-adjoint on smooth functions with compact support. We refer to [**LL16**, **LR15**, **Lun16**] and references therein for a detailed discussion of extended anyon models.

1.1. The model: average-field approximation. The physical system we plan to describe in this note is a many-body gas of a large number N of non-interacting anyons confined in a trap. When N gets very large, the quantum mechanical description above becomes very inefficient and the complexity of the corresponding equations calls for a simplified effective model. Such a problem has already been faced in the physics literature [**IL92**, **Wil90**], where it was noted that, if the statistic parameter α is close to 0 (bosons[2]), the mean-field behavior of the anyonic gas is expected to be suitably approximated by the *average-field energy functional*[3], which reads

$$\boxed{\mathcal{E}_\beta^{\mathrm{af}}[u] := \int_{\mathbb{R}^2} \mathrm{d}\mathbf{r} \, \left\{ \left| \left(-i\nabla + \beta \mathbf{A}[|u|^2] \right) u \right|^2 + V(\mathbf{r}) |u|^2 \right\}} \tag{1.7}$$

where V is the trapping potential and

$$\mathbf{A}[\rho] := \nabla^\perp w_0 * \rho, \qquad w_0(\mathbf{r}) := \log |\mathbf{r}|, \tag{1.8}$$

and $\nabla^\perp := (-\partial_y, \partial_x)$. Since w_0 is the fundamental solution of the Laplace equation $\Delta w_0 = 2\pi\delta(\mathbf{r})$, the magnetic field associated to $\mathbf{A}[\rho]$ is

$$\left(\operatorname{curl} \mathbf{A}[\rho] \right)(\mathbf{r}) = 2\pi \rho(\mathbf{r}), \tag{1.9}$$

i.e., its intensity is proportional to the particle density $\rho(\mathbf{r}) = |u(\mathbf{r})|^2$. The parameter $\beta \in \mathbb{R}$ is related to the statistics parameter α in a way that will be made precise below. The energy (1.7) can indeed be easily obtained by evaluating the expectation value of the many-body Schrödinger operator (1.4) on a product state $\Psi(\mathbf{r}_1, \ldots, \mathbf{r}_N) = u(\mathbf{r}_1) \cdot \cdots \cdot u(\mathbf{r}_N)$ and dropping any term vanishing in the limit $N \to \infty$. The parameter β is obtained as the limit of $N\alpha$ and therefore one is forced to assume that $\alpha = \mathcal{O}(N^{-1})$ to obtain a finite energy in the limit, i.e., $\alpha \to 0$ and

---

[2] A similar approximation is often applied in the opposite regime of α close to 1, i.e., for almost fermionic anyons. A one-particle effective model is expected to emerge in this case too, although the expression of the function may be different.

[3] In fact the explicit expression of the average-field functional has never appeared in the physics literature, where a further approximation is made (see Section 2). In this respect $\mathcal{E}_\beta^{\mathrm{af}}$ was first introduced in [**LR15**].

we are considering almost-bosonic anyons. A similar behavior can be recovered in the opposite regime of almost-fermionic anyons, i.e., when $\alpha \to 1$ [**Wil90**].

This heuristic derivation of the average-field functional was in fact made rigorous in [**LR15**, Theorem 1.1], where it is proven that, assuming

$$\alpha = \frac{\beta}{N-1}, \tag{1.10}$$

in the limit $N \to \infty$, the ground state energy per particle of a gas of non-interacting anyons converges to the infimum of the average-field functional (1.7). The result is actually proven for extended anyons but with a particle radius $\sim N^{-\gamma}$, for some $\gamma > 0$, i.e., the limit $N \to \infty$ describes also the convergence to point-like particles. In addition the k-th particle reduced density matrix associated to any quasi-ground state converges to a convex combination of projectors onto the tensor product of minimizers of $\mathcal{E}^{\mathrm{af}}_{\beta}$. Hence one can say that the minimization of the average-field functional (see below) provides a good approximation for the ground state behavior of the many-body anyon gas in this particular limit.

The effective theory provided by the average-field energy depends on only one physical parameter $\beta \in \mathbb{R}$, which in fact can be assumed to be positive

$$\beta \geqslant 0, \tag{1.11}$$

thanks to the symmetry under exchange $u \to u^*$ (complex conjugation) of the energy. In spite of a certain analogy with other effective nonlinear theories applied to the description of, e.g., Bose-Einstein condensates (Gross-Pitaevskii theory) or superconductors (Ginzburg-Landau theory), the specificity of the average-field functional lies in the special form of the nonlinearity, which appears only in the magnetic potential $\mathbf{A}[|u|^2]$ and therefore affects mostly the phase of the effective wave function u. In addition, the nonlinearity in $\mathcal{E}^{\mathrm{af}}_{\beta}[u]$ is actually nonlocal and, as it becomes apparent in its Euler-Lagrange equation (1.25), it generates cubic quasi-linear and quintic semi-linear terms.

Given the physical meaning of the parameter β and (1.10), we are thus describing a model of anyonic behavior which is a "small perturbation" of the usual bosonic one. In the regime $\beta \to 0$ one thus expects to recover conventional bosons, and this can actually be proven rigorously, at least for the ground state properties [**LR15**, Proposition 3.8]. More interesting and unexplored is the opposite regime

$$\beta \gg 1, \tag{1.12}$$

where the anyonic features of the model should emerge. This is indeed the asymptotic regime we are going to discuss in this note in the framework of the average-field approximation.

1.2. Minimization of the average-field functional. Let us now discuss closer the average-field functional from the mathematical view point. First of all we assume that the trapping potential is positive

$$V(\mathbf{r}) \geqslant 0, \tag{1.13}$$

which does not imply any loss of generality but simply a change of energy scale. In the following we will restrict our attention to a smaller class of smooth homogeneous potentials such that

$$V(\lambda \mathbf{r}) = \lambda^s V(\mathbf{r}), \qquad s > 1, \tag{1.14}$$

for any $\lambda > 0$. The potential must also be trapping and therefore we assume that

$$\lim_{R\to\infty} \min_{|\mathbf{r}|\geqslant R} V(\mathbf{r}) = +\infty. \tag{1.15}$$

A typical case is the (anisotropic) harmonic potential $V(\mathbf{r}) = ax^2 + by^2$, $a, b \in \mathbb{R}^+$, but we are also going to consider the flat case, i.e., formally $s = +\infty$ or

$$V(\mathbf{r}) = \begin{cases} 0, & \text{in } \Omega \subset \mathbb{R}^2, \\ +\infty & \text{otherwise,} \end{cases} \tag{1.16}$$

for some simply connected domain Ω with Lipschitz boundary. The average-field functional in this case simply becomes

$$\mathcal{E}^{\mathrm{af}}_{\beta,\Omega}[u] := \int_\Omega \mathrm{d}\mathbf{r}\ \left|\left(-i\nabla + \beta\mathbf{A}[|u|^2]\right) u\right|^2. \tag{1.17}$$

Note that the choice (1.16) would naturally lead to consider Dirichlet conditions on the boundary of Ω, while the Neumann case seems to be more appropriate to describe the homogeneous setting. As a matter of fact we are going to prove that boundary conditions do not matter at all in the limit $\beta \to \infty$ and we keep referring to both cases as the *homogenous* anyonic gas.

The energy functional (1.8) is well defined for any $u \in H^1(\mathbb{R}^2)$ such that $V|u|^2 \in L^1(\mathbb{R}^2)$, as it can be easily seen thanks to the inequality [**LR15**, Lemma 3.4]

$$\int_{\mathbb{R}^2} \mathrm{d}\mathbf{r}\ \left|\mathbf{A}\left[|u|^2\right]\right|^2 |u|^2 \leqslant \tfrac{3}{2} \|u\|_2^4 \left\|\nabla|u|\right\|_2^2,$$

which allows to bound the most singular term in the energy by the kinetic term. We thus consider the minimization domain

$$\mathscr{D}^{\mathrm{af}} := \left\{u \in H^1(\mathbb{R}^2) \,\middle|\, V|u|^2 \in L^1(\mathbb{R}^2), \|u\|_2 = 1\right\}, \tag{1.18}$$

where the L^2 normalization has the usual meaning of the quantum mechanical probability conservation, and set

$$E^{\mathrm{af}}_\beta := \inf_{u\in\mathscr{D}^{\mathrm{af}}} \mathcal{E}^{\mathrm{af}}_\beta[u]. \tag{1.19}$$

The boundedness from below of E^{af}_β, in fact its positivity, is a simple consequence of the assumption on V. The existence of a corresponding minimizing u^{af}_β does not immediately follow but can be proven exploiting the estimate above [**LR15**, Proposition 3.7] together with the following magnetic bounds [**LR15**, Lemma 3.4]: for any $\beta \in \mathbb{R}$ and $u \in H^1(\mathbb{R}^2)$,

$$\int_{\mathbb{R}^2} \mathrm{d}\mathbf{r}\ \left|\left(-i\nabla + \beta\mathbf{A}[|u|^2]\right) u\right|^2 \geqslant \int_{\mathbb{R}^2} \mathrm{d}\mathbf{r}\ \left|\nabla|u|\right|^2, \tag{1.20}$$

$$\int_{\mathbb{R}^2} \mathrm{d}\mathbf{r}\ \left|\left(-i\nabla + \beta\mathbf{A}[|u|^2]\right) u\right|^2 \geqslant 2\pi|\beta| \int_{\mathbb{R}^2} \mathrm{d}\mathbf{r}\ |u|^4. \tag{1.21}$$

Both inequalities can actually be generalized [**CLR17**, Lemma 3.2] to any domain $\Omega \subset \mathbb{R}^2$, but the second one requires the additional assumption $u \in H^1_0(\Omega)$, i.e., the support of u must be strictly contained in Ω. Note also that the first bound is simply the extension to the nonlinear vector potential $\mathbf{A}[|u|^2]$ of the usual diamagnetic inequality [**LL01**, Theorem 7.21].

For the homogeneous gas we have to specify the dependence on boundary conditions and we thus set

$$\mathscr{D}^{\mathrm{af}}_{\mathrm{N}} := \left\{ u \in H^1(\Omega) \,\middle|\, \|u\|_2 = 1 \right\}, \tag{1.22}$$

$$\mathscr{D}^{\mathrm{af}}_{\mathrm{D}} := \left\{ u \in H^1_0(\Omega) \,\middle|\, \|u\|_2 = 1 \right\}, \tag{1.23}$$

and correspondingly

$$E^{\mathrm{af}}_{\beta,\mathrm{N/D}} := \inf_{u \in \mathscr{D}^{\mathrm{af}}_{\mathrm{N/D}}} \mathcal{E}^{\mathrm{af}}_{\beta,\Omega}[u]. \tag{1.24}$$

For any trapping potential V satisfying the above assumptions, one can derive the variational equation associated to the minimization of $\mathcal{E}^{\mathrm{af}}_{\beta}$ [**CLR17**, Appendix A], which reads

$$\left[\left(-i\nabla + \beta \mathbf{A}[|u|^2] \right)^2 + V - 2\beta \nabla^{\perp} w_0 * \left(\beta \mathbf{A}[|u|^2] |u|^2 + \mathbf{j}[u] \right) \right] u = \mu^{\mathrm{af}}_{\beta} u, \tag{1.25}$$

where the current $\mathbf{j}[u]$ is given by

$$\mathbf{j}[u] := \tfrac{i}{2} \left(u \nabla u^* - u^* \nabla u \right). \tag{1.26}$$

The chemical potential $\mu^{\mathrm{af}}_{\beta}$ can be expressed in terms of the ground state energy E^{af}_{β} and the corresponding minimizer as

$$\begin{aligned} \mu^{\mathrm{af}}_{\beta} &= E^{\mathrm{af}}_{\beta} + \int_{\mathbb{R}^2} \mathrm{d}\mathbf{r} \, \left\{ 2\beta \mathbf{A}\left[|u^{\mathrm{af}}_{\beta}|^2 \right] \cdot \mathbf{j}\left[u^{\mathrm{af}}_{\beta} \right] + 2\beta^2 \left| \mathbf{A}\left[|u^{\mathrm{af}}_{\beta}|^2 \right] \right|^2 |u^{\mathrm{af}}_{\beta}|^2 \right\} \\ &= \int_{\mathbb{R}^2} \mathrm{d}\mathbf{r} \, \left\{ \left| \nabla u^{\mathrm{af}}_{\beta} \right|^2 + V |u^{\mathrm{af}}_{\beta}|^2 + 4\beta \mathbf{A}\left[|u^{\mathrm{af}}_{\beta}|^2 \right] \cdot \mathbf{j}\left[u^{\mathrm{af}}_{\beta} \right] + 3\beta^2 \left| \mathbf{A}\left[|u^{\mathrm{af}}_{\beta}|^2 \right] \right|^2 |u^{\mathrm{af}}_{\beta}|^2 \right\}. \end{aligned} \tag{1.27}$$

2. Main Results

As anticipated we are going to study the ground state properties of a trapped gas of non-interacting anyons in the average-field approximation, i.e., we will investigate the minimization of the effective energy functional (1.7) (or its homogenous counterpart (1.17)) in the asymptotic regime $\beta \gg 1$.

In physics literature the average-field approximation is usually performed by simplifying further the problem and assuming that a local anyonic density $\rho(\mathbf{r})$ would generate a magnetic field

$$\mathbf{B}(\mathbf{r}) \simeq 2\pi N |\alpha| \rho(\mathbf{r}) \simeq 2\pi \beta \rho(\mathbf{r})$$

proportional to the density itself. This is clearly inspired by the self-generated magnetic potential appearing in the functional $\mathcal{E}^{\mathrm{af}}$ and is very often applied to the homogenous case, i.e., when $\rho = \mathrm{const.}$ in some bounded region. In the lowest energy state (lowest Landau level) the energy per particle should then be given by the magnetic energy $|\mathbf{B}|$, to be averaged over the local density $\rho(\mathbf{r})$ itself. Taking into account the potential energy, one thus recovers the Thomas-Fermi (TF)-like functional

$$\int_{\mathbb{R}^2} \mathrm{d}\mathbf{r} \, \left\{ 2\pi\beta\rho^2(\mathbf{r}) + V\rho(\mathbf{r}) \right\}. \tag{2.1}$$

It is interesting to remark that the form of the energy above is very close to the one obtained via the conventional Thomas-Fermi approximation for two-dimensional fermions: recalling that ρ is normalized to 1, the TF energy per particle of N fermions is

$$\int_{\mathbb{R}^2} \mathrm{d}\mathbf{r}\ \left\{2\pi N\rho^2(\mathbf{r}) + V\rho(\mathbf{r})\right\}.$$

Since fermions are identified by the choice $\alpha = 1$, it is not surprising that the energy per particle of non-interacting anyons would be given by

$$\int_{\mathbb{R}^2} \mathrm{d}\mathbf{r}\ \left\{2\pi\alpha N\rho^2(\mathbf{r}) + V\rho(\mathbf{r})\right\},$$

which is close to (2.1), when $\alpha \sim \beta/N$.

2.1. Homogeneous anyonic gas. When the gas is confined to some bounded region Ω, where V is constant (e.g., zero), the optimal density minimizing the energy (2.1) is simply the constant function $\rho(\mathbf{r}) = |\Omega|^{-1}$ and the corresponding energy is thus const. β. According to (2.1) the precise prefactor should be $2\pi|\Omega|^{-1}$, which simply amounts to 2π when Ω has unit area $|\Omega| = 1$.

In our main result about the homogenous gas we are going to recover the linear dependence of $E^{\mathrm{af}}_{\beta,\mathrm{N/D}}$ on β and prove that the expression is independent of boundary conditions and of the shape of the domain. However, whether the prefactor is 2π or larger remains an open question (see the discussion below).

THEOREM 2.1 (Energy asymptotics). *Let $\Omega \subset \mathbb{R}^2$ be a bounded simply connected domain with Lipschitz boundary, then*

$$\frac{E^{\mathrm{af}}_{\beta,\mathrm{N}}}{\beta} = \frac{E^{\mathrm{af}}_{\beta,\mathrm{D}}}{\beta} + \mathcal{O}(\beta^{-1/7+\varepsilon}), \tag{2.2}$$

and the limits of both quantities coincide and are finite. Moreover we have

$$|\Omega| \lim_{\beta\to+\infty} \frac{E^{\mathrm{af}}_{\beta,\mathrm{N/D}}}{\beta} =: e(1,1) \geqslant 2\pi. \tag{2.3}$$

The reason why we define the coefficient $e(1,1)$ as a function depending on two parameters is that one can think of a more general limit in which the rescaled statistic parameter β is set equal to γ/ε, $\gamma > 0$ and $0 \leqslant \varepsilon \ll 1$, and define (the dependence on $|\Omega|^{-1/2}$ is chosen for further convenience)

$$e\big(\gamma, |\Omega|^{-1/2}\big) := \lim_{\varepsilon\to 0^+} \varepsilon E^{\mathrm{af}}_{\frac{\gamma}{\varepsilon},\mathrm{N/D}}, \tag{2.4}$$

which reduces to (2.2) when Ω has unit area and $\gamma = 1$. Equivalently, one can define $e(\beta,\rho)$ as the energy per unit area in a thermodynamic limit $L \to \infty$ on a rescaled domain $L\Omega$, under normalization $\|u\|_2^2 = \lambda L^2|\Omega|$, $\lambda > 0$ (see also Section 3.1). The two quantities, which are a priori different, equal because of the following scaling property of the functional: let $\lambda, \mu \in \mathbb{R}^+$ and set $u_{\lambda,\mu}(\mathbf{r}) := \lambda u(\mathbf{r}/\mu)$, then

$$\mathcal{E}^{\mathrm{af}}_{\beta,\mu\Omega}[u_{\lambda,\mu}] = \lambda^2 \mathcal{E}^{\mathrm{af}}_{\lambda^2\mu^2\beta,\Omega}[u], \tag{2.5}$$

where $\mu\Omega$ stands for the dilated domain $\{\mu\mathbf{r} \,|\, \mathbf{r} \in \Omega\}$. Once applied to $e\left(\gamma, |\Omega|^{-1}\right)$, this scaling law implies that

$$e\big(\gamma, |\Omega|^{-1/2}\big) = \gamma|\Omega|^{-1} e(1,1), \tag{2.6}$$

and therefore $e(1,1)$ determines the function entirely.

The bound (2.3) is not expected to be optimal, i.e., we conjecture that the strict inequality holds true

$$e(1,1) > 2\pi. \tag{2.7}$$

The estimate (2.3) is indeed a consequence of (1.21), which yields for any $u \in H_0^1(\Omega)$, i.e., satisfying Dirichlet boundary conditions,

$$\mathcal{E}^{\mathrm{af}}_{\beta,\Omega}[u] \geqslant 2\pi\beta \|u\|_4^4 . \tag{2.8}$$

The r.h.s. is saturated by the constant function $u(\mathbf{r}) = |\Omega|^{-1/2}$ (although it does not satisfy Dirichlet boundary conditions), yielding the estimate (2.3), via (2.2). Even neglecting the problem of boundary conditions, there are strong indications that the constant function is very far from a minimizer of the functional for large β: in order to minimize the contribution of the magnetic field and its huge circulation, it is indeed much more convenient to distribute more or less uniformly a large number of vortices, whose fluxes compensate $\mathbf{A}[|u|^2]$. The picture would resemble then what is expected for fast rotating Bose-Einstein condensates [**CY08**, **CPRY12**, **CR13**] or for superconductors in strong magnetic fields [**SS07**], i.e., the occurrence of vortices on a regular lattice (Abrikosov lattice). In this case, u must vanish at the center of each vortex and, even though u can still be close to a constant at larger scales in a weak sense (e.g., in L^p, $p < \infty$), the vortex core being very small, the interaction energy between vortices is expected to make the inequality (2.3) strict. It is interesting to note that such a behavior for anyons has already been conjectured in [**CWWH89**, p. 1012], although the consequence on $e(1,1)$ has not been noticed. We plan to investigate further this question in a future work.

The homogeneity of the system is confirmed by the following result about the density, which is independent of boundary conditions. In view of the discussion above, it is worth remarking that our next estimate (2.9) is perfectly compatible with the presence of a huge number of vortices with small core.

THEOREM 2.2 (Density asymptotics). *Under the same hypothesis of Theorem 2.1 and for any minimizer $u^{\mathrm{af}}_{\beta,\mathrm{N/D}}$ of $\mathcal{E}^{\mathrm{af}}_{\beta,\Omega}$ in $\mathscr{D}^{\mathrm{af}}_{\mathrm{N/D}}$,*

$$|\Omega|^{1/2} \left| u^{\mathrm{af}}_{\beta,\mathrm{N/D}} \right| \xrightarrow[\beta\to\infty]{\left(C_0^{0,1}(\Omega)\right)^*} 1, \tag{2.9}$$

where $C_0^{0,1}(\Omega)$ is the space of Lipschitz functions vanishing on $\partial\Omega$.

The above result guarantees that on scale one, i.e., on the scale of the domain $|\Omega|$, any minimizer of the average-field functional (1.17) can be well approximated by a constant. In fact, we expect this to be true even at finer scales, which are much larger than the vortex spacing. This latter characteristic length is of order $\beta^{-1/2}$ for the homogeneous gas, since the circulation to compensate is of order β and the optimal vortex distribution is thought to be given by a regular lattice of singly quantized vortices, thus leading to an average spacing of order $\beta^{-1/2}$. Hence, on any length scale much larger than $\beta^{-1/2}$ the approximation by a constant should be accurate. Next result, which is certainly not optimal (one would expect it to hold true up to $\eta < 1/2$), shows however an instance of this behavior. We denote by Ω° the interior of the domain Ω.

PROPOSITION 2.3 (Local density approximation). *Let* $\mathbf{r}_0 \in \Omega^\circ$, *and* $R > 0$ *finite, then*[4] *under the same hypothesis of Theorem* 2.1,

$$\left| u^{\mathrm{af}}_{\beta,\mathrm{N/D}}\left(\mathbf{r}_0 + \beta^{-\eta}\,\cdot\right)\right| \xrightarrow[\beta\to+\infty]{\left(C^{0,1}_0(\mathcal{B}_R)\right)^*} |\Omega|^{-1/2}, \qquad \textit{for any } 0 < \eta < \tfrac{1}{14}. \tag{2.10}$$

2.2. Trapped anyonic gas. We now consider a gas of non-interacting anyons which is trapped by a more general potential $V(\mathbf{r})$ satisfying the assumptions (1.13)–(1.15). As anticipated in the Introduction, the paradigm is given by the anisotropic harmonic oscillator $V(\mathbf{r}) = ax^2 + by^2$, which contains the usual symmetric oscillator as a special case. Indeed, we do not require radial symmetry of V but only homogeneity of degree $s > 1$ as a function of $\mathbf{r}$. This last assumption can be relaxed as well and our result applies to the more general class of asymptotically homogeneous potentials as defined in [**LSY01**, Definition 1.1]. We stick however to the hypothesis above for the sake of concreteness. The relevance of the scaling property $V(\lambda\mathbf{r}) = \lambda^s V(\mathbf{r})$ will become apparent when we will discuss the effective TF theory describing the behavior of the average-field functional in the regime $\beta \gg 1$: the so-obtained TF functional indeed admits a scaling property which allows to factor out the dependence on β, provided V is a homogeneous function (or an asymptotically homogeneous function for β large).

The heuristics for the asymptotics $\beta \to \infty$ of the functional $\mathcal{E}^{\mathrm{af}}_\beta$ relies on the results proven in the previous Section: it is reasonable to assume that locally, on a suitable fine scale, the inhomogeneity generated by $V(\mathbf{r})$ does not play any role and therefore the energy on that scale is given by the one of the homogeneous gas, i.e., $\simeq e(\beta,\rho) = e(1,1)\beta\rho^2$. Hence, the TF effective functional we expect to recover in the regime $\beta \gg 1$ must have the form

$$\mathcal{E}^{\mathrm{TF}}_\beta[\rho] := \int_{\mathbb{R}^2} \mathrm{d}\mathbf{r}\,\left\{e(1,1)\beta\rho^2 + V\rho\right\}, \tag{2.11}$$

where $\rho := |u|^2$ is the gas density. The ground state energy of the TF functional is defined as

$$E^{\mathrm{TF}}_\beta := \min_{\rho\in\mathscr{D}^{\mathrm{TF}}} \mathcal{E}^{\mathrm{TF}}_\beta[\rho] \tag{2.12}$$

with

$$\mathscr{D}^{\mathrm{TF}} := \left\{\rho \in L^2(\mathbb{R}^2)\;\middle|\; \rho \geqslant 0, \|\rho\|_1 = 1\right\}. \tag{2.13}$$

The associated (unique) minimizer is denoted by ρ^{TF}_β. Under the hypothesis (1.13)–(1.15), the minimization (2.12) is actually explicit and one can extract the β dependence simply by rescaling the lengths. The result is

$$E^{\mathrm{TF}}_\beta = \beta^{s/(s+2)} E^{\mathrm{TF}}_1, \qquad \rho^{\mathrm{TF}}_\beta(\mathbf{r}) = \beta^{-2/(2+s)}\rho^{\mathrm{TF}}_1\left(\beta^{-1/(s+2)}\mathbf{r}\right), \tag{2.14}$$

where ρ^{TF}_1 minimizes the TF energy with $\beta = 1$ and is given by the compactly supported function

$$\rho^{\mathrm{TF}}_1(\mathbf{r}) = \frac{1}{2e(1,1)}\left[\lambda^{\mathrm{TF}}_1 - V(\mathbf{r})\right]_+, \tag{2.15}$$

with chemical potential $\lambda^{\mathrm{TF}}_1 = E^{\mathrm{TF}}_1 + e(1,1)\left\|\rho^{\mathrm{TF}}_1\right\|_2^2 > 0$. If the potential V was radial, then the support of the rescaled TF minimizer ρ^{TF}_1 would be a disc of radius

[4]We denote by $\mathcal{B}_R := \mathcal{B}_R(0)$ a ball of radius $R > 0$ centered at the origin. An analogous ball centered at $\mathbf{r}$ will be denoted by $\mathcal{B}_R(\mathbf{r})$.

$\propto (\lambda_1^{\mathrm{TF}})^{1/s}$. In the general case the support of ρ_1^{TF} is more complicated but still contained in a ball of finite radius R_0, so that

$$\mathrm{supp}\left(\rho_\beta^{\mathrm{TF}}\right) \subset \mathcal{B}_{R_0 \beta^{1/(2+s)}}. \tag{2.16}$$

Without loss of generality we also assume that $\partial_{\mathbf{n}} V \neq 0$ a.e. along $\partial\, \mathrm{supp}\, \rho_1^{\mathrm{TF}}$, with $\mathbf{n}$ the outward normal to the boundary.

The following result shows that the TF functional (2.11) provides indeed a fine approximation of the average-field energy:

THEOREM 2.4 (Energy asymptotics). *Let V satisfy the assumptions* (1.13)–(1.15), *then as $\beta \to +\infty$*

$$\beta^{-s/(s+2)} E_\beta^{\mathrm{af}} = E_1^{\mathrm{TF}} + o(1). \tag{2.17}$$

The analogue of Theorem 2.2 describing the asymptotics of the average-field density as $\beta \to +\infty$ is:

THEOREM 2.5 (Local density approximation). *Under the same hypothesis of Theorem 2.4 and for any minimizer u_β^{af} of $\mathcal{E}_\beta^{\mathrm{af}}$ in $\mathscr{D}^{\mathrm{af}}$, we have for any $\mathbf{r} \in \mathbb{R}^2$ and $R > 0$*

$$\beta^{2/(s+2)} \left| u_\beta^{\mathrm{af}} \left(\beta^{1/(s+2)} \,\cdot\, \right) \right|^2 \xrightarrow[\beta\to\infty]{\left(C_0^{0,1}(\mathcal{B}_R)\right)^*} \rho_1^{\mathrm{TF}}(\,\cdot\,). \tag{2.18}$$

As for the homogeneous gas, the above estimate is expected to holds true on any finer scale, which is much larger than the vortex one, i.e., the rescaled average-field density should be close to ρ_1^{TF} even locally in any small ball of radius $\beta^{-\eta}$, $\eta < s/(2(s+2))$, which is the mean spacing of the conjectured vortex lattice. In fact, by exploiting the explicit remainder of the energy asymptotics (2.17) derived in the proofs, we could prove that (2.18) holds on a scale shorter than the one of ρ_β^{TF} but still much larger than the optimal one, exactly as in Proposition 2.3. We skip the statement for the sake of brevity.

3. Sketch of the Proofs

We present here a synthetic exposition of the main arguments used in the proofs of the results stated in the previous Section. We refer to [**CLR17**] for further details. The starting point is the discussion of the homogeneous gas (see Section 2.1), which will be used as a key tool to take into account the inhomogeneity introduced by the trapping potential V.

3.1. Homogeneous gas. Exploiting the scaling property (2.5), it is possible to show that the large β limit in a fixed domain Ω is equivalent to a thermodynamic limit $L \to \infty$ of a domain $L\Omega$ with normalization $\|u\|_2^2 \propto L^2 |\Omega|$. Explicitly, for any $\gamma \geqslant 0$,

$$\lim_{\beta\to+\infty} \frac{E_{\gamma\beta,\mathrm{N}}^{\mathrm{af}}}{\beta} = \lim_{L\to+\infty} \frac{1}{\lambda^2 L^2 |\Omega|^2} \inf_{u \in H^1(L\Omega), \|u\|_2^2 = \lambda L^2 |\Omega|} \mathcal{E}_{\gamma, L\Omega}^{\mathrm{af}}[u], \tag{3.1}$$

where $\lambda > 0$ is a positive parameter, which is kept fixed as $L \to +\infty$, i.e., we are considering a large volume limit with fixed density

$$\lambda = \frac{\|u\|_{L^2(L\Omega)}^2}{|\Omega| L^2}.$$

The relation between β and L in the identity above is $\beta =: \lambda|\Omega|L^2$ and γ plays the role of a rescaled statistic parameter. In the Dirichlet case the definition is perfectly analogous and the only difference is that $H^1(L\Omega)$ must be replaced with $H_0^1(L\Omega)$ on the r.h.s..

In [**CLR17**] Theorem 2.1 (and consequently Theorem 2.2) is proven by a direct inspection of the large L limit of the r.h.s. of (3.1). The main steps are the following:

(i) a priori bound on the r.h.s. of (3.1), showing that it is a bounded quantity, which allows to define $e(\gamma, \lambda)$ at least as

$$e(\gamma, \lambda) := \liminf_{L \to +\infty} \frac{1}{L^2|\Omega|} \inf_{u \in H^1(L\Omega), \|u\|_2^2 = \lambda L^2|\Omega|} \mathcal{E}^{\mathrm{af}}_{\gamma, L\Omega}[u], \tag{3.2}$$

since we do not know at this stage whether the limit does exist;

(ii) use of the scaling property (2.5), which yields

$$e(\gamma, \lambda) = \gamma\lambda^2 e(1, 1); \tag{3.3}$$

(iii) proof of the existence of the thermodynamic limit $L \to \infty$ when Ω is a (unit) square;

(iv) comparison of the Dirichlet and Neumann energies for squares showing that they coincide in the limit;

(v) extension of the existence of the thermodynamic limit to any domain Ω. For this last step it suffices to prove that the limit of the Dirichlet energy equals $e(\gamma, \lambda)$, because the result then follows from the definition (3.2) and the usual bound $E^{\mathrm{af}}_{\beta,\mathrm{N}} \leqslant E^{\mathrm{af}}_{\beta,\mathrm{D}}$.

We are not going to discuss all the details of the steps above but the first one (see Proposition 3.1 below). We just point out that the most non-trivial result is the comparison Dirichlet/Neumann, whose fundamental tool is an IMS-type localization formula [**CLR17**, Lemma 3.3] for a suitable partition of unity, together with the estimates (1.20) and (1.21). The initial restriction to squares is motivated by the simplicity in constructing suitable partitions and coverings.

The a priori bound showing that the energy $E^{\mathrm{af}}_{\beta,\Omega}$ grows at most linearly in β is proven in [**CLR17**, Lemma 3.1] in the same thermodynamic setting discussed above. We state it here in a different but equivalent form:

PROPOSITION 3.1 (Trial upper bound). *Under the same hypothesis as in Theorem 2.1, there exists a finite constant $C > 0$ such that as $\beta \to \infty$*

$$\frac{E^{\mathrm{af}}_{\beta,\mathrm{N}}}{\beta} \leqslant \frac{E^{\mathrm{af}}_{\beta,\mathrm{D}}}{\beta} \leqslant C. \tag{3.4}$$

PROOF. The inequality between the Neumann and Dirichlet energies is trivial, since the minimization domain in the latter case is smaller. Assume then without loss of generality (thanks to the scaling property (2.5)) that $|\Omega| = 1$ and fill Ω with[5] $N = \beta \gg 1$ disjoint balls of radius $1/\sqrt{\beta}$ centered at points $\mathbf{r}_j \in \Omega$,

$$\mathcal{B}_j := \mathcal{B}_{1/\sqrt{\beta}}(\mathbf{r}_j), \qquad j = 1, \ldots, N.$$

Let $f \in C_0^1(\mathcal{B}_1(0))$ be a radial function so that $\|f\|_2^2 = 1$, set

$$u_j(\mathbf{r}) := f\big(\sqrt{\beta}(\mathbf{r} - \mathbf{r}_j)\big) \in C_0^1(\mathcal{B}_j), \tag{3.5}$$

[5]Let us assume for the sake of simplicity that β is an integer number, otherwise one would have to take a number of balls equal to, e.g., the integer part of β and the computation would then become more involved, while the core of the proof would be unaffected.

which clearly satisfies $\|u_j\|_2^2 = 1/\beta$. Then the trial state we are going to use is

$$u(\mathbf{r}) := \sum_{j=1}^N u_j(\mathbf{r}) e^{-i\sum_{k\neq j}\arg(\mathbf{r}-\mathbf{r}_k)} = \sum_{j=1}^N u_j(\mathbf{r}) \prod_{k\neq j} \frac{z^*-z_k^*}{|z-z_k|}, \tag{3.6}$$

where for any point $\mathbf{r} = (x,y)$ in the plane we have used the complex notation $z = x+iy$ and $\arg z = \arctan\frac{y}{x}$. The properties of f and in particular its compact support contained in the unit ball imply that

$$|u(\mathbf{r})|^2 = \sum_{j=1}^N |u_j(\mathbf{r})|^2 = \begin{cases} |u_j(\mathbf{r})|^2, & \text{on } \mathcal{B}_j, \\ 0, & \text{otherwise.} \end{cases}$$

Then

$$\begin{aligned}
\mathcal{E}^{\mathrm{af}}_{\beta,\Omega}[u] &= \sum_{j=1}^N \int_{\mathcal{B}_j} \mathrm{d}\mathbf{r}\, \left|\left(-i\nabla + \beta\textstyle\sum_{k=1}^N \mathbf{A}[|u_k|^2]\right) e^{-i\sum_{k\neq j}\arg(\mathbf{r}-\mathbf{r}_k)} u_j\right|^2 \\
&= \sum_{j=1}^N \int_{\mathcal{B}_j} \mathrm{d}\mathbf{r}\, \left|\left(-i\nabla + \beta\mathbf{A}[|u_j|^2] + \textstyle\sum_{k\neq j}\left(\beta\mathbf{A}[|u_k|^2] - \nabla\arg(\mathbf{r}-\mathbf{r}_k)\right)\right) u_j\right|^2 \\
&= \sum_{j=1}^N \int_{\mathcal{B}_j} \mathrm{d}\mathbf{r}\, \left|(-i\nabla + \beta\mathbf{A}[|u_j|^2])u_j\right|^2,
\end{aligned}$$

where we used that by 2D Newton's theorem

$$\begin{aligned}
\mathbf{A}[|u_k|^2](\mathbf{r}) = \nabla^\perp \int_{\mathcal{B}_k} \mathrm{d}\mathbf{r}'\, \ln|\mathbf{r}-\mathbf{r}'||u_k(\mathbf{r}')|^2 &= \nabla^\perp \ln|\mathbf{r}-\mathbf{r}_k| \int_{\mathcal{B}_k} \mathrm{d}\mathbf{r}'\, |u_k(\mathbf{r}')|^2 \\
&= \frac{1}{\beta}\nabla\arg(\mathbf{r}-\mathbf{r}_k)
\end{aligned}$$

for $\mathbf{r} \notin \mathcal{B}_k$. Now note that for all $\mathbf{r} \in \mathbb{R}^2$ and for all $j = 1,\dots,N$,

$$\begin{aligned}
\mathbf{A}[|u_j|^2](\mathbf{r}) &= \int_{\mathcal{B}_j} \mathrm{d}\mathbf{r}'\, \frac{(\mathbf{r}-\mathbf{r}')^\perp}{|\mathbf{r}-\mathbf{r}'|^2} |u_j(\mathbf{r}')|^2 \\
&= \frac{1}{\sqrt{\beta}} \int_{\mathcal{B}_1} \mathrm{d}\mathbf{r}'\, \frac{(\sqrt{\beta}(\mathbf{r}-\mathbf{r}_j)-\mathbf{r}')^\perp}{|\sqrt{\beta}(\mathbf{r}-\mathbf{r}_j)-\mathbf{r}'|^2} |f(\mathbf{r}')|^2 = \beta^{-\frac{1}{2}}\mathbf{A}[|f|^2](\sqrt{\beta}(\mathbf{r}-\mathbf{r}_j))
\end{aligned}$$

and thus

$$\begin{aligned}
\mathcal{E}^{\mathrm{af}}_{\beta,\Omega}[u] &= \sum_{j=1}^N \int_{\mathcal{B}_j} \mathrm{d}\mathbf{r}\, \left|(-i\nabla u_j(\mathbf{r}) + \beta\mathbf{A}[|u_j|^2](\mathbf{r})u_j(\mathbf{r}))\right|^2 \\
&= N\beta^{-1} \int_{\mathcal{B}_1} \mathrm{d}\mathbf{r}'\, \left|\left(-i\sqrt{\beta}\nabla f(\mathbf{r}') + \sqrt{\beta}\mathbf{A}[|f|^2](\mathbf{r}')f(\mathbf{r}')\right)\right|^2 \\
&= N \int_{\mathcal{B}_1} \mathrm{d}\mathbf{r}\, \left|(-i\nabla + \mathbf{A}[|f|^2])f\right|^2 = N\mathcal{E}^{\mathrm{af}}_{1,\mathcal{B}_1}[f].
\end{aligned}$$

Since $\mathcal{E}^{\mathrm{af}}_{1,\mathcal{B}_1}[f]$ does not depend on β we obtain the desired bound

$$E_0(\Omega,\beta,M) \leqslant CN = C\beta$$

by the variational principle. □

The bounds proven in Proposition 3.1 and its proof are particularly interesting because of the form of the trial state (3.6): unlike the constant function heuristic mentioned in Section 2.1 (see the discussion below Theorem 2.1 on $e(1,1)$), the trial state has a very large phase circulation generated by singly-quantized vortices sitting at the centers of the balls $\mathcal{B}_j$. As it is apparent from the energy computation, this huge circulation is in fact crucial in order to obtain an energy which depends linearly on β. Note indeed that a naive approach would give an estimate of order β^2 for the energy $E^{\mathrm{af}}_{\beta,\Omega}$ due to the square of the vector potential. This justifies once more the conjecture about the value of $e(1,1)$.

The density estimate (2.9) is a straightforward consequence of the energy convergence. We skip most of the details and focus on the proof of Proposition 2.3:

PROOF OF PROPOSITION 2.3. Acting as in [**CLR17**, Proof of Lemma 4.1] one can show that the energy estimate (2.2) implies the following bound

$$\left\| \bar{\rho}^{\mathrm{af}}_{\beta} - |\Omega|^{-1} \right\|_{L^2(\Omega)} = \mathcal{O}(\beta^{-1/14+\varepsilon}), \tag{3.7}$$

where $\bar{\rho}^{\mathrm{af}}_{\beta}$ is a suitable piecewise approximation of $u^{\mathrm{af}}_{\beta,\mathrm{N/D}}$, i.e., the result does not depend on boundary conditions. More precisely, $\bar{\rho}^{\mathrm{af}}_{\beta}$ is constructed as follows: let $\{Q_j\}_{j\in J}$ a tiling covering Ω made of squares Q_j of side length $\beta^{-\nu}$, with

$$0 < \nu < \tfrac{1}{2}, \tag{3.8}$$

then we set

$$\bar{\rho}^{\mathrm{af}}_{\beta}(\mathbf{r}) := \sum_{j\in\tilde{J}} \rho_j \mathbb{1}_{Q_j}(\mathbf{r}), \qquad \rho_j := \beta^{2\nu} \int_{Q_j} \mathrm{d}\mathbf{r} \left| u^{\mathrm{af}}_{\beta,\mathrm{N/D}}(\mathbf{r}) \right|^2, \tag{3.9}$$

where for some $0 < \mu < 1 - 2\nu$

$$\tilde{J} := \left\{ j \in J \,\middle|\, \rho_j \geqslant \beta^{2\nu-1+\mu} \right\}, \tag{3.10}$$

i.e., the mass of $u^{\mathrm{af}}_{\beta,\mathrm{N/D}}$ in the cells Q_j, $j \in \tilde{J}$, is not too small.

Let now ϕ be a Lipschitz function with compact support contained in $\mathcal{B}_R(0)$, then for any

$$\eta < \nu, \tag{3.11}$$

one has

$$\begin{aligned}
&\int_{\mathcal{B}_R(0)} \mathrm{d}\mathbf{r}\, \phi(\mathbf{r}) \left| u^{\mathrm{af}}_{\beta,\mathrm{N/D}}\left(\mathbf{r}_0 + \beta^{-\eta}\mathbf{r}\right) \right|^2 \\
&= \beta^{2\eta} \sum_{Q_j \subset \mathcal{B}_{\beta^{-\eta}R}(\mathbf{r}_0),\, j\in\tilde{J}} \int_{Q_j} \mathrm{d}\mathbf{r}\, \phi\left(\beta^{\eta}(\mathbf{r}-\mathbf{r}_0)\right) \left| u^{\mathrm{af}}_{\beta,\mathrm{N/D}}(\mathbf{r}) \right|^2 + \mathcal{O}(\beta^{2(\eta-\mu)}) \\
&\quad = \beta^{2(\eta-\nu)} \sum_{Q_j \subset \mathcal{B}_{\beta^{-\eta}R}(\mathbf{r}_0)} \phi\left(\beta^{\eta}(\mathbf{r}_j-\mathbf{r}_0)\right) \rho_j + \mathcal{O}(\beta^{2(\eta-\nu)}) \\
&\qquad = \int_{\mathcal{B}_R(0)} \mathrm{d}\mathbf{r}\, \phi(\mathbf{r}) \left| \bar{\rho}^{\mathrm{af}}_{\beta}\left(\mathbf{r}_0 + \beta^{-\eta}\mathbf{r}\right) \right|^2 + \mathcal{O}(\beta^{2(\eta-\nu)}).
\end{aligned} \tag{3.12}$$

Then

$$\begin{aligned}(3.13)\quad & \left| \int_{\mathcal{B}_R(0)} \mathrm{d}\mathbf{r}\, \phi(\mathbf{r}) \left[\left| u^{\mathrm{af}}_{\beta,\mathrm{N/D}}\left(\mathbf{r}_0 + \beta^{-\eta}\mathbf{r}\right)\right|^2 - |\Omega|^{-1} \right] \right| \\ & = \left| \int_{\mathcal{B}_R(0)} \mathrm{d}\mathbf{r}\, \phi(\mathbf{r}) \left[\bar{\rho}^{\mathrm{af}}_{\beta}\left(\mathbf{r}_0 + \beta^{-\eta}\mathbf{r}\right) - |\Omega|^{-1} \right] \right| + o(1) \\ & \leqslant C\beta^{\eta} \left\|\phi\right\|_{\mathrm{Lip}} \left(\int_{\mathcal{B}_{\beta^{-\eta}R}(\mathbf{r}_0)} \mathrm{d}\mathbf{r} \left| \bar{\rho}^{\mathrm{af}}_{\beta}(\mathbf{r}) - |\Omega|^{-1} \right|^2 \right)^{1/2} + o(1) \\ & \leqslant C\beta^{\eta} \left\|\phi\right\|_{\mathrm{Lip}} \left\| \bar{\rho}^{\mathrm{af}}_{\beta} - |\Omega|^{-1} \right\|_{L^2(\Omega)} \leqslant C\beta^{\eta - 1/14 + \varepsilon} \left\|\phi\right\|_{\mathrm{Lip}} + o(1),\end{aligned}$$

and taking the supremum over Lipschitz functions with norm bounded by 1, one obtains the result, provided $\eta < 1/14$, which is allowed since ν can be larger than $1/14$. □

3.2. Trapped gas. The proof of the local density approximation for the trapped gas relies heavily on the result for the homogeneous system discussed in the previous Section. The first step is as usual the energy estimate (1.19), which is proven by deriving suitable upper and lower bounds. In both cases the proof scheme is rather simple: by tiling the plane with squares of suitable length smaller than the TF scale, i.e., $\beta^{1/(s+2)}$, but also much larger that the expected fine scale of the vortex lattice, i.e., $\beta^{-s/(2(s+2))}$, one can use locally the result for the homogeneous gas.

In the upper bound one has to get rid of the "interaction" between different cells, i.e., the magnetic field generated by the density in other cells, which is done by exploiting a trial state inspired by (3.6). The kinetic energy thus results in the sum of Dirichlet homogeneous energies inside the cells, up to remainder terms. The asymptotics (2.2) together with a Riemann sum approximation of the potential term then yields the upper bound.

The key step in the lower bound is on the other hand the use of the gauge invariance of the functional in order to cancel inside one given cell the magnetic potential generated by the density in all the other cells. The kinetic energy can then be bounded from below by the Neumann energy in a square. Riemann sum estimates and the scaling properties of the energy are then sufficient to complete the proof. Note that it is crucial to know that the Dirichlet and Neumann energies have the same limit as $\beta \to \infty$.

The density estimate is proven by first deriving an L^2 estimate of the form (3.7) for a piecewise approximation of $|u^{\mathrm{af}}_{\beta}|^2$ and then using Cauchy-Schwarz as in (3.13).

Acknowledgments: This work is supported by MIUR through the FIR grant 2013 "Condensed Matter in Mathematical Physics (Cond-Math)" (code RBFR13WAET), the Swedish Research Council (grant no. 2013-4734) and the ANR (Project Mathostaq ANR-13-JS01-0005-01). We thank Jan Philip Solovej for insightful suggestions and Romain Duboscq for inspiring numerical simulations. M.C. is also grateful to the organizing committee of the conference "QMath13: Mathematical Results in Quantum Physics", session "Many-body Systems and Statistical Mechanics", for the invitation to present this work there.

References

[AT98] R. Adami and A. Teta, *On the Aharonov-Bohm Hamiltonian*, Lett. Math. Phys. **43** (1998), no. 1, 43–53, DOI 10.1023/A:1007330512611. MR1607550

[BS92] M. Bourdeau and R. D. Sorkin, *When can identical particles collide?*, Phys. Rev. D (3) **45** (1992), no. 2, 687–696, DOI 10.1103/PhysRevD.45.687. MR1144230

[CLR17] Michele Correggi, Douglas Lundholm, and Nicolas Rougerie, *Local density approximation for the almost-bosonic anyon gas*, Anal. PDE **10** (2017), no. 5, 1169–1200, DOI 10.2140/apde.2017.10.1169. MR3668588

[CPRY12] M. Correggi, F. Pinsker, N. Rougerie, and J. Yngvason, *Critical rotational speeds for superfluids in homogeneous traps*, J. Math. Phys. **53** (2012), no. 9, 095203, 45, DOI 10.1063/1.3697418. MR2905785

[CR13] M. Correggi and N. Rougerie, *Inhomogeneous vortex patterns in rotating Bose-Einstein condensates*, Comm. Math. Phys. **321** (2013), no. 3, 817–860, DOI 10.1007/s00220-013-1697-y. MR3070038

[CWWH89] Yi-Hong Chen, Frank Wilczek, Edward Witten, and Bertrand I. Halperin, *On anyon superconductivity*, Internat. J. Modern Phys. B **3** (1989), no. 7, 1001–1067, DOI 10.1142/S0217979289000725. MR1070254

[CY08] Michele Correggi and Jakob Yngvason, *Energy and vorticity in fast rotating Bose-Einstein condensates*, J. Phys. A **41** (2008), no. 44, 445002, 19, DOI 10.1088/1751-8113/41/44/445002. MR2453175

[DFT97] Gianfausto Dell'Antonio, Rodolfo Figari, and Alessandro Teta, *Statistics in space dimension two*, Lett. Math. Phys. **40** (1997), no. 3, 235–256, DOI 10.1023/A:1007361832622. MR1453763

[IL92] Roberto Iengo and Kurt Lechner, *Anyon quantum mechanics and Chern-Simons theory*, Phys. Rep. **213** (1992), no. 4, 179–269, DOI 10.1016/0370-1573(92)90039-3. MR1170980

[Kha05] Avinash Khare, *Fractional statistics and quantum theory*, World Scientific Publishing Co., Inc., River Edge, NJ, 1997. MR1795025

[LL01] Elliott H. Lieb and Michael Loss, *Analysis*, 2nd ed., Graduate Studies in Mathematics, vol. 14, American Mathematical Society, Providence, RI, 2001. MR1817225

[LL16] Simon Larson and Douglas Lundholm, *Exclusion bounds for extended anyons*, Arch. Ration. Mech. Anal. **227** (2018), no. 1, 309–365, DOI 10.1007/s00205-017-1161-9. MR3740376

[LR15] Douglas Lundholm and Nicolas Rougerie, *The average field approximation for almost bosonic extended anyons*, J. Stat. Phys. **161** (2015), no. 5, 1236–1267, DOI 10.1007/s10955-015-1382-y. MR3422925

[LR16] ———, *Emergence of fractional statistics for tracer particles in a Laughlin liquid*, Phys. Rev. Lett. **116** (2016), 170401.

[LSY01] Elliott H. Lieb, Robert Seiringer, and Jakob Yngvason, *A rigorous derivation of the Gross-Pitaevskii energy functional for a two-dimensional Bose gas*, Comm. Math. Phys. **224** (2001), no. 1, 17–31, DOI 10.1007/s002200100533. Dedicated to Joel L. Lebowitz. MR1868990

[LS13] Douglas Lundholm and Jan Philip Solovej, *Local exclusion and Lieb-Thirring inequalities for intermediate and fractional statistics*, Ann. Henri Poincaré **15** (2014), no. 6, 1061–1107, DOI 10.1007/s00023-013-0273-5. MR3205745

[Lun16] D. Lundholm, *Many-anyon trial states*, preprint arXiv:1608.05067, 2016.

[Myr99] J. Myrheim, *Anyons*, Aspects topologiques de la physique en basse dimension/Topological aspects of low dimensional systems (Les Houches, 1998), EDP Sci., Les Ulis, 1999, pp. 265–413, DOI 10.1007/3-540-46637-1_4. MR1850419

[SS07] Etienne Sandier and Sylvia Serfaty, *Vortices in the magnetic Ginzburg-Landau model*, Progress in Nonlinear Differential Equations and their Applications, vol. 70, Birkhäuser Boston, Inc., Boston, MA, 2007. MR2279839

[Wil82] Frank Wilczek, *Quantum mechanics of fractional-spin particles*, Phys. Rev. Lett. **49** (1982), no. 14, 957–959, DOI 10.1103/PhysRevLett.49.957. MR675053

[Wil90] Frank Wilczek, *Fractional statistics and anyon superconductivity*, World Scientific Publishing Co., Inc., Teaneck, NJ, 1990. MR1081990

Dipartimento di Matematica "G. Castelnuovo", "Sapienza" Università di Roma, P.le Aldo Moro, 5, 00185, Rome, Italy
Email address: michele.correggi@gmail.com
URL: http://www1.mat.uniroma1.it/people/correggi/

Department of Mathematics, KTH Royal Institute of Technology, SE-100 44 Stockholm, Sweden
Email address: dogge@math.kth.se

CNRS & Université Grenoble Alpes, LPMMC (UMR 5493), B.P. 166, F-38042 Grenoble, France
Email address: nicolas.rougerie@lpmmc.cnrs.fr

Contemporary Mathematics
Volume **717**, 2018
https://doi.org/10.1090/conm/717/14443

Lieb-Robinson bounds, the spectral flow, and stability of the spectral gap for lattice fermion systems

Bruno Nachtergaele, Robert Sims, and Amanda Young

ABSTRACT. We prove Lieb-Robinson bounds for a general class of lattice fermion systems. By making use of a suitable conditional expectation onto subalgebras of the CAR algebra, we can apply the Lieb-Robinson bounds much in the same way as for quantum spin systems. We preview how to obtain the spectral flow automorphisms and to prove stability of the spectral gap for frustration-free gapped systems satisfying a Local Topological Quantum Order condition.

1. Introduction

In the past dozen years, a considerable number of mathematical results on quantum spin systems made use of Lieb-Robinson Bounds [**26**] in an essential way. These include extensions of the Lieb-Schultz-Mattis theorem to higher dimensions [**18**, **37**], the exponential decay of correlations in gapped ground states [**20**, **36**], Area Laws for the entanglement entropy [**9**, **19**], construction of the spectral flow and adiabatic theorems [**5**, **6**, **8**], stability results for gapped ground states [**11**, **30**] and more.

It is quite clear that in many cases it should be possible to answer the same type of questions for lattice fermion systems based on the same principles that apply to quantum spin systems. Indeed, there are several special cases where the close analogy between lattice systems of spins and of fermions allowed for the successful application of Lieb-Robinson bounds. For example, a proof of the quantized Hall effect for interacting lattice fermions [**16**, **21**], and linear response theory for lattice fermion systems [**13**].

In this paper, we discuss several crucial ingredients which enter the proof of stability for gapped, frustration-free models of lattice fermions. First, we prove a Lieb-Robinson bound for a general class of models. Variants of this quasi-locality estimate enter the proof of stability in many different stages. Next, we introduce a conditional expectation which enables strictly local approximations of quasi-local observables; this can be seen as an analogue of the normalized partial trace familiar in the context of quantum spin systems. Then, we provide a version of the martingale method suitable to prove lower bounds for the spectral gap of fermion

2010 *Mathematics Subject Classification.* Primary 82B10; Secondary 82C10, 82C20.
Based on work supported by the National Science Foundation under Grant DMS-1515850.

models. Equipped with this, a wealth of potential unperturbed models can be explored. After giving a few specific classes of gapped fermion models, we overview the main tool of analysis for stability: the spectral flow. More detailed estimates and additional results will be provided in forth-coming papers [**32**, **41**].

The methods discussed in this paper can be applied to the Aubry-André model studied by Mastropietro in a recent series of papers [**27**–**29**]. Both the molecular and the free fermion limit can be shown to be stable under general, uniformly small, short-range perturbations. This is more general than the class of perturbations studied by Mastropietro, but the general result is weaker. In particular the renormalization group method of Mastropietro takes into account the quasi-periodicity of the potential, in a way that allows for an estimate of the correlation length that is indicative of many-body localization. The details of the interactions are ignored by the general approach discussed here and adapting the method to study many-body localization will require further research.

2. Lattice fermion systems

Spinless fermions on a countable set Γ, which is often referred to as 'the lattice', are described by the CAR algebra $\mathcal{A}_\Gamma = \mathrm{CAR}(\ell^2(\Gamma))$. $\mathcal{A}_\Gamma$ is the C^*-algebra generated by creation and annihilation operators a_x^*, a_x, $x \in \Gamma$, which satisfy the canonical anti-commutation relations, i.e. the CAR:

$$\{a_x, a_y\} = \{a_x^*, a_y^*\} = 0 \quad \text{and} \quad \{a_x, a_y^*\} = \delta_{x,y}\mathbb{1} \quad \text{for any } x, y \in \Gamma\,. \tag{2.1}$$

Here $\{A, B\} = AB + BA$ denotes the anti-commutator of A and B. As is discussed in detail in [**10**], this CAR algebra can be represented as the collection of bounded linear operators over the Hilbert space corresponding to the anti-symmetric Fock space generated by $\ell^2(\Gamma)$. Furthermore, we note that spin and/or band indices can be included in this description by extending Γ, for example, to $\tilde{\Gamma} = \Gamma \times \{1, \dots, n\}$.

For $X \subset \Gamma$, $\mathcal{A}_X$ is naturally embedded as a subalgebra of $\mathcal{A}_\Gamma$ by identifying $\ell^2(X)$ with the subspace of $\ell^2(\Gamma)$ consisting of the functions that vanish on $\Gamma \setminus X$. Let $\mathcal{P}_0(\Gamma)$ denote the set of finite subsets of Γ. For any $X \in \mathcal{P}_0(\Gamma)$, it is useful to define the parity automorphism of $\mathcal{A}_X$, which we denote by Θ_X, by setting

$$\Theta_X(A) = (-1)^{N_X} A (-1)^{N_X} \quad \text{for any } A \in \mathcal{A}_X\,. \tag{2.2}$$

Here $N_X = \sum_{x \in X} a_x^* a_x$ is the local number operator. Using the quasi-local structure of $\mathcal{A}_\Gamma$, we see that there is a unique automorphism Θ of $\mathcal{A}_\Gamma$ for which $\Theta\restriction_{\mathcal{A}_X} = \Theta_X$ for any $X \in \mathcal{P}_0(\Gamma)$. It is clear that $\Theta^2 = \mathrm{id}$. The *even* and *odd* elements of $\mathcal{A}_\Gamma$ are the eigenvectors of Θ with eigenvalue 1 and -1, respectively. By $\mathcal{A}_\Gamma^+$ and $\mathcal{A}_\Gamma^-$, we denote the corresponding eigenspaces. Similarly, for any $\Lambda \in \mathcal{P}_0(\Gamma)$, we set $\mathcal{A}_\Lambda^+$ and $\mathcal{A}_\Lambda^-$ to be the even and odd eigenspaces of Θ_Λ on the subalgebra $\mathcal{A}_\Lambda$. Note that $\mathcal{A}_\Lambda^+$ is a C^*-subalgebra of $\mathcal{A}_\Lambda$, but $\mathcal{A}_\Lambda^-$ is not a subalgebra. In fact, it is immediate that $(\mathcal{A}_\Lambda^-)^2 \subset \mathcal{A}_\Lambda^+$.

A convenient basis for the local subalgebra $\mathcal{A}_\Lambda$ is the one consisting of all monomials. Recall that $A \in \mathcal{A}_\Lambda$ is a monomial if

$$A = \prod_{x \in \Lambda} A_x \quad \text{with} \quad A_x \in \{\mathbb{1}, a_x, a_x^*, a_x^* a_x\}\,. \tag{2.3}$$

Since each monomial is either even or odd, we conclude that any $A \in \mathcal{A}_\Lambda$ can be written as $A = A^+ + A^-$ where $A^\pm \in \mathcal{A}_\Lambda^\pm$ and with each of A^+, resp. A^-, being a linear combination of even, resp. odd, monomials. Many of the results we will present depend on the support of the observable under consideration. The following proposition describes some useful commutation properties related to support.

PROPOSITION 2.1. *Let* $X, Y \in \mathcal{P}_0(\Gamma)$ *with* $X \cap Y = \emptyset$.
(i) If $A \in \mathcal{A}_X^+$ *and* $B \in \mathcal{A}_Y$*, then* $[A, B] = 0$*. Moreover, if* $A \in \mathcal{A}_X$*,* $B \in \mathcal{A}_Y$*, and* $[A, B] = 0$*, then either* $A \in \mathcal{A}_X^+$ *or* $B \in \mathcal{A}_Y^+$*.*
(ii) If $A \in \mathcal{A}_X^-$ *and* $B \in \mathcal{A}_Y^-$*, then* $\{A, B\} = 0$*. Moreover, if* $A \in \mathcal{A}_X$*,* $B \in \mathcal{A}_Y$*, and* $\{A, B\} = 0$*, then either* A *or* B *is identically zero, or* $A \in \mathcal{A}_X^-$ *and* $B \in \mathcal{A}_Y^-$*.*

PROOF. The first statements in (i) and (ii) above are easy to check for monomials, and they extend to general observables by linearity. Now, suppose $A \in \mathcal{A}_X$, $B \in \mathcal{A}_Y$, and $[A, B] = 0$. Using the first part of (i), it is clear that

$$0 = [A, B] = [A^+ + A^-, B^+ + B^-] = [A^-, B^-] = 2A^-B^- - \{A^-, B^-\} \tag{2.4}$$

and therefore, $A^-B^- = 0$ by the first part of (ii). The claim now follows by expanding A^- and B^- in the monomial basis. Proving the second part of (ii) is similar. □

An interaction Φ for a system of fermions on Γ is defined similarly to that of an interaction for a quantum spin system. Specifically, a map $\Phi : \mathcal{P}_0(\Gamma) \to \mathcal{A}_\Gamma$ is an interaction if $\Phi(X)^* = \Phi(X) \in \mathcal{A}_X$ for all $X \in \mathcal{P}_0(\Gamma)$. For the results we are interested in here, we will restrict our attention to *even* interactions. An interaction Φ is said to be even if $\Phi(X) \in \mathcal{A}_X^+$ for all $X \in \mathcal{P}_0(\Gamma)$. As such, each term $\Phi(X)$ is itself a sum of terms of the form

$$a_Y^* h(Y, Z) a_Z + a_Z^* \overline{h(Y, Z)} a_Y, \tag{2.5}$$

where $Y \cup Z = X$, $|Y| + |Z|$ is even, $h(Y, Z) \in \mathbb{C}$, and $a_Y = a_{y_1} \cdots a_{y_k}$ for a suitable enumeration of $Y = \{y_1, \ldots, y_k\}$. This assumption of even interactions has a physical justification; it follows from the conservation of angular momentum that the parity of the number of particles with half-integral spin is preserved. In other words, fermions can only be created or annihilated in an even number at a time.

Given an interaction Φ and $\Lambda \in \mathcal{P}_0(\Gamma)$, a local Hamiltonian H_Λ^Φ is defined by

$$H_\Lambda^\Phi = \sum_{X \subset \Lambda} \Phi(X). \tag{2.6}$$

When the interaction Φ is understood, we often drop its dependence in the local Hamiltonians. These finite-volume Hamiltonians generate a local Heisenberg dynamics which is a one-parameter group of automorphisms of $\mathcal{A}_\Lambda$:

$$\tau_t^\Lambda(A) = e^{itH_\Lambda} A e^{-itH_\Lambda}, \quad t \in \mathbb{R}, A \in \mathcal{A}_\Lambda. \tag{2.7}$$

If a dynamics τ_t^Λ is generated by an even interaction, it leaves $\mathcal{A}_\Lambda^\pm$ invariant. In the next section we will show that for such dynamics that Lieb-Robinson bounds identical to the well-known bounds for quantum spin systems hold with essentially the same proof. One could also consider lattice fermion systems with an infinite number of bands, but for simplicity we will not do this here. In the finite band case, it is not a loss of generality to assume that the time-dependence of each interaction term is continuous in the operator norm, and this allows for a more straightforward presentation.

3. Lieb-Robinson bounds

Lieb-Robinson bounds provide an upper bound on the speed of propagation of disturbances in an extended many-body system. Such bounds can be proved under quite general conditions on the many-body interactions. In fact, the argument we describe below applies to time-dependent interactions as well, and so we introduce this now. Let $I \subset \mathbb{R}$ be an interval; often $I = \mathbb{R}$, but this is not necessary. An even, time-dependent interaction Φ is a mapping $\Phi : \mathcal{P}_0(\Gamma) \times I \to \mathcal{A}_\Gamma$ for which

$$\Phi(X,t)^* = \Phi(X,t) \in \mathcal{A}_X^+ \quad \text{for all } X \in \mathcal{P}_0(\Gamma) \text{ and } t \in I, \tag{3.1}$$

and moreover,

$$t \mapsto \Phi(X,t) \text{ is continuous for each } X \in \mathcal{P}_0(\Gamma)\,. \tag{3.2}$$

Associated to any even, time-dependent interaction Φ and each $\Lambda \in \mathcal{P}_0(\Gamma)$, there is a corresponding finite-volume, time-dependent Hamiltonian

$$H_\Lambda(t) = \sum_{X \subset \Lambda} \Phi(X,t) \quad \text{for all } t \in I\,. \tag{3.3}$$

As is well-known, see e.g. Theorem X.69 in [**43**], the solution of

$$\frac{d}{dt}U_\Lambda(t,s) = -iH_\Lambda(t)U_\Lambda(t,s) \quad \text{with } U_\Lambda(s,s) = \mathbb{1} \quad \text{and } s,t \in I, \tag{3.4}$$

produces a two-parameter family of unitary propagators $U_\Lambda(t,s) \in \mathcal{A}_\Lambda$, and in terms of these unitaries, a co-cycle of automorphisms of $\mathcal{A}_\Lambda$ is defined by

$$\tau_{t,s}^\Lambda(A) = U_\Lambda(t,s)^* A U_\Lambda(t,s) \quad \text{for all } A \in \mathcal{A}_\Lambda \text{ and } t,s \in I\,. \tag{3.5}$$

These automorphisms $\tau_{t,s}^\Lambda$ are commonly referred to as the finite-volume Heisenberg dynamics associated to Φ, and it is to these dynamics that the Lieb-Robinson bounds apply. As we previously observed in the time-independent case, since $H_\Lambda(t)$ in (3.4) is even, $U_\Lambda(t,s)$ is even and so the map $\tau_{t,s}^\Lambda$ leaves $\mathcal{A}_\Lambda^+$ and $\mathcal{A}_\Lambda^-$ invariant.

Lieb-Robinson bounds are valid for interactions that decay sufficiently fast. A precise formulation of these bounds requires a notion of distance. For this reason, we will further assume that the countable set Γ is equipped with a metric d. In many physically interesting models, $\Gamma = \mathbb{Z}^\nu$ for some integer $\nu \geq 1$, and we may, for instance, take d to be the ℓ^1-metric.

It is convenient to express the decay of interactions in terms of a function $G : \Gamma \times \Gamma \to (0,\infty)$ with the following properties:

(i) for all $x, y \in \Gamma$, $G(x,y) = G(y,x)$;
(ii) for all $x, y \in \Gamma$, $\sum_{z\in\Gamma} G(x,z)G(z,y) \leq G(x,y)$;
(iii) $x \mapsto \sum_{z\in\Gamma} G(x,z)$ is a uniformly bounded function on Γ.

We will denote the supremum of the bounded function described in (iii) by $\|G\|$.

Given such a function G, we define a set of even interactions, denoted $\mathcal{B}_G^+(I)$, for which we can prove a Lieb-Robinson bound: an even interaction Φ belongs to $\mathcal{B}_G^+(I)$ if there is a locally integrable function $\|\Phi\|_G : I \to [0,\infty)$ for which

$$\sum_{\substack{Z\in\mathcal{P}_0(\Gamma)\\ x,y\in Z}} \|\Phi(Z,t)\| \leq \|\Phi\|_G(t)G(x,y) \quad \text{for all } x,y \in \Gamma \text{ and } t \in I. \tag{3.6}$$

Here $\|\Phi\|_G(t)$ plays the role of a time-dependent norm on the space of interactions.

Two further definitions are useful in the statement of the Lieb-Robinson bounds. For $X \subset \Lambda \subset \Gamma$, we define a collection of surface sets associated to X in Λ by

$$S_\Lambda(X) = \{Z \in \mathcal{P}_0(\Lambda) : Z \cap X \neq \emptyset \text{ and } Z \cap (\Lambda \setminus X) \neq \emptyset\}. \tag{3.7}$$

Furthermore, if Φ is a time-dependent interaction, then the Φ-boundary of a set $X \in \mathcal{P}_0(\Gamma)$ is defined by

$$\partial_\Phi X = \{x \in X \mid \exists Z \in S_\Gamma(X), t \in I \text{ s.t. } x \in Z, \text{ and } \Phi(Z,t) \neq 0\}. \tag{3.8}$$

If Φ is long-range, then it is often the case that $\partial_\Phi X = X$, and $|\partial_\Phi X|$ is not a good measure of surface effects on the dynamics. However, when Φ is finite-range, the above definition is of more consequence.

THEOREM 3.1 (Lieb-Robinson Bound for Fermions). *Let Γ be a countable set equipped with a metric d and a function G satisfying (i) - (iii) above, and let $\Phi \in \mathcal{B}_G^+(I)$. For $\Lambda \in \mathcal{P}_0(\Gamma)$, let $\tau_{t,s}^\Lambda$ be a finite-volume dynamics associated to Φ, as defined in (3.5). Let $X, Y \subset \Lambda$ with $X \cap Y = \emptyset$.*
(i) If $A \in \mathcal{A}_X$, $B \in \mathcal{A}_Y$, and $[A,B] = 0$, then

$$\left\| \left[\tau_{t,s}^\Lambda(A), B\right] \right\| \leq 2\|A\|\|B\| \left(\exp\left[2\int_s^t \|\Phi\|_G(r)\,dr \right] - 1 \right) \sum_{x \in \partial_\Phi X} \sum_{y \in Y} G(x,y) \tag{3.9}$$

for all $t, s \in I$, $s \leq t$.
(ii) If $A \in \mathcal{A}_X$, $B \in \mathcal{A}_Y$, and $\{A,B\} = 0$, then for any $s \leq t$, one has that

$$\left\| \left\{\tau_{t,s}^\Lambda(A), B\right\} \right\| \leq 2\|A\|\|B\| \left(\exp\left[2\int_s^t \|\Phi\|_G(r)\,dr \right] - 1 \right) \sum_{x \in \partial_\Phi X} \sum_{y \in Y} G(x,y). \tag{3.10}$$

A number of remarks are in order.

First, the proof of this theorem mimics techniques that are well-known in the context of quantum spin systems, i.e. tensor product algebras. In fact, given Lemma 3.2 and Lemma 3.3, both proved below, the proof of Lieb-Robinson bounds in both settings proceeds almost identically. This was noted already in [**13**], where a variant of Theorem 3.1 (i) is proved along the same lines as what we present here (see [**13**][Theorem 5.1-Corollary 5.2(ii)]. It has been known for some time that to treat odd observables for fermionic systems, one should consider anti-commutators in addition to commutators. See, e.g., the discussion in [**20**].

Next, as explained in Proposition 2.1, the assumptions on the observables above can be re-formulated: for (3.9), one has assumed that either A or B is even, whereas for (3.10), excepting trivialities, one has assumed that both A and B are odd.

Further, the existence of a function G satisfying properties (i)-(iii) above is, implicitly, a condition on Γ. In many applications, the function G is defined as a function of the distance between sites. More precisely, let $F : [0,\infty) \to (0,\infty)$ be a non-increasing function with the following two properties:

(iv) F is uniformly integrable on Γ, i.e.

$$\|F\| = \sup_{x \in \Gamma} \sum_{y \in \Gamma} F(d(x,y)) < \infty, \tag{3.11}$$

and

(v) F satisfies the convolution condition

$$C = \sup_{x,y\in\Gamma} \sum_{z\in\Gamma} \frac{F(d(x,z))F(d(z,y))}{F(d(x,y))} < \infty. \tag{3.12}$$

Any function F as above is called an F-function on Γ. Moreover, it is clear that, in terms of any such F, we can define a function G with the properties (i)-(iii), by setting $G(x,y) = C^{-1}F(d(x,y))$.

For technical estimates, it is often convenient to consider classes of decay functions on Γ. Note that if F is an F-function on Γ, then for any subadditive function $f : [0,\infty) \to [0,\infty)$, i.e., $f(r+s) \leq f(r) + f(s)$ for all $r, s \in [0,\infty)$, the function $F_f(r) = e^{-f(r)}F(r)$ also satisfies (iv) and (v) with $\|F_f\| \leq \|F\|$ and $C_f \leq C$. Similarly, given G satisfying (i)-(iii) and any $g : \Gamma \to (0,1]$, the function $G_g(x,y) = g(x)g(y)G(x,y)$ also satisfies (i)-(iii). In this latter case, the function g can be used to introduce a spatial dependence in the decay of the interaction.

For $\Gamma = \mathbb{Z}^\nu$ and $d(x,y) = |x-y|$, i.e. the ℓ^1-distance, a typical example of an F-function is given by

$$F(r) = \frac{1}{(1+r)^{\nu+\epsilon}} \tag{3.13}$$

where $\epsilon > 0$ can be arbitrary. In fact, it is clear that this F-function is uniformly integrable, i.e. (3.11) holds, and moreover, for (3.12) one may take

$$C = 2^{\nu+\epsilon}\|F\|\,. \tag{3.14}$$

In combination with $f(r) = ar$, $a > 0$, we obtain a useful family of F-functions F_a given by $F_a(r) = e^{-ar}/(1+r)^{\nu+\epsilon}$.

Before proving Theorem 3.1, we state two simple lemmas. First, in Lemma 3.2 below, we prove a basic estimate for solutions of certain $\mathcal{B}(\mathcal{H})$-valued differential equations. Next, Lemma 3.3 summarizes an application of Lemma 3.2 demonstrating a one-step locality estimate for the dynamically evolved quantities of interest. Theorem 3.1 will then follow by iterating the result from Lemma 3.3.

We start with the following solution estimate.

LEMMA 3.2. *Let $\mathcal{H}$ be a complex Hilbert space, $I \subset \mathbb{R}$ an interval, and $A, B : I \to \mathcal{B}(\mathcal{H})$, be norm continuous with A self-adjoint, i.e. $A(t)^* = A(t)$ for all $t \in I$. Then, for any $t_0 \in I$, the solution of the initial value problem*

$$\frac{d}{dt}f(t) = i[A(t), f(t)] + B(t) \quad \textit{with} \quad f(t_0) = f_0 \in \mathcal{B}(\mathcal{H}) \tag{3.15}$$

satisfies the estimate

$$\|f(t)\| \leq \|f(t_0)\| + \int_{t_-}^{t_+} \|B(s)\|\, ds \quad \textit{for any } t \in I. \tag{3.16}$$

Here we have set $t_+ = \max\{t, t_0\}$ and $t_- = \min\{t, t_0\}$.

The assumption of norm continuity above is convenient because it guarantees that the mapping $s \mapsto \|B(s)\|$ is continuous and thus measurable. A variant of this result in the case that A and B are merely strongly continuous is proved in [**39**].

PROOF. Since A is self-adjoint, the unitary propagator corresponding to

$$\frac{d}{dt}U(t,t_0) = iA(t)U(t,t_0) \quad \text{with } U(t_0,t_0) = \mathbb{1} \tag{3.17}$$

is well-defined for any $t \in I$. In this case, the function $g : I \to \mathcal{B}(\mathcal{H})$ given by

$$g(t) = U(t,t_0)g_0U(t,t_0)^* \tag{3.18}$$

is the unique solution of the initial value problem

$$\frac{d}{dt}g(t) = i[A(t), g(t)] \quad \text{with} \quad g(t_0) = g_0 \in \mathcal{B}(\mathcal{H}). \tag{3.19}$$

From this, one readily checks that

$$f(t) = U(t,t_0)\left(f_0 + \int_{t_0}^{t} U(s,t_0)^*B(s)U(s,t_0)\,ds\right)U(t,t_0)^* \tag{3.20}$$

is the unique solution of (3.15) from which (3.16) follows. □

We now use Lemma 3.2 to provide estimates on two families of operators. Let Φ be an even, time-dependent interaction. For any $\Lambda \in \mathcal{P}_0(\Gamma)$, consider the dynamics $\tau_{t,s}^{\Lambda}$ on $\mathcal{A}_\Lambda$, as in (3.5) above. Given $X \subset \Lambda$ and any $B \in \mathcal{A}_\Lambda$, we define maps $g_{t,s}^{X,B}, h_{t,s}^{X,B} : \mathcal{A}_X \to \mathcal{A}_\Lambda$ as follows:

$$g_{t,s}^{X,B}(A) = [\tau_{t,s}^{\Lambda}(A), B] \quad \text{and} \quad h_{t,s}^{X,B}(A) = \{\tau_{t,s}^{\Lambda}(A), B\} \quad \text{for all } A \in \mathcal{A}_X. \tag{3.21}$$

LEMMA 3.3. *Let Φ be an even, time-dependent interaction and take $\Lambda \in \mathcal{P}_0(\Gamma)$. For $X \subset \Lambda$, $A \in \mathcal{A}_X$, and $s,t \in I$, with $s \le t$, the mappings $g_{t,s}^{X,B}$ and $h_{t,s}^{X,B}$: $\mathcal{A}_X \to \mathcal{A}_\Lambda$, defined in (3.21), satisfy the bounds*

$$\|g_{t,s}^{X,B}(A)\| \le \|[\tau_{t,s}^{X}(A), B]\| + 2\|A\| \sum_{Z \in S_\Lambda(X)} \int_s^t \|g_{r,s}^{Z,B}(\Phi(Z,r))\|\,dr \tag{3.22}$$

$$\|h_{t,s}^{X,B}(A)\| \le \|\{\tau_{t,s}^{X}(A), B\}\| + 2\|A\| \sum_{Z \in S_\Lambda(X)} \int_s^t \|g_{r,s}^{Z,B}(\Phi(Z,r))\|\,dr, \tag{3.23}$$

where $\tau_{t,s}^{X}(A)$ is the finite-volume dynamics associated to $H_X(t)$, see (3.3) and (3.5).

As will be the case in our applications, these bounds are particularly useful when the first term on the right-hand-sides above vanish.

PROOF OF LEMMA 3.3. Let $A \in \mathcal{A}_X, B \in \mathcal{A}_\Lambda$ and $s \in I$ be fixed. To derive the two bounds in parallel, define two functions:

$$f_1(t) = [\tau_{t,s}^{\Lambda} \circ \hat{\tau}_{t,s}^{X}(A), B] \quad \text{and} \quad f_2(t) = \{\tau_{t,s}^{\Lambda} \circ \hat{\tau}_{t,s}^{X}(A), B\}. \tag{3.24}$$

Here, $\hat{\tau}_{t,s}^{X}$ denotes the inverse dynamics of the system restricted to X, i.e.

$$\hat{\tau}_{t,s}^{X}(A) = U_X(t,s)AU_X(t,s)^*, \tag{3.25}$$

which is the inverse of (3.5). One readily checks that for f_1:

$$\begin{aligned}
\frac{d}{dt}f_1(t) &= i\left[\tau_{t,s}^{\Lambda}\left([H_\Lambda(t) - H_X(t), \hat{\tau}_{t,s}^{X}(A)]\right), B\right] \\
&= i \sum_{Z \in S_\Lambda(X)} \left[\left[\tau_{t,s}^{\Lambda}(\Phi(Z,t)), \tau_{t,s}^{\Lambda} \circ \hat{\tau}_{t,s}^{X}(A)\right], B\right] \\
&= i \sum_{Z \in S_\Lambda(X)} \left[\tau_{t,s}^{\Lambda}(\Phi(Z,t)), f_1(t)\right] - i \sum_{Z \in S_\Lambda(X)} \left[\tau_{t,s}^{\Lambda} \circ \hat{\tau}_{t,s}^{X}(A), \left[\tau_{t,s}^{\Lambda}(\Phi(Z,t)), B\right]\right].
\end{aligned}$$

In the first equality above, we use that the adjoints of the unitary propagators satisfy the adjoint of (3.4). For the second, we use that $\operatorname{supp}(\hat{\tau}_{t,s}^X(A)) \subset X$, and for the final equality we use the Jacobi identity.

An almost identical calculation for $f_2(t)$ gives:

$$\begin{aligned}
\frac{d}{dt} f_2(t) &= i\left\{\tau_{t,s}^\Lambda\left(\left[H_\Lambda(t) - H_X(t), \hat{\tau}_{t,s}^X(A)\right]\right), B\right\} \\
&= i \sum_{Z \in S_\Lambda(X)} \left\{\left[\tau_{t,s}^\Lambda(\Phi(Z,t)), \tau_{t,s}^\Lambda \circ \hat{\tau}_{t,s}^X(A)\right], B\right\} \\
&= i \sum_{Z \in S_\Lambda(X)} \left[\tau_{t,s}^\Lambda(\Phi(Z,t)), f_2(t)\right] - i \sum_{Z \in S_\Lambda(X)} \left\{\tau_{t,s}^\Lambda \circ \hat{\tau}_{t,s}^X(A), \left[\tau_{t,s}^\Lambda(\Phi(Z,t)), B\right]\right\}.
\end{aligned}$$

We note that the only change here is that, instead of the Jacobi identity for commutators, we use the following identity which holds for any three elements x, y, z in an associative algebra:

$$\{[x,y],z\} - \{[z,x],y\} + [\{y,z\},x] = 0. \tag{3.26}$$

Both differential equations above are of the form required to apply Lemma 3.2. The claimed bounds follow from this and the substitution $A \mapsto \tau_{t,s}^X(A)$. $\square$

We can now present the proof of the Lieb-Robinson bounds for lattice fermions for which the following notation will be useful: for $X, Y \subset \Gamma$, set

$$\delta_Y(X) = \begin{cases} 0 & \text{if } X \cap Y = \emptyset \\ 1 & \text{if } X \cap Y \neq \emptyset. \end{cases} \tag{3.27}$$

PROOF OF THEOREM 3.1. We first prove (3.9). Lemma 3.3 implies

$$\|[\tau_{t,s}^\Lambda(A), B]\| \leq \|[\tau_{t,s}^X(A), B]\| + 2\|A\| \sum_{Z \in S_\Lambda(X)} \int_s^t \|[\tau_{r,s}^\Lambda(\Phi(Z,r)), B]\| \, dr. \tag{3.28}$$

Since the observables A and B have disjoint support and one of them is even, it is clear that $[\tau_{t,s}^X(A), B] = 0$. More generally, if $A \in \mathcal{A}_Z^+$, then $\tau_{r,s}^Z(A) \in \mathcal{A}_Z^+$ for all $r, s \in I$ and so the bound $\|[\tau_{r,s}^Z(A), B]\| \leq 2\|A\|\|B\|\delta_Y(Z)$ follows. In this case, by iterating (3.28) $N \geq 1$ times, we obtain

$$\|[\tau_{t,s}^\Lambda(A), B]\| \leq 2\|A\|\|B\| \left(\delta_Y(X) + \sum_{n=1}^N a_n(t)\right) + R_{N+1}(t) \tag{3.29}$$

where

$$\begin{aligned}
a_n(t) = 2^n \sum_{Z_1 \in S_\Lambda(X)} \sum_{Z_2 \in S_\Lambda(Z_1)} \cdots \sum_{Z_n \in S_\Lambda(Z_{n-1})} \delta_Y(Z_n) \int_s^t \int_s^{r_1} \cdots \int_s^{r_{n-1}} \times \\
\times \left(\prod_{j=1}^n \|\Phi(Z_j, r_j)\|\right) dr_n dr_{n-1} \cdots dr_1
\end{aligned} \tag{3.30}$$

and

$$R_{N+1}(t) = 2^{N+1} \sum_{Z_1 \in S_\Lambda(X)} \sum_{Z_2 \in S_\Lambda(Z_1)} \cdots \sum_{Z_{N+1} \in S_\Lambda(Z_N)} \int_s^t \int_s^{r_1} \cdots \int_s^{r_N} \times$$

$$\left(\prod_{j=1}^N \|\Phi(Z_j, r_j)\| \right) \|[\tau^\Lambda_{r_{N+1},s}(\Phi(Z_{N+1}, r_{N+1})), B]\| dr_{N+1} dr_N \cdots dr_1. \tag{3.31}$$

The remainder term $R_{N+1}(t)$ is estimated as follows. First, we observe that

$$\|[\tau^\Lambda_{r_{N+1},s}(\Phi(Z_{N+1}, r_{N+1})), B]\| \leq 2\|B\| \, \|\Phi(Z_{N+1}, r_{N+1})\| \, . \tag{3.32}$$

Next, we note that the sums above are actually sums over chains of sets $Z_1, \cdots Z_{N+1}$ that satisfy $Z_1 \cap \partial_\Phi X \neq \emptyset$ and $Z_j \cap Z_{j-1} \neq \emptyset$ for $2 \leq j \leq N+1$. As such, there are points $w_1, w_2, \cdots, w_{N+1} \in \Lambda$ with $w_1 \in Z_1 \cap \partial_\Phi X$ and $w_j \in Z_j \cap Z_{j-1}$ for all $2 \leq j \leq N+1$. A simple upper bound on these sums is then obtained by over counting:

$$\sum_{Z_1 \in S_\Lambda(X)} \sum_{Z_2 \in S_\Lambda(Z_1)} \cdots \sum_{Z_{N+1} \in S_\Lambda(Z_N)} * \leq \sum_{w_1 \in \partial_\Phi X} \sum_{w_2, \ldots, w_{N+2} \in \Lambda} \sum_{\substack{Z_1, \ldots, Z_{N+1} \subset \Lambda: \\ w_k, w_{k+1} \in Z_k, k=1,\ldots,N+1}} * \tag{3.33}$$

where $*$ denotes an arbitrary non-negative quantity. Note that we have also used that the last set Z_{N+1} must contain more than one point since $Z_{N+1} \in S_\Lambda(Z_N)$. Now, from (3.6) we have that

$$\sum_{\substack{Z_k \subset \Lambda: \\ w_k, w_{k+1} \in Z_k}} \|\Phi(Z_k, r_k)\| \leq \|\Phi\|_G(r_k) G(w_k, w_{k+1}) \tag{3.34}$$

holds for each $1 \leq k \leq N+1$. We conclude that

$$\begin{aligned}
R_{N+1}(t) &\leq 2\|B\| 2^{N+1} \int_s^t \cdots \int_s^{r_N} \sum_{w_1 \in \partial_\Phi X} \sum_{w_2, \ldots, w_{N+2} \in \Lambda} \sum_{\substack{Z_1, \ldots, Z_{N+1} \subset \Lambda: \\ w_k, w_{k+1} \in Z_k, k=1,\ldots,N+1}} \\
&\quad \times \prod_{j=1}^{N+1} \|\Phi(Z_j, r_j)\| dr_{N+1} \cdots dr_1 \\
&\leq 2\|B\| 2^{N+1} \int_s^t \cdots \int_s^{r_N} \sum_{w_1 \in \partial_\Phi X} \sum_{w_2, \ldots, w_{N+2} \in \Lambda} \\
&\quad \times \prod_{j=1}^{N+1} \|\Phi\|_G(r_j) G(w_j, w_{j+1}) dr_{N+1} \cdots dr_1 \\
&\leq 2\|B\| 2^{N+1} \sum_{w_1 \in \partial_\Phi X} \sum_{w_{N+2} \in \Lambda} G(w_1, w_{N+2}) \\
&\quad \times \int_s^t \cdots \int_s^{r_N} \prod_{j=1}^{N+1} \|\Phi\|_G(r_j) dr_{N+1} \cdots dr_1 \\
&\leq 2\|B\| |\partial_\Phi X| \|G\| \frac{\left(2 \int_s^t \|\Phi\|_G(r) \, dr\right)^{N+1}}{(N+1)!}.
\end{aligned}$$

Since $\|\Phi\|_G(\cdot)$ is locally integrable on I, this remainder clearly goes to 0 as $N \to \infty$.

A similar estimate can be applied to the terms $a_n(t)$. Note that these terms are also sums over chains of sets. However, there is a restriction: only those chains whose final link Z_n satisfies $Z_n \cap Y \neq \emptyset$ contribute to the sum. The bound

$$a_n(t) \leq \frac{\left(2\int_s^t \|\Phi\|_G(r)\,dr\right)^n}{n!} \sum_{x\in\partial_\Phi X}\sum_{y\in Y} G(x,y). \tag{3.35}$$

follows as above. Since $\delta_Y(X) = 0$ and $n \geq 1$, the bound in (3.9) is now clear.

The proof of (3.10) proceeds similarly. In fact, Lemma 3.3 implies

$$\|\{\tau_{t,s}^\Lambda(A), B\}\| \leq \|\{\tau_{t,s}^X(A), B\}\| + 2\|A\| \sum_{Z\in S_\Lambda(X)} \int_s^t \|[\tau_{r,s}^\Lambda(\Phi(Z,r)), B]\|\,dr \tag{3.36}$$

Since $\{A, B\} = 0$, the first term on the left-hand-side above vanishes. Using (3.28) to estimate the second term and iterating in exactly the same way as above, yields (3.10) as claimed. □

Theorem 3.1 gives an estimate of $\left\|\left[\tau_{t,s}^\Lambda(A), B\right]\right\|$ for commuting observables A and B with disjoint supports. As a function of $|t-s|$, this estimate grows exponentially and for small $|t-s|$ vanishes linearly. A few additional comments are in order. First, note that when the supports of A and B have non-empty intersection, in general, one cannot expect to improve on the trivial bound: $\left\|\left[\tau_{t,s}^\Lambda(A), B\right]\right\| \leq 2\|A\|\|B\|$. On the other hand, given a lower bound on the distance between the supports of A and B, and if the interaction is of finite range, a slight modification to the proof of Theorem 3.1 shows that the behavior for small $|t-s|$ is $o(|t-s|^n)$, where n is the minimum number of interactions terms necessary to connect the supports of A and B with a chain of sets, see e.g. (3.30). For similar reasons, one can also show that single-site terms in the Hamiltonian do not contribute to the estimate of Lieb-Robinson velocity. This can easily be seen using the interaction picture as is done, e.g., in [**38**]. Of course, this is not the statement that single-site terms in the Hamiltonian do not affect the velocity associated to certain time-evolved observables; rather it is the fact that this general upper bound is insensitive to such terms. For example, in specific models with a random external field, the speed of propagation has been shown to vanish [**17**].

4. Conditional expectation and local approximations

In many applications to quantum spin systems, the commutator estimates provided by Lieb-Robinson bounds are used to approximate quasi-local observables by strictly local ones. These local approximations are given by a conditional expectation with respect to a suitable product state (see, e.g., [**12, 34, 35**]). In the setting of lattice fermions, the conditional expectations that come to mind are those with respect to a product state such as, e.g., the tracial state, which is also the quasi-free state on $\mathcal{A}_\Gamma$ determined by $\omega^{\mathrm{tr}}(a_x a_y) = 0$, and $\omega^{\mathrm{tr}}(a_x^* a_y) = \frac{1}{2}\delta_{x,y}$, for all $x, y \in \Gamma$. This state has the following product property [**1**, Theorem 6.12]: for any finite set of distinct $x_1, \ldots, x_k \in \Gamma$, and $A_{x_i} \in \mathcal{A}_{\{x_i\}}$, we have

$$\omega^{\mathrm{tr}}(A_{x_1}\cdots A_{x_k}) = \prod_{i=1}^k \omega^{\mathrm{tr}}(A_{x_i}). \tag{4.1}$$

It is not difficult to see that for all $\Lambda \in \mathcal{P}_0(\Gamma)$, $\omega^{\rm tr}$ restricted to $\mathcal{A}_\Lambda$ is the state of maximal entropy. It was shown in [**4**][Theorem 4.7] that for each finite $X \subset \Gamma$, there is a unique conditional expectation $\mathbb{F}_X : \mathcal{A}_\Gamma \to \mathcal{A}_X$ that leaves $\omega^{\rm tr}$ invariant, meaning

$$\omega^{\rm tr}(\mathbb{F}_X(AB)) = \omega^{\rm tr}(\mathbb{F}_X(A)B), \quad A \in \mathcal{A}_\Gamma, B \in \mathcal{A}_X. \tag{4.2}$$

The same result for arbitrary even product states ω was proved in [**3**]. The family of conditional expectations $\{\mathbb{F}_X : \mathcal{A}_\Gamma \to \mathcal{A}_X \mid X \in \mathcal{P}_0(\Gamma)\}$ determined by (4.2), satisfies the commuting diagram:

$$\begin{array}{ccc} \mathcal{A}_{X\cup Y} & \xrightarrow{\mathbb{F}_X} & \mathcal{A}_X \\ \downarrow{\scriptstyle \mathbb{F}_Y} & & \downarrow{\scriptstyle \mathbb{F}_{X\cap Y}} \\ \mathcal{A}_Y & \xrightarrow{\mathbb{F}_{X\cap Y}} & \mathcal{A}_{X\cap Y} \end{array} . \tag{4.3}$$

The commutativity of this diagram is the essential property that allows one to use the maps $\mathbb{F}_X$ in the same role as the partial trace in the case of of quantum spin systems. We note however, that for applications one often only needs these relations for even observables, in which case there are other options for the local algebras and the conditional expectations. For our purposes, it will be useful to introduce a slightly different set of conditional expectations, which we will now explain.

Under the isomorphism $\mathcal{A}_\Lambda \cong M_{2^{|\Lambda|}}$, $\omega^{\rm tr}$ coincides with the normalized trace on $M_{2^{|\Lambda|}}$. For even observables, the conditional expectation acts exactly as in the situation of spin system. The anticommutation properties of odd observables, however, introduce a small twist, which implies that the Krauss form of the map $\mathbb{F}_X$ contains global operators (see (4.19) below). This creates a complication for the proof of Lemma 4.3. Therefore, we introduce another family of conditional expectations which, as we will show, also satisfy a commuting diagram similar to (4.3), and which coincide with the maps $\mathbb{F}_X$ on even observables.

The remainder of this section is devoted to describing the conditional expectations of interest for finite volume systems. Given a C^*-algebra $\mathcal{A}$, and a subalgebra $\mathcal{B}$ of $\mathcal{A}$, by a conditional expectation of $\mathcal{A}$ onto $\mathcal{B}$, we mean a unity-preserving, completely positive map $\mathbb{E} : \mathcal{A} \to \mathcal{A}$ with $\operatorname{ran}(\mathbb{E}) = \mathcal{B}$, such that $\mathbb{E}(BAC) = B\mathbb{E}(A)C$ for all $A \in \mathcal{A}$ and $B, C \in \mathcal{B}$. Let $X \subseteq \Lambda \subset \Gamma$ with X and Λ finite. We will show that the range of the relevant conditional expectation $\mathbb{E}_X^\Lambda : \mathcal{A}_\Lambda \to \mathcal{A}_\Lambda$ is given by the C^* -subalgebra:

$$\mathcal{A}_X^\Lambda = \{A + B\theta_\Lambda \,:\, A \in \mathcal{A}_X^+,\, B \in \mathcal{A}_X^-\},$$

where $\theta_\Lambda = (-1)^{N_\Lambda}$ is the parity operator used to define Θ_Λ in (2.2). Since we only consider Hamiltonians defined by even interactions, in applications it will be sufficient to only consider the restriction of $\mathbb{E}_X^\Lambda$ to $\mathcal{A}_\Lambda^+$. In the restricted case $\mathbb{E}_X^\Lambda : \mathcal{A}_\Lambda^+ \to \mathcal{A}_\Lambda$ will be a conditional expectation with range equal to $\mathcal{A}_X^+$.

We define $\mathbb{E}_X^\Lambda$ by giving its Krauss form. For each site $x \in \Lambda$, define

$$u_x^{(0)} = \mathbb{1}, \quad u_x^{(1)} = a_x^* + a_x, \quad u_x^{(2)} = a_x^* - a_x, \quad u_x^{(3)} = \mathbb{1} - 2a_x^*a_x. \tag{4.4}$$

It follows from the CAR that these are unitary. Clearly, $u_x^{(i)} \in \mathcal{A}_{\{x\}}^+$ for $i = 0, 3$, and $u_x^{(i)} \in \mathcal{A}_{\{x\}}^-$ for $i = 1, 2$. Therefore, $u_x^{(i)}$ commutes with the elements of $\mathcal{A}_{\Lambda\setminus\{x\}}$ for $i = 0, 3$, and $u_x^{(i)}$ commutes with $\mathcal{A}_{\Lambda\setminus\{x\}}^+$ and anti-commutes with $\mathcal{A}_{\Lambda\setminus\{x\}}^-$ for $i = 1, 2$.

For a subset $X \subseteq \Lambda$, let $I_{\Lambda\setminus X} = \{0,1,2,3\}^{\Lambda\setminus X}$ and fix an ordering of the sites of $\Lambda \setminus X = \{x_1, \dots, x_n\}$. Then, for each $\alpha \in I_{\Lambda\setminus X}$ define the unitary operator $u(\alpha) \in \mathcal{A}_{\Lambda\setminus X}$ by

$$u(\alpha) = u_{x_1}^{(\alpha(x_1))} \cdots u_{x_n}^{(\alpha(x_n))}. \tag{4.5}$$

We will show that the following unity-preserving, completely positive map $\mathbb{E}_X^\Lambda : \mathcal{A}_\Lambda \to \mathcal{A}_\Lambda$ is the conditional expectation of interest associated with X on Λ:

$$\mathbb{E}_X^\Lambda(A) = \frac{1}{4^{|\Lambda\setminus X|}} \sum_{\alpha \in I_{\Lambda\setminus X}} u(\alpha)^* A u(\alpha), \quad A \in \mathcal{A}_\Lambda. \tag{4.6}$$

It is important to notice that the map $\mathbb{E}_X^\Lambda$ does not depend on the ordering chosen for the site in $\Lambda \setminus X$. Since the unitaries $u_x^{(k)}, u_y^{(l)}$, $x \neq y$, $k, l \in \{0,1,2,3\}$, either commute or anti-commute, any reordering $\tilde{u}(\alpha)$ of $u(\alpha)$ equals either $u(\alpha)$ or $-u(\alpha)$. Either way, the α-term in (4.6) is not affected.

The first important property for $\mathbb{E}_X^\Lambda$ is that

$$\mathbb{E}_X^\Lambda(A) = \omega^{\mathrm{tr}}(A)\mathbb{1} \ \text{ for all } \ A \in \mathcal{A}_{\Lambda\setminus X}. \tag{4.7}$$

It is easy to verify that $\mathbb{E}_{\Lambda\setminus\{x\}}^\Lambda(A) = \omega^{\mathrm{tr}}(A)\mathbb{1}$ for a monomial $A \in \mathcal{A}_{\{x\}}$. Using this, the product property of ω^{tr} given in (4.1), and the CAR it follows that (4.7) holds for any monomial $A \in \mathcal{A}_{\Lambda\setminus X}$. The property then extends to $\mathcal{A}_{\Lambda\setminus X}$ by linearity.

To establish that $\mathbb{E}_X^\Lambda$ is a conditional expectation, it is left to verify that

$$\mathbb{E}_X^\Lambda(BAC) = B\mathbb{E}_X^\Lambda(A)C \tag{4.8}$$

for all $A \in \mathcal{A}_\Lambda$ and $B, C \in \mathrm{ran}(\mathbb{E}_X^\Lambda)$. In the situation that $\mathrm{ran}(\mathbb{E}_X^\Lambda)$ is a C^*-subalgebra of $\mathcal{A}_\Lambda$, a theorem by Tomiyama [**44**] shows that (4.8) is satisfied if $\mathbb{E}_X^\Lambda$ is a norm-1 projection. In the next lemma we establish that Tomiyama's result applies to both $\mathbb{E}_X^\Lambda : \mathcal{A}_\Lambda \to \mathcal{A}_\Lambda$ and the restriction $\mathbb{E}_X^\Lambda : \mathcal{A}_\Lambda^+ \to \mathcal{A}_\Lambda$.

LEMMA 4.1. *For $X, \Lambda \in \mathcal{P}_0(\Gamma)$ with $X \subset \Lambda$, the map $\mathbb{E}_X^\Lambda : \mathcal{A}_\Lambda \to \mathcal{A}_\Lambda$ defined in (4.6) satisfies*
(i) $\|\mathbb{E}_X^\Lambda\| = 1$;
(ii) $(\mathbb{E}_X^\Lambda)^2 = \mathbb{E}_X^\Lambda$;
(iii) $\mathbb{E}_X^\Lambda(\mathcal{A}_\Lambda) = \mathcal{A}_X^\Lambda$;
(iv) $\mathbb{E}_X^\Lambda(\mathcal{A}_\Lambda^+) = \mathcal{A}_\Lambda^+$.

PROOF. Property (i) follows immediately from the fact that $\mathbb{E}_X^\Lambda$ is an average of unitary conjugations.

To prove (ii)-(iv) we consider observables of the form $A = \prod_{x\in\Lambda} A_x$ where $A_x \in \mathcal{A}_{\{x\}}^\pm$, for all $x \in \Lambda$. We refer to such operators as product observables. Since the monomials defined in (2.3) satisfy this condition, computing $\mathbb{E}_X^\Lambda(A)$ for product observables completely describes the map.

By (anti-)commuting the factors A_x, A can be written in the form $A = BC$ where B and C are product observables in $\mathcal{A}_X$ and $\mathcal{A}_{\Lambda\setminus X}$, respectively. Then, using the CAR:

$$\mathbb{E}_X^\Lambda(A) = \begin{cases} B\frac{1}{4^{|\Lambda\setminus X|}} \sum_{\alpha\in I_{\Lambda\setminus X}} u(\alpha)^* C u(\alpha) & \text{if } B \in \mathcal{A}_\Lambda^+ \\ B\frac{1}{4^{|\Lambda\setminus X|}} \sum_{\alpha\in I_{\Lambda\setminus X}} \pi(u(\alpha)) u(\alpha)^* C u(\alpha) & \text{if } B \in \mathcal{A}_\Lambda^-. \end{cases} \tag{4.9}$$

Here, $\pi(u(\alpha)) = \pm 1$ denotes the parity of $u(\alpha)$. For all $\alpha \in I_{\Lambda\setminus X}$

$$\pi(u(\alpha))u(\alpha)^* C u(\alpha) = u(\alpha)^* C \theta_{\Lambda\setminus X} u(\alpha) \theta_{\Lambda\setminus X},$$

so applying (4.7) to (4.9) we find:

$$\mathbb{E}_X^\Lambda(A) = \begin{cases} B\omega^{\mathrm{tr}}(C) & \text{if } B \in \mathcal{A}_\Lambda^+ \\ B\omega^{\mathrm{tr}}(C\theta_{\Lambda\setminus X})\theta_{\Lambda\setminus X} & \text{if } B \in \mathcal{A}_\Lambda^-. \end{cases} \tag{4.10}$$

Applying (4.10) a second time shows $(\mathbb{E}_X^\Lambda)^2(A) = \mathbb{E}_X^\Lambda(A)$ for any product observable A, and so property (ii) holds. Alternatively, (ii) also follows from the observation that $\mathbb{E}_X^\Lambda$ is defined as the average of the adjoint actions $\{\mathrm{Ad}_{u(\alpha)} \mid \alpha \in I_{\Lambda\setminus X}\}$, which form a group.

For property (iii), we see from (4.10) that $\mathbb{E}_X^\Lambda(\mathcal{A}_\Lambda) \subseteq \mathcal{A}_X^\Lambda$ by noting that $\theta_{\Lambda\setminus X} = \theta_X\theta_\Lambda$. The opposite containment follows from observing that

$$\mathbb{E}_X^\Lambda(A + B\theta_\Lambda) = A + B\theta_\Lambda$$

for any pair of product observables $A \in \mathcal{A}_X^+$ and $B \in \mathcal{A}_X^-$.

For (iv), if $A \in \mathcal{A}_\Lambda^+$ then the factors B and C above are either both even or both odd. If they are both odd, then $\mathbb{E}_X^\Lambda(A) = 0$ since ω^{tr} is zero for odd observables. It follows that $\mathbb{E}_X^\Lambda(\mathcal{A}_\Lambda^+) \subseteq \mathcal{A}_X^+$. The opposite containment holds since $\mathbb{E}_X^\Lambda(A) = A$ for all $A \in \mathcal{A}_X^+$. □

Several comments are in order. First, an important consequence of (4.10) is that

$$\mathcal{A}_{\Lambda\setminus X}^- \subseteq \ker(\mathbb{E}_X^\Lambda). \tag{4.11}$$

Second, we will want to consider the family of all conditional expectations $\mathbb{E}_X^\Lambda$ such that $X \subseteq \Lambda \in \mathcal{P}_0(\Gamma)$. Since the definition of $\mathbb{E}_X^\Lambda$ is independent of the ordering of the sites $x \in \Lambda \setminus X$, given any two sets $Y, Z \subseteq \Lambda$ such that $\Lambda \setminus X$ is the disjoint union of $\Lambda \setminus Y$ and $\Lambda \setminus Z$ it immediately follows that

$$\mathbb{E}_X^\Lambda = \mathbb{E}_Y^\Lambda \circ \mathbb{E}_Z^\Lambda = \mathbb{E}_Z^\Lambda \circ \mathbb{E}_Y^\Lambda. \tag{4.12}$$

LEMMA 4.2. *The family of conditional expectations $\{\mathbb{E}_X^\Lambda : X \subseteq \Lambda \in \mathcal{P}_0(\Gamma)\}$ defined as in (4.6) satisfy the following properties:*
(i) For any X, $Y \subseteq \Lambda$,

$$\mathbb{E}_X^\Lambda \circ \mathbb{E}_Y^\Lambda = \mathbb{E}_{X\cap Y}^\Lambda. \tag{4.13}$$

(ii) For any $X \subseteq \Lambda$ and $A \in \mathcal{A}_Z^+$ and $B \in \mathcal{A}_Y^+$ with $Z \cap Y = \emptyset$,

$$\mathbb{E}_X^\Lambda(AB) = \mathbb{E}_{X\cup Y}^\Lambda(A) \cdot \mathbb{E}_{X\cup Y^c}^\Lambda(B) = \mathbb{E}_{X\cup Z^c}^\Lambda(A) \cdot \mathbb{E}_{X\cup Z}^\Lambda(B). \tag{4.14}$$

(iii) Given $X \subseteq \Lambda_1 \subseteq \Lambda_2$ and $A \in \mathcal{A}_{\Lambda_1}^+$,

$$\mathbb{E}_X^{\Lambda_1}(A) = \mathbb{E}_X^{\Lambda_2}(A). \tag{4.15}$$

PROOF. For (i), since $\Lambda\setminus X$ is the disjoint union of $\Lambda\setminus(X\cup Y)$ and $\Lambda\setminus(X\cup Y^c)$, (4.12) implies

$$\mathbb{E}_X^\Lambda = \mathbb{E}_{X\cup Y^c}^\Lambda \circ \mathbb{E}_{X\cup Y}^\Lambda. \tag{4.16}$$

Analogously, $\mathbb{E}_Y^\Lambda = \mathbb{E}_{X\cup Y}^\Lambda \circ \mathbb{E}_{X^c\cup Y}^\Lambda$. Using that $\mathbb{E}_{X\cup Y}^\Lambda$ is a projection, we find

$$\mathbb{E}_X^\Lambda \circ \mathbb{E}_Y^\Lambda = \mathbb{E}_X^\Lambda \circ \mathbb{E}_{X^c\cup Y}^\Lambda.$$

The result follows from noticing that $\Lambda \setminus (X \cap Y)$ is the disjoint union of $\Lambda \setminus X$ and $\Lambda \setminus (X^c \cup Y)$.

For (ii), we first use (4.16) to rewrite $\mathbb{E}_X(AB)$. Then, using the commutation properties of even observables we find

$$\begin{aligned}\mathbb{E}_X^\Lambda(AB) &= \frac{1}{4^{|\Lambda\setminus X|}} \sum_{\substack{\alpha\in I_{X^c\cap Y^c}\\ \beta\in I_{X^c\cap Y}}} u(\beta)^* u(\alpha)^* A u(\alpha)\, B u(\beta) \\ &= \frac{1}{4^{|\Lambda\setminus X|}} \sum_{\substack{\alpha\in I_{X^c\cap Y^c}\\ \beta\in I_{X^c\cap Y}}} u(\alpha)^* A u(\alpha)\, u(\beta)^* B u(\beta) \\ &= \mathbb{E}_{X\cup Y}^\Lambda(A)\cdot \mathbb{E}_{X\cup Y^c}^\Lambda(B).\end{aligned}$$

An analogous argument holds for showing the final equality in (4.14).

Finally, (iii), since $A' = \mathbb{E}_X^{\Lambda_1}(A) \in \mathcal{A}_X^+$, and $\mathcal{A}_X^+ \subseteq \mathcal{A}_{\Lambda_1}^+ \subseteq \mathcal{A}_{\Lambda_2}^+$, it is clear that

$$\mathbb{E}_X^{\Lambda_2}(A) = \mathbb{E}_{\Lambda_1}^{\Lambda_2}(A') = A'.$$

□

Recall that the motivation for introducing these conditional expectations was to produce local approximations of global observables. Since we consider even interactions, it is sufficient to just localize even observables. The final result we provide shows that $\mathbb{E}_X^\Lambda$ does indeed produce local approximations of even observables.

LEMMA 4.3. *Let $A \in \mathcal{A}_\Lambda^+$, $X \subset \Lambda$, and $\epsilon > 0$. If $\|[A,B]\| \leq \epsilon\|B\|$ for all $B \in \mathcal{A}_{X^c}$, then there exists $A' \in \mathcal{A}_X^+$ such that $\|A - A'\| \leq \epsilon$.*

This lemma is proved by a straightforward estimate using $A' = \mathbb{E}_X^\Lambda(A)$:

$$\|A - \mathbb{E}_X^\Lambda(A)\| = \left\| \frac{1}{4^{|\Lambda\setminus X|}} \sum_{\alpha\in I_{\Lambda\setminus X}} u(\alpha)^*[u(\alpha), A] \right\|. \tag{4.17}$$

Since $u(\alpha) \in \mathcal{A}_{\Lambda\setminus X}$ and is unitary, for all $\alpha \in I_{\Lambda\setminus X}$, by the assumption in the lemma we have the bound

$$\|[A, u(\alpha)]\| \leq \epsilon, \tag{4.18}$$

which is sufficient to prove lemma.

Since A is even, using the Krauss forms of $\mathbb{E}_X^\Lambda$, see (4.6), and of $\mathbb{F}_X^\Lambda$, see (4.19) below, one can easily check $\mathbb{E}_X^\Lambda(A) = \mathbb{F}_X^\Lambda(A)$. Note, however, that the Krauss form for $\mathbb{F}_X^\Lambda$ on all of $\mathcal{A}_\Lambda$ involves global operators and, as such, is not suitable for the above argument. To present a Krauss form for $\mathbb{F}_X^\Lambda$, we first define $\pi(0) = \pi(3) = 1$, $\pi(1) = \pi(2) = -1$ and for $\alpha \in I_Y, \pi(\alpha) = \prod_{y\in Y} \pi(\alpha_y)$, and denote by $I_Y^\pm$ the sets of $\alpha \in I_Y$ for which $\pi(\alpha) = \pm 1$. Then we have

$$\mathbb{F}_X^\Lambda(A) = \frac{1}{4^{|\Lambda\setminus X|}} \sum_{\alpha\in I_{\Lambda\setminus X}} \tilde{u}(\alpha)^* A \tilde{u}(\alpha), \quad A \in \mathcal{A}_\Lambda, \tag{4.19}$$

with

$$\tilde{u}(\alpha) = \begin{cases} u(\alpha) & \text{if } \alpha \in I_{\Lambda\setminus X}^+ \\ \theta_X u(\alpha) & \text{if } \alpha \in I_{\Lambda\setminus X}^- \end{cases}. \tag{4.20}$$

Clearly, the support of $\theta_X u(\alpha)$ is all of Λ and there is no obvious way to apply a Lieb-Robinson bound to estimate a commutator with it.

5. Martingale method for lattice fermions

Before we address the question of stability of a spectral gap in the presence of perturbations, we would like to demonstrate the existence of a wealth of models for which one can prove the existence of a spectral gap. These models can then serve as the 'unperturbed model.'

As in the situation of quantum spin systems, it is helpful to consider frustration-free models. Examples of frustration-free fermion models in one dimension are easily found by making use of the Jordan-Wigner transformation [**25**], to translate the abundance of frustration-free quantum spin chains into fermion Hamiltonians. An example of this type is Kitaev's Majorana chain [**24**] and also the path of frustration-free fermion chains connecting Kitaev's Majorana chain to a family of interacting fermion chains introduced by Katsura et al. [**23**].

In Section 6 we will also introduce a class of frustration-free fermion models in arbitrary dimension that have a quasi-free ground state and a spectral gap above it. In the latter case, for the quadratic Hamiltonians associated to these models, one can often determine a lower bound for the spectral gap by inspection. For models with interactions of a more general form, however, proving a volume-independent lower bound for the spectral gap above the ground state is generally a challenging problem. For spin systems the martingale method has often been applied to establish a nonzero spectral gap above the ground state energy [**7, 15, 33**]. In this section we give a formulation of the method suited for lattice fermions. Although the tensor product structure of quantum spin systems is not available, under general conditions, one can still obtain the commutation relations required for the method. The approach we introduce here was first used in [**45**].

Let Λ be finite and $H_\Lambda \in \mathcal{A}_\Lambda^+$ be the lattice fermion Hamiltonian acting on a fermionic Fock space. Note further that for any self-adjoint, even observable $A \in \mathcal{A}_\Lambda^+$ that the spectral projections associated to A also belong to $\mathcal{A}_\Lambda^+$. This follows immediately from the fact that the spectral projections can be expressed as a polynomial of A. From this, we conclude that the spectral projections of two self-adjoint even observables with disjoint support commute.

Now consider an increasing sequence of non-negative Hamiltonians

$$H_0 \leq H_1 \leq \ldots \leq H_N$$

with $H_0 = 0$, $H_N = H_\Lambda$, and $H_n \in \mathcal{A}_\Lambda^+$ for all $0 \leq n \leq N$. More generally, it is sufficient to consider an increasing, non-negative sequence of Hamiltonians for which

$$cH_N \leq H_\Lambda - E_\Lambda \mathbb{1} \leq CH_N \tag{5.1}$$

where $c, C > 0$ and E_Λ is the ground state energy of H_Λ. When the conditions of the martingale method are met, it produces a lower bound for the spectral gap above the ground state energy of H_N.

To apply the martingale method, it is necessary that the Hamiltonians H_n each have a non-trivial kernel. Since these Hamiltonians are non-negative and increasing, the kernels $\mathcal{G}_n = \ker H_n$ form a decreasing sequence of subspaces:

$$\mathcal{H}_\Lambda = \mathcal{G}_0 \supset \mathcal{G}_1 \supset \mathcal{G}_2 \cdots \supset \mathcal{G}_N = \ker(H_N).$$

In order to state the assumptions of the martingale method we define

$$h_n = H_n - H_{n-1}, \text{ for } n = 1, \ldots, N. \tag{5.2}$$

Clearly, (5.2) implies $H_n = \sum_{k=1}^{n} h_k$, and H_n is increasing if and only if $h_k \geq 0$, for all $k = 1, \ldots, N$. Furthermore, define G_n and g_n to be the orthogonal projections onto $\ker H_n$ and $\ker h_n$, respectively, and let

$$E_n = \begin{cases} \mathbb{1} - G_1 & \text{if } n = 0 \\ G_n - G_{n+1} & \text{if } 1 \leq n \leq N-1 \\ G_N & \text{if } n = N \end{cases} . \tag{5.3}$$

It is easy to verify that the E_n are a mutually orthogonal family of orthogonal projections that form a resolution of the identity, i.e., $E_n^* = E_n$, $E_n E_m = \delta_{n,m} E_n$, and $\sum_{n=0}^{N} E_n = \mathbb{1}$.

Assumptions for the Martingale Method:

(i) There is a constant $\gamma > 0$ such that $h_n \geq \gamma(\mathbb{1} - g_n)$ for $1 \leq n \leq N$.

(ii) There is an integer $\ell \geq 0$ such that whenever $0 \leq n \leq N-1$ and $k \notin [n-\ell, n]$, $[E_k, g_{n+1}] = 0$.

(iii) There exists a positive $\epsilon < 1/\sqrt{\ell + 1}$, such that $E_n g_{n+1} E_n \leq \epsilon^2 E_n$, for $0 \leq n \leq N-1$.

THEOREM 5.1 (Martingale Method). *Suppose that Assumptions (i)–(iii) hold for a sequence of Hamiltonians H_n, $n = 0, \ldots, N$ as described above. If $\psi \in \mathcal{H}_\Lambda$ such that $G_N \psi = 0$, then*

$$\langle \psi, H_N \psi \rangle \geq \gamma (1 - \epsilon\sqrt{1+\ell})^2 \|\psi\|^2.$$

PROOF. By assumption $E_N \psi = G_N \psi = 0$. Hence

$$\|\psi\|^2 = \sum_{n=0}^{N-1} \|E_n \psi\|^2. \tag{5.4}$$

Given this, for any $0 \leq n \leq N-1$, we have that

$$\begin{aligned} \|E_n \psi\|^2 &= \langle \psi, (\mathbb{1} - g_{n+1}) E_n \psi \rangle + \langle \psi, g_{n+1} E_n \psi \rangle \\ &= \langle \psi, (\mathbb{1} - g_{n+1}) E_n \psi \rangle + \langle \psi, \left(\sum_{k=n_\ell}^{n} E_k \right) g_{n+1} E_n \psi \rangle, \end{aligned} \tag{5.5}$$

where we have set $n_\ell = \max(0, n - \ell)$. For the last equality above, we inserted the resolution of the identity, used Assumption (ii), and applied the mutual orthogonality of the projections E_n. Two applications of the inequality

$$|\langle \varphi_1, \varphi_2 \rangle| \leq \frac{1}{2c} \|\varphi_1\|^2 + \frac{c}{2} \|\varphi_2\|^2, \quad \varphi_1, \varphi_2 \in \mathcal{H}, c > 0, \tag{5.6}$$

now show that

$$\begin{aligned} \|E_n \psi\|^2 \leq\ & \frac{1}{2c_1} \langle \psi, (\mathbb{1} - g_{n+1}) \psi \rangle + \frac{c_1}{2} \langle \psi, E_n \psi \rangle \\ & + \frac{1}{2c_2} \langle \psi, E_n g_{n+1} E_n \psi \rangle + \frac{c_2}{2} \langle \psi, \left(\sum_{k=n_\ell}^{n} E_k \right)^2 \psi \rangle. \end{aligned}$$

We estimate each of the four terms above as follows. For the first term, we apply Assumption (i), whereas the second term is immediately combined with the left-hand-side. With the third term, we use Assumption (iii), and for the fourth term, we again use the mutual orthogonality of the E_n. After some reordering, we have

$$(5.7)\qquad (1-\frac{c_1}{2}-\frac{\epsilon^2}{2c_2})\|E_n\psi\|^2-\frac{c_2}{2}\sum_{k=n_\ell}^{n}\|E_k\psi\|^2\le\frac{1}{2c_1\gamma}\langle\psi,h_{n+1}\psi\rangle.$$

Summing both sides of (5.7) from $n=0,\ldots,N-1$, yields

$$(5.8)\qquad \langle\psi,H_N\psi\rangle\ge 2c_1\gamma\left[1-\frac{c_1}{2}-\frac{\epsilon^2}{2c_2}-\frac{c_2(1+\ell)}{2}\right]\|\psi\|^2.$$

Maximizing this lower bound leads to the choice of $c_1=1-\epsilon\sqrt{1+\ell}$ and $c_2=\epsilon/\sqrt{1+\ell}$. This produces the inequality stated in the theorem. □

6. Discussion: stability, examples, and the spectral flow

Recently, methods were introduced, see e.g. [**11**, **30**], which allow for a proof of stability for gapped ground states of frustration-free quantum spin systems satisfying a Local Topological Quantum Order (LTQO) condition. Here, the term stability refers to the property that the there is a lower bound of the spectral gap, uniform in Λ, for finite-volume Hamiltonians of the form

$$H_\Lambda(s)=\sum_{X\subset\Lambda}\Phi(X)+s\Psi(X),$$

where Ψ is any other short-range interaction and $|s|<s_0$ for some $s_0>0$. Roughly speaking, this LTQO condition on Φ amounts to a precise formulation of the notion that, in this situation, degenerate ground states cannot be distinguished by local operations. In a forth-coming work [**40**], a systematic refinement of these techniques will be presented which generalize previous results. For example, a stability result for gapped ground state phases of quantum spin models with a spontaneously broken discrete symmetry is contained in [**40**].

In this section, we give an indication of how analogous results extend to models of lattice fermions. More details on this extension will be given in [**41**]. We begin by defining general frustration-free fermion models. Next, we give a large class of examples of gapped, frustration-free fermion which may serve as unperturbed models for stability results. Then, we introduce several types of symmetry, which may or may not be broken in a given model, and discuss briefly how the LTQO condition should be modified in this context. Lastly, we discuss the construction of the spectral flow automorphism which is main tool of analysis in proofs of stability.

In the same way as for quantum spin systems, an even fermion interaction Φ on Γ is defined to be *frustration-free* if Φ is finite-range and for $\Lambda\in\mathcal{P}_0(\Gamma)$ we have

$$(6.1)\qquad \inf\operatorname{spec}(H_\Lambda)=\sum_{X\subset\Lambda}\inf\operatorname{spec}(\Phi(X)).$$

It is standard to normalize the interactions so that each interaction term is non-negative, i.e. $\Phi(X)\ge 0$ for each $X\in\mathcal{P}_0(\Gamma)$, and further that the ground state energy of H_Λ vanishes. In this case, a ground state eigenvector for H_Λ is necessarily in the kernel of each of its terms. Since we do not assume the existence of simultaneous eigenvectors for other energy values, this set-up does not imply that the interaction terms commute.

We now consider two classes of gapped frustration-free fermion models with finite-range interactions. As already indicated in Section 5, a first class of examples are those one-dimensional fermion systems with finite-range interactions that are mapped to a gapped, frustration-free quantum spin chain by the Jordan-Wigner transformation. Note that even interaction terms $\Phi([a,b]) \in \mathcal{A}^+_{[a,b]}$ are mapped into $\tilde{\Phi}([a,b]) \in \mathcal{A}^{\rm spin}_{[a,b]}$. Here *evenness* of the interactions is important to preserve its finite-range property. It also implies a local discrete symmetry for the quantum spin model; more on this below. The properties of the ground states of such models can be used to verify the conditions of the martingale method discussed in Section 5. This provides a wide class of gapped fermion systems in one-dimension. Another approach for constructing frustration-free one-dimensional fermion systems defines fermionic Matrix Product States using graded vector spaces [**14**].

A second class of interesting frustration-free fermion models with a spectral gap is obtained by considering quasi-free systems with two (or more) bands separated by a gap $\gamma > 0$. For clarity, we will use two sets of fermion creation and annihilation operators $\{b_k, b_k^* \mid k \in B\}$ and $\{c_l, c_l^* \mid l \in C\}$, labeled by index sets B and C, respectively, which together span $\mathcal{A}_\Gamma$, where Γ is a lattice of arbitrary dimension. These operators are defined as follows. There are two subsets of $\ell^2(\Gamma)$, $\{f_k | k \in B\}$, and $\{g_l \mid l \in C\}$ such that:

(i) $\|f_k\| = 1, \|g_l\| = 1$, for all $k \in B$ and $l \in C$;

(ii) $\operatorname{span}(\{f_k \mid k \in B\} \cup \{g_l \mid l \in C\})$ is dense in $\ell^2(\Gamma)$;

(iii) $\langle f_k, g_l \rangle = 0$, for all $k \in B$ and $l \in C$;

(iv) there exists $R \geq 0$, and $x_k, y_l \in \Gamma$, such that $\operatorname{supp} f_k \subset B_{x_k}(R)$, and $\operatorname{supp} g_l \subset B_{y_l}(R)$, for all $k \in B$ and $l \in C$.

Then, the new creation and annihilation operators are defined by

$$b_k^* = \sum_{x \in \Gamma} f_k(x) a_x^* \qquad b_k = \sum_{x \in \Gamma} \overline{f}_k(x) a_x \tag{6.2}$$

$$c_l^* = \sum_{x \in \Gamma} g_l(x) a_x^* \qquad c_l = \sum_{x \in \Gamma} \overline{g}_l(x) a_x. \tag{6.3}$$

Next, define an interaction Φ by setting

$$\Phi(B_{x_k}(R)) = \mathbb{1} - b_k^* b_k, \quad \Phi(B_{y_l}(R)) = c_l^* c_l, \tag{6.4}$$

and $\Phi(X) = 0$ if X is not a ball of radius R centered at a site x_k or y_l, for any $k \in B$, or $l \in C$. In a standard application, the functions f_k and g_l are the orbitals in the valence and conduction band, respectively. It is then straightforward to construct a quasi-free state ω_Λ on $\mathcal{A}_\Lambda$ satisfying

$$\omega_\Lambda(b_k^* b_k) = 1, \quad \omega_\Lambda(c_l^* c_l) = 0. \tag{6.5}$$

Clearly, the Hamiltonians

$$H_\Lambda = \sum_{X \in \subset \Lambda} \Phi(X) \tag{6.6}$$

are non-negative. Since (6.5) implies that $\omega_\Lambda(H_\Lambda) = 0$, these models are frustration-free. The orthogonal complement to the ground state given by ω_Λ, is spanned by the Fock space vectors with at least one hole in the valence band or one particle in the conduction band. In this case, the gap above the ground state energy is $\gamma = 1$.

An interesting class of examples of models of this type are the so-called flat-band Hamiltonians studied by Mielke and Tasaki [**31**]. A thorough study of the conditions under which there exists a spanning set of compactly supported orbitals for a band structure was recently carried out by Read [**42**]. He shows that certain band structures in two or more dimensions, which yield ground states with certain types of topological order, cannot be spanned by compactly supported orbitals. It is also worth noting that if the goal is to show stability of the gap under small perturbations, one can also deal with some cases in which the bands are spanned by orbitals that are not compactly supported. If the orbitals are well approximated by compactly supported functions and the gap is not too small, then one can treat the error as part of the perturbation.

We now turn to symmetry. In the classification of gapped ground state phases it is often important to consider symmetries of the model. These may or may not be spontaneously broken in the ground state. Common examples of the symmetries we will consider include the following:

(1) *Parity of the fermion number:* We will always assume that all terms in the interaction are even: $\Phi(X) \in \mathcal{A}_X^+$. This means that all interaction terms supported in $X \in \mathcal{P}_0(\Gamma)$ commute with $(-1)^{N_X}$.
(2) *Local symmetries:* A local symmetry is described by a representation of a finite group G consisting of automorphisms, β_g^x for $g \in G$, acting on $\mathcal{A}_{\{(x,k)|k=1,\dots,n\}}$, for each $x \in \Gamma$. This invariance can equivalently be described by the commutation of the interaction terms with a unitary representation of G: $U_X(g)\Phi(X) = \Phi(X)U_X(g)$. The parity of the fermion number is an example of such a discrete local symmetry.
(3) *Translation invariance:* Often Γ is a lattice such as $\mathcal{L} = \mathbb{Z}^\nu$ or $\mathcal{L} = \mathbb{Z}^\nu/(L\mathbb{Z}^\nu)$, or Γ contains a lattice as a factor: $\Gamma = \mathcal{L} \times \Gamma_1$. In either case, there is a natural action of $\mathcal{L}$ on Γ which, for simplicity, we will denote by addition. Then, $\mathcal{L}$ acts on $\mathcal{A}_\Gamma$ as a group of translation automorphisms $\beta_x, x \in \mathcal{L}$. Translation invariance of the interaction is then expressed by $\beta_x(\Phi(X)) = \Phi(X + x)$.
(4) *Space inversion and other lattice symmetries:* Besides translations, Γ may often possess other discrete symmetries, such as inversion ($x \mapsto -x$) or rotation by certain angles.
(5) *Time reversal invariance:* If there is a basis in Fock space with respect to which the matrix of the Hamiltonian is real, meaning that there is a complex conjugation with which it commutes, we have a symmetry between the forward and backward dynamics, i.e., $\tau_t(\beta(A)) = \beta(\tau_{-t}(A))$, for the corresponding anti-automorphism.

When the goal is to prove stability of the gapped ground state phases in cases that allow for a spontaneously broken discrete symmetry, one is led to assume a slight modification of the LTQO condition introduced in [**11**]. For example, it no longer makes sense to assume LTQO for arbitrary local observables; one should restrict attention to those observables that preserve the symmetry. Mimicking methods in [**40**], one can prove that the spectral gap and the structure of the ground state phases are stable under sufficiently small perturbations of the interaction, if one additionally assumes that perturbations preserve the symmetry. A precise statement of this result will appear in [**41**].

Finally, we turn to the main tool used in the recent proofs of stability of gapped phases [**11, 30**], the so-called *spectral flow* or *quasi-adiabatic continuation* [**8, 22**]. To define the spectral flow, consider a one-parameter family of interactions Φ_s : $\mathcal{P}_0(\Gamma) \to \mathcal{A}_\Gamma$, such that for each $X \in \mathcal{P}_0(\Gamma)$, $\Phi_s(X)$ is differentiable with respect to $s \in [0, 1]$. In this case, Hamiltonians of the form $H_\Lambda(s) = \sum_{X \subseteq \Lambda} \Phi_s(X)$ are defined on any finite volume $\Lambda \in \mathcal{P}_0(\Gamma)$.

The situation of interest is where the spectrum of $H_\Lambda(s)$ is composed of two parts, $\Sigma_1(s)$ and $\Sigma_2(s)$, separated by a gap bounded below by a constant $\gamma > 0$, for all $s \in [0,1]$. Let $P(s)$ denote the spectral projection of $H_\Lambda(s)$ corresponding to the set $\Sigma_1(s)$. Due to the spectral gap assumption, general results imply that $P(s)$ is unitarily equivalent to $P(0)$ for all $s \in [0,1]$, i.e., there exists a curve of unitaries $U(s)$ satisfying

$$P(s) = U(s)P(0)U(s)^*. \tag{6.7}$$

The spectral flow is constructed in terms of a particular choice of unitaries satisfying (6.7).

As discussed e.g. in [**8**], one choice for a family of unitaries satisfying (6.7) is obtained as the solution of the Schrödinger equation for unitaries $U(s)$, with s playing the role of time. The Hamiltonian for this Schrödinger equation, $D_\Lambda(s)$, is defined by an expression of the following form:

$$D_\Lambda(s) = \int_{-\infty}^{\infty} \tau_t^{H_\Lambda(s)}(H'_\Lambda(s))\, W_\gamma(t)\, dt, \tag{6.8}$$

where $W_\gamma(t) \in L^1(\mathbb{R})$ is a well-chosen function which decays faster than any power law as $t \to \pm\infty$. In terms of the corresponding unitaries, the spectral flow automorphism is then defined as $\alpha_s^\Lambda(A) = U(s)^* AU(s)$ for all $A \in \mathcal{A}_\Lambda$.

A crucial technical result, used in all proofs of stability, is the fact that this spectral flow satisfies a Lieb-Robinson bound with a decay function that is explicit. Since the Hamiltonian terms corresponding to (6.8) are not strictly local, such an estimate is not a direct application of known results. With some effort, one can show that if the family of interactions Φ_s and Φ'_s decay sufficiently fast (usually expressed in terms of an F-function that decays at an exponential rate), then $D_\Lambda(s)$ can be realized as a local Hamiltonian associated to an interaction Ψ_Λ

$$D_\Lambda(s) = \sum_{X \subseteq \Lambda} \Psi_\Lambda(X, s). \tag{6.9}$$

Moreover, there exists an F-function, denoted by F_Ψ, for which $\|\Psi_\Lambda(s)\|_{F_\Psi} < \infty$ uniformly in the finite volume Λ. As the explicit F_Ψ decays sub-exponentially, one obtains Lieb-Robinson bounds, i.e. locality estimates, for the spectral flow that decay as fast.

In the construction of the interaction $\Psi_\Lambda(\cdot, s)$, for fermions, one uses the conditional expectation discussed in Section 4 in combination with the Lieb-Robinson bounds of Section 3. As far as we are aware, the first use of a conditional expectation to construct an interaction Φ from a set of local Hamiltonians of a quantum system appeared in [**2**]. To decompose $D_\Lambda(s)$ into strictly local terms, it is crucial that the initial interaction terms $\Phi(X, s)$ be even. In this case, $\Psi_\Lambda(\cdot, s)$ is also even. Of course, for models with other symmetries, one must check that this localizing operation preserves the symmetry. Given that this is the case, the proof of stability then proceeds in the same way as in the case of quantum spin systems. Explicit estimates and further applications will be given in [**41**].

Acknowledgements

We thank Detlev Buchholz for pointing out to us the use of a conditional expectation in the work of Araki and Moriya [**4**].

References

[1] R. Alicki and M. Fannes, *Quantum dynamical systems*, Oxford University Press, Oxford, 2001. MR2011940

[2] H. Araki, *On KMS states of a C^* dynamical system*, C^*-algebras and applications to physics (Proc. Second Japan-USA Sem., Los Angeles, Calif., 1977), Lecture Notes in Math., vol. 650, Springer, Berlin, 1978, pp. 66–84. MR504753

[3] H. Araki, *Conditional expectations relative to a product state and the corresponding standard potentials*, Comm. Math. Phys. **246** (2004), no. 1, 113–132, DOI 10.1007/s00220-003-1028-9. MR2044893

[4] H. Araki and H. Moriya, *Equilibrium statistical mechanics of fermion lattice systems*, Rev. Math. Phys. **15** (2003), no. 2, 93–198, DOI 10.1142/S0129055X03001606. MR1972301

[5] S. Bachmann, W. De Roeck, and M. Fraas, *The adiabatic theorem and linear response theory for extended quantum systems*, arXiv:1705.02838, 2017.

[6] S. Bachmann, W. De Roeck, and M. Fraas, *Adiabatic theorem for quantum spin systems*, Phys. Rev. Lett. **119** (2017), 060201, arXiv:1612.01505.

[7] S. Bachmann, E. Hamza, B. Nachtergaele, and A. Young, *Product vacua and boundary state models in d-dimensions*, J. Stat. Phys. **160** (2015), no. 3, 636–658, DOI 10.1007/s10955-015-1260-7. MR3366096

[8] S. Bachmann, S. Michalakis, B. Nachtergaele, and R. Sims, *Automorphic equivalence within gapped phases of quantum lattice systems*, Comm. Math. Phys. **309** (2012), no. 3, 835–871, DOI 10.1007/s00220-011-1380-0. MR2885611

[9] F. G. S. L. Brandão, T. S. Cubitt, A. Lucia, S. Michalakis, and D. Perez-Garcia, *Area law for fixed points of rapidly mixing dissipative quantum systems*, J. Math. Phys. **56** (2015), no. 10, 102202, 15, DOI 10.1063/1.4932612. MR3407511

[10] O. Bratteli and D. W. Robinson, *Operator algebras and quantum statistical mechanics. 2*, 2nd ed., Texts and Monographs in Physics, Springer-Verlag, Berlin, 1997. Equilibrium states. Models in quantum statistical mechanics. MR1441540

[11] S. Bravyi, M. B. Hastings, and S. Michalakis, *Topological quantum order: stability under local perturbations*, J. Math. Phys. **51** (2010), no. 9, 093512, 33, DOI 10.1063/1.3490195. MR2742836

[12] S. Bravyi, M. B. Hastings, and F. Verstraete, *Lieb-Robinson bounds and the generation of correlations and topological quantum order*, Phys. Rev. Lett. **97** (2006), 050401.

[13] J.-B. Bru and W. de Siqueira Pedra, *Lieb-Robinson bounds for multi-commutators and applications to response theory*, SpringerBriefs in Mathematical Physics, vol. 13, Springer, Cham, 2017. MR3586437

[14] N. Bultinck, D.J. Williamson, J. Haegeman, and F. Verstraete, *Fermionic matrix product states and one-dimensional topological phases*, Phys. Rev. B **95** (2017), 075108, arXiv:1610.07849v1.

[15] M. Fannes, B. Nachtergaele, and R. F. Werner, *Finitely correlated states on quantum spin chains*, Comm. Math. Phys. **144** (1992), no. 3, 443–490. MR1158756

[16] A. Giuliani, V. Mastropietro, and M. Porta, *Universality of the Hall conductivity in interacting electron systems*, Comm. Math. Phys. **349** (2017), no. 3, 1107–1161, DOI 10.1007/s00220-016-2714-8. MR3602824

[17] E. Hamza, R. Sims, and G. Stolz, *Dynamical localization in disordered quantum spin systems*, Comm. Math. Phys. **315** (2012), no. 1, 215–239, DOI 10.1007/s00220-012-1544-6. MR2966945

[18] M. B. Hastings, *Lieb-Schultz-Mattis in higher dimensions*, Phys. Rev. B **69** (2004), 104431.

[19] M. B. Hastings, *An area law for one-dimensional quantum systems*, J. Stat. Mech. Theory Exp. **8** (2007), P08024, 14. MR2338267

[20] M. B. Hastings and T. Koma, *Spectral gap and exponential decay of correlations*, Comm. Math. Phys. **265** (2006), no. 3, 781–804, DOI 10.1007/s00220-006-0030-4. MR2231689

[21] M. B. Hastings and S. Michalakis, *Quantization of Hall conductance for interacting electrons on a torus*, Comm. Math. Phys. **334** (2015), no. 1, 433–471, DOI 10.1007/s00220-014-2167-x. MR3304282

[22] M. B. Hastings and X. G. Wen, *Quasi-adiabatic continuation of quantum states: The stability of topological ground-state degeneracy and emergent gauge invariance*, Phys. Rev. B **72** (2005), 045141.

[23] H. Katsura, D. Schuricht, and M. Takahashi, *Exact ground states and topological order in interacting Kitaev/Majorana chains*, Phys. Rev. B **92** (2015), 115137.

[24] A. Yu. Kitaev, *Unpaired Majorana fermions in quantum wires*, Phys. Usp. **44** (2001), 131.

[25] E. Lieb, T. Schultz, and D. Mattis, *Two soluble models of an antiferromagnetic chain*, Ann. Physics **16** (1961), 407–466, DOI 10.1016/0003-4916(61)90115-4. MR0136392

[26] E. H. Lieb and D. W. Robinson, *The finite group velocity of quantum spin systems*, Comm. Math. Phys. **28** (1972), 251–257. MR0312860

[27] V. Mastropietro, *Localization of interacting fermions in the aubry-andre' model*, Phys. Rev. Lett. **115** (2015), 180401.

[28] V. Mastropietro, *Localization in the ground state of an interacting quasi-periodic fermionic chain*, Comm. Math. Phys. **342** (2016), no. 1, 217–250, DOI 10.1007/s00220-015-2498-2. MR3455150

[29] V. Mastropietro, *Localization in interacting fermionic chains with quasi-random disorder*, Comm. Math. Phys. **351** (2017), no. 1, 283–309, DOI 10.1007/s00220-016-2773-x. MR3613505

[30] S. Michalakis and J. P. Zwolak, *Stability of frustration-free Hamiltonians*, Comm. Math. Phys. **322** (2013), no. 2, 277–302, DOI 10.1007/s00220-013-1762-6. MR3077916

[31] A. Mielke and H. Tasaki, *Ferromagnetism in the Hubbard model. Examples from models with degenerate single-electron ground states*, Comm. Math. Phys. **158** (1993), no. 2, 341–371. MR1249599

[32] A. Moon and B. Nachtergaele, *Stability of the spectral gap for open chains of spins or fermions*, In preparation.

[33] B. Nachtergaele, *The spectral gap for some spin chains with discrete symmetry breaking*, Comm. Math. Phys. **175** (1996), no. 3, 565–606. MR1372810

[34] B. Nachtergaele, Y. Ogata, and R. Sims, *Propagation of correlations in quantum lattice systems*, J. Stat. Phys. **124** (2006), no. 1, 1–13, DOI 10.1007/s10955-006-9143-6. MR2256615

[35] B. Nachtergaele, V. B. Scholz, and R. F. Werner, *Local approximation of observables and commutator bounds*, Operator methods in mathematical physics, Oper. Theory Adv. Appl., vol. 227, Birkhäuser/Springer Basel AG, Basel, 2013, pp. 143–149, DOI 10.1007/978-3-0348-0531-5_8. MR3050163

[36] B. Nachtergaele and R. Sims, *Lieb-Robinson bounds and the exponential clustering theorem*, Comm. Math. Phys. **265** (2006), no. 1, 119–130, DOI 10.1007/s00220-006-1556-1. MR2217299

[37] B. Nachtergaele and R. Sims, *A multi-dimensional Lieb-Schultz-Mattis theorem*, Comm. Math. Phys. **276** (2007), no. 2, 437–472, DOI 10.1007/s00220-007-0342-z. MR2346396

[38] B. Nachtergaele and R. Sims, *Locality estimates for quantum spin systems*, New Trends in Mathematical Physics. Selected contributions of the XVth International Congress on Mathematical Physics (V. Sidovaricius, ed.), Springer Verlag, 2009, pp. 591–614.

[39] B. Nachtergaele and R. Sims, *On the dynamics of lattice systems with unbounded on-site terms in the Hamiltonian*, arXiv:1410.8174, 2014.

[40] B. Nachtergaele, R. Sims, and A. Young, *Quasi-locality bounds for quantum lattice systems and perturbations of gapped ground states*, in preparation.

[41] B. Nachtergaele, R. Sims, and A. Young, *Stability of gapped phases of fermionic lattice systems*, in preparation.

[42] N. Read, *Compactly-supported Wannier functions and algebraic K-theory*, Phys. Rev. B **95** (2017), 115309.

[43] M. Reed and B. Simon, *Methods of modern mathematical physics. II. Fourier analysis, self-adjointness*, Academic Press [Harcourt Brace Jovanovich, Publishers], New York-London, 1975. MR0493420

[44] J. Tomiyama, *On the projection of norm one in W^*-algebras*, Proc. Japan Acad. **33** (1957), 608–612. MR0096140

[45] A. M. Young, *Spectral Properties of Multi-Dimensional Quantum Spin Systems*, ProQuest LLC, Ann Arbor, MI, 2016. Thesis (Ph.D.)–University of California, Davis. MR3611147

Department of Mathematics and Center for Quantum Mathematics and Physics, University of California, Davis, Davis, California 95616
Email address: bxn@math.ucdavis.edu

Department of Mathematics, University of Arizona, Tuscon, Arizona 85721
Email address: rsims@math.arizona.edu

Department of Mathematics, University of Arizona, Tuscon, Arizona 85721
Email address: amyoung@math.arizona.edu

Contemporary Mathematics
Volume **717**, 2018
https://doi.org/10.1090/conm/717/14444

Solvable models of interacting n-particle systems on quantum graphs

Jens Bolte and George Garforth

ABSTRACT. We introduce n-particle quantum graphs with singular two-particle interactions in such a way that eigenfunctions can be given in the form of a Bethe ansatz. We show that this leads to a secular equation characterising eigenvalues of the Hamiltonian that is based on a finite-dimensional determinant. These findings generalise previous results about two-particle quantum graphs.

1. Introduction

A quantum graph is metric graph equipped with a differential operator that typically is a self-adjoint realisation of the differential Laplacian. Spectral properties of quantum graphs recently attracted much attention (see, e.g., [**EKK+08**, **BK13a**]). They often mimic corresponding properties of Laplacians on manifolds and have become widely used models in quantum chaos [**KS99b**, **GS06**]. The popularity of quantum graph models is partly due to the fact that eigenvalues of compact quantum graphs can be characterised in terms of a secular equation involving a finite-dimensional determinant [**KS99b**]. This leads to a Gutzwiller-like trace formula that is an identity rather than an asymptotic relation [**KS99b**, **KN05**, **BE09**]. Moreover, numerical calculations, with the aim of analysing the distribution of eigenvalues, can be performed very efficiently [**KS99b**].

More recently, many-particle quantum systems on graphs with singular interactions have been introduced [**BK13c**, **BK13b**], primarily with the intention to study the existence or absence of Bose-Einstein condensation [**BK14**, **BK16**]. The latter mostly requires knowledge about low-lying eigenvalues, which can be gained without a secular equation. As the classical configuration space of $n \geq 2$ particles on a graph is n-dimensional, one would, in general, not expect to find a secular equation based on a finite-dimensional determinant, but rather an approximate scheme involving a truncated determinant as, e.g., developed in [**DS92**]. However, it is well known that particular many-particle quantum systems, such as the Lieb-Liniger gas on an interval or a circle [**LL63**, **Gau71**], are solvable by making use of an n-particle Bethe ansatz for the eigenfunctions [**Bet31**], which is a *finite* sum of plane waves. In such a case only finitely many unknown coefficients need to be calculated in order to determine an eigenfunction. Although intervals and circles can be viewed as very simple graphs, an important step to generalising the Bethe ansatz

2010 *Mathematics Subject Classification.* Primary 81Q35.

to quantum graphs was done in [**CC07**], where the punctured real line, i.e. a graph with one vertex and two infinite edges, was equipped with a singular two-particle interaction that was specially designed to make a Bethe ansatz possible. In [**BG17**] we extended this construction to two particles on arbitrary compact graphs, with explicit constructions for equilateral star graphs as well as for tetrahedra. In all cases the Bethe ansatz led to a secular equation with a finite-dimensional determinant. We numerically calculated eigenvalues and analysed their distribution finding that, on the scale of the mean level spacing, eigenvalues appear to be uncorrelated. As such an observation would normally be attributed to an integrable classical limit [**BT77**], we see it as a hint of an underlying quantum integrable field theory (as it is the case for the Lieb-Liniger model).

In what follows we generalise the approach to n-particle graphs. We maintain the singular two-particle interactions between any pairs of the n particles and show that the Bethe ansatz can be extended to this situation. We then derive a secular equation, which is our main result.

2. Preliminaries

2.1. One-particle quantum graphs. We denote by Γ a metric graph with finite vertex set $\mathcal{V}$ and finite edge set $\mathcal{E}$. Each edge $e_j \in \mathcal{E}$ is assigned a (finite) length l_j. The one particle Hilbert space is

$$L^2(\Gamma) = \bigoplus_{j=1}^{|\mathcal{E}|} L^2(0, l_j), \tag{2.1}$$

and the Laplacian acting on $\Psi = (\psi_j)_{j=1}^{|\mathcal{E}|} \in H^2(\Gamma)$, with the Sobolev space $H^2(\Gamma)$ defined in analogy to $L^2(\Gamma)$, is

$$-\Delta_1 \Psi = \left(-\psi_j''\right)_{j=1}^{|\mathcal{E}|}. \tag{2.2}$$

Self-adjoint realisations of the Laplacian can be characterised in terms of boundary conditions at the vertices [**KS99a**]. Let

$$\Psi_{\text{bv}} = \left(\psi_1(0), \ldots, \psi_{|\mathcal{E}|}(0), \psi_1(l_1), \ldots, \psi_{|\mathcal{E}|}(l_{|\mathcal{E}|})\right), \tag{2.3}$$

$$\Psi'_{\text{bv}} = \Big(\psi'_1(0), \ldots, \psi'_{|\mathcal{E}|}(0), -\psi'_1(l_1), \ldots, -\psi'_{|\mathcal{E}|}(l_{|\mathcal{E}|})\Big), \tag{2.4}$$

be the vectors of boundary values and let A, B be complex $|\mathcal{E}| \times |\mathcal{E}|$-matrices, such that AB^* is self-adjoint and (A, B) has maximal rank. Then the boundary conditions

$$A\Psi_{\text{bv}} + B\Psi'_{\text{bv}} = 0 \tag{2.5}$$

define a domain of a self-adjoint realisation of the Laplacian as a subspace of $H^2(\Gamma)$. Further details can be found in, e.g., [**KS99a, GS06, BK13a**].

The secular equation for Laplace eigenvalues requires two unitary $2|\mathcal{E}| \times 2|\mathcal{E}|$ matrices. One is the vertex S-matrix

$$S_v(k) = -(A + ikB)^{-1}(A - ikB), \tag{2.6}$$

with $k \in \mathbb{R}$, and

$$T(k; \mathbf{l}) = \begin{pmatrix} 0 & e^{ik\mathbf{l}} \\ e^{ik\mathbf{l}} & 0 \end{pmatrix}, \tag{2.7}$$

where

$$e^{ik\mathbf{l}} = \operatorname{diag}(e^{ikl_j})_{j=1}^{|\mathcal{E}|}. \tag{2.8}$$

Then, $k^2 > 0$ is an eigenvalue of the Laplacian with boundary conditions (2.5), iff

$$\det\left(\mathbb{I} - S_v(k)T(k;\mathbf{l})\right) = 0, \tag{2.9}$$

see [**KS99b, KS06**].

2.2. Weyl group. Before we proceed it is useful to define the symmetric group S_n and the Weyl group $\mathcal{W}_n$, which we use to characterise the symmetries of exactly solvable n-particle systems.

Elements Q of the symmetric group S_n of order $n!$ will be written in terms of generators $T_1, \dots, T_{n-1}$ which satisfy the conditions

(1) $T_iT_i = I$;
(2) $T_iT_j = T_jT_i$ for $|i-j| > 1$;
(3) $T_iT_{i+1}T_i = T_{i+1}T_iT_{i+1}$.

The $n-1$ generators of S_n act on wave numbers $(k_1, \dots, k_n)$ according to

$$T_i(k_1, \dots, k_n) \equiv (k_{T_i(1)}, \dots, k_{T_i(n)}) = (k_1, \dots, k_{i-1}, k_{i+1}, k_i, k_{i+2}, \dots, k_n). \tag{2.10}$$

DEFINITION 2.1. The Weyl group $\mathcal{W}_n$ is defined as a semidirect product of $(\mathbb{Z}/2\mathbb{Z})^n$ with the symmetric group S_n,

$$\mathcal{W}_n := (\mathbb{Z}/2\mathbb{Z})^n \rtimes S_n. \tag{2.11}$$

Some useful and well-known properties of the Weyl group are the following:

PROPOSITION 2.2. *The Weyl group $\mathcal{W}_n$ has order $2^n n!$ and is generated by the elements $T_1, \dots, T_n, R_1$ that satisfy the conditions*

(1) $R_1R_1 = I$;
(2) $T_iT_i = I$;
(3) $T_iT_j = T_jT_i$ *for* $|i-j| > 1$;
(4) $T_iT_{i+1}T_i = T_{i+1}T_iT_{i+1}$;
(5) $R_1T_1R_1T_1 = T_1R_1T_1R_1$;
(6) $R_1T_i = T_iR_1$ *for* $i > 1$.

The n generators of $\mathcal{W}_n$ act on wave numbers $(k_1, \dots, k_n)$ according to (2.10) and

$$R_1(k_1, \dots, k_n) = (-k_1, k_2, \dots, k_n). \tag{2.12}$$

Finally it will be useful to relate the Weyl groups $\mathcal{W}_n$ and $\mathcal{W}_{n-1}$. To this end, it is convenient to define elements

$$R_i = T_{i-1} \dots T_1R_1T_1 \dots T_{i-1} \tag{2.13}$$

so that

$$R_i(k_1, \dots, k_n) = (k_1, \dots, k_{i-1}, -k_i, k_{i+1}, \dots, k_n), \tag{2.14}$$

and the cyclic permutation

$$C_n = T_{n-1}T_{n-2} \dots T_1 \tag{2.15}$$

so that

$$C_n(k_1, k_2, \dots, k_{n-1}, k_n) = (k_n, k_1, \dots, k_{n-2}, k_{n-1}), \tag{2.16}$$

where we note the relation

$$R_n = C_n R_1 C_n^{-1}. \tag{2.17}$$

The Weyl group $\mathcal{W}_n$ can then be written in terms of $\mathcal{W}_{n-1}$ according to

$$\mathcal{W}_n = \left\{ C_n^d (R_n)^j X;\ X \in \mathcal{W}_{n-1},\ d = 0, \ldots, n-1,\ j = 0, 1 \right\}. \tag{2.18}$$

To simplify the notation we denote the action of any $P \in \mathcal{W}_n$ in analogy to (2.10) as

$$P(k_1, \ldots, k_n) = (k_{P(1)}, \ldots, k_{P(n)}). \tag{2.19}$$

2.3. Bosons in an interval. In order to establish some key concepts in the n-particle setting, we begin by presenting the model of n δ-interacting bosons confined to an interval of length l first solved by Gaudin [**Gau71**]. The problem is formulated as a search for solutions of the formal Schrödinger equation

$$\left(-\Delta_n + 2\alpha \sum_{i \neq j} \delta(x_i - x_j) \right) \psi(x_1, \ldots, x_n) = E\psi(x_1, \ldots, x_n) \tag{2.20}$$

with $\alpha \in \mathbb{R}$ and particle positions $x_1, \ldots, x_n$ on the half-line $\mathbb{R}_+ = (0, \infty)$. Here the n-particle Laplacian acts according to

$$-\Delta_n \psi = -\sum_{j=1}^{n} \frac{\partial^2 \psi}{\partial x_j{}^2}. \tag{2.21}$$

For a suitable choice of a domain, the Laplacian becomes self-adjoint and provides a rigorous realisation of the formal operator in (2.20). Imposing bosonic symmetry

$$\psi(x_1, \ldots, x_n) = \psi(x_{Q(1)}, \ldots, x_{Q(n)}), \tag{2.22}$$

for all $Q \in S_n$, this domain can be shown to consist of the set of $n-1$ jump conditions in the derivatives,

$$\left(\frac{\partial}{\partial x_{j+1}} - \frac{\partial}{\partial x_j} \right) \psi|_{x_{j+1}=x_j^+} = \alpha \psi|_{x_{j+1}=x_j^+}, \tag{2.23}$$

for $j \in \{1, \ldots, n-1\}$, and the Dirichlet condition

$$\psi|_{x_1=0} = 0. \tag{2.24}$$

Due to the bosonic symmetry, ψ can be restricted to the subspace

$$d^I = \{(x_1, \ldots, x_n) \in \mathbb{R}_+^n;\ x_1 < \cdots < x_n\}. \tag{2.25}$$

By applying a permutation $Q \in S_n$, it is also defined in all subspaces

$$d^Q = \{(x_1, \ldots, x_n) \in \mathbb{R}_+^n;\ x_{Q(1)} < \cdots < x_{Q(n)}\}, \tag{2.26}$$

and thus in all of $\mathbb{R}_+^n$. In the two-particle setting, interactions between particles means $\mathbb{R}_+^2$ is naturally dissected, along the line $x_1 = x_2$, into two subspaces d^I and d^{T_1}. Here, in the n-particle setting, appropriate dissections result in $n!$ subspaces labeled by elements $Q \in S_n$. Bosonic symmetry establishes equivalence between each of these subspaces so that we need only consider one.

The task is then to construct explicit Laplace eigenfunctions ψ in d^I which satisfy conditions (2.23) and (2.24). The Bethe ansatz method in this context is the assumption that the appropriate ansatz is the sum of possible constituent plane wave states,

$$\psi(x_1,\dots,x_n)=\sum_{P\in\mathcal{W}_n}\mathcal{A}^P e^{i(k_{P(1)}x_1+\dots+k_{P(n)}x_n)}. \tag{2.27}$$

A function of this form is an eigenfunction of the Laplacian with eigenvalue

$$E=\sum_{j=1}^{n}k_j^2. \tag{2.28}$$

The δ-type boundary conditions (2.23) which characterise interactions between particles, imply the relations

$$\mathcal{A}^{PT_i}=s_p(k_{P(i)}-k_{P(i+1)})\mathcal{A}^P, \tag{2.29}$$

for $i\in\{1,\dots,n-1\}$, with

$$s_p(k)=\frac{k-i\alpha}{k+i\alpha} \tag{2.30}$$

for all $P\in\mathcal{W}_n$. The Dirichlet condition (2.24) at the end-point of the half line implies the relation

$$\mathcal{A}^{PR}=-\mathcal{A}^P \tag{2.31}$$

for all $P\in\mathcal{W}_n$. Exact solvability is then assured if relations (2.29) and (2.31) are compatible with the properties of $\mathcal{W}_n$ as prescribed in Proposition 2.2. This amounts only to the requirement $s_p(u)s_p(-u)=1$ which is easily verified.

Until this point, particle positions have been defined on the half-line $\mathbb{R}^+$. Of course, in order to restrict the particles to an interval of length l we must impose the further Dirichlet condition at the point $x=l$. Since bosonic symmetry allows us to restrict our attention to the domain $d^I\subset\mathbb{R}^n_+$, the appropriate boundary condition is given by

$$\psi(x_1,\dots,x_{n-1},l)=0 \tag{2.32}$$

which implies the relation

$$\begin{aligned}\mathcal{A}^P&=-\,e^{-2ik_{P(n)}l}\mathcal{A}^{PR_n}\\&=-\,e^{-2ik_{P(n)}l}\mathcal{A}^{PC_nR_1C_n^{-1}},\end{aligned} \tag{2.33}$$

where, for the latter equality, we have used (2.17). Finally, applying (2.29), (2.31) and (2.33) successively, we arrive at the condition

$$e^{-2ik_{P(n)}l}=\prod_{i=1}^{n-1}s_p(k_{P(n)}+k_{P(i)})s_p(k_{P(n)}-k_{P(i)}) \tag{2.34}$$

for all $P\in\mathcal{W}_n$. We note here that the form of $s_p(k)$ is such that, if (2.34) is satisfied for some $P\in\mathcal{W}_n$ then it is necessarily satisfied for elements

$$PT_1,\dots,PT_{n-2},PR_1 \text{ and } PR_n \tag{2.35}$$

in $\mathcal{W}_n$ and thus for every

$$PX \text{ and } PR_nX \tag{2.36}$$

in $\mathcal{W}_n$ with $X \in \mathcal{W}_{n-1}$. Using (2.18), we then have that the condition (2.34) need only be satisfied for elements

$$P \in \{I, C_n, C_n^2, \dots, C_n^{n-1}\}. \tag{2.37}$$

This is equivalent to the n quantisation conditions

$$e^{-2ik_j l} = \prod_{i \neq j} s_p(k_j + k_i) s_p(k_j - k_i), \tag{2.38}$$

with $j \in \{1, \dots, n\}$. Solutions $(k_1, \dots, k_n) \neq (0, \dots, 0)$, such that $0 \leq k_1 \dots \leq k_n$, then constitute energies (2.28).

3. General graphs with $\tilde{\delta}$-interactions

Now we have established how to construct exactly solvable n-particle systems on an interval, we would like to extend the approach to general graphs. It turns out that systems of δ-interacting particles on graphs with more than a single edge, in general, are not compatibile with the Bethe ansatz method. In [**BG17**], which itself is based on a construction in [**CC07**], graphs were equipped with singular, non-local interactions, referred to as $\tilde{\delta}$-type, designed to make a Bethe ansatz possible. In what follows we extend this approach to n-particle quantum graphs. Defining an appropriate n-particle Bethe ansatz, we show exact solvability and calculate a quantisation condition, in the form of a collection of n secular equations, which provide the exact spectra.

Let us begin by viewing the compact graph Γ in what we call its star representation $\Gamma^{(s)}$ by cutting all edges of Γ to produce $|\mathcal{V}|$ star graphs and extending all the edges of the star graphs to infinity. The appropriate n-particle Hilbert space on $\Gamma^{(s)}$ is then

$$\mathcal{H}_n^{(s)} = \bigotimes_{i=1}^{n} \left(\bigoplus_{j=1}^{|\mathcal{E}|} L^2(0,\infty) \right). \tag{3.1}$$

We remark that here $\mathcal{E}$ is now the union of the edge sets of all infinite star graphs. Its order is twice the number of edges of the compact graph Γ. Vectors

$$\Psi = \left(\psi_{j_1 \dots j_n}^{(s)} \right)_{j_1, \dots, j_n = 1}^{|\mathcal{E}|} \tag{3.2}$$

in $\mathcal{H}_n^{(s)}$ are then lists of n-particle functions

$$\psi_{j_1 \dots j_n}^{(s)} : D_{j_1 \dots j_n}^{(s)} \to \mathbb{C} \tag{3.3}$$

in $L^2(D_{j_1 \dots j_n}^{(s)})$ with infinite subdomains defined as

$$D_{j_1 \dots j_n}^{(s)} = (0,\infty)^n. \tag{3.4}$$

The total configuration space for n particles on $\Gamma^{(s)}$ is the disjoint union

$$D_\Gamma^{(s)} = \bigsqcup_{j_1, \dots, j_n = 1}^{|\mathcal{E}|} D_{j_1 \dots j_n}^{(s)} \tag{3.5}$$

of these subdomains. The n-particle Hilbert space can then be written $\mathcal{H}_n^{(s)} = L^2(D_\Gamma^{(s)})$.

In the two-particle setting (see [**BG17**]), interactions take place along the diagonals $x_1 = x_2$ of two-dimensional configuration spaces $D^{(s)}_{mn}$. In the n-particle setting, we wish to impose interactions at the boundaries of subdomains

$$D^{(s,Q)}_{j_1\cdots j_n} = \{(x_1,\dots,x_n) \in D^{(s)}_{j_1\cdots j_n};\ x_{Q(1)} < \cdots < x_{Q(n)}\}, \tag{3.6}$$

with $Q \in S_n$. The appropriate total dissected configuration space is then

$$D^{(s,*)}_\Gamma = \bigsqcup_{j_1,\dots,j_n=1}^{|\mathcal{E}|} \left(\bigsqcup_{Q\in S_n} D^{(s,Q)}_{j_1\cdots j_n} \right), \tag{3.7}$$

with the total dissected two-particle Hilbert space $\mathcal{H}^{(s,*)}_n = L^2(D^{(s,*)}_\Gamma)$. Thus vectors

$$\Psi = \left(\left(\psi^{(s,Q)}_{j_1\cdots j_n} \right)^{|\mathcal{E}|}_{j_1,\dots,j_n=1} \right)_{Q\in S_n} \tag{3.8}$$

in $\mathcal{H}^{(s,*)}_n$ are lists of square-integrable functions $\psi^{(s,Q)}_{j_1\cdots j_n} : D^{(s,Q)}_{j_1\cdots j_n} \to \mathbb{C}$. The corresponding Sobolev space $H^2(D^{(s,*)}_\Gamma)$ is the set of $\Psi \in \mathcal{H}^{(s,*)}_n$ consisting of functions $\psi^{(s,Q)}_{j_1\cdots j_n} \in H^2(D^{(s,Q)}_{j_1\cdots j_n})$.

Boundary conditions will be imposed on eigenfunctions $\Psi \in H^2(D^{(s,*)}_\Gamma)$ of the n-particle Laplacian $-\Delta_n$. We reiterate here that these will be n-particle analogues of the boundary conditions imposed in the two-particle setting in [**BG17**]. Before we proceed with establishing these conditions, it is convenient to define the permutation matrices $\mathbb{Q}$ as representations of $Q \in S_n$ on

$$\bigotimes_{j=1}^{n} \mathbb{C}^{|\mathcal{E}|} \tag{3.9}$$

such that

(1) $\mathbb{I} = \mathbb{I}_{|\mathcal{E}|^n}$ is the representation of I;
(2) $\mathbb{T}^{(i)} = \mathbb{I}_{|\mathcal{E}|^{i-1}} \otimes \mathbb{T}_{|\mathcal{E}|^2} \otimes \mathbb{I}_{|\mathcal{E}|^{n-i-1}}$ is the representation of T_i.

Here

$$\mathbb{T}_{|\mathcal{E}|^2} = \begin{pmatrix} \mathbb{I}_{|\mathcal{E}|} \otimes m_1 \\ \vdots \\ \mathbb{I}_{|\mathcal{E}|} \otimes m_{|\mathcal{E}|} \end{pmatrix}, \tag{3.10}$$

with the $|\mathcal{E}|$-dimensional row vectors

$$m_j = (0,\dots,0,1,0\dots,0) \tag{3.11}$$

in which the 1 is in j-th position. We note the properties

$$\mathbb{T}^{(i)} \left(\mathcal{A}_{j_1\cdots j_n}\right)^{|\mathcal{E}|}_{j_1,\dots,j_n=1} = \left(\mathcal{A}_{j_1\cdots j_n}\right)^{|\mathcal{E}|}_{j_1,\dots,j_{i-1},j_{i+1},j_i,j_{i+2},\dots,j_n=1} \tag{3.12}$$

for $|\mathcal{E}|^n$-dimensional column vectors $\mathcal{A}$ and that

$$\begin{aligned} &\mathbb{T}^{(i)}(M_1 \otimes \cdots \otimes M_{i-1} \otimes M_i \otimes M_{i+1} \otimes M_{i+2} \otimes \cdots \otimes M_n)\mathbb{T}^{(i)} \\ &= M_1 \otimes \cdots \otimes M_{i-1} \otimes M_{i+1} \otimes M_i \otimes M_{i+2} \otimes \cdots \otimes M_n \end{aligned} \tag{3.13}$$

for any $|\mathcal{E}| \times |\mathcal{E}|$ matrices M_j. Finally, it is convenient to note the property

$$\mathbb{Q}\left(M \otimes \mathbb{I}_{|\mathcal{E}|^{n-1}}\right)\mathbb{Q}^{-1} = \mathbb{I}_{|\mathcal{E}|} \otimes \cdots \otimes \mathbb{I}_{|\mathcal{E}|} \otimes M \otimes \mathbb{I}_{|\mathcal{E}|} \otimes \cdots \otimes \mathbb{I}_{|\mathcal{E}|} \tag{3.14}$$

where on the right hand side, the matrix M is the $Q(1)$-th position.

Let us begin by establishing boundary conditions which prescribe single-particle interactions with the vertices. These will be given as simple n-particle lifts of those given in (2.5) imposed in the one-particle setting. The values of $\Psi \in H^2(D_\Gamma^{(s,*)})$ at the vertices, along with corresponding derivatives, are given by boundary vectors

$$\begin{aligned} \Psi_{\mathrm{bv}}^{(v)} &= \left((\psi_{j_1\cdots j_n}^{Q}(x_1,\ldots,x_n)|_{x_{Q(1)}=0})_{j_1,\ldots,j_n=1}^{|\mathcal{E}|} \right)_{Q\in S_n}; \\ \Psi_{\mathrm{bv}}^{(v)\prime} &= \left((\psi_{j_1\cdots j_n,Q(1)}^{Q}(x_1,\ldots,x_n)|_{x_{Q(1)}=0})_{j_1,\ldots,j_n=1}^{|\mathcal{E}|} \right)_{Q\in S_n}, \end{aligned} \tag{3.15}$$

where $\psi_{j_1\cdots j_n,Q(1)}^{Q}$ are inward derivatives normal to the lines $x_{Q(1)}=0$. Then, using (3.14), the appropriate boundary condition is given by

$$\left(\mathbb{I}_{n!}\otimes \mathbb{Q}\left(A\otimes \mathbb{I}_{|\mathcal{E}|^{n-1}}\right)\mathbb{Q}^{-1}\right)\Psi_{\mathrm{bv}}^{(v)} + \left(\mathbb{I}_{n!}\otimes \mathbb{Q}\left(B\otimes \mathbb{I}_{|\mathcal{E}|^{n-1}}\right)\mathbb{Q}^{-1}\right)\Psi_{\mathrm{bv}}^{(v)\prime} = 0. \tag{3.16}$$

Here the matrices A, B are of the same form as in (2.5) .

Next, we would like to impose $\tilde{\delta}$-type interactions between pairs of particles located on the same infinite star and impose continuity across dissections otherwise. Such interactions are prescribed in the n-particle setting according to the conditions

$$\begin{aligned} &\psi^{Q}_{j_{Q^{-1}(1)}\cdots j_{Q^{-1}(n)}}(x_1,\ldots,x_n)|_{x_{Q(i)}=x_{Q(i+1)}} \\ =&\psi^{QT_i}_{j_{T_iQ^{-1}(1)}\cdots j_{T_iQ^{-1}(n)}}(x_1,\ldots,x_n)|_{x_{Q(i)}=x_{Q(i+1)}}; \\ &\left(\frac{\partial}{\partial x_{Q(i+1)}}-\frac{\partial}{\partial x_{Q(i)}}-2\alpha\right)\psi^{Q}_{j_{Q^{-1}(1)}\cdots j_{Q^{-1}(n)}}(x_1,\ldots,x_n)|_{x_{Q(i)}=x_{Q(i+1)}} \\ =&\left(\frac{\partial}{\partial x_{Q(i+1)}}-\frac{\partial}{\partial x_{Q(i)}}\right)\psi^{QT_i}_{j_{T_iQ^{-1}(1)}\cdots j_{T_iQ^{-1}(n)}}(x_1,\ldots,x_n)|_{x_{Q(i)}=x_{Q(i+1)}}, \end{aligned} \tag{3.17}$$

if the edges e_{j_i} and $e_{j_{i+1}}$ belong to the same star graph, and

$$\begin{aligned} &\psi^{Q}_{j_{Q^{-1}(1)}\cdots j_{Q^{-1}(n)}}(x_1,\ldots,x_n)|_{x_{Q(i)}=x_{Q(i+1)}} \\ =&\psi^{QT_i}_{j_{Q^{-1}(1)}\cdots j_{Q^{-1}(n)}}(x_1,\ldots,x_n)|_{x_{Q(i)}=x_{Q(i+1)}}; \\ &\left(\frac{\partial}{\partial x_{Q(i+1)}}-\frac{\partial}{\partial x_{Q(i)}}\right)\psi^{Q}_{j_{Q^{-1}(1)}\cdots j_{Q^{-1}(n)}}(x_1,\ldots,x_n)|_{x_{Q(i)}=x_{Q(i+1)}} \\ =&\left(\frac{\partial}{\partial x_{Q(i+1)}}-\frac{\partial}{\partial x_{Q(i)}}\right)\psi^{QT_i}_{j_{Q^{-1}(1)}\cdots j_{Q^{-1}(n)}}(x_1,\ldots,x_n)|_{x_{Q(i)}=x_{Q(i+1)}}, \end{aligned} \tag{3.18}$$

if the edges e_{j_i} and $e_{j_{i+1}}$ belong to different star graphs.

The task is now to specify eigenvectors $\Psi \in H^2(D_\Gamma^{(s,*)})$ which satisfy boundary conditions (3.16)-(3.18). Taking care to distinguish between subdomains $D_{j_1\cdots j_n}^{(s,Q)}$, the vector Ψ will be described by the collection of functions

$$\psi_{j_1\cdots j_n}^{Q} = \sum_{P\in\mathcal{W}_n} \mathcal{A}_{j_1\cdots j_n}^{(P,Q)} e^{i(k_{P(1)}x_1+\cdots+k_{P(n)}x_n)}. \tag{3.19}$$

This form obviously leads to eigenfunctions of $-\Delta_n$ with Laplace eigenvalues (2.28).

Let us define the $|\mathcal{E}|^n$-dimensional vectors

$$\mathcal{A}^{(P,Q)} = \left(\mathcal{A}_{j_1\cdots j_n}^{(P,Q)}\right)_{j_1,\ldots,j_n=1}^{|\mathcal{E}|} \tag{3.20}$$

and then the $n!|\mathcal{E}|^n$-dimensional vectors

$$\mathcal{A}^P = \left(\mathbb{Q}^{-1}\mathcal{A}^{(PQ^{-1},Q)}\right)_{Q\in S_n}. \tag{3.21}$$

It is convenient at this point to impose an ordering on (3.21) by associating with each element Q the number $[Q] \in (1,\dots,n!)$ so that

$$\mathbb{Q}^{-1}\mathcal{A}^{(PQ^{-1},Q)} \tag{3.22}$$

is the $[Q]^{\text{th}}$ block in the list $\mathcal{A}^P$.

Boundary conditions at the vertices (3.16) imply the relations

$$\mathbb{Q}^{-1}\mathcal{A}^{(PR_{Q(1)},Q)} = \left(S_v(-k_{PQ(1)})\otimes \mathbb{I}_{|\mathcal{E}|^{n-1}}\right)\mathbb{Q}^{-1}\mathcal{A}^{(P,Q)}. \tag{3.23}$$

Noting then, that the properties of $\mathcal{W}_n$ imply

$$R_{Q(1)} = QR_1Q^{-1}, \tag{3.24}$$

we have that

$$\mathcal{A}^{PR_1} = \mathbb{I}_{n!}\otimes S_v(-k_{P(1)})\otimes \mathbb{I}_{|\mathcal{E}|^{n-1}}\mathcal{A}^P. \tag{3.25}$$

At this point, it is convenient to define the diagonal matrices

$$\boldsymbol{c}_i = \operatorname{diag}(c^{(i)}_{j_1\cdots j_n})^{|\mathcal{E}|}_{j_1,\dots,j_n=1}, \tag{3.26}$$

where $c^{(i)}_{j_1\cdots j_n} = 1$, if the edges e_{j_i} and $e_{j_{i+1}}$ belong to the same star graph, and $c^{(i)}_{j_1\cdots j_n} = 0$ otherwise. These matrices distinguish domains with $\tilde{\delta}$-type interactions from those which are continuous across dissections. We notice here the relations

$$\boldsymbol{c}_i = \mathbb{I}_{|\mathcal{E}|^{i-1}}\otimes \boldsymbol{c}\otimes \mathbb{I}_{|\mathcal{E}|^{n-i-1}}, \tag{3.27}$$

where $\boldsymbol{c}$ is defined for two-particle quantum graphs in [**BG17**]. The $\tilde{\delta}$-type conditions (3.17) and continuity conditions (3.18) imply the relations

$$\begin{aligned}&(\mathbb{I}_2\otimes \boldsymbol{c}_i)\begin{pmatrix}\mathbb{Q}^{-1}\mathcal{A}^{(PT_iQ^{-1},Q)}\\ \mathbb{T}^{(i)}\mathbb{Q}^{-1}\mathcal{A}^{(PQ^{-1},QT_i)}\end{pmatrix}\\ &\quad= \left(S_p(k_{P(i)}-k_{P(i+1)})\otimes \mathbb{I}_{|\mathcal{E}|^n}\right)(\mathbb{I}_2\otimes \boldsymbol{c}_i)\begin{pmatrix}\mathbb{Q}^{-1}\mathcal{A}^{(PQ^{-1},Q)}\\ \mathbb{T}^{(i)}\mathbb{Q}^{-1}\mathcal{A}^{(PT_iQ^{-1},QT_i)}\end{pmatrix}\end{aligned} \tag{3.28}$$

and

$$(\mathbb{I}_{|\mathcal{E}|^n}-\boldsymbol{c}_i)\mathbb{Q}^{-1}\mathcal{A}^{(PQ^{-1},Q)} = (\mathbb{I}_{|\mathcal{E}|^n}-\boldsymbol{c}_i)\mathbb{Q}^{-1}\mathcal{A}^{(PQ^{-1},QT_i)} \tag{3.29}$$

respectively. We then have that

$$\mathcal{A}^{PT_i} = Y_i(k_{P(i)}-k_{P(i+1)})\mathcal{A}^P, \tag{3.30}$$

where

$$(Y_i(k))_{[Q][Q']} = \frac{-i\alpha}{k+i\alpha}\boldsymbol{c}_i\delta_{[Q][Q']} + \left(\frac{k}{k+i\alpha}\boldsymbol{c}_i + \mathbb{T}^{(i)}\left(\mathbb{I}_{|\mathcal{E}|^n}-\boldsymbol{c}_i\right)\right)\delta_{[QT_i][Q']}. \tag{3.31}$$

Exact solvability is assured if relations (3.25) and (3.30) are compatible with the properties of $\mathcal{W}_n$. This amounts to the following consistency relations:

(1) $S_v(u)S_v(-u) = \mathbb{I}_{|\mathcal{E}|}$;
(2) $Y_i(u)Y_i(-u) = \mathbb{I}_{n!|\mathcal{E}|^n}$;
(3) $Y_i(u)Y_j(v) = Y_j(v)Y_i(u)$ for $|i-j|>1$;
(4) $Y_{i+1}(u)Y_i(u+v)Y_{i+1}(v) = Y_i(v)Y_{i+1}(u+v)Y_i(u)$;

(5) $\left(\mathbb{I}_{n!} \otimes S_v(u) \otimes \mathbb{I}_{|\mathcal{E}|^{n-1}}\right) Y_1(u+v) \left(\mathbb{I}_{n!} \otimes S_v(v) \otimes \mathbb{I}_{|\mathcal{E}|^{n-1}}\right) Y_1(v-u)$
$= Y_1(v-u) \left(\mathbb{I}_{n!} \otimes S_v(v) \otimes \mathbb{I}_{|\mathcal{E}|^{n-1}}\right) Y_1(u+v) \left(\mathbb{I}_{n!} \otimes S_v(u) \otimes \mathbb{I}_{|\mathcal{E}|^{n-1}}\right)$;

(6) $Y_i(u) \left(\mathbb{I}_{n!} \otimes S_v(v) \otimes \mathbb{I}_{|\mathcal{E}|^{n-1}}\right) = \left(\mathbb{I}_{n!} \otimes S_v(v) \otimes \mathbb{I}_{|\mathcal{E}|^{n-1}}\right) Y_i(u)$ for $i > 1$.

These conditions can be verified by the explicit forms of $S_v(k)$ and $Y_i(k)$.

In order to turn the eigenfunctions in the star representation into eigenfunctions on the compact graph, we must impose appropriate joining conditions. These are the n-particle analogues of the joining conditions established in [**BG17**] and are given by

$$\psi^Q_{j_1\cdots j_n}(x_1,\dots,x_n) = \psi^Q_{j'_1\cdots j'_n}(x'_1,\dots,x'_n) \tag{3.32}$$

for all $Q \in S_n$, where

$$(x'_{Q(1)},\dots,x'_{Q(n)}) = (x_{Q(1)},\dots,x_{Q(n-1)}, l_{j_{Q(n)}} - x_{Q(n)}) \tag{3.33}$$

and

$$(j'_{Q(1)},\dots,j'_{Q(n)}) = (j_{Q(1)},\dots,j_{Q(n-1)}, j_{Q(n)} + |\mathcal{E}|). \tag{3.34}$$

We then have

$$\begin{aligned} \mathcal{A}^P &= E(-k_{P(n)})\mathcal{A}^{PR_n} \\ &= E(-k_{P(n)})\mathcal{A}^{PC_nR_1C_n^{-1}}, \end{aligned} \tag{3.35}$$

where

$$E(k) = \mathbb{I}_{n!|\mathcal{E}|^{n-1}} \otimes \left(\begin{smallmatrix} 0 & 1 \\ 1 & 0 \end{smallmatrix}\right) \otimes e^{ik\boldsymbol{l}}. \tag{3.36}$$

Applying (3.25), (3.30) and (3.35) successively we have that the relation

$$Z(k_{P(1)},\dots,k_{P(n)}) = 0, \tag{3.37}$$

with

$$\begin{aligned} Z(k_1,\dots,k_n) = \det\Big[&\mathbb{I}_{n!|\mathcal{E}|^n} - E(k_n)Y_{n-1}(k_n-k_{n-1})\dots Y_1(k_n-k_1) \\ &(\mathbb{I}_{n!} \otimes S_v(k_n) \otimes \mathbb{I}_{|\mathcal{E}|^{n-1}})Y_1(k_1+k_n)\dots Y_{n-1}(k_{n-1}+k_n)\Big], \end{aligned} \tag{3.38}$$

is satisfied for all $P \in \mathcal{W}_n$. By using properties of determinants, it can be shown that the explicit forms of $Y_i(k)$, $S_v(k)$ and $E(k)$ are such that if (3.37) is satisfied for some $P \in \mathcal{W}_n$, then it is necessarily satisfied for elements

$$PT_1,\dots,PT_{n-2}, R_1 \text{ and } R_n \tag{3.39}$$

in $\mathcal{W}_n$ and thus for every

$$PX \text{ and } PR_nX \tag{3.40}$$

in $\mathcal{W}_n$ with $X \in \mathcal{W}_{n-1}$. Using (2.18), we can state the main result of this section.

THEOREM 3.1. *Non-zero eigenvalues of a self-adjoint n-particle Laplacian $-\Delta_n$ defined on Γ with local vertex interactions specified through A, B and $\tilde{\delta}$-type interactions between particles when they are located on neighbouring edges, are the values $E = k_1^2 + \cdots + k_n^2 \neq 0$ with multiplicity m, where $(k_1,\dots,k_n)$, such that $0 \le k_1 \le \cdots \le k_n$, are solutions to the n secular equations*

$$Z(k_{i_1},\dots,k_{i_n}) = 0, \tag{3.41}$$

for $(i_1,\dots,i_n) \in \{C_n^d(1,\dots,n)\}_{d=0}^{n-1}$, with multiplicity m.

4. Recovering specific results

In this final section, by choosing particular parameters, we show how to recover established results from the general n-particle quantisation condition prescribed by Theorem 3.1.

4.1. Non-interacting particles. Non-interacting systems are recovered by reestablishing continuity across all dissected domains. This is achieved by setting $\boldsymbol{c}_i = \boldsymbol{0}$ for all $i \in \{1, \dots, n-1\}$. Matrices

$$Y_i(k)|_{\boldsymbol{c}=\boldsymbol{0}} \tag{4.1}$$

are then composed of blocks

$$(Y_i(k)|_{\boldsymbol{c}=\boldsymbol{0}})_{[Q][Q']} = \mathbb{T}^{(i)} \delta_{[QT_i][Q']}. \tag{4.2}$$

By substituting into (3.38) we recover the secular equation (2.9) for the one-particle quantum graph.

4.2. Two-particle graphs. Simply by choosing $n = 2$, one immediately recovers the spectra of two-particle quantum graphs as established in [**BG17**] (see Theorem 5.1 therein).

References

[BE09] Jens Bolte and Sebastian Endres, *The trace formula for quantum graphs with general self adjoint boundary conditions*, Ann. Henri Poincaré **10** (2009), no. 1, 189–223, DOI 10.1007/s00023-009-0399-7. MR2496307

[Bet31] H. Bethe, *Zur Theorie der Metalle. i. Eigenwerte und Eigenfunktionen der linearen Atomkette (German) [On the theory of metals. i. Eigenvalues and eigenfunctions of a linear chain of atoms]*, Z. Phys. **71** (1931), 205–226.

[BG17] Jens Bolte and George Garforth, *Exactly solvable interacting two-particle quantum graphs*, J. Phys. A **50** (2017), no. 10, 105101, 27. MR3609089

[BK13a] Gregory Berkolaiko and Peter Kuchment, *Introduction to quantum graphs*, Mathematical Surveys and Monographs, vol. 186, American Mathematical Society, Providence, RI, 2013. MR3013208

[BK13b] Jens Bolte and Joachim Kerner, *Quantum graphs with singular two-particle interactions*, J. Phys. A **46** (2013), no. 4, 045206, 25, DOI 10.1088/1751-8113/46/4/045206. MR3029862

[BK13c] ———, *Quantum graphs with two-particle contact interactions*, J. Phys. A: Math. Theor. **46** (2013), 045207.

[BK14] ———, *Many-particle quantum graphs and Bose-Einstein condensation*, J. Math. Phys. **55** (2014), 061901.

[BK16] Jens Bolte and Joachim Kerner, *Instability of Bose-Einstein condensation into the one-particle ground state on quantum graphs under repulsive perturbations*, J. Math. Phys. **57** (2016), no. 4, 043301, 9, DOI 10.1063/1.4946044. MR3487209

[BT77] M. V. Berry and M. Tabor, *Level clustering in the regular spectrum*, Proc. R. Soc. London. Series A, Math. and Phys. Sci. **356** (1977), 375–394.

[CC07] V. Caudrelier and N. Crampé, *Exact results for the one-dimensional many-body problem with contact interaction: including a tunable impurity*, Rev. Math. Phys. **19** (2007), no. 4, 349–370, DOI 10.1142/S0129055X07002973. MR2325347

[DS92] Eyal Doron and Uzy Smilansky, *Semiclassical quantization of chaotic billiards: a scattering theory approach*, Nonlinearity **5** (1992), no. 5, 1055–1084. MR1187738

[EKK+08] P. Exner, J. P. Keating, P. Kuchment, T. Sunada, and A. Teplyaev, *Vladimir A. Geyler, April 29, 1943–April 2, 2007*, Analysis on graphs and its applications, Proc. Sympos. Pure Math., vol. 77, Amer. Math. Soc., Providence, RI, 2008, pp. xiii, DOI 10.1090/pspum/077. MR2459861

[Gau71] M. Gaudin, *Boundary energy of a Bose gas in one dimension*, Phys. Rev. A. **4** (1971), 386–394.

[GS06] S. Gnutzmann and U. Smilansky, *Quantum graphs: Applications to quantum chaos and universal spectral statistics*, Advances in Physics **55** (2006), 527–625.

[KN05] Pavel Kurasov and Marlena Nowaczyk, *Inverse spectral problem for quantum graphs*, J. Phys. A **38** (2005), no. 22, 4901–4915, DOI 10.1088/0305-4470/38/22/014. MR2148632

[KS99a] V. Kostrykin and R. Schrader, *Kirchhoff's rule for quantum wires*, J. Phys. A **32** (1999), no. 4, 595–630, DOI 10.1088/0305-4470/32/4/006. MR1671833

[KS99b] Tsampikos Kottos and Uzy Smilansky, *Periodic orbit theory and spectral statistics for quantum graphs*, Ann. Physics **274** (1999), no. 1, 76–124, DOI 10.1006/aphy.1999.5904. MR1694731

[KS06] Vadim Kostrykin and Robert Schrader, *Laplacians on metric graphs: eigenvalues, resolvents and semigroups*, Quantum graphs and their applications, Contemp. Math., vol. 415, Amer. Math. Soc., Providence, RI, 2006, pp. 201–225, DOI 10.1090/conm/415/07870. MR2277618

[LL63] Elliott H. Lieb and Werner Liniger, *Exact analysis of an interacting Bose gas. I. The general solution and the ground state*, Phys. Rev. (2) **130** (1963), 1605–1616. MR0156630

Department of Mathematics, Royal Holloway, University of London, Egham, TW20 0EX, United Kingdom

Email address: jens.bolte@rhul.ac.uk

Department of Mathematics, Royal Holloway, University of London, Egham, TW20 0EX, United Kingdom

Email address: George.Garforth.2012@live.rhul.ac.uk

Contemporary Mathematics
Volume **717**, 2018
https://doi.org/10.1090/conm/717/14445

Nonlinear Sturm oscillation: From the interval to a star

Ram Band and August J. Krueger

Abstract. The Sturm oscillation property, i.e. that the n-th eigenfunction of a Sturm-Liouville operator on an interval has $n-1$ zeros (nodes), has been well studied. This result is known to hold when the interval is replaced by a metric (quantum) tree graph. We prove that the solutions of the real stationary nonlinear Schrödinger equation on an interval satisfy a nonlinear version of the Sturm oscillation property. However, we show that unlike for the linear theory, the nonlinear version of Sturm oscillation breaks down already for a star graph. We point out conditions under which this violation can be assured.

1. Introduction

The linear theory of Sturm-Liouville operators, and the associated oscillation theorems, that began in [**19, 20**] has lead to an extensive and robust field of ideas and results. See, for example, [**4**] for a broad review of the classical and modern theory. Put simply, Sturm oscillation theorem states that if the eigenvalues of an operator are indexed increasingly by $\mathbb{N}$, then the n-th eigenfunction has $n-1$ interior zeros. The theory of Sturm oscillation may be extended in many different directions, one of which is to consider differential operators on collections of line segments joined at their endpoints with suitable matching conditions. Theses networks, or graphs, are called tree graphs if they admit no closed cycles. For the cases where the differential operator is the Laplacian with Robin matching conditions, the Sturm oscillation property for a tree graph has been established, see e.g. recent results in [**6, 17, 18**] and review in [**5**].

We consider the generalization of oscillation theory to nonlinear differential equations on line segments and star graphs. This extension immediately prohibits a direct appeal to linear spectral theory and therefore new definitions must be given not only for the operators encountered but also of a suitable notion of nonlinear Sturm oscillation.

In the linear theory, one may rescale eigenfunctions with impunity. The nonlinear theory, however, lacks such a trivial scale factor. Therefore in order to characterize all stationary solutions of the nonlinear Schrödinger equation one needs to introduce an additional parameter. Such a parameter is usually taken to be some norm of the solution and we take it here to be the L^∞ norm. In the two-dimensional

2010 *Mathematics Subject Classification.* Primary 34A34, 81Q35, 34C10, 34B45.
Key words and phrases. Nonlinear Schrödinger equation, quantum graph, Sturm oscillation, spectral curve.

space that is parametrized by the spectral parameter and the norm, one may represent families of stationary solutions as simple curves, which we call spectral curves. We prove a kind of nonlinear Sturm oscillation property for the interval, where the spectral curves may be indexed and each solution lying on the n-th curve has $n-1$ interior zeros (Theorem 2.4). Following this, the nonlinear Schrödinger equation on a star graph is considered. The full nonlinear spectrum for the star graph is beyond the scope of this paper, however local properties of spectral curves are explored. We prove that the nodal count is not constant in general along each spectral curve (Theorem 2.9) and show how to construct star graphs and solutions for which such a nodal count change occurs. Therefore, in distinction to the linear theory, the analogous form of nonlinear Sturm oscillation property does not generically hold already for the simplest tree graphs.

We refer the interested reader to the book [**7**] and review [**8**] on the linear theory of metric (quantum) graphs. The nonlinear Schrödinger equation on metric graphs is addressed in many recent works. We mention here only those works which deal with stationary solutions and are closer in spirit to those discussed in the current paper. With regard to stationary solutions of the nonlinear Schrödinger equation we note the study of general scattering in [**9,10**], of general solutions on star graphs in [**1–3**], and of bifurcation and stability properties of solutions on various graphs in [**13,15,16**]. Finally, a framework to aid in the solving of the stationary nonlinear Schrödinger equation on metric graphs was presented recently in [**11,12**].

2. Main results

2.1. Nonlinear Schrödinger equation on an interval.

DEFINITION 2.1. *Let the* real, stationary, nonlinear Schrödinger equation *on an interval of length* $0 < l \in \mathbb{R}$ *be given by*

$$\mu\phi = -\partial_x^2\phi - (\sigma+1)\nu\phi^{2\sigma+1}, \quad \mu \in \mathbb{R},\ 0 \neq \nu \in \mathbb{R},\ \sigma \in \mathbb{N}, \tag{1}$$

where $\phi \in C^2([0,l],\mathbb{R})$ *and subject to boundary conditions that can be either of Dirichlet type,* $\phi(x_j) = 0$, *or of Neumann type,* $\partial_x\phi(x_j) = 0$, *where* $j = 1,2$, $x_1 = 0$, $x_2 = l$.

One may analogously define (1) *on a ray, e.g.* $\phi \in C^2([0,\infty),\mathbb{R})$, *by eliminating the second boundary condition, and alternatively on a line, e.g.* $\phi \in C^2(\mathbb{R},\mathbb{R})$, *by eliminating both boundary conditions.*

It is easiest to classify the solutions of (1) on an interval and ray by first considering those on a line. Furthermore integrating (1) yields

$$h = (\partial_x\phi)^2 + \mu\phi^2 + \nu\phi^{2(\sigma+1)}, \quad h \in \mathbb{R}, \tag{2}$$

which can be interpreted as a Hamiltonian energy conservation constraint (see the derivation leading to (27)). Here the constant h takes the role of the energy of a particle moving on a line in the time coordinate x, whose trajectory is represented by $\phi(x)$. One may therefore visualize the solutions by considering the effective potential energy $h_{\mathrm{p}}(\phi) = \mu\phi^2 + \nu\phi^{2(\sigma+1)}$. We will distinguish between four special cases:

- Case I, $(\nu > 0, \mu \geq 0)$. Shown in Figure 1 (a). $h_{\mathrm{p}}(\phi)$ has a global minimum at $\phi = 0$ and $\lim_{\phi\to\pm\infty} h_{\mathrm{p}}(\phi) = \infty$.

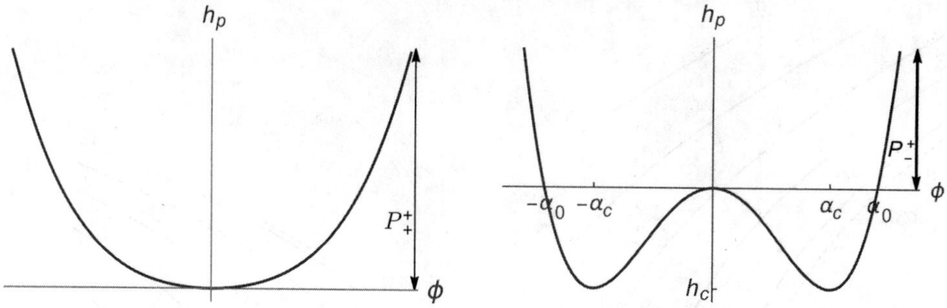

(a) Case I, ($\nu > 0, \mu \geq 0$). (b) Case II, ($\nu > 0, \mu < 0$).

FIGURE 1. $\nu > 0$.

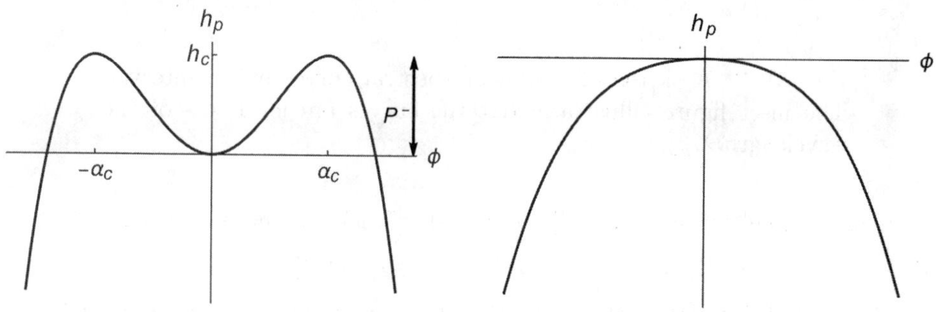

(a) Case III, ($\nu < 0, \mu > 0$). (b) Case IV, ($\nu < 0, \mu \leq 0$).

FIGURE 2. $\nu < 0$.

- Case II, ($\nu > 0, \mu < 0$). Shown in Figure 1 (b). $h_{\rm p}(\phi)$ has a local maximum at $\phi = 0$, global minima at $\phi = \pm|\mu/(\sigma+1)\nu|^{1/2\sigma}$, and $\lim_{\phi\to\pm\infty} h_{\rm p}(\phi) = \infty$.

- Case III, ($\nu < 0, \mu > 0$). Shown in Figure 2 (a). $h_{\rm p}(\phi)$ has a local minimum at $\phi = 0$, global maxima at $\phi = \pm|\mu/(\sigma+1)\nu|^{1/2\sigma}$, and $\lim_{\phi\to\pm\infty} h_{\rm p}(\phi) = -\infty$.

- Case IV, ($\nu < 0, \mu \leq 0$). Shown in Figure 2 (b). $h_{\rm p}(\phi)$ has a global maximum at $\phi = 0$ and $\lim_{\phi\to\pm\infty} h_{\rm p}(\phi) = -\infty$.

DEFINITION 2.2. *We associate with each bounded solution of* (1) *on a line a parameter* $\alpha = ||\phi||_\infty$. *We define the following distinguished subsets of* $\mathbb{R}^2$. *For* $\nu > 0$*:*

$$P_+^+ := \{(\mu,\alpha) \in \mathbb{R}^2 : \mu > 0, \alpha > 0\}, \quad P_-^+ := \{(\mu,\alpha) \in \mathbb{R}^2 : \mu \leq 0, \alpha > \alpha_0\}, \tag{3}$$

and for $\nu < 0$*:*

$$P^- := \{(\mu,\alpha) \in \mathbb{R}^2 : \mu > 0, 0 < \alpha < \alpha_{\rm c}\}, \tag{4}$$

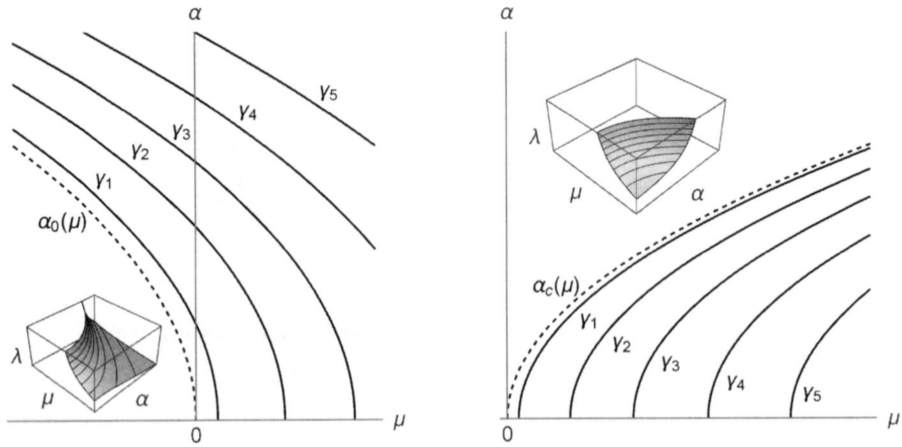

(a) $\nu > 0$ (b) $\nu < 0$

FIGURE 3. A sketch of nonlinear spectral curves on an interval. The inset figures illustrate that the curves are level sets of the wavelength $\lambda_{(\mu,\alpha)}$.

where $\alpha_0 := |\mu/\nu|^{1/2\sigma}$ *and* $\alpha_c := |\mu/(\sigma+1)\nu|^{1/2\sigma}$. *Furthermore we take*

$$P^+ := P^+_+ \cup P^+_-, \quad P := P^+ \cup P^-. \tag{5}$$

We would like to study the properties of solutions that oscillate symmetrically through zero, as these are most similar to the solutions of the analogous linear system, i.e. $\nu = 0$. Let $\nu \neq 0$ and fix a Dirichlet or Neumann boundary condition at each endpoint of an interval. We denote

$$\Phi_{\text{int}} := \{\phi \neq 0 \text{ solves (1) on an interval,} \tag{6}$$

$$\text{with the prescribed boundary conditions,} \tag{7}$$

$$\text{such that } \phi \text{ attains at least one zero on } [0,\ell]\}. \tag{8}$$

The next lemma parametrizes Φ_{int} by P.

LEMMA 2.3. *Fix* $\nu \neq 0$. *Fix an interval of length* $l > 0$ *with Dirichlet or Neumann boundary conditions at each of its endpoints. The following holds.*

(1) *Given* $\phi \in \Phi_{\text{int}}$, *there is a unique value of* $\mu \in \mathbb{R}$ *such that* ϕ *is a solution of* (1) *on this interval with the given values* μ, ν. *This allows one to define the map*

$$\Lambda_{\text{int}} : \Phi_{\text{int}} \to P, \quad \Lambda_{\text{int}} : \phi \mapsto (\mu, ||\phi||_\infty). \tag{9}$$

(2) Λ_{int} *is two to one since* $\Lambda_{\text{int}}(\phi_1) = \Lambda_{\text{int}}(\phi_2)$ *if and only if* $\phi_1 = \zeta\phi_2$, *where* $\zeta = \pm 1$.

The above lemma allows one to parametrize solutions of (1) with points $(\mu, \alpha) \in P$, which we write as $\phi = \phi_{(\mu,\alpha)}$. Those solutions lie on curves in the (μ, α) half plane, as is demonstrated in Figure 3. This is stated in an exact manner in the next theorem, which also establishes a nonlinear form of the Sturm oscillation property for the interval.

THEOREM 2.4. *Fix an interval of length $l > 0$ and $\nu \neq 0$. The following hold*

(1) $\Lambda_{\text{int}}(\Phi_{\text{int}}) = \bigsqcup_{n\in\mathscr{N}} \gamma_n$, *where $\bigsqcup$ represents a disjoint union, each γ_n is a connected, non self intersecting curve, and $\mathscr{N} = \mathbb{N}$ if at least one boundary condition is Dirichlet and $\mathscr{N} = \mathbb{N} \setminus \{1\}$ if both are Neumann.*
(2) *If $\Lambda_{\text{int}}(\phi) \in \gamma_n$, then ϕ has $n-1$ interior zeros.*
(3) *Let $0 < \alpha \in \mathbb{R}$ be fixed. Each γ_n intersects the line $\{(\mu, \alpha)\}_{\mu\in\mathbb{R}}$ only once. Furthermore these intersection points occur for $\mu = \mu_n$, where the μ_n are monotonically strictly increasing in n.*
(4) $\lim_{\alpha\to 0} \gamma_n = (\mu_n^{\text{lin}}, 0)$, *where μ_n^{lin} is the n-th eigenvalue of the linear problem on an interval.*

The first two parts of the theorem above show that the solutions of the nonlinear Schrödinger equation on an interval are naturally arranged in a sequential order and that all the solutions corresponding to the n-th set (curve) possess $n-1$ internal nodal points. We treat this as the nonlinear analogue of the Sturm oscillation property. The third part shows that once the solution norm, α, is fixed one obtains a discrete spectrum of solutions which obeys Sturm oscillation. The fourth part connects this with the linear spectrum, which is obtained as $\alpha \to 0$.

2.2. Nonlinear Schrödinger equation on a star graph.

DEFINITION 2.5. *Consider a set of $d > 2$ intervals with edge lengths $0 < l_j \in \mathbb{R}$, $j = 1, \ldots, d$. Join one endpoint of each of these intervals, hereafter termed* edges, *at a single point, hereafter termed the* central vertex, *and denote the resulting set a* star graph, Γ, *of degree d, whose endpoints are hereafter termed boundary vertices. We endow Γ with a fixed coordinate system where $x_j \in [0, l_j]$ is a coordinate on edge j such that $x_j = 0$ at the central vertex along each edge and $x_j = l_j$ at each boundary vertex. For any function $\phi : \Gamma \to \mathbb{R}$, we take ϕ_j be its restriction to edge j.*

Let the real stationary nonlinear Schrödinger equation on a degree $d > 2$ star graph Γ be given by

$$\mu\phi_j = -\partial_x^2\phi_j - (\sigma + 1)\nu\phi_j^{2\sigma+1}, \tag{10}$$

$$\mu \in \mathbb{R},\ 0 \neq \nu \in \mathbb{R},\ \sigma \in \mathbb{N},\ \phi_j \in C^2([0, l_j], \mathbb{R}), \tag{11}$$

for $j = 1, \ldots, d$, with a Neumann condition at the central vertex

$$\phi_1(0) = \phi_j(0), \quad j = 2, \ldots, d, \tag{12}$$

$$\sum_{j=1}^{d} \partial_x\phi_j(0) = 0, \tag{13}$$

and a Dirichlet or Neumann condition at each boundary vertex, $x_j = l_j$.

Let $\nu \neq 0$ and fix a Dirichlet or Neumann boundary condition at each endpoint of a graph. We denote

$$\Phi_{\text{star}} := \{\phi \neq 0 \text{ solves (10), with the prescribed boundary conditions,} \tag{14}$$

$$\text{such that } \phi \text{ attains at least one zero}\}, \tag{15}$$

and parametrize Φ_{star}, in a similar manner as was done for the solutions on the interval.

DEFINITION 2.6. *Let $Q \subset \mathbb{R}^{d+1}$ be the space of points $q = (\mu, \alpha_1, \ldots, \alpha_d)$ such that $(\mu, \alpha_j) \in P$, for all $j = 1, \ldots, d$ and fixed $\nu \neq 0$.*

LEMMA 2.7. *Fix $\nu \neq 0$. Fix a star graph with edges of length $l_j > 0$, $j = 1, \ldots, d$, with boundary conditions as described in Definition 2.5. Given $\phi \in \Phi_{\text{star}}$, there is a unique value of $\mu \in \mathbb{R}$ such that ϕ is a solution of (10) on this graph with the given values μ, ν. This allows us to define the map*

$$\Lambda_{\text{star}} : \Phi_{\text{star}} \to Q, \quad \Lambda_{\text{star}} : \phi \mapsto (\mu, ||\phi_1||_\infty, \ldots, ||\phi_d||_\infty), \tag{16}$$

for which $\Lambda_{\text{star}}(\phi^{(1)}) = \Lambda_{\text{star}}(\phi^{(2)})$ implies $\phi_j^{(1)} = \zeta_j \phi_j^{(2)}$, where $\zeta_j = \pm 1$ for all j.

The above lemma allows one to parametrize solutions of (10) with points $q \in Q$, which we write as $\phi = \phi_{(q)}$.

DEFINITION 2.8. *If there exists a continuous map $\gamma : \mathbb{R} \to Q$ such that $\phi_{(\gamma(\tau))} \in \Phi_{\text{star}}$ for all τ then we term γ a* local spectral curve.

For the interval, we managed to decompose Φ_{int} as a union of spectral curves, such that the nodal count of the solution is fixed along each curve. For the star we do not address the global structure of spectral curves but rather show that locally the nodal count may change along the spectral curves, thereby preventing a nonlinear Sturm oscillation property from being satisfied on the star.

THEOREM 2.9. *Let $\phi_{(q_*)} \in \Phi_{\text{star}}$ for some $q_* \in Q$. If $\phi_{(q_*)}$ vanishes at the central vertex then:*

1. *There exists a local spectral curve $\gamma : \mathbb{R} \to Q$ which passes through $q_* \in Q$.*
2. *For all $q \in \gamma$ sufficiently close to $q_* \in \gamma$, we have that the change in nodal count between the solutions at q and q_* is given by*

$$Z(\phi_{(q)}) - Z(\phi_{(q_*)}) \tag{17}$$

$$= \text{sgn}^2(\phi_{(q)}) \left[-1 + 2^{-1}d - 2^{-1}\text{sgn}(\phi_{(q)}) \sum_{j=1}^{d} \text{sgn}(\partial_x \phi_{(q_*),j}) \right] \Big|_{x=0}, \tag{18}$$

where $Z(\phi) \in \mathbb{N}$ is the number of zeros (nodes) of ϕ in the interior of Γ.

3. *If there are only Dirichlet conditions on exterior vertices, $\nu > 0$, and the edge lengths $\ell_1, \ldots, \ell_d$ satisfy $\sum_{j=1}^d \zeta_j (n_j/l_j)^{1+1/\sigma} = 0$ for some set of $\zeta_j \in \{-1, 1\}$ and $n_j \in \mathbb{N}$ for all j, then there exists a $\phi_{(q_*)} \in \Phi_{\text{star}}$ with $\mu_* = 0$, $\phi_{(q_*)}(0) = 0$, $\zeta_j = \text{sgn}(\partial_x \phi_{(q_*),j}(0))$, and interior nodal count $Z(\phi_{(q_*)}) = 1 - d + \sum_{j=1}^d n_j$.*
4. *In addition to assumptions in (3), we further have that if*

$$\sum_{j=1}^{d} \zeta_j (n_j/l_j)^{-1+1/\sigma} \neq 0,$$

then through this q_* passes a local spectral curve γ such that for all $q_+, q_- \in \gamma$ sufficiently close to q_*, where $\mu_+ > 0 > \mu_-$, one has the interior nodal count change $|Z(\phi_{(q_+)}) - Z(\phi_{(q_-)})| = |\sum_{j=1}^d \zeta_j|$.

REMARK 2.10. *Parts (3) and (4) of Theorem 2.9 can be made slightly more general and also apply to a star with either a Dirichlet or Neumann condition on each exterior vertex. The statements would be modified as follows. In (3), we change*

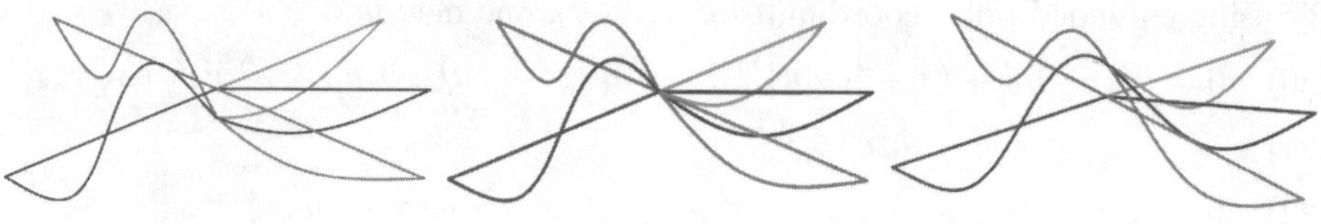

(a) 4 interior zeros (b) 3 interior zeros (c) 5 interior zeros

FIGURE 4. Sketch of nodal count change along a local spectral curve

the condition $\sum_{j=1}^{d} \zeta_j (n_j/l_j)^{1+1/\sigma} = 0$ *in a way such that each term corresponding to an edge with a Neumann condition becomes* $\zeta_j[(n_j - 1/2)/l_j]^{1+1/\sigma}$*. A similar change is done for the condition of* (4).

The theorem above demonstrates that for a star graph we cannot obtain a Sturm oscillation property similar to the one we got for the interval. In order for such a property to hold, we need to have that the nodal count is constant along spectral curves. Part (1) of the theorem shows the existence of a local curve. Then Part (2) of the theorem shows what is the nodal count change between a solution vanishing at the central vertex and a neighboring solution. Finally, the last two parts of the theorem show how to construct neighboring solutions which exhibit a nodal count change. We note that in such a construction, the nodal count change differs from zero for all star graphs with odd number of edges, an example of which is illustrated in Figure 4.

The paper is structured as follows. Sections 3 and 4 provide some work that is required for the proof of Theorems 2.4 and 2.9. Section 5 presents the proof of Theorem 2.4. Section 6 presents the proof of Theorem 2.9.

3. Preliminaries

Due to the importance, for general theory as well as applications, of the fact that solutions of the standard stationary nonlinear Scrödinger equation are complex valued, we choose to first couch the real stationary solutions in the context of the larger theory of complex stationary solutions. To this end we introduce a convenient means of coordinate decomposition.

We denote by *extended polar coordinates* the pair (ϕ, θ) where $\phi \in \mathbb{R}$ and $0 \leq \theta < 2\pi$ such that for each $z \in \mathbb{C}$ there exists at least one pair (ϕ, θ) such that $z = \phi e^{i\theta}$. To $z = 0$ one may associate any pair of the form $(0, \theta)$ and to $z \neq 0$ one always has the two equivalent associated pairs (ϕ, θ) and $(-\phi, \theta + \pi)$. These coordinates are useful for representing motion of point particles in a plane as influenced by central forces. There are no problems with the algebra represented by such coordinates so long as one is consistent about representation and it will be seen that we need consider no possible subtleties nor issues.

Consider the stationary nonlinear Schrödinger equation on a line

$$\mu\psi = -\partial_x^2\psi - (\sigma + 1)\nu|\psi|^{2\sigma}\psi, \quad \mu \in \mathbb{R},\ 0 \neq \nu \in \mathbb{R},\ \sigma \in \mathbb{N},\ \psi \in C^2(\mathbb{R}, \mathbb{C}). \tag{19}$$

By using extended polar coordinates $\psi = (\phi, \theta)$, one may find

$$0 = \partial_x^2\psi + \mu\psi + (\sigma+1)\nu|\psi|^{2\sigma}\psi = e^{i\theta}[\partial_x^2\phi + 2i\partial_x\phi\partial_x\theta - \phi(\partial_x\theta)^2 + i\phi\partial_x^2\theta] \tag{20}$$

$$+ e^{i\theta}\mu\phi + 2e^{i\theta}\nu\phi^{2\sigma}, \tag{21}$$

$$0 = [\partial_x^2\phi + 2i\partial_x\phi\partial_x\theta - \phi(\partial_x\theta)^2 + i\phi\partial_x^2\theta] + \mu\phi + (\sigma+1)\nu\phi^{2\sigma+1}. \tag{22}$$

By taking the imaginary and real parts of this equation one arrives at equations which are respectively equivalent to the angular and radial equations of motion of a Newtonian point particle moving in a planar, anharmonic, central force

$$(\text{Im.}):\quad 0 = 2\partial_x\phi\partial_x\theta + \phi\partial_x^2\theta, \tag{23}$$

$$(\text{Re.}):\quad 0 = \partial_x^2\phi - \phi(\partial_x\theta)^2 + \mu\phi + (\sigma+1)\nu\phi^{2\sigma+1}. \tag{24}$$

Integrating these respectively gives analogues of conservation of angular momentum and energy:

$$(\text{Im.}):\quad \text{const.} = \omega = \phi^2\partial_x\theta, \quad \omega \in \mathbb{R} \tag{25}$$

$$(\text{Re.}):\quad \text{const.} = h = (\partial_x\phi)^2 + \omega\phi^{-2} + \mu\phi^2 + \nu\phi^{2(\sigma+1)}, \quad h \in \mathbb{R}. \tag{26}$$

The system is equivalent to that of a particle moving in the plane, with the exception that the particle might transition from one plane to the adjoined one, i.e. $\phi \mapsto -\phi$, if it passes through the origin. If $\omega \neq 0$ then the centrifugal potential energy becomes arbitrarily large as the particle moves closer to $\phi = 0$. If $|h| < \infty$, then the centrifugal potential energy cannot be overcome. Solutions with $\omega = 0$ are different from those with $\omega \neq 0$ in that the former are not differentiable in standard polar coordinates, hence our introduction of the extended polar coordinates. These observations can be summarized as follows.

REMARK 3.1. *If $\psi = (\phi, \theta)$ is a solution of* (19) *on a line, then $\psi(x)$ can vanish for some x only if $[\phi(x)]^2\partial_x\theta(x) = \omega = 0$ for all $x \in \mathbb{R}$.*

We henceforth take $\omega = 0$ and $\theta = 0$ everywhere and consider only the real solutions and then (19) becomes (1) on a line. This restricts our focus to all solutions that can attain zeros, and possibly a few more, at the cost of a wide class of solutions that feature nontrivial complex oscillation without attaining zeros. With respect to the effective particle total energy, one then has

$$h = (\partial_x\phi)^2 + \mu\phi^2 + \nu\phi^{2(\sigma+1)}, \quad h, \mu \in \mathbb{R},\ 0 \neq \nu \in \mathbb{R},\ \sigma \in \mathbb{N},\ \phi \in C^2(\mathbb{R}, \mathbb{R}). \tag{27}$$

One may partition the effective particle total energy h into kinetic and potential parts, respectively h_{k} and h_{p}, via

$$h = h_{\mathrm{k}} + h_{\mathrm{p}}, \quad h_{\mathrm{k}}(\phi) := (\partial_x\phi)^2, \quad h_{\mathrm{p}}(\phi) := \mu\phi^2 + \nu\phi^{2(\sigma+1)}. \tag{28}$$

We have now reduced the system to that of a classical point particle constrained to move along a potential energy surface with constant total energy. This allows us to classify all solutions of (1) on a line. Fix $\nu \neq 0$, we denote

$$\Phi_{\text{line}} := \{\phi \neq 0 \text{ solves (1) on a line}, \tag{29}$$

$$\text{such that } \phi \text{ is periodic and attains zeros}\}. \tag{30}$$

PROPOSITION 3.2. *Let $\nu \neq 0$. Given $\phi \in \Phi_{\text{line}}$, there is a unique value of $\mu \in \mathbb{R}$ such that ϕ is a solution of* (1) *on the line with the given values μ, ν. This allows us to define the surjective map*

$$\Lambda_{\text{line}} : \Phi_{\text{line}} \to P, \quad \Lambda_{\text{line}} : \phi \mapsto (\mu, ||\phi||_\infty). \tag{31}$$

Furthermore $\Lambda_{\text{line}}(\phi_1) = \Lambda_{\text{line}}(\phi_2)$ *if and only if* $\phi_1(x) = \zeta\phi_2(x+x_0)$ *for some fixed* $\zeta = \pm 1$, *some fixed* $x_0 \in \mathbb{R}$, *and all* x.

PROOF. *Solution theory via energy conservation*

First we prove that Λ_{line} is onto and study the degree of freedom in its preimage. At the end of the proof we show the uniqueness of μ. Solutions of (1) follow from conservation of effective particle total energy and qualitative analysis of dynamics through determination of critical points of the effective particle motion as follows. From (27) we get

$$\widehat{x}(\widehat{\phi}_0, \widehat{\phi}) = x_0 + \zeta \int_{\widehat{\phi}_0}^{\widehat{\phi}} \mathrm{d}w\ [h - \mu w^2 - \nu w^{2(\sigma+1)}]^{-1/2}, \tag{32}$$

where $\zeta = \pm 1$, $x_0 \in \mathbb{R}$ is an initial value of x along a trajectory and

$$\widehat{x}(\widehat{\phi}_0, \widehat{\phi}_0) = x_0, \quad \widehat{x}(\widehat{\phi}_0, \widehat{\phi}) = x, \quad \phi(x_0) = \widehat{\phi}_0, \quad \phi(x) = \widehat{\phi}. \tag{33}$$

The map $\widehat{x}(\widehat{\phi}_0, \widehat{\phi})$ presents an inverse function for the solution, $\phi : x \mapsto \phi(x)$, that is defined piecewise between the obstructions $\partial_x \phi = 0$. Since $\widehat{x}(\widehat{\phi}_0, \widehat{\phi})$ is necessarily monotone in $\widehat{\phi}$ between the obstructions, the function may be inverted on these intervals to recover $\phi(x)$ piecewise. The solutions can be continued past the obstructions by adjoining the piecewise solutions in a manner that satisfies (1) and effective particle energy conservation appropriately.

The turning values are specified by the values of ϕ, which we denote by $\phi(x) = \beta$, and satisfy

$$h = \mu\beta^2 + \nu\beta^{2(\sigma+1)}. \tag{34}$$

These are illustrated in Figures 5-2. For $\sigma = 1$, one may find

$$\beta_n^2 = -2^{-1}\nu^{-1}[\mu + (-1)^n(\mu^2 + 4h\nu)^{1/2}], \quad n = 1, 2. \tag{35}$$

For other values of σ, calculation of the β_n might not be so straightforward but they can be assured to exist due to the simple local monotonicity properties of $h_{\text{p}}(\cdot)$ and thereby are also qualitatively similar to the values for $\sigma = 1$ in that they appear in pairs that are both real, both imaginary, or otherwise accordingly.

To prove that Λ_{line} is onto and study its preimage, it is helpful to first exhaustively classify the solutions of (1) on a line up to translation $x \mapsto x + x_0$, which may also be seen in [**11**], and relate the results to the auxiliary parameter α where possible. We do so by considering the distinguished parameter regions for the effective particle potential energy while recalling that the effective particle kinetic energy $h_{\text{k}}(\phi)$ is necessarily nonnegative.

Case I: $(\nu > 0, \mu \geq 0)$. $h_{\text{p}}(\phi)$ has a global minimum at $\phi = 0$ and $\lim_{\phi \to \pm\infty} h_{\text{p}}(\phi) = \infty$. There are three notable ranges of h:

(1) $h < 0$. There are no solutions.
(2) $h = 0$. There is only the constant solution $\phi(x) = 0$.
(3) $h > 0$. Shown in Figure 5. There is only the solution, which oscillates as $-\beta_1 \leq \phi(x) \leq \beta_1$. This solution is bounded, periodic, attains zeros, and satisfies $\alpha = \beta_1 > 0$.

Then for Case I, elements of Φ_{line} may belong only to the sub-case (3) above, for which $h > 0$, and $(\mu, \alpha) \in P_+^+$.

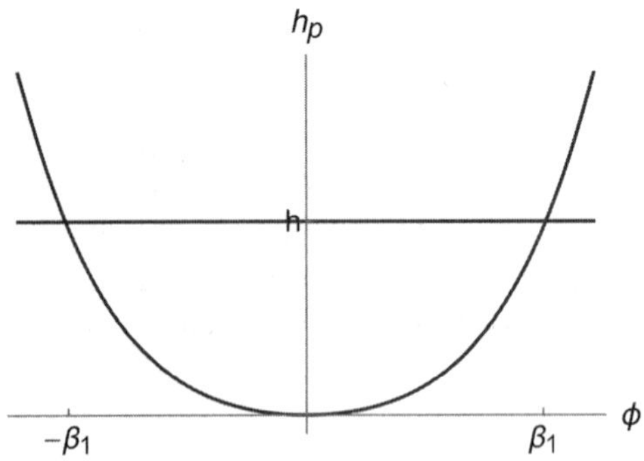

FIGURE 5. Case I: $(\nu > 0, \mu \geq 0)$: $h > 0$

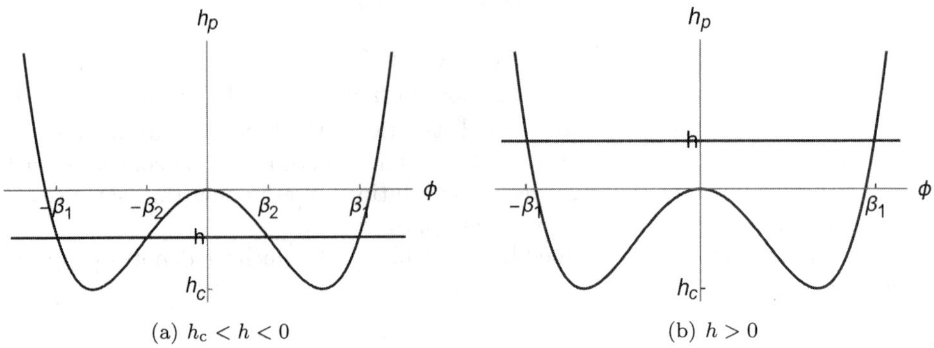

FIGURE 6. Case II: $(\nu > 0, \mu < 0)$

Case II: $(\nu > 0, \mu < 0)$. $h_p(\phi)$ has a local maximum at $\phi = 0$, global minima at $\phi = \pm\alpha_c$, and $\lim_{\phi\to\pm\infty} h_p(\phi) = \infty$. There are five notable ranges of h:

(1) $h < h_c$. There are no solutions.
(2) $h = h_c$. There are only the two constant solutions $\phi = \pm\alpha_c$.
(3) $h_c < h < 0$. Shown in Figure 6 (a). There are two solutions. Each has definite sign and are negatives of one another. The positive solution oscillates as $\beta_2 \leq \phi(x) \leq \beta_1$. These solutions are bounded, periodic, and attain no zeros.
(4) $h = 0$. There are two solutions. They are "soliton solutions" and are negatives of one another. They have the maximum absolute value $||\phi||_\infty = \alpha_0$. One is strictly positive and for $\phi(0) = \alpha_0$ it satisfies $\phi(x) \to 0$ monotonically as $0 < x \to \infty$ and $\phi(-x) = -\phi(x)$ since by (32) one has

$$\left[\lim_{\epsilon\searrow 0} \widehat{x}(\epsilon, \widehat{\phi}) - x_0\right] / \zeta = \lim_{\epsilon\searrow 0} \int_\epsilon^{\widehat{\phi}} \mathrm{d}w\ (|\mu| w^2 - |\nu| w^{2(\sigma+1)})^{-1/2} \tag{36}$$

$$= \lim_{\epsilon\searrow 0} \int_\epsilon^{\widehat{\phi}} \mathrm{d}w\ w^{-1}(|\mu| - |\nu| w^{2\sigma})^{-1/2} = \infty. \tag{37}$$

These solutions are bounded, not periodic, and attain no zeros.

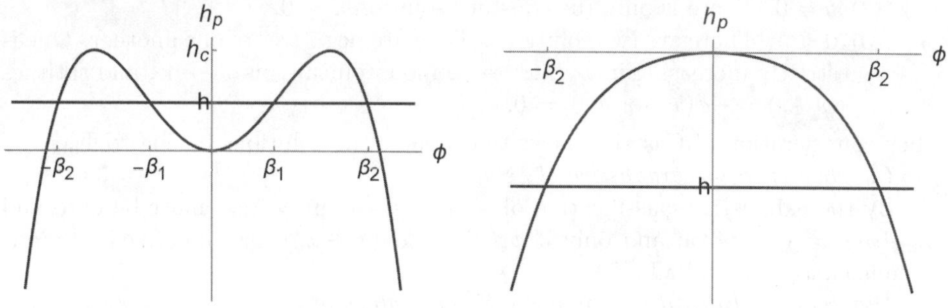

(a) Case III: $(\nu < 0, \mu \geq 0)$: $0 < h < h_c$ (b) Case IV: $(\nu < 0, \mu < 0)$: $h < 0$

FIGURE 7. Some subcases of Cases III and IV

(5) $h > 0$. Shown in Figure 6 (b). There is only the solution, which oscillates as $-\beta_1 \leq \phi(x) \leq \beta_1$. This solution is bounded, periodic, attains zeros, and satisfies $\alpha = \beta_1 > \alpha_0$.

Then for Case II, elements of Φ_{line} may belong only to the sub-case (5) above, for which $h > 0$, and $(\mu, \alpha) \in P_-^+$.

Case III: $(\nu < 0, \mu \geq 0)$. $h_p(\phi)$ has a local minimum at $\phi = 0$, global maxima at $\phi = \pm\alpha_c$, and $\lim_{\phi\to\pm\infty} h_p(\phi) = -\infty$. There are five notable ranges of h:

(1) $h < 0$. There are two solutions. Each have definite sign and are negatives of one another. The positive solution has a minimum value $\phi = \beta_2$. For $\phi(0) = \beta_2$, one has that $\phi(x) \to \infty$ monotonically as $0 \leq x \to \infty$ and $\phi(-x) = -\phi(x)$.

(2) $h = 0$. There are three solutions. Two are analogous to those for $h < 0$ and the remaining one is the constant solution $\phi = 0$.

(3) $0 < h < h_c$. Shown in Figure 7 (a). There are three solutions. Two are analogous to those for $h < 0$. The remaining one oscillates as $-\beta_1 \leq \phi(x) \leq \beta_1$. The oscillating solution is bounded, periodic, attains zeros, and satisfies $\alpha = \beta_1 < \alpha_c$.

(4) $h = h_c$. There is are two solutions. They are "kink solutions" and are negatives of one another. One is strictly increasing in x, $\phi(x) \to \alpha_c$ monotonically as $x \to \infty$, and satisfies $\phi(-x) = -\phi(x)$ for $\phi(0) = 0$, which can be shown by a calculation similar to that of (36) and (37). These solutions are bounded, not periodic, and attain only one zero.

(5) $h_c < h$. There are two solutions. They are negatives of one another. One is strictly increasing in x, $\phi(x) \to \infty$ monotonically as $x \to \infty$, and satisfies $\phi(-x) = -\phi(x)$ for $\phi(0) = 0$.

Then for Case III, elements of Φ_{line} may belong only to the sub-case (3) above, for which $0 < h < h_c$, and $(\mu, \alpha) \in P_-$.

Case IV: $(\nu < 0, \mu < 0)$. $h_p(\phi)$ has a global maximum at $\phi = 0$ and $\lim_{\phi\to\pm\infty} h_p(\phi) = -\infty$. There are three notable ranges of h:

(1) $h < 0$. Shown in Figure 7 (b). There are two solutions. Each have definite sign and are negatives of one another. The positive solution has a minimum value $\phi(x) = \beta_2$. Without loss of generality take $\phi(0) = \beta_2$. One has that $\phi(x) \to \infty$ monotonically as $0 \leq x \to \infty$ and $\phi(-x) = -\phi(x)$.

(2) $h = 0$. There is only the constant solution $\phi = 0$.
(3) $0 < h$. There are two solutions. They are negatives of one another. One is strictly increasing in x, $\phi(x) \to \infty$ monotonically as $x \to \infty$, and satisfies $\phi(-x) = -\phi(x)$ for $\phi(0) = 0$.

Then consideration of Case IV shows that none of its solutions belong to Φ_{line}.

Conclusion to the exhaustive classification

By the exhaustive classification of solutions, the map Λ_{line} must be onto and $\Lambda_{\text{line}}(\phi_1) = \Lambda_{\text{line}}(\phi_2)$ if and only if $\phi_1(x) = \zeta\phi_2(x + x_0)$ for some fixed $\zeta = \pm 1$, some fixed $x_0 \in \mathbb{R}$ and all x.

Different solutions if and only if different values of μ

Assume that $\phi \in \Phi$ and take $x \in \mathbb{R}$ to satisfy at least one of $\phi(x), \partial_x\phi(x)$ differs from zero. Such an x exists by the classification done above. By observing the RHS of (1) one can see that a given ϕ uniquely specifies the μ with which it is associated. This proves that Λ_{line} is well defined.

□

The classification made above for solutions in Φ_{line} allows us to study the map Λ_{int}. We may now prove Lemma 2.3.

PROOF OF LEMMA 2.3. Part (1). Proving that Λ_{int} is well defined follows by the same argument as that which was used for Λ_{line}, as was implemented above in the proof of Proposition 3.2.

Part (2). Let $\phi_1, \phi_2 \in \Phi_{\text{int}}$ such that $\Lambda_{\text{int}}(\phi_1) = \Lambda_{\text{int}}(\phi_2)$. By Part (1) of the Lemma, ϕ_1, ϕ_2 share the same values of μ, α. Hence both of them correspond to trajectories of a classical particle moving in the same potential, which belongs to one of the four cases in the proof of Proposition 3.2. Having the same value of α means that both trajectories have the same energy h as seen in (27). Pick one boundary point of the interval. If the boundary condition at this point is Dirichlet, then both trajectories start at $\phi(x) = 0$ and since they have equal energies their initial velocities are the same up to a sign, from which we conclude that the trajectories are equal up to a sign, i.e. $\phi_1 = \zeta\phi_2$ where $\zeta = \pm 1$. Alternatively, if the boundary condition is Neumann, then both trajectories start at $\phi(x) = \pm\alpha$ with zero velocity and once again this implies that they are equal up sign, i.e. $\phi_1 = \zeta\phi_2$ where $\zeta = \pm 1$. The proof is finished once we note that $\Lambda_{\text{int}}(\phi) = \Lambda_{\text{int}}(-\phi)$. □

Next is an easy but important Lemma that establishes a connection between solutions of (1) on an interval and on a line.

LEMMA 3.3. *Every solution* $\phi \in \Phi_{\text{int}}$ *is a restriction of a solution* $\widehat{\phi} = \Lambda_{\text{line}}^{-1} \circ \Lambda_{\text{int}}(\phi)$ *from the line to an appropriate interval, where* $||\widehat{\phi}||_\infty = ||\phi||_\infty$. *Note that* Λ_{line} *is not injective and hence* $\Lambda_{\text{line}}^{-1}$ *is not uniquely defined however, for the sake of the statement, any preimage of* Λ_{line} *can be chosen as the image of* $\Lambda_{\text{line}}^{-1}$.

PROOF. Given a solution $\phi \in \Phi_{\text{int}}$ we apply Λ_{int} to get the corresponding (μ, α). By the classification of solutions done in the proof of Proposition 3.2, this (μ, α) corresponds to a particle trajectory on the line, i.e. $\widehat{\phi} \in \Phi_{\text{line}}$. Our solution ϕ serves as a sub-trajectory and hence can be obtained as a restriction of $\widehat{\phi}$. The equality $||\widehat{\phi}||_\infty = ||\phi||_\infty$ follows as the trajectory ϕ always attains the maximal absolute value of the trajectory $\widehat{\phi}$ either at an endpoint if there is a Neumann condition there or somewhere in between if both boundary conditions are Dirichlet. □

4. The wavelength λ

The elements of Φ_{line} are periodic on the line. We call this period the *wavelength* and denote it by λ. In this section we study the dependence of λ on the parameters μ, α, which would allow the classification of elements of Φ_{int}.

DEFINITION 4.1. *Fix $\nu \neq 0$, for $(\mu,\alpha) \in P$ denote:*

$$\kappa_{(\mu,\alpha)}(w) := [\mu(1-w^2) + \nu\alpha^{2\sigma}(1-w^{2(\sigma+1)})]^{-1/2}, \quad w \in [0,1]. \tag{38}$$

PROPOSITION 4.2. *For each $(\mu,\alpha) \in P$, the solution $\phi = \phi_{(\mu,\alpha)} \in \Phi_{\text{line}}$ is periodic with wavelength (period) $\lambda \equiv \lambda_{(\mu,\alpha)}$ of the form*

$$\lambda_{(\mu,\alpha)} = 4\int_0^1 dw\ \kappa_{(\mu,\alpha)}(w) \tag{39}$$

and that satisfies the following properties.

$\underline{(\mu,\alpha) \in P^+_+}$:

$$\lim_{\mu\to\infty} \lambda_{(\mu,\alpha)} = 0, \quad \lim_{\alpha\to 0} \lambda_{(\mu,\alpha)} = 2\pi\mu^{-1/2}, \quad \partial_\mu\lambda_{(\mu,\alpha)} < 0, \quad \partial_\alpha\lambda_{(\mu,\alpha)} < 0, \tag{40}$$

$\underline{(\mu,\alpha) \in P^+_-}$:

$$\lim_{\mu\searrow\mu_0} \lambda_{(\mu,\alpha)} = \infty, \quad \partial_\mu\lambda_{(\mu,\alpha)} < 0, \quad \partial_\alpha\lambda_{(\mu,\alpha)} < 0, \tag{41}$$

$\underline{(\mu,\alpha) \in P^-}$:

$$\lim_{\mu\to\infty} \lambda_{(\mu,\alpha)} = 0, \quad \lim_{\alpha\to 0} \lambda_{(\mu,\alpha)} = 2\pi\mu^{-1/2}, \quad \lim_{\mu\searrow\mu_c} \lambda_{(\mu,\alpha)} = \infty, \tag{42}$$

$$\partial_\mu\lambda_{(\mu,\alpha)} < 0, \quad \partial_\alpha\lambda_{(\mu,\alpha)} > 0, \tag{43}$$

where $\mu_0 = -|\nu|\alpha^{2\sigma}$, $\mu_c = (\sigma+1)|\nu|\alpha^{2\sigma}$.

PROOF. Proof of representation of λ in (39). For solutions of the form $\phi = \phi_{(\mu,\alpha)} \in \Phi_{\text{line}}$ for, $(\mu,\alpha) \in P$, one may calculate the quarter wavelength through

$$\lambda_{(\mu,\alpha)}/4 = \int_0^\alpha dw\ [h - \mu w^2 - \nu w^{2(\sigma+1)}]^{-1/2} \tag{44}$$

$$= \int_0^1 dw\ \alpha[h - \mu\alpha^2 w^2 - \nu\alpha^{2(\sigma+1)}w^{2(\sigma+1)}]^{-1/2} \tag{45}$$

$$= \int_0^1 dw\ [\alpha^{-2}h - \mu w^2 - \nu\alpha^{2\sigma}w^{2(\sigma+1)}]^{-1/2}. \tag{46}$$

By following the analogy to particle dynamics, one may consider λ to be the particle period of oscillation in time.

Since the effective particle total energy satisfies

$$h = (\partial_x\phi)^2 + \mu\phi^2 + \nu\phi^{2(\sigma+1)} \tag{47}$$

one has

$$h = \mu\alpha^2 + \nu\alpha^{2(\sigma+1)} \tag{48}$$

and therefore

$$\lambda_{(\mu,\alpha)}/4 = \int_0^1 dw\ [\mu(1-w^2) + \nu\alpha^{2\sigma}(1-w^{2(\sigma+1)})]^{-1/2} = \int_0^1 dw\ \kappa_{(\mu,\alpha)}(w), \tag{49}$$

which proves the representation of λ in (39).

Proof of regular limits in (40), (41), (42). Consider that $\nu > 0$. By inspection we have that $\lim_{\mu\to\infty} \kappa_{(\mu,\alpha)} = 0$ uniformly, and therefore $\lim_{\mu\to\infty} \lambda_{(\mu,\alpha)} = 0$, which proves the first relations of (40) and (42).

One may write

$$\kappa_{(\mu,\alpha)}(w) = \left(\mu + \nu\alpha^{2\sigma}\sum_{j=0}^{\sigma} w^{2j}\right)^{-1/2} (1-w^2)^{-1/2}, \tag{50}$$

which is a form that is very useful for analysis of the integral. For $\nu > 0$ one has

$$\kappa_{(\mu,\alpha)}(w) \tag{51}$$

$$= \left(\mu + \nu\alpha^{2\sigma}\sum_{j=0}^{\sigma} w^{2j}\right)^{-1/2} (1-w^2)^{-1/2} \le (\mu+\nu\alpha^{2\sigma})^{-1/2}(1-w^2)^{-1/2} \tag{52}$$

for all $w \in (0,1)$ and for $\nu < 0$ one has

$$\kappa_{(\mu,\alpha)}(w) \le [\mu + (\sigma+1)\nu\alpha^{2\sigma}]^{-1/2}(1-w^2)^{-1/2} \tag{53}$$

for all $w \in (0,1)$, both bounds of $\kappa_{(\mu,\alpha)}(w)$ are positive and integrable. These bounds justify the application of dominated convergence theorem to evaluate

$$\mu > 0: \lim_{\alpha\to 0} \lambda_{(\mu,\alpha)} = \int_0^1 \lim_{\alpha\to 0} \kappa_{(\mu,\alpha)} = 4\mu^{-1/2}\int_0^1 (1-w^2)^{-1/2} = 2\pi\mu^{-1/2}, \tag{54}$$

which proves the second relations in (40) and (42).

Proof of singular limits in (41), (42). Take $\nu > 0$ and $\mu < 0$. We recall that $\mu_0 = -|\nu|\alpha^{2\sigma}$ and observe that for $\mu = \mu_0 + \epsilon$, $\epsilon > 0$, one has

$$\kappa_{(\mu,\alpha)}(w) = [\mu(1-w^2) + \nu\alpha^{2\sigma}(1-w^{2(\sigma+1)})]^{-1/2} \tag{55}$$

$$= (\mu - \mu w^2 + \nu\alpha^{2\sigma} - \nu\alpha^{2\sigma}w^{2(\sigma+1)})^{-1/2} \tag{56}$$

$$= (\epsilon - \mu_0 w^2 - \epsilon w^2 - \nu\alpha^{2\sigma}w^{2(\sigma+1)})^{-1/2} \tag{57}$$

$$\ge (\epsilon + |\mu_0|w^2)^{-1/2} = |\mu_0|^{-1/2}(\epsilon/|\mu_0| + w^2)^{-1/2} \tag{58}$$

and then by direct calculation

$$\lim_{\epsilon\to 0} \lambda_{(\mu,\alpha)}/4 \ge \lim_{\epsilon\to 0}\int_0^1 dw\ |\mu_0|^{-1/2}(\epsilon/|\mu_0| + w^2)^{-1/2} = \infty, \tag{59}$$

which proves the first relation in (41).

Take $\nu < 0$ and $\mu > 0$. We recall that $\mu_c = (\sigma+1)|\nu|\alpha^{2\sigma}$ and observe that for $\mu = \mu_c(1-\epsilon)^{-2\sigma}$, $\epsilon > 0$, one has

$$\kappa_{(\mu,\alpha)}(w) = \left(\mu + \nu\alpha^{2\sigma}\sum_{j=0}^{\sigma} w^{2j}\right)^{-1/2} (1-w^2)^{-1/2} \tag{60}$$

$$= \left[\mu_c(1-\epsilon)^{-2\sigma} - |\nu|\alpha^{2\sigma}\sum_{j=0}^{\sigma} w^{2j}\right]^{-1/2} (1-w^2)^{-1/2} \tag{61}$$

$$\ge \left[\mu_c(1-\epsilon)^{-2\sigma} - (\sigma+1)|\nu|\alpha^{2\sigma}w^{2\sigma}\right]^{-1/2}(1-w^2)^{-1/2} \tag{62}$$

$$= [(1-\epsilon)^{2\sigma}/\mu_c]^{1/2}[1-(1-\epsilon)^{2\sigma}w^{2\sigma}]^{-1/2}(1-w^2)^{-1/2} \tag{63}$$

$$\geq \mu_c^{-1/2}(1-\epsilon)^\sigma[1-(1-\epsilon)^{2\sigma}w^{2\sigma}]^{-1/2}[1-(1-\epsilon)^{2\sigma}w^{2\sigma}]^{-1/2} \tag{64}$$

$$= \mu_c^{-1/2}(1-\epsilon)^\sigma[1-(1-\epsilon)^{2\sigma}w^{2\sigma}]^{-1} \tag{65}$$

$$= \mu_c^{-1/2}(1-\epsilon)^\sigma \left\{\sum_{j=0}^{2\sigma-1}[(1-\epsilon)w]^j\right\}^{-1}[1-(1-\epsilon)w]^{-1} \tag{66}$$

$$\geq (2\sigma)^{-1}\mu_c^{-1/2}(1-\epsilon)^\sigma[1-(1-\epsilon)w]^{-1}, \tag{67}$$

and then by direct calculation

$$\lim_{\epsilon\to 0}\lambda_{(\mu,\alpha)}/4 \geq \lim_{\epsilon\to 0}\int_0^1 dw\ (2\sigma)^{-1}\mu_c^{-1/2}(1-\epsilon)^\sigma[1-(1-\epsilon)w]^{-1} \tag{68}$$

$$= (2\sigma)^{-1}\mu_c^{-1/2}\lim_{\epsilon\to 0}\int_0^1 dw\ [1-(1-\epsilon)w]^{-1} \tag{69}$$

$$= (2\sigma)^{-1}\mu_c^{-1/2}\lim_{\epsilon\to 0}[-(1-\epsilon)^{-1}\log(1-\{1-\epsilon\}w)]_0^1 = \infty, \tag{70}$$

which proves the third relation in (42).

Proof of monotonicity in (40), (41), (43). By inspection one can see that $\kappa_{(\mu,\alpha)}(w)$, $\partial_\mu\kappa_{(\mu,\alpha)}(w)$, $\partial_\alpha\kappa_{(\mu,\alpha)}(w)$ are continuous in μ, α, w, and therefore by Leibniz integral rule:

$$\partial_\mu\int_0^1 dw\ \kappa_{(\mu,\alpha)}(w) = \int_0^1 dw\ \partial_\mu\kappa_{(\mu,\alpha)}(w), \tag{71}$$

$$\partial_\alpha\int_0^1 dw\ \kappa_{(\mu,\alpha)}(w) = \int_0^1 dw\ \partial_\alpha\kappa_{(\mu,\alpha)}(w). \tag{72}$$

One may thereby determine the signs of $\partial_\mu\lambda_{(\mu,\alpha)}$ and $\partial_\alpha\lambda_{(\mu,\alpha)}$ from

$$\partial_\mu\kappa_{(\mu,\alpha)}(w) = -2^{-1}\left(\mu+\nu\alpha^{2\sigma}\sum_{j=0}^{\sigma}w^{2j}\right)^{-3/2}(1-w^2)^{-1/2} \tag{73}$$

$$\partial_\alpha\kappa_{(\mu,\alpha)}(w) = -\sigma\nu\alpha^{2\sigma-1}\sum_{j=0}^{\sigma}w^{2j}\left(\mu+\nu\alpha^{2\sigma}\sum_{j=0}^{\sigma}w^{2j}\right)^{-3/2}(1-w^2)^{-1/2}, \tag{74}$$

which proves the last two relations in (40), (41), (43). □

5. Proof of Theorem 2.4

We are now prepared to directly prove our results on an interval.

PROOF OF THEOREM 2.4. Proof of Parts (1), (2). By Lemma 3.3, elements of Φ_{int} may be formed by restricting elements of Φ_{line} to functions on an interval. Such restrictions must be made so that the endpoints are zeros or local extrema as needed to satisfy the Dirichlet or Neumann boundary points. All elements of Φ_{line} are periodic with wavelength (period) λ. Therefore $\phi = \phi_{(\mu,\alpha)} \in \Phi_{\text{int}}$ and ϕ has a given number of $n-1$ internal zeros if one can ensure that $\lambda_{(\mu,\alpha)} = \lambda_n$, $n \in \mathcal{N}$, is chosen appropriately:

- For one matching condition of Dirichlet type and one of Neumann type one requires $\lambda_n = 4l/(2n-1)$ where $n \in \mathbb{N}$.

- For both boundary conditions of Dirichlet type one requires $\lambda_n = 2l/n$ where $n \in \mathbb{N}$.
- For both boundary conditions of Neumann type one requires $\lambda_n = 2l/n$ where $n \in \mathbb{N} \setminus \{1\}$.

We note that if both boundary conditions are Neumann, then by our construction there are no stationary solutions without zeros, hence $\mathscr{N} = \mathbb{N} \setminus \{1\}$ for this case.

In each case, we must ensure that the wavelength takes on the appropriate value, $\lambda_{(\mu,\alpha)} = \lambda_n$, which guarantees that $\phi_{(\mu,\alpha)}$ has $n-1$ internal zeros. Now we will show that this can be done for each $\alpha > 0$. Take to be any fixed values $\alpha > 0$ and $\nu \neq 0$. By the monotonicity properties of Proposition 4.2, $f_\alpha : \mu \mapsto f_\alpha(\mu) = \lambda_{(\mu,\alpha)}$ is a monotone strictly decreasing function and, due to the limiting properties seen in Proposition 4.2, that is surjective on $(0, \infty)$. Therefore there must exist only one $\mu = f_\alpha^{-1}(\lambda_n)$ for which $\lambda_{(\mu,\alpha)} = \lambda_n$ and one may write

$$\gamma_n = \bigcup_{\alpha \in (0,\infty)} \{(f_\alpha^{-1}(\lambda_n), \alpha)\}, \tag{75}$$

which must be a connected curve because of continuity of $f_\alpha^{-1}(\lambda_n)$ in $\alpha \in (0, \infty)$. We also observe that, due to the above representation, each γ_n is a level curve of $\lambda_{(\mu,\alpha)}$ along $\lambda = \lambda_n$. This implies that the γ_n are mutually disjoint. The gradient of $\lambda_{(\mu,\alpha)}$ on the (μ, α) half plane is nonvanishing by the monotonicity properties of Proposition 4.2 and this implies that the level curves γ_n are non self intersecting.

Proof of Part (3). From above, one may construct each spectral curve γ_n by taking the union of points obtained from $(f_\alpha^{-1}(\lambda_n), \alpha)$ by allowing α to vary on $(0, \infty)$. Therefore, through this construction, fixing α furnishes exaclty one point $(f_\alpha^{-1}(\lambda_n), \alpha) \in \gamma_n$. Since λ_n and $f_\alpha^{-1}(\lambda)$ are monotonically strictly decreasing respectively in n and λ, it follows that $\mu_n = f_\alpha^{-1}(\lambda_n)$ is monotonically strictly increasing in n.

Proof of Part (4). By the small α limit properties of Proposition 4.2 in (40) and (42) and the fact that $\alpha > \alpha_0$ for $(\mu, \alpha) \in P_-^+$, one has that $\lim_{\alpha \to 0} \lambda_{(\mu,\alpha)}$ exists only for $\mu > 0$ and for this case one has $\lim_{\alpha \to 0} \lambda_{(\mu,\alpha)} = 2\pi\mu^{-1/2}$, which is the wavelength of the linear system. Matching the wavelength in this limit gives the eigenvalue $\mu_1^{\text{lin}} = 0$ of the linear problem. We note that μ_1^{lin} is an eigenvalue of the linear problem with Neumann conditions at both ends of an interval. Yet it is not obtained as a limit of any spectral curve as γ_1 is missing from Φ_{int} and in this case we actually do not have any solutions without zeros at all. $\square$

6. Proof of Theorem 2.9

We prove Theorem 2.9 by considering a set of d semi-infinite rays, each of which corresponds to one of the star edges. We equip each ray with the same boundary condition as its corresponding edge. Next we identify d points, one on each ray, and require matching conditions on them such that combining the solutions of (1) on the rays yields a solution of (10). Hence we start by describing a set of solutions of (1) on the ray.

Let $\nu \neq 0$ and fix a Dirichlet or Neumann boundary condition on the endpoint of a ray, we denote

$$\Phi_{\text{ray}} := \{\phi \neq 0 \text{ solves (1) on a ray, such that } \phi \text{ is periodic and attains zeros}\}. \tag{76}$$

We next follow the same path as for the line and the interval and parametrize solutions of (1) on a ray with points $(\mu, \alpha) \in P$, which we write as $\phi = \phi_{(\mu,\alpha)}$

LEMMA 6.1. *The following holds.*

(1) *Let $\nu \neq 0$. Given $\phi \in \Phi_{\text{ray}}$, there is a unique value of $\mu \in \mathbb{R}$ such that ϕ is a solution of (1) on a ray with the given values μ, ν. This allows us to define the map*

$$\Lambda_{\text{ray}} : \Phi_{\text{ray}} \to P, \quad \Lambda_{\text{ray}} : \phi \mapsto (\mu, ||\phi||_\infty), \tag{77}$$

which is onto and two to one since $\Lambda_{\text{ray}}(\phi_1) = \Lambda_{\text{ray}}(\phi_2)$ if and only if $\phi_1 = \zeta\phi_2$ for some fixed $\zeta = \pm 1$.

(2) *Every solution $\phi \in \Phi_{\text{ray}}$ is a restriction of the solution $\widehat{\phi} = \Lambda_{\text{line}}^{-1} \circ \Lambda_{\text{ray}}(\phi)$ from the line to a ray, where $||\widehat{\phi}||_\infty = ||\phi||_\infty$. Note that Λ_{line} is not injective and hence $\Lambda_{\text{line}}^{-1}$ is not uniquely defined however, for the sake of the statement, any preimage of Λ_{line} can be chosen as the image of $\Lambda_{\text{line}}^{-1}$.*

PROOF. Part (1). The proofs of uniqueness of μ for a ray and that Λ_{ray} is onto are the same as that for the line, seen in the proof of Proposition 3.2. The proof that Λ_{ray} is two to one is the same as that of the proof of Lemma 2.3 Part (1).

Part (2). The proof follows from the same argument as that of the proof of Lemma 2.3 Part (2). □

We have just established the existence and properties of Λ_{ray}. The same was done for Λ_{star} in Lemma 2.7, whose proof follows the same lines as Lemma 2.3.

PROOF OF THEOREM 2.9 PART (1). Take a set of $d > 2$ rays of the form $L_j = (-\infty, l_j]$ and take $\Phi_{\text{ray},j}$ to be the space of solutions of (1) on each such ray with the same boundary condition as the j-th edge of the star. Furthermore, define $\Lambda_{\text{ray},j} : \Phi_{\text{ray},j} \to P$ to act as $\Lambda_{\text{ray},j} : \phi_j \mapsto (\mu, \alpha_j = ||\phi_j||_\infty)$. Take $\phi_{(q_*)} \in \Phi_{\text{star}}$ as in the statement of the theorem. Now for each j we choose $\widetilde{\Phi}_{\text{ray},j} \subset \Phi_{\text{ray},j}$ such that $\Lambda_{\text{ray},j} \restriction_{\widetilde{\Phi}_{\text{ray},j}}$ is one to one. Explicitly, this choice is made as follows. If the j-th ray has a Neumann condition, then

$$\widetilde{\Phi}_{\text{ray},j} = \{\phi_j \in \Phi_{\text{ray},j} : \text{sgn}(\phi_j(l_j)) = \text{sgn}(\phi_{(q_*),j}(l_j))\} \tag{78}$$

and alternatively if the j-th ray has a Dirichlet condition, then

$$\widetilde{\Phi}_{\text{ray},j} = \{\phi_j \in \Phi_{\text{ray},j} : \text{sgn}(\partial_x\phi_j(l_j)) = \text{sgn}(\partial_x\phi_{(q_*),j}(l_j))\}. \tag{79}$$

We further define the vector of constraints $N : Q \to \mathbb{R}^d$ to act as

$$N_1(q) := \sum_{j=1}^{d} \partial_x\phi_j(0), \tag{80}$$

$$N_j(q) := \phi_j(0) - \phi_1(0), \quad j = 2, \ldots, d, \tag{81}$$

where $\phi_j \in \widetilde{\Phi}_{\text{ray},j}$ is uniquely obtained from $q = (\mu, \alpha_1, \ldots, d)$ by requiring $(\mu, \alpha_j) = \Lambda_{\text{ray},j}(\phi_j)$ for all j. We introduce the following useful notation. For each $\phi_{(q)} \in \Phi_{\text{star}}$, denote by $\lambda_{(q)} \in \mathbb{R}^d$ the vector whose j-th component is the wavelength of the $\phi_j \in \widetilde{\Phi}_{\text{ray},j}$ corresponding to the j-th edge component of $\phi_{(q)}$, i.e. $\lambda_{(q),j} = \lambda_{(\mu,\alpha_j)}$.

We observe that zeros of N correspond to solutions of (10) in the following sense. For each $q = (\mu, \alpha_1, \ldots, \alpha_d)$ that satisfies $N(q) = 0$, take $\phi_j \in \widetilde{\Phi}_{\text{ray},j}$ such that $(\mu, \alpha_j) = \Lambda_{\text{ray},j}(\phi_j)$ for all j. Then there exists a solution of (10), $\psi \in \Phi_{\text{star}}$, such that $\psi_j = \phi_j \restriction_{[0,l_j]}$ for all j. The validity of this solution indeed holds since the central vertex condition is satisfied as $N(q) = 0$. This motivates the application of Implicit Function Theorem to show the existence of a local spectral curve through $q_* \in Q$. In order to do so we need to assure the continuous differentiability of $N(q)$ in $q \in Q$, which is guaranteed by the following lemma, whose proof is postponed.

LEMMA 6.2. *For* $\phi_{(\mu,\alpha)} \in \Phi_{\text{ray}}$, $\phi_{(\mu,\alpha)}$ *and* $\partial_x \phi_{(\mu,\alpha)}$ *are continuously differentiable in* μ *and* α.

Now we aim to show that the Jacobian matrix $\partial_{\alpha_k} N_j(q_*)$ has a nonzero determinant. Let the linear operator $J : \mathbb{R}^d \to \mathbb{R}^d$ be defined by $J_{j,k} = \partial_{\alpha_k} N_j(q_*)$. We utilize the block decomposition

$$J = \begin{bmatrix} A & B \\ C & D \end{bmatrix}, \tag{82}$$

where

$$J_{1,1} = A := \partial_{\alpha_1} \partial_x \phi_{(q_*),1}(0), \quad J_{1,j} = B_j := \partial_{\alpha_j} \partial_x \phi_{(q_*),j}(0), \tag{83}$$

$$J_{j,1} = C_j := -\partial_{\alpha_1} \phi_{(q_*),1}(0), \quad J_{j,j} = D_{j,j} := \partial_{\alpha_j} \phi_{(q_*),j}(0), \tag{84}$$

and where $j = 2, \ldots, d$ and $J_{j,k} = 0$ otherwise.

We note that the diagonal components of D are nonvanishing by the following argument. We now make use of Lemma 6.5, which appears at the end of this section, and this is where the assumption that ϕ vanishes at the central vertex is being used. By (127) there, one has that

$$D_{j,j} = \partial_\alpha \phi_{(q_*),j}(0) = \zeta \xi \alpha_j (\mu + \nu \alpha_j^{2\sigma})^{1/2} \partial_\alpha \lambda_{(q),j} \restriction_{q=q_*} . \tag{85}$$

The above RHS vanishes only if one of α_j, $(\mu + \nu\alpha_j^{2\sigma})^{1/2}$, or $\partial_\alpha \lambda_{(q),j}$ vanishes. By the definition of P, neither α_j nor $(\mu + \nu\alpha_j^{2\sigma})^{1/2}$ can vanish. By Proposition 4.2 one has that $\partial_\alpha \lambda_{(q),j} \neq 0$ for all $\alpha > 0$, and therefore cannot vanish. Thus, the tentative assumption cannot hold and $D_{j,j} \neq 0$ for $j = 2, \ldots, d$.

Since $\det D = \prod_{j=2}^d D_{j,j} \neq 0$ one may find

$$J = \begin{bmatrix} A & B \\ C & D \end{bmatrix} = \begin{bmatrix} I & 0 \\ 0 & D \end{bmatrix} \begin{bmatrix} A & B \\ D^{-1}C & I \end{bmatrix}. \tag{86}$$

then

$$\det(A - BD^{-1}C) = \det J / \det D = A - \sum_{j=2}^d B_j C_j / D_{j,j} \tag{87}$$

$$= \{\partial_{\alpha_1} \partial_x \phi_{(q),1} - \sum_{j=2}^d \partial_{\alpha_j} \partial_x \phi_{(q),j} [-\partial_{\alpha_1} \phi_{(q),1}] / \partial_{\alpha_j} \phi_{(q),j}\} \restriction_{x=0,q=q_*} \tag{88}$$

$$= \partial_{\alpha_1} \phi_{(q),1} \sum_{j=1}^d \partial_{\alpha_j} \partial_x \phi_{(q),j} / \partial_{\alpha_j} \phi_{(q),j} \restriction_{x=0,q=q_*} . \tag{89}$$

Since $\partial_{\alpha_1} \phi_{(q),1} \restriction_{x=0,q=q_*} \neq 0$ by the same argument that gives $D_{j,j} \neq 0$, if we denote

$$S := \sum_{j=1}^d \partial_{\alpha_j} \partial_x \phi_{(q),j} / \partial_{\alpha_j} \phi_{(q),j} \restriction_{x=0,q=q_*} \tag{90}$$

then $\det J = 0$ if and only if $S = 0$. We will now show that $S \neq 0$. We define

$$t_j := [\mu + (\sigma+1)\nu\alpha_j^{2\sigma}]/(\mu + \nu\alpha_j^{2\sigma}) \tag{91}$$

and note that due to (127) and (128) of Lemma 6.5 one may write

$$S = \sum_{j=1}^{d} t_j/(\xi_j\alpha_j\partial_\alpha\lambda_{(q),j}) \restriction_{q=q_*} . \tag{92}$$

We recall from Proposition 4.2 that one has $\partial_\alpha\lambda_{(q),j} < 0$ for $\nu > 0$ and $\partial_\alpha\lambda_{(q),j} > 0$ for $\nu < 0$. Therefore $S \neq 0$ if $\mathrm{sgn}(t_j)$ is nonvanishing and constant for all j.

First consider $(\mu, \alpha_j) \in P_+^+$ for all j; clearly since $\nu, \mu, \alpha > 0$ it must be true that $t_j > 0$. Now consider $(\mu, \alpha_j) \in P_-^+$ for all j; one has

$$t_j = [\mu + (\sigma+1)\nu\alpha_j^{2\sigma}]/(\mu + \nu\alpha_j^{2\sigma}) = [(\sigma+1)|\nu/\mu|\alpha_j^{2\sigma} - 1]/(|\nu/\mu|\alpha_j^{2\sigma} - 1), \tag{93}$$

and since $\alpha_j \in (\alpha_0, \infty)$ and $\alpha_0 = |\mu/\nu|^{1/2\sigma}$ we get that $t_j > 2 > 0$. Lastly consider $(\mu, \alpha_j) \in P^-$ for all j; one has

$$t_j = [\mu + (\sigma+1)\nu\alpha_j^{2\sigma}]/(\mu + \nu\alpha_j^{2\sigma}) = [1 - (\sigma+1)|\nu/\mu|\alpha_j^{2\sigma}]/(1 - |\nu/\mu|\alpha_j^{2\sigma}), \tag{94}$$

and since $\alpha_j \in (0, \alpha_c)$ and $\alpha_c = |\mu/(\sigma+1)\nu|^{1/2\sigma}$ we get that $t_j > 0$. Therefore it must be true that $t_j > 0$ for all j. Thus $\det S \neq 0$ and in turn $\det J \neq 0$.

We have ensured that requirements for Implicit Function Theorem are assured. Therefore its application, as stated in the beginning of the proof, guarantees the existence of a local spectral curve through q_*. □

PROOF OF THEOREM 2.9 PART (2). Along a local spectral curve the solution always satisfies the matching conditions and is continuous. Although the shape of the oscillations can change slightly, the most dramatic change occurs at the central vertex. If $\phi_{(q_*)}(0) = 0$, then $\mu \mapsto \mu + \delta\mu$ can yield only three results: $\phi_{(q)}(0) = 0$, $\phi_{(q)}(0) > 0$, or $\phi_{(q)}(0) < 0$. One can always find a $\delta\mu$ small enough so that the variation does not move through more than one of these cases and in this sense one must take q to be appropriately close to q_*. If the central value remains zero, then no change in nodal count can occur, hence the need for the factor of $\mathrm{sgn}^2(\phi_{(q)})$ in the RHS of (17). If the central value increases from zero, then the zero at the center vanishes and a zero forms on each edge for which $\partial_x\phi_{(q_*),j}(0) < 0$. The contribution to the nodal count change is then a sum of a term -1 for the vanishing of the zero at the center and a term $2^{-1}[1 - \mathrm{sgn}(\phi_{(q)}\partial_x\phi_{(q_*),j})] \restriction_{x=0}$ for each edge. If the central value decreases from zero, then the zero at the center vanishes and a zero forms on each edge for which $\partial_x\phi_{(q_*),j}(0) > 0$. The contribution to the nodal count change is the same as before, i.e. a sum of a term -1 for the vanishing of the zero at the center and a term $2^{-1}[1 - \mathrm{sgn}(\phi_{(q)}\partial_x\phi_{(q_*),j})] \restriction_{x=0}$ for each edge. Combining these contributions and considerations for each case, one has

$$Z(\phi_{(q)}) - Z(\phi_{(q_*)}) = \mathrm{sgn}^2(\phi_{(q)}) \left\{ -1 + 2^{-1} \sum_{j=1}^{d} [1 - \mathrm{sgn}(\phi_{(q)}\partial_x\phi_{(q_*),j})] \right\} \restriction_{x=0} \tag{95}$$

$$= \mathrm{sgn}^2(\phi_{(q)}) \left[-1 + 2^{-1}d - 2^{-1}\mathrm{sgn}(\phi_{(q)}) \sum_{j=1}^{d} \mathrm{sgn}(\partial_x\phi_{(q_*),j}) \right] \restriction_{x=0} . \tag{96}$$

□

Proof of Theorem 2.9 Part (3). We are at liberty to construct an appropriate $\phi_{(q_*)} \in \Phi_{\text{star}}$ so that it automatically satisfies the desired properties. For only Dirichlet conditions on exterior vertices and $\phi(0) = 0$ we must have $l_j = n_j \lambda_{(q),j}/2$ for some $n_j \in \mathbb{N}$ and all j. Since for $\mu = 0$ one has from equation (144) of Lemma 6.6, which appears at the end of this section, that

$$\lambda_{(0,\alpha_j)} = 4c_1 \nu^{-1/2} \alpha_j^{-\sigma}, \tag{97}$$

where c_1 is given in (147) of Lemma 6.6, it must then be the case that

$$\alpha_j = (2c_1 \nu^{-1/2} n_j / l_j)^{1/\sigma}. \tag{98}$$

Since $\phi_{(q_*),j}(0) = 0$ for all j, it is the case that $0 = N_j(q_*)$, as required by (80) and (81), is automatically satisfied for $j = 2, \ldots, d$. We now check what is required for the matching condition at the central vertex. We take

$$\zeta_j = \operatorname{sgn}(\partial_x \phi_{(q_*),j}(0)) \tag{99}$$

for all j. We recall from the definition of N_1 in (80), the expression of $\partial_x \phi$ in (27), and $\mu_* = 0$ in turn

$$0 = N_1(q_*) = \sum_{j=1}^{d} \partial_x \phi_{(q_*),j}(0) = \sum_{j=1}^{d} \zeta_j [\mu \alpha_j^2 + \nu \alpha_j^{2(\sigma+1)}]^{1/2} \restriction_{q=q_*} \tag{100}$$

$$= \sum_{j=1}^{d} \zeta_j \nu^{1/2} \alpha_j^{\sigma+1} \restriction_{q=q_*} = \sum_{j=1}^{d} \zeta_j \nu^{1/2} (2c_1 \nu^{-1/2} n_j / l_j)^{1+1/\sigma} \tag{101}$$

$$\Leftrightarrow \quad 0 = \sum_{j=1}^{d} \zeta_j (n_j / l_j)^{1+1/\sigma}, \tag{102}$$

which is automatically satisfied by assumption for appropriately chosen ζ_j, n_j. Since we have specified α_j and $\operatorname{sgn}(\partial_x \phi_{(q_*),j}(0))$, for all j, the solution $\phi_{(q_*)}$ has been uniquely determined. The nodal count $Z(\phi_{(q_*)}) = 1 - d + \sum_{j=1}^{d} n_j$ follows by inspection of the hereto constructed solution $\phi_{(q_*)}$, which completes the proof. □

Remark 6.3. *As mentioned after the statement of the theorem, the proof above generalizes to the case where some (or all) of the Dirichlet boundary conditions are replaced by Neumann ones. This may be obtained by replacing, in the assumed condition and proof, n_j with $n_j - 1/2$ for each edge that possesses a Neumann boundary condition at its end. Nevertheless, the nodal count expression is unchanged.*

Proof of Theorem 2.9 Part (4). We first combine the conclusions of Parts (1) and (3) of the Theorem and get the existence of a local spectral curve $\gamma(\mu) = (\mu, \gamma_1(\mu), \ldots, \gamma_d(\mu))$ through q_* given in (3). We now tentatively assume for the sake of contradiction that

$$\frac{\mathrm{d}}{\mathrm{d}\mu} \phi_{(\gamma(\mu_*)),j}(0) = 0, \tag{103}$$

where the complete derivative is taken along the spectral curve γ mentioned above. From $\mathrm{d}\phi_{(\gamma(\mu_*)),j}/\mathrm{d}\mu = 0$ we also conclude that $\mathrm{d}\lambda_{(\gamma(\mu_*)),j}/\mathrm{d}\mu = 0$, for which one must have

$$0 = \frac{\mathrm{d}}{\mathrm{d}\mu}\lambda_{(q_*),j} = \partial_\mu \lambda_{(q_*),j} + \partial_{\alpha_j}\lambda_{(q_*),j}\frac{\mathrm{d}}{\mathrm{d}\mu}\gamma_j(\mu_*) \tag{104}$$

$$\Rightarrow \frac{\mathrm{d}}{\mathrm{d}\mu}\gamma_j(\mu_*) = -\partial_\mu\lambda_{(q_*),j}/\partial_{\alpha_j}\lambda_{(q_*),j}, \tag{105}$$

for all $j = 1, \dots, d$ and then for $\mu_* = 0$ one has by Lemma 6.6 that

$$\frac{\mathrm{d}}{\mathrm{d}\mu}\gamma_j(\mu_*) = -(-2c_2\nu^{-3/2}\alpha_j^{-3\sigma})/[-4\sigma c_1\nu^{-1/2}\alpha_j^{-(\sigma+1)}] \lfloor_{q=q_*} \tag{106}$$

$$= -2^{-1}\sigma^{-1}c_3\nu^{-1}\alpha_j^{-(2\sigma-1)} \lfloor_{q=q_*}, \tag{107}$$

where c_1, c_2, c_3 are given respectively by (147), (148), (149), of Lemma 6.6.

We recall from (126) and (128) of Lemma 6.5 that for $\mu_* = 0$ one has

$$\partial_\mu\partial_x\phi_{(q_*),j}(0) = 2^{-1}\zeta_j\nu^{-1/2}\alpha_j^{1-\sigma} \lfloor_{q=q_*} \tag{108}$$

$$\partial_\alpha\partial_x\phi_{(q_*),j}(0) = (\sigma+1)\zeta_j\nu^{1/2}\alpha_j^{\sigma} \lfloor_{q=q_*}, \tag{109}$$

and since $N_1(q) = \sum_{j=1}^d \partial_x\phi_{(q),j}(0)$, as defined by (80), one has for $\mu_* = 0$ that

$$\frac{\mathrm{d}}{\mathrm{d}\mu}N_1(\gamma(\mu_*)) = \left[\partial_\mu N_1(q_*) + \sum_{j=1}^d \partial_{\alpha_j}N_1(q_*)\frac{\mathrm{d}}{\mathrm{d}\mu}\gamma_j(\mu_*)\right] \tag{110}$$

$$= \sum_{j=1}^d \left[\partial_\mu\partial_x\phi_{(q_*),j}(0) + \partial_{\alpha_j}\partial_x\phi_{(q_*),j}(0)\frac{\mathrm{d}}{\mathrm{d}\mu}\gamma_j(\mu_*)\right] \tag{111}$$

$$= \sum_{j=1}^d \{2^{-1}\zeta_j\nu^{-1/2}\alpha_j^{1-\sigma} + [(\sigma+1)\zeta_j\nu^{1/2}\alpha_j^{\sigma}][-2^{-1}\sigma^{-1}c_3\nu^{-1}\alpha_j^{-(2\sigma-1)}]\} \lfloor_{q=q_*} \tag{112}$$

$$= \sum_{j=1}^d \zeta_j[2^{-1}\nu^{-1/2}\alpha_j^{1-\sigma} - 2^{-1}\sigma^{-1}(\sigma+1)c_3\nu^{-1/2}\alpha_j^{1-\sigma}] \lfloor_{q=q_*}. \tag{113}$$

Then by (98) of the previous part of the proof

$$\frac{\mathrm{d}}{\mathrm{d}\mu}N_1(\gamma(\mu_*)) = 2^{-1}\nu^{-1/2}[1-(1+1/\sigma)c_3]\sum_{j=1}^d \zeta_j(2c_1\nu^{-1/2}n_j/l_j)^{-1+1/\sigma} \tag{114}$$

$$= 2^{-1}\nu^{-1/2}[1-(1+1/\sigma)c_3](2c_1\nu^{-1/2})^{1/\sigma-1}\sum_{j=1}^d \zeta_j(n_j/l_j)^{-1+1/\sigma} \neq 0, \tag{115}$$

which contradicts γ being a local spectral curve passing through q_*. By this contradiction, it must be the case that our tentative assumption is false and therefore $\mathrm{d}\phi_{(\gamma_1(\mu_*)),j}(0)/\mathrm{d}\mu \neq 0$, for all j. As a consequence hereof, taken with the fact that $\phi_{(q)}(0)$ is continuous in q thanks to Lemma 6.2, it must be the case that through this q_* passes a local spectral curve γ such that for all $q_+, q_- \in \gamma$ sufficiently close to q_*, where $\mu_+ > 0 > \mu_-$, one has $\mathrm{sgn}(\phi_{(q_+)}) = -\mathrm{sgn}(\phi_{(q_-)})$. Then using (17) and

(99), we find

$$Z(\phi_{(q_+)}) - Z(\phi_{(q_-)}) = [Z(\phi_{(q_+)}) - Z(\phi_{(q_*)})] - [Z(\phi_{(q_-)}) - Z(\phi_{(q_*)})] \tag{116}$$

$$= \left\{ \operatorname{sgn}^2(\phi_{(q_+)}) \left[-1 + 2^{-1}d - 2^{-1}\operatorname{sgn}(\phi_{(q_+)}) \sum_{j=1}^{d} \operatorname{sgn}(\partial_x \phi_{(q_*),j}) \right] \lfloor_{x=0} \right\} \tag{117}$$

$$- \left\{ \operatorname{sgn}^2(\phi_{(q_-)}) \left[-1 + 2^{-1}d - 2^{-1}\operatorname{sgn}(\phi_{(q_-)}) \sum_{j=1}^{d} \operatorname{sgn}(\partial_x \phi_{(q_*),j}) \right] \lfloor_{x=0} \right\} \tag{118}$$

$$= \operatorname{sgn}(\phi_{(q_-)}) \sum_{j=1}^{d} \zeta_j, \tag{119}$$

which gives $|Z(\phi_{(q_+)}) - Z(\phi_{(q_-)})| = |\sum_{j=1}^{d} \zeta_j|$ and completes the proof. □

We end this section by proving Lemma 6.2, which was stated earlier in the proof of part (1) of Theorem 2.9, and stating and proving Lemmata 6.5 and 6.6, which were referenced above.

Proof of Lemma 6.2. We recall from (32) of the proof Proposition 3.2 that the solutions of (1) may be formed as the inverse functions of

$$\widehat{x}(\widehat{\phi}_0, \widehat{\phi}) = x_0 + \zeta \int_{\widehat{\phi}_0}^{\widehat{\phi}} \mathrm{d}w \; [h - \mu w^2 - \nu w^{2(\sigma+1)}]^{-1/2} \tag{120}$$

$$= x_0 + \zeta \int_{\widehat{\phi}_0}^{\widehat{\phi}} \mathrm{d}w \; \{\mu(\alpha^2 - w^2) + \nu[\alpha^{2(\sigma+1)} - w^{2(\sigma+1)}]\}^{-1/2}. \tag{121}$$

For $\phi = \phi_{(p)} \in \Phi_{\mathrm{ray}}$ one may write part of the inverse solution on each quarter wavelength, starting at a zero of ϕ, as

$$\widehat{x}(\widehat{\phi}) = \zeta \int_{0}^{\widehat{\phi}} \mathrm{d}w \; \{\mu(\alpha^2 - w^2) + \nu[\alpha^{2(\sigma+1)} - w^{2(\sigma+1)}]\}^{-1/2}. \tag{122}$$

Since the full solution can be composed of appropriately gluing together partial solutions, it is sufficient to confirm that each partial solution is continuously differentiable in μ and α. Let

$$F(x, \mu, \alpha, \widehat{\phi}) := x - \zeta \int_{0}^{\widehat{\phi}} \mathrm{d}w \; \{\mu(\alpha^2 - w^2) + \nu[\alpha^{2(\sigma+1)} - w^{2(\sigma+1)}]\}^{-1/2} \tag{123}$$

so that the formula for the inverse function of the partial solution may be expressed as $F(x, \mu, \alpha, \widehat{\phi}) = 0$.

It is easy to see that F is continuously differentiable in x and $\widehat{\phi}$. F can be shown to be continuously differentiable in in μ and α by arguments similar to those used to study $\partial_\mu \lambda$ and $\partial_\alpha \lambda$ in the proof of the monotonicity properties of Proposition 4.2. Furthermore, since $\partial_{\widehat{\phi}} F(x, \mu, \alpha, \widehat{\phi}) \neq 0$ for all x, μ, α, Implicit Function Theorem gives that there must exist a $\phi_{(\mu,\alpha)}(x)$ that is continuously differentiable in μ, α, and x in an open neighborhood of any point $(\mu, \alpha, x) \in P \times \mathbb{R}$. This continuous differentiability also holds in the same manner for $\partial_x \phi_{(\mu,\alpha)}(x)$ since by (27) one has

$$\partial_x \phi_{(\mu,\alpha)}(x) = \zeta\{\mu[\alpha^2 - \phi^2_{(\mu,\alpha)}(x)] + \nu[\alpha^{2(\sigma+1)} - \phi^{2(\sigma+1)}_{(\mu,\alpha)}(x)]\}^{1/2}, \tag{124}$$

which completes the proof. □

REMARK 6.4. *The lemma above could also be proven, for the case of $\sigma = 1$, by referring to Jacobi elliptic functions, which are the explicit solutions of (1) in this case.*

LEMMA 6.5. *Let $\phi_{(\mu_*,\alpha_*)} \in \Phi_{\mathrm{ray}}$ where the ray has the form $(-\infty, l]$. If $\phi_{(\mu_*,\alpha_*)}(0) = 0$ then*

$$\partial_\mu \phi_{(\mu_*,\alpha_*)}(0) = \zeta\xi\alpha(\mu + \nu\alpha^{2\sigma})^{1/2}\partial_\mu \lambda_{(\mu,\alpha)} \restriction_{(\mu,\alpha)=(\mu_*,\alpha_*)}, \tag{125}$$

$$\partial_\mu \partial_x \phi_{(\mu_*,\alpha_*)}(0) = 2^{-1}\zeta(\mu + \nu\alpha^{2\sigma})^{-1/2}\alpha \restriction_{(\mu,\alpha)=(\mu_*,\alpha_*)}, \tag{126}$$

$$\partial_\alpha \phi_{(\mu_*,\alpha_*)}(0) = \zeta\xi\alpha(\mu + \nu\alpha^{2\sigma})^{1/2}\partial_\alpha \lambda_{(\mu,\alpha)} \restriction_{(\mu,\alpha)=(\mu_*,\alpha_*)}, \tag{127}$$

$$\partial_\alpha \partial_x \phi_{(\mu_*,\alpha_*)}(0) = \zeta(\mu + \nu\alpha^{2\sigma})^{-1/2}[\mu + (\sigma+1)\nu\alpha^{2\sigma}] \restriction_{(\mu,\alpha)=(\mu_*,\alpha_*)}, \tag{128}$$

where $\zeta = \pm 1$ and $l = \xi\lambda_{(\mu_*,\alpha_*)}$ for some fixed $\xi \in \mathbb{N}/4$.

PROOF. Let $\widehat{x}_{(\mu,\alpha)}$ satisfy $\widehat{x}_{(\mu_*,\alpha_*)} = 0$ and $\phi_{(\mu,\alpha)}(\widehat{x}_{(\mu,\alpha)}) = 0$ for all (μ,α). Namely $\widehat{x}_{(\mu,\alpha)}$ denotes the position of some nodal point of the solution $\phi_{(\mu,\alpha)}$. Therefore the variation of $\widehat{x}_{(\mu,\alpha)}$ with respect to μ must take the form $\partial_\mu \widehat{x}_{(\mu,\alpha)} = -\xi\partial_\mu\lambda_{(\mu,\alpha)}$ for all (μ,α) and for some fixed $\xi \in \mathbb{N}/4$. Therefore

$$0 = \phi_{(\mu,\alpha)}(\widehat{x}_{(\mu,\alpha)}) \tag{129}$$

$$\Rightarrow \quad 0 = \frac{\mathrm{d}}{\mathrm{d}\mu}\phi_{(\mu,\alpha)}(\widehat{x}_{(\mu,\alpha)}) = \partial_\mu\phi_{(\mu,\alpha)}(\widehat{x}_{(\mu,\alpha)}) + \partial_x\phi_{(\mu,\alpha)}(\widehat{x}_{(\mu,\alpha)})\partial_\mu\widehat{x}_{(\mu,\alpha)}, \tag{130}$$

$$\begin{aligned} \Rightarrow \quad \partial_\mu\phi_{(\mu_*,\alpha_*)}(0) &= \xi\partial_x\phi_{(\mu_*,\alpha_*)}(0)\partial_\mu\lambda_{(\mu_*,\alpha_*)} \\ &= \zeta\xi\alpha(\mu + \nu\alpha^{2\sigma})^{1/2}\partial_\mu\lambda_{(\mu,\alpha)} \restriction_{(\mu,\alpha)=(\mu_*,\alpha_*)}, \end{aligned} \tag{131}$$

where we have used

$$\partial_x\phi_{(\mu_*,\alpha_*)}(0) = \zeta\alpha(\mu + \nu\alpha^{2\sigma})^{1/2} \restriction_{(\mu,\alpha)=(\mu_*,\alpha_*)}, \tag{132}$$

which holds thanks to (124) and $\phi_{(\mu_*,\alpha_*)}(\widehat{x}_{(\mu_*,\alpha_*)}) = 0$. Furthermore, by differentiating the expression of $\partial_x\phi$ in (124) with respect to μ, one has

$$[\partial_\mu\partial_x\phi_{(\mu,\alpha)}(\widehat{x}_{(\mu,\alpha)})] \restriction_{(\mu,\alpha)=(\mu_*,\alpha_*)} \tag{133}$$

$$= 2^{-1}\zeta\{\mu[\alpha^2 - \phi^2_{(\mu,\alpha)}(\widehat{x}_{(\mu,\alpha)})] + \nu[\alpha^{2(\sigma+1)} - \phi^{2(\sigma+1)}_{(\mu,\alpha)}(\widehat{x}_{(\mu,\alpha)})]\}^{-1/2} \tag{134}$$

$$\times \{[\alpha^2 - \phi^2_{(\mu,\alpha)}(\widehat{x}_{(\mu,\alpha)})] - 2\phi_{(\mu,\alpha)}(\widehat{x}_{(\mu,\alpha)})\partial_\mu[\phi_{(\mu,\alpha)}(\widehat{x}_{(\mu,\alpha)})] \tag{135}$$

$$- 2(\sigma+1)\nu\phi^{2\sigma+1}_{(\mu,\alpha)}(\widehat{x}_{(\mu,\alpha)})\partial_\mu[\phi_{(\mu,\alpha)}(\widehat{x}_{(\mu,\alpha)})]\} \restriction_{(\mu,\alpha)=(\mu_*,\alpha_*)} \tag{136}$$

$$= 2^{-1}\zeta[\mu\alpha^2 + \nu\alpha^{2(\sigma+1)}]^{-1/2}\alpha^2 \restriction_{(\mu,\alpha)=(\mu_*,\alpha_*)} \tag{137}$$

$$= 2^{-1}\zeta(\mu + \nu\alpha^{2\sigma})^{-1/2}\alpha \restriction_{(\mu,\alpha)=(\mu_*,\alpha_*)}. \tag{138}$$

The variation of $\widehat{x}_{(\mu,\alpha)}$ with respect to α must take the form $\partial_\alpha\widehat{x}_{(\mu,\alpha)} = -\xi\partial_\alpha\lambda_{(\mu,\alpha)}$ for all (μ,α). Therefore we find similarly to (132) above that

$$\partial_\alpha\phi_{(\mu_*,\alpha_*)}(0) \tag{139}$$

$$= \xi\partial_x\phi_{(\mu_*,\alpha_*)}(0)\partial_\alpha\lambda_{(\mu_*,\alpha_*)} = \zeta\xi\alpha(\mu + \nu\alpha^{2\sigma})^{1/2}\partial_\alpha\lambda_{(\mu,\alpha)} \restriction_{(\mu,\alpha)=(\mu_*,\alpha_*)}. \tag{140}$$

Furthermore, by differentiating the expression of $\partial_x \phi$ in (124) with respect to α, one has, by a similar calculation to that of (138) above, that

$$\partial_\alpha \partial_x \phi_{(\mu_*,\alpha_*)}(0) \tag{141}$$
$$= 2^{-1}\zeta[\mu\alpha^2 + \nu\alpha^{2(\sigma+1)}]^{-1/2}[2\mu\alpha + 2(\sigma+1)\nu\alpha^{2\sigma+1}] \downharpoonright_{(\mu,\alpha)=(\mu_*,\alpha_*)} \tag{142}$$
$$= \zeta(\mu + \nu\alpha^{2\sigma})^{-1/2}[\mu + (\sigma+1)\nu\alpha^{2\sigma}] \downharpoonright_{(\mu,\alpha)=(\mu_*,\alpha_*)} . \tag{143}$$

□

LEMMA 6.6. *For $\mu = 0$ one has*

$$\lambda_{(0,\alpha)} = 4c_1\nu^{-1/2}\alpha^{-\sigma}, \tag{144}$$
$$\partial_\alpha \lambda_{(0,\alpha)} = -4\sigma c_1\nu^{-1/2}\alpha^{-(\sigma+1)}, \tag{145}$$
$$\partial_\mu \lambda_{(0,\alpha)} = -2c_2\nu^{-3/2}\alpha^{-3\sigma}, \tag{146}$$

where

$$c_1 := \int_0^1 dw\ (1 - w^{2(\sigma+1)})^{-1/2} > 0, \tag{147}$$
$$c_2 := \int_0^1 dw\ (1 - w^{2(\sigma+1)})^{-3/2}(1 - w^2) > 0, \tag{148}$$
$$c_3 := c_2/c_1 > 0. \tag{149}$$

PROOF. Due to (39) of Proposition 4.2, one has

$$\lambda_{(0,\alpha)} = 4\int_0^1 dw\ [\nu\alpha^{2\sigma}(1 - w^{2(\sigma+1)})]^{-1/2} \tag{150}$$
$$= 4\nu^{-1/2}\alpha^{-\sigma}\int_0^1 dw\ (1 - w^{2(\sigma+1)})^{-1/2} \tag{151}$$
$$= 4c_1\nu^{-1/2}\alpha^{-\sigma}. \tag{152}$$

Differentiating $\lambda_{(\mu,\alpha)}$, as given in (39), one has

$$\partial_\alpha \lambda_{(\mu,\alpha)}/4 \tag{153}$$
$$= -2^{-1}\int_0^1 dw\ [\mu(1 - w^2) + \nu\alpha^{2\sigma}(1 - w^{2(\sigma+1)})]^{-3/2} 2\sigma\nu\alpha^{2\sigma-1}(1 - w^{2(\sigma+1)}) \tag{154}$$
$$\partial_\alpha \lambda_{(0,\alpha)} = -4\sigma\nu^{-1/2}\alpha^{-(\sigma+1)}\int_0^1 dw\ (1 - w^{2(\sigma+1)})^{-3/2}(1 - w^{2(\sigma+1)}) \tag{155}$$
$$= -4\sigma\nu^{-1/2}\alpha^{-(\sigma+1)}\int_0^1 dw\ (1 - w^{2(\sigma+1)})^{-1/2} = -4\sigma c_1\nu^{-1/2}\alpha^{-(\sigma+1)}, \tag{156}$$

and

$$\partial_\mu \lambda_{(\mu,\alpha)}/4 = -2^{-1}\int_0^1 dw\ [\mu(1 - w^2) + \nu\alpha^{2\sigma}(1 - w^{2(\sigma+1)})]^{-3/2}(1 - w^2) \tag{157}$$
$$\partial_\mu \lambda_{(0,\alpha)} = -2\nu^{-3/2}\alpha^{-3\sigma}\int_0^1 dw\ (1 - w^{2(\sigma+1)})^{-3/2}(1 - w^2) = -2c_2\nu^{-3/2}\alpha^{-3\sigma}. \tag{158}$$

□

Acknowledgements

We are indebted to Sven Gnutzmann for inspiring this work. We thank Sebastian Egger for his careful reading and useful direction. We thank Lior Alon and Michael Bersudsky for their helpful comments.

References

[1] R. Adami, C. Cacciapuoti, D. Finco, and D. Noja, *Stationary states of NLS on star graphs*, Europhys. Lett. 100, 10003 (2012).

[2] Riccardo Adami and Diego Noja, *Stability and symmetry-breaking bifurcation for the ground states of a NLS with a δ' interaction*, Comm. Math. Phys. **318** (2013), no. 1, 247–289, DOI 10.1007/s00220-012-1597-6. MR3017068

[3] Riccardo Adami, Claudio Cacciapuoti, Domenico Finco, and Diego Noja, *Variational properties and orbital stability of standing waves for NLS equation on a star graph*, J. Differential Equations **257** (2014), no. 10, 3738–3777, DOI 10.1016/j.jde.2014.07.008. MR3260240

[4] Werner O. Amrein, Andreas M. Hinz, and David B. Pearson (eds.), *Sturm-Liouville theory*, Birkhäuser Verlag, Basel, 2005. Past and present; Including papers from the International Colloquium held at the University of Geneva, Geneva, September 15–19, 2003. MR2132131

[5] Ram Band, *The nodal count* $\{0, 1, 2, 3, \ldots\}$ *implies the graph is a tree*, Philos. Trans. R. Soc. Lond. Ser. A Math. Phys. Eng. Sci. **372** (2014), no. 2007, 20120504, 24, DOI 10.1098/rsta.2012.0504. MR3151080

[6] Gregory Berkolaiko, *A lower bound for nodal count on discrete and metric graphs*, Comm. Math. Phys. **278** (2008), no. 3, 803–819, DOI 10.1007/s00220-007-0391-3. MR2373444

[7] Gregory Berkolaiko and Peter Kuchment, *Introduction to quantum graphs*, Mathematical Surveys and Monographs, vol. 186, American Mathematical Society, Providence, RI, 2013. MR3013208

[8] S. Gnutzmann, U. Smilansky, *Quantum graphs: Applications to quantum chaos and universal spectral statistics*, Advances in Physics **55** (2006), 527-625.

[9] S. Gnutzmann, U. Smilansky, S. Derevyanko, *Soliton scattering from a nonlinear network*, Phys Rev A **83** (2011), 033831.

[10] S. Gnutzmann, H. Schanz, U. Smilansky, *Topological Resonances in Scattering on Networks (Graphs)*, Phys. Rev. Lett. 110 (2013), 094101.

[11] Sven Gnutzmann and Daniel Waltner, *Stationary waves on nonlinear quantum graphs: general framework and canonical perturbation theory*, Phys. Rev. E **93** (2016), no. 3, 032204, 19. MR3652709

[12] Sven Gnutzmann and Daniel Waltner, *Stationary waves on nonlinear quantum graphs: general framework and canonical perturbation theory*, Phys. Rev. E **93** (2016), no. 3, 032204, 19. MR3652709

[13] Jeremy L. Marzuola and Dmitry E. Pelinovsky, *Ground state on the dumbbell graph*, Appl. Math. Res. Express. AMRX **1** (2016), 98–145, DOI 10.1093/amrx/abv011. MR3483843

[14] Diego Noja, *Nonlinear Schrödinger equation on graphs: recent results and open problems*, Philos. Trans. R. Soc. Lond. Ser. A Math. Phys. Eng. Sci. **372** (2014), no. 2007, 20130002, 20, DOI 10.1098/rsta.2013.0002. MR3151078

[15] Diego Noja, Dmitry Pelinovsky, and Gaukhar Shaikhova, *Bifurcations and stability of standing waves in the nonlinear Schrödinger equation on the tadpole graph*, Nonlinearity **28** (2015), no. 7, 2343–2378, DOI 10.1088/0951-7715/28/7/2343. MR3366647

[16] Dmitry Pelinovsky and Guido Schneider, *Bifurcations of standing localized waves on periodic graphs*, Ann. Henri Poincaré **18** (2017), no. 4, 1185–1211, DOI 10.1007/s00023-016-0536-z. MR3626301

[17] Yu. V. Pokornyĭ, V. L. Pryadiev, and A. Al′-Obeĭd, *On the oscillation of the spectrum of a boundary value problem on a graph* (Russian), Mat. Zametki **60** (1996), no. 3, 468–470, DOI 10.1007/BF02320380; English transl., Math. Notes **60** (1996), no. 3-4, 351–353 (1997). MR1428867

[18] Philipp Schapotschnikow, *Eigenvalue and nodal properties on quantum graph trees*, Waves Random Complex Media **16** (2006), no. 3, 167–178, DOI 10.1080/17455030600702535. MR2253041

[19] C. Sturm, *Mémoire sur les équations différentielles linéaires du second ordre*, J. Math. Pures Appl. **1** (1836), 106-186.

[20] C. Sturm, *Mémoire sur une classe d'équations à différences partielles*, J. Math. Pures Appl. **1**, 373-444, (1836).

MATHEMATICS DEPARTMENT, TECHNION, AMADO BUILDING, HAIFA, 32000, ISRAEL
Email address: `ramband@tx.technion.ac.il`

CHEMISTRY AND BIOMOLECULAR ENGINEERING DEPARTMENT, JOHNS HOPKINS UNIVERSITY, MARYLAND HALL, BALTIMORE, MARYLAND 21218
Email address: `akruegl1@jhu.edu`

Contemporary Mathematics
Volume **717**, 2018
https://doi.org/10.1090/conm/717/14446

Existence of the ground state for the NLS with potential on graphs

Claudio Cacciapuoti

ABSTRACT. We review and extend several recent results on the existence of the ground state for the nonlinear Schrödinger (NLS) equation on a metric graph. By ground state we mean a minimizer of the NLS energy functional constrained to the manifold of fixed L^2-norm. In the energy functional we allow for the presence of a potential term, of delta-interactions in the vertices of the graph, and of a power-type focusing nonlinear term. We discuss both subcritical and critical nonlinearity. Under general assumptions on the graph and the potential, we prove that a ground state exists for sufficiently small mass, whenever the constrained infimum of the quadratic part of the energy functional is strictly negative.

1. Introduction

Analysis on metric graphs and networks is a very well established research field, potentially with many physical and technological applications. From a mathematical point of view the interest in these structures lies in the fact that, despite being essentially simple one-dimensional objects, they still exhibit several intriguing features due to nontrivial connectivity and topology.

For an introduction to metric graphs and an extended list of references we refer to one of the many monographs on the subject, see, e.g., [**14, 21, 30, 34**].

The study of nonlinear equations on graphs is still at its beginning yet quickly developing. A monograph on quasilinear wave equations on one-dimensional networks, mostly dealing with the problem of the well-posedness, is [**12**] (see also [**13, 15**]).

Concerning the NLS equation on simple networks (e.g., the Y-junction or star-graph, see Fig. 2) a certain amount of work has been recently carried on: for the scattering and transmission properties of simple networks, see, e.g., [**2, 18, 38, 41**]; the inverse scattering method has recently been applied to the cubic NLS on a star-graph in [**19**]; the shrinking limit for the dynamics in a thin network (a relevant problem from the point of view of applications) has been studied in [**27, 39, 44**]. For a review on recent results and open problems related to the NLS equation on graphs we refer to [**31**].

2010 *Mathematics Subject Classification.* Primary 35Q55, 81Q35, 35R02, 49J40.

The author acknowledges the support of the FIR 2013 project "Condensed Matter in Mathematical Physics", Ministry of University and Research of Italian Republic (code RBFR13WAET).

In what follows we shall focus attention on the problem of the existence of the ground state. We shall discuss several related works at the end of the introduction.

The problem we are interested in is the minimization of the nonlinear Schrödinger energy functional

$$E[\Psi] := \|\Psi'\|^2 + (\Psi, W\Psi) + \sum_{v\in\mathcal{V}} \alpha_v |\Psi(v)|^2 - \frac{1}{\mu+1}\|\Psi\|_{2\mu+2}^{2\mu+2} \qquad 0<\mu\leq 2 \tag{1.1}$$

defined on a metric graph $\mathcal{G}$, where W is a potential on the graph, $\mathcal{V}$ is the set of vertices of the graph, and α_v are some real constants that take into account possible delta-interactions in the vertices. We shall focus attention on the minimization problem

$$-\nu_\mu(m) := \inf\{E[\Psi] \mid \Psi \in H^1(\mathcal{G}),\ \|\Psi\|^2 = m\}. \tag{1.2}$$

The parameter m is called *mass*, and will play an important role in our analysis. It is easy to check that the functional $E[\Psi]$, defined on the space $H^1(\mathcal{G})$ (see Sec. 2 for a precise definition of $H^1(\mathcal{G})$) is unbounded from below; to this aim it is enough to consider the behavior of $E[\lambda\Psi]$ for large λ. It is well known that imposing the constraint $\|\Psi\|^2 = m$ may solve this issue; indeed, the first question that we try to answer concerns the existence of a lower bound for the infimum in Eq. (1.2).

Whenever such a lower bound exists, we will be concerned about the existence of a minimizer. We shall use the following definition

Definition 1.1 (Ground state). A minimizer of problem (1.2), i.e., a function $\hat{\Psi} \in H^1(\mathcal{G})$, such that $\|\hat{\Psi}\|^2 = m$, and $E[\hat{\Psi}] = -\nu_\mu(m)$ (if it exists), is called *ground state* (of mass m).

We recall that a metric graph can be understood as a metric space made up of a set of segments, referred to as *edges*, and a set of points, called *vertices*. In our analysis, we shall always assume that the graph has a finite number of edges and vertices. The edges can be of finite or infinite length; in the first case each edge is identified with a segment $[0,\ell_e]$ (ℓ_e being the length of the edge), in the latter with a copy of the half-line $[0,+\infty)$. The edges of the graph are glued together according to a connection map which identifies each endpoint (0 or ℓ_e) of each edge with a vertex v of the graph, see Fig. 1.

As for the standard NLS with power-type nonlinearity on the real-line, the case $0<\mu<2$ is called *subcritical* while the case $\mu = 2$ is called *critical.* The terminology is associated to the scaling properties of the *kinetic* and *nonquadratic* terms in the energy functional ($\|\Psi'\|^2$ and $\|\Psi\|_{2\mu+2}^{2\mu+2}$ respectively). In fact, under the mass invariant transformation $\Psi \in L^2(\mathcal{G}) \to \Psi_\lambda \in L^2(\lambda^{-1}\mathcal{G})$, defined by $\Psi_\lambda(x) := \sqrt{\lambda}\Psi(\lambda x)$, the scaling relations are

$$\|\Psi_\lambda'\|_{L^2(\lambda^{-1}\mathcal{G})}^2 = \lambda^2\|\Psi'\|_{L^2(\mathcal{G})}^2 \quad \text{and} \quad \|\Psi_\lambda\|_{L^{2\mu+2}(\lambda^{-1}\mathcal{G})}^{2\mu+2} = \lambda^\mu\|\Psi\|_{L^{2\mu+2}(\mathcal{G})}^{2\mu+2}. \tag{1.3}$$

When $\mathcal{G}$ is itself scale invariant, i.e., $\mathcal{G}$ coincides with the real-line, or the half-line, or N half lines with a common vertex (a *star-graph*), the scaling (1.3) clearly implies that the kinetic (positive) term dominates for $0<\mu<2$ and λ large enough, thus suggesting that the infimum (1.2) is bounded from below. For $\mu = 2$ the two terms scale in the same way, and it turns out that the infimum $-\nu_\mu(m)$ is bounded from below only for small mass. The same behavior can be observed for generic graphs.

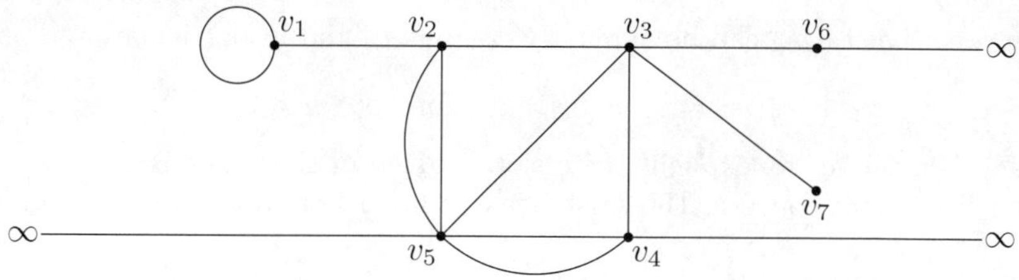

FIGURE 1. A graph with 7 vertices and 14 edges: 3 external, and 11 internal (of which 1 terminal).

Our main results are stated in Ths. 1 and 2 below. We remark that the subcritical case was discussed in [**17**], here we include the case $\mu = 2$. When $\mathcal{G}$ is a star-graph (and $W = 0$), the critical case was already discussed in [**5**], to extend the analysis to generic graphs we shall use several results from [**10**].

We make the following assumptions.

ASSUMPTION A. $\mathcal{G}$ is a finite, connected graph, with at least one external edge.

We call *finite* a graph that has a finite number of edges and vertices; a graph $\mathcal{G}$ is *connected* if given any two points of the graph there is always a path in $\mathcal{G}$ connecting them; and we call *external* an edge of infinite length.

ASSUMPTION B. $W = W_+ - W_-$ with $W_\pm \geq 0$, $W_+ \in L^1(\mathcal{G}) + L^\infty(\mathcal{G})$, and $W_- \in L^r(\mathcal{G})$ for some $r \in [1, 1 + 1/\mu]$.

We recall that $L^r(\mathcal{G}) \subset L^1(\mathcal{G}) + L^\infty(\mathcal{G})$ for all $r \in [1, +\infty]$. Hence, under Ass. B, one has $W \in L^1(\mathcal{G}) + L^\infty(\mathcal{G})$. We remark that we shall use the stronger assumption $W_- \in L^r(\mathcal{G})$ only in Th. 2. More precisely, the additional constraint $r \in [1, 1+1/\mu]$ is needed only in the bound (4.7); in the remaining part of the proof it is enough to assume $1 \leq r < \infty$.

Denote by $E^{lin}[\Psi]$ the quadratic part of energy functional $E[\Psi]$,

$$E^{lin}[\Psi] := \|\Psi'\|^2 + (\Psi, W\Psi) + \sum_{v \in \mathcal{V}} \alpha_v |\Psi(v)|^2, \tag{1.4}$$

and by $-E_0$ the infimum

$$-E_0 := \inf \left\{ E^{lin}[\Psi] \mid \Psi \in H^1(\mathcal{G}),\ \|\Psi\|^2 = 1 \right\}. \tag{1.5}$$

We shall assume that this infimum is negative.

ASSUMPTION C. $E_0 > 0$.

In the statements of our main results, some constraints on μ and m, with an interplay between the two parameters, are needed. Two quantities will enter the constraints. For the critical case it will be relevant the best constant $K_{6,2}(\mathcal{G})$ satisfying the Gagliardo-Nirenberg inequality (2.2). In the subcritical case, instead, the constraint will depend on a constant, denoted by γ_μ, which is related to the infimum of the "free" nonlinear NLS energy functional on the real-line

$$-t_\mu(m) := \inf \left\{ E_{\mathbb{R}}[\psi] \,\middle|\, \psi \in H^1(\mathbb{R}),\ \|\psi\|^2_{L^2(\mathbb{R})} = m \right\} \qquad 0 < \mu \leq 2,$$

with

$$E_{\mathbb{R}}[\psi] := \|\psi'\|^2_{L^2(\mathbb{R})} - \frac{1}{\mu+1} \|\psi\|^{2\mu+2}_{L^{2\mu+2}(\mathbb{R})}.$$

It turns out that $t_\mu(m)$ can be explicitly computed, and in particular

$$t_\mu(m) = \gamma_\mu m^{1+\frac{2\mu}{2-\mu}} \qquad \text{for} \quad 0<\mu<2, \tag{1.6}$$

where γ_μ is positive (its explicit expression is given in Eq. (4.12)).

Our first result concerns the existence of a lower bound for the infimum in Eq. (1.2).

THEOREM 1. *Let Assumption* A *hold true and assume* $W \in L^1(\mathcal{G}) + L^\infty(\mathcal{G})$. *If* $0<\mu<2$ *then* $\nu_\mu(m) < +\infty$ *for any* $m>0$. *If* $\mu = 2$ *then* $\nu_\mu(m) < +\infty$ *for any* $0 < m < \sqrt{3}/K_{6,2}^3(\mathcal{G})$.

The second result concerns the existence of the ground state.

THEOREM 2. *Let Assumptions* A, B, *and* C *hold true. Then* $-\nu_\mu(m) \le -mE_0$. *Moreover, let*

$$m_\mu^* := \begin{cases} (E_0/\gamma_\mu)^{\frac{1}{\mu}-\frac{1}{2}} & \textit{if} \quad 0<\mu<2 \\ \sqrt{3}/K_{6,2}^3(\mathcal{G}) & \textit{if} \quad \mu = 2 \end{cases} \tag{1.7}$$

Then the ground state $\hat{\Psi}$ *exists for all* $0<m<m_\mu^*$.

We remark that it is possible to show that the upper bound for $-\nu_\mu(m)$ is strict, i.e., $-\nu_\mu(m) < -mE_0$. However, since we do not need this additional information, we will not pursue this goal.

A main tool to prove Th. 2 is the concentration-compactness lemma (Lem. 2.2 below). We remark that concentration-compactness methods are standard in $\mathbb{R}^d$ (see, e.g., [**20**]). Here we adapt the technique to a setting where there is no translation invariance.

Once that the existence of a lower bound for the infimum (1.2) is granted by Th. 1, Lemma 2.2 is used to show that a sufficient condition for the existence of a minimizer is

$$-\nu_\mu(m) < -t_\mu(m), \tag{1.8}$$

both in the subcritical and in the critical case. Condition (1.8) can be used together with the fact that $t_\mu(m)$ can be explicitly computed and that it must be $-\nu_\mu(m) \le -E_0 m$. In the subcritical case, the threshold mass m_μ^* follows by requiring $-E_0 m < -t_\mu(m)$. In the critical case the situation is slightly different because $t_2(m)$ has a sharp transition depending on the value of m, precisely

$$-t_2(m) = \begin{cases} 0 & \text{if } m \le \pi\sqrt{3}/2 \\ -\infty & \text{if } m > \pi\sqrt{3}/2 \end{cases} \tag{1.9}$$

Note that $\pi\sqrt{3}/2 = \sqrt{3}/K_{6,2}^3(\mathbb{R})$, where $K_{6,2}(\mathbb{R})$ is the best Gagliardo-Nirenberg constant for the real-line. In this case the value of m_2^* arises from the thresholds in Th. 1 and Eq. (1.9), together with the fact that $K_{6,2}(\mathcal{G}) \ge K_{6,2}(\mathbb{R})$, see [**10**].

We remark that, in the subcritical case and for $W \in L^r(\mathcal{G})$ for some $r \in [1,+\infty)$, one can prove that the infimum $-\nu_\mu(m)$ cannot exceed $-t_\mu(m)$, see Prop. A.1 below.

We conclude the introduction with several remarks and a discussion on the related literature.

Conservation laws and well-posedness. The nonlinear energy functional (1.1), defined on $D(E) = H^1(\mathcal{G})$, is the conserved energy associated to the NLS equation compactly written as

$$i\frac{d}{dt}\Psi(t) = H\Psi(t) - |\Psi(t)|^{2\mu}\Psi(t). \tag{1.10}$$

Where H is the linear self-adjoint operator on $\mathcal{G}$ associated to the quadratic form E^{lin} with domain $D(E^{lin}) = H^1(\mathcal{G})$, we refer to Sec. 2, Eqs. (2.4) - (2.5) for its rigorous definition. Here we just remark that H encodes both the presence of the potential W and the presence of delta-interactions of strength α_v in the vertices. By the definition of E_0, one has $-E_0 = \inf \sigma(H)$, $\sigma(H)$ being the spectrum of H. The nonlinear term $|\Psi|^{2\mu}\Psi$ in Eq. (1.10) must be understood componentwise as $(|\Psi|^{2\mu}\Psi)_e = |\psi_e|^{2\mu}\psi_e$ for every edge e, where ψ_e is the component of the wavefunction Ψ on the edge e (see Sec. 2 for the details). Hence, Eq. (1.10) is understood as a single-particle equation on a one-dimensional ramified structure. On each branch (edge), the dynamics is governed by: a dispersive term (the second order spatial derivative in Eq. (2.5)); plus a potential term; plus delta-type interactions in the vertices (which can be understood as singular potentials); plus a (focusing) nonlinear power-type term.

We recall that, for any initial datum $\Psi(0) = \Psi_0 \in H^1(\mathcal{G})$, the energy $E[\Psi(t)]$ and the mass $\|\Psi(t)\|^2$ are conserved along the flow associated to Eq. (1.10). Moreover: if $0 < \mu < 2$ then Eq. (1.10) (in weak form) is globally well-posed in $H^1(\mathcal{G})$; if $\mu = 2$ global well-posedness holds true for small enough mass ($\|\Psi_0\|^2 < \sqrt{3}/K_{6,2}^3(\mathcal{G})$). We refer to [**17**] for the precise statements and the proofs.

Stationary states and bifurcations. It is well known that, whenever a ground state exists, it is a stationary solution of Eq. (1.10), i.e., a solution of the form $\Psi(t) = e^{i\omega t}\Psi(\omega)$, with $\omega \in \mathbb{R}$. The function $\Psi(\omega)$ satisfies the stationary equation

$$H\Psi(\omega) - |\Psi(\omega)|^{2\mu}\Psi(\omega) = -\omega\Psi(\omega), \tag{1.11}$$

where the parameter $\omega \in \mathbb{R}$ must be chosen in order to satisfy the mass constraint. Solutions of Eq. (1.11) are called *stationary states.*

In the subcritical case, for mass small enough (possibly smaller than m_μ^* in Eq. (1.7)), one can prove that the ground state $\hat{\Psi}$ is the solution of Eq. (1.11) bifurcating from the null state along the direction of the eigenvector of H corresponding to the eigenvalue $-E_0$. The bifurcation occurs for $\omega = E_0$. We refer to [**17**] for the details.

Eq. (1.11) has an interest in its own as its solutions identify the critical points of the energy functional (1.1) subject to the mass constraint.

Star-graph with delta-interaction in the vertex. A first rigorous analysis of Eq. (1.11) was performed for the case of a star-graph with N edges, when $W = 0$, and $\alpha \in \mathbb{R}$, in [**4**] (see also [**6**]). In such a case all the stationary states can be explicitly computed.

In the same setting, the minimization problem (1.2), for $0 < \mu \leq 2$, was studied in [**5**]. One main result in [**5**] is that, for $\alpha < 0$ (attractive interaction in the vertex), by exploiting the explicit form of the stationary states, it is possible to identify the ground state with the unique (up to phase multiplication) symmetric stationary state. Several techniques and ideas, that can also be applied to generic graphs, such as the interplay between the concentration-compactness lemma and condition

(1.8), were first used in [**5**]. We remark that in Th. 2 the value of the threshold mass m_μ^* is slightly improved with respect to the one given in [**5**].

In general it is not possible to say what happens when m crosses the mass threshold m_μ^*. For a star-graph with $0 < \mu < 2$, $W = 0$, and $\alpha < 0$, it was proved in [**7**] that for mass large enough the ground state does not exist. Despite that, one can show that when the mass is larger than a certain threshold, the symmetric stationary state is a local minimum of the energy functional (1.1) constrained to the manifold of fixed mass, see [**7**].

Free Laplacian with Kirchhoff conditions in the vertices. The case $W = 0$ and $\alpha_v = 0$ for all $v \in \mathcal{V}$ requires a separate discussion, as in this case Ass. C is not satisfied. We shall use the following notation $E_{(0,0)}[\Psi] = E_{W=0,\alpha_v=0}[\Psi]$, and $E^{lin}_{(0,0)}[\Psi] = E^{lin}_{W=0,\alpha_v=0}[\Psi]$.

We recall that the Hamiltonian $H_{(0,0)}$ associated to the quadratic form $E^{lin}_{(0,0)}$ is still defined as in Eqs. (2.4) and (2.5), with $W = 0$ and $\alpha_v = 0$. When $\alpha_v = 0$, the gluing conditions encoded in the definition of $D(H)$ are usually referred to as *Kirchhoff* (or *standard*) conditions.

As a first remark we note that

$$\inf\left\{E^{lin}_{(0,0)}[\Psi] \mid \Psi \in H^1(\mathcal{G}),\ \|\Psi\|^2 = 1\right\} = 0,$$

hence, Ass. C is not satisfied and Th. 2 does not give any information on the existence of the ground state (Th. 1 still hods true though).

A first result on the minimization problem (1.2) for $E_{(0,0)}[\Psi]$ was given in [**3**], where it was shown that on a star-graph and in the cubic case ($\mu = 1$), the minimizer does not exist for any value of the mass. The stationary states were explicitly computed in [**4**], see also [**6**].

A systematic analysis of the minimization problem for $E_{(0,0)}$ on generic graphs has been performed in [**8, 9, 11**] for the subcritical case (see [**1**] for the analysis of the cubic case), and in [**10**] for the critical case. One main result in the series of works [**8, 9, 11**] is the identification of a topological condition (called Assumption H) that excludes (apart for very specific examples of graphs) the existence of the ground state for any value of the mass m. Assumption H can be stated as: the graph $\mathcal{G}$ can be covered by cycles (here the ∞-points of the external edges, see. Figs. 1 and 2, are regarded as a single vertex). One example of graph to which Assumption H applies is the star-graph.

It is worth noticing that this result is very unstable under perturbations of the energy functional, in the sense that adding any arbitrarily small negative potential may turn the infimum in (1.5) into strictly negative. Hence, as a consequence of Th. 2, the ground state would exist for small mass, despite the topological condition. This is exactly what happens for a star-graph with N edges: for arbitrary $\alpha < 0$ one has $E_0 = |\alpha|^2/N^2$ see [**5**]; on the other hand, for $\alpha = 0$ and $N \geq 3$ no ground exists for any $m > 0$.

Concerning the critical case, we remark that in [**10**], among other results, the authors prove that for a large class of graphs (e.g., the ones that do not satisfy Assumption H, have no terminal edge, and have at least two external edges) the ground state exists if and only if $m \in [\sqrt{3}/K^3_{6,2}(\mathcal{G}), \sqrt{3}/K^3_{6,2}(\mathbb{R})]$. In view of Ths. 1 and 2, the existence of the lower bound for the mass parameter might be surprising, so we briefly comment on it.

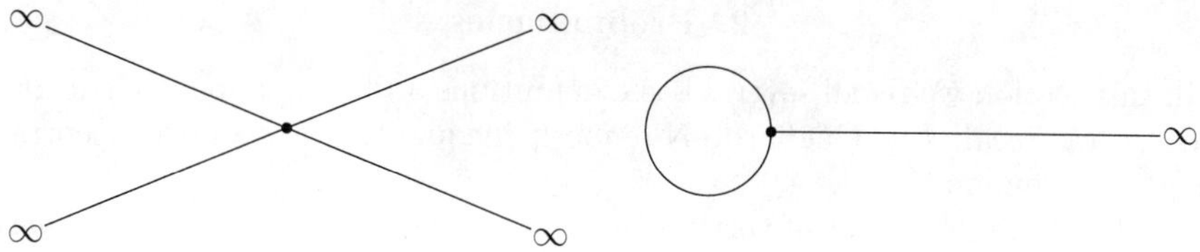

FIGURE 2. On the right, a star-graph with 4 edges. On the left a tadpole graph.

One issue in the minimization of $E_{(0,0)}$ for small mass, is that the infimum might be zero, and never attained because minimizing sequences are *vanishing* in the sense of Lemma 2.2. This is indeed the case when $\mathcal{G} = \mathbb{R}$, the mass threshold being $\pi\sqrt{3}/2$, see Eq. (1.9). In the presence of potential terms or delta-interactions, if Ass. C is satisfied then $-\nu_\mu(m)$ is strictly negative by Th. 2; hence, vanishing cannot occur and there is no lower bound on m.

On the other hand, for large mass, exactly larger than $\sqrt{3}/K^3_{6,2}(\mathcal{G})$, one has a different issue: the Gagliardo-Nirenberg inequality does not guarantee that the infimum in (1.2) is lower bounded. In [**10**], the authors show that by the topological assumptions, $-\nu_\mu(m)$ is indeed lower bounded (and strictly negative due to the large mass) thus implying that the ground state exists.

Other related works. For several specific examples of graphs Eq. (1.11) can be explicitly solved. One interesting case is the tadpole-graph, see Fig. 2, for $E = E_{(0,0)}$. The stationary states for the tadpole-graph have been completely characterized in [**16**]. Certain families of solutions can be understood in terms of bifurcation theory from embedded eigenvalues and threshold resonances. Bifurcations and stability properties have been further analyzed in [**32**]. The existence of the ground state for the tadpole-graph for any $m > 0$ has been proved in [**9**], see also [**11**]. A general approach to the study of the stationary solutions has been recently proposed in [**24, 25**]. The stationary solutions on a compact star-graph, in a setting in which the nonlinear term changes from edge to edge has been studied in [**35, 40**].

A similar analysis for stationary states on periodic graphs is in [**33**] (see also [**22**]). While the ground state for the dumbbell-graph is studied in [**29**].

Existence/nonexistence of the ground state in a slightly different setting, i.e., when the nonlinearity is supported only on a compact region of the graph, has been investigated in [**36, 37, 43**]. The same model was first proposed in [**23**], to study the scattering through a nonlinear network.

The paper is structured as follows. In Section 2 we set up the model and recall several preliminary results, included the Concentration-Compactness Lemma 2.2. Section 3 is devoted to the proof of Th. 1. Section 4 is devoted to the proof of Th. 2. We conclude the paper with a short appendix (App. A) in which we prove that in the subcritical case, if the potential W decays at infinity, $-t_\mu(m)$ is an upper bound for the infimum in (1.2).

2. Preliminaries

In this section we recall several basic definitions and facts about metric graphs, moreover we recall the Gagliardo-Nirenberg inequalities and the concentration-compactness lemma.

We denote by $\mathcal{V}$ the set of vertices of the graph $\mathcal{G}$, and by $\mathcal{V}_-$ the set of vertices for which the coupling constant α_v is strictly negative. We denote by $\mathcal{E}$ the set of edges of the graph and decompose it as $\mathcal{E} = \mathcal{E}^{in} \cup \mathcal{E}^{ext}$. The set $\mathcal{E}^{ext}$ denotes the set of external edges, these are the edges that start from one vertex of the graph and extend to infinity; which means that every element of $\mathcal{E}^{ext}$ can be identified with the half-line $[0, +\infty)$. The set $\mathcal{E}^{in}$ denotes the set of internal edges of the graph, these are the edges of finite length; each edge $e \in \mathcal{E}^{in}$ can be identified with a segment $[0, \ell_e]$, with ℓ_e denoting its length.

In what follows we shall use the notation $I_e \equiv [0, +\infty)$ for $e \in \mathcal{E}^{ext}$, and $I_e \equiv [0, \ell_e]$ for $e \in \mathcal{E}^{in}$.

A (complex valued) function on $\mathcal{G}$ is a map $\Psi : \mathcal{G} \to \mathbb{C}^{|\mathcal{E}|}$, to be understood as $\Psi = \bigoplus_{e \in \mathcal{E}} \psi_e$ with $\psi_e : I_e \to \mathbb{C}$, ψ_e denoting the wave function component on the edge e.

For $p \in [1, +\infty]$, one has $L^p(\mathcal{G}) = \bigoplus_{e \in \mathcal{E}} L^p(I_e)$, and we denote by $\|\cdot\|_p$ the corresponding norm

$$\|\Psi\|_p^p = \sum_{e \in \mathcal{E}} \|\psi_e\|_{L^p(I_e)}^p, \quad p \in [1, +\infty); \qquad \|\Psi\|_\infty = \max_{e \in \mathcal{E}} \|\psi_e\|_{L^\infty(I_e)}.$$

For $p = 2$ we shall denote the norm in $L^2(\mathcal{G})$ simply by $\|\cdot\|$.

We also recall the definition of the Sobolev spaces $H^1(\mathcal{G})$ and $H^2(\mathcal{G})$. Denote by $C(\mathcal{G})$ the set of continuous functions on $\mathcal{G}$, then

$$H^1(\mathcal{G}) = \left\{\Psi \in C(\mathcal{G}) \,|\, \psi_e \in H^1(I_e)\ \forall e \in \mathcal{E}\right\}, \tag{2.1}$$

equipped with the norm

$$\|\Psi\|_{H^1(\mathcal{G})}^2 = \sum_{e \in \mathcal{E}} \|\psi_e\|_{H^1(I_e)}^2;$$

and

$$H^2(\mathcal{G}) = \left\{\Psi \in H^1(\mathcal{G}) \,|\, \psi_e \in H^2(I_e)\ \forall e \in \mathcal{E}\right\}$$

equipped with the norm

$$\|\Psi\|_{H^2(\mathcal{G})}^2 = \sum_{e \in \mathcal{E}} \|\psi_e\|_{H^2(I_e)}^2.$$

We recall the Gagliardo-Nirenberg inequalities on graphs.

PROPOSITION 2.1 (Gagliardo-Nirenberg inequalities on graphs). *For any connected graph $\mathcal{G}$ such that $|\mathcal{E}| < \infty$ and $|\mathcal{V}| < \infty$, and for any $p, q \in [2, +\infty]$, with $p \geq q$, and $\alpha = \frac{2}{2+q}(1 - q/p)$, there exist two constants $K_{p,q}(\mathcal{G})$ and $\mathcal{K}_{p,q}(\mathcal{G})$ such that*

$$\|\Psi\|_p \leq K_{p,q}(\mathcal{G}) \|\Psi'\|^\alpha \|\Psi\|_q^{1-\alpha} \qquad \text{if } |\mathcal{E}^{ext}| \geq 1, \tag{2.2}$$

$$\|\Psi\|_p \leq \mathcal{K}_{p,q}(\mathcal{G}) \|\Psi\|_{H^1}^\alpha \|\Psi\|_q^{1-\alpha} \qquad \text{if } |\mathcal{E}^{ext}| = 0, \tag{2.3}$$

for all $\Psi \in H^1(\mathcal{G})$.

A proof of inequality (2.2) is in [**11**] (see also [**5, 6, 26, 43**]). In the case of compact graphs ($|\mathcal{E}^{ext}| = 0$) inequality (2.2) cannot hold true (it is clearly violated by the constant function). Nevertheless, it can be replaced by the weaker inequality (2.3), for a proof we refer to [**30**].

We recall some facts about the quadratic form $E^{lin}[\Psi]$ defined in Eq. (1.4). We always consider E^{lin} on the domain $D(E^{lin}) = H^1(\mathcal{G})$. We note that on such domain, $\Psi(v)$ (the value of the wave function in a vertex of the graph) is well defined due to the global continuity condition, see Eq. (2.1).

Moreover by using Gagliardo-Nirenberg inequalities it is easy to prove (see [**17**]) that for $W \in L^1(\mathcal{G}) + L^\infty(\mathcal{G})$

$$\Big|(\Psi, W\Psi) + \sum_{v\in\mathcal{V}} \alpha(v)|\Psi(v)|^2\Big| \le a\|\Psi'\|^2 + b\|\Psi\|^2, \qquad \text{with } 0 < a < 1,\ b > 0,$$

which, by KLMN theorem, implies that the form E^{lin} is lower bounded ($E_0 < +\infty$) and closed, hence defines a selfadjoint operator. This a standard result for Schrödinger operators on the real-line, see, e.g., [**28**, Ch. 11.3].

It is easy to prove that the self-adjoint operator corresponding to the quadratic form E^{lin} coincides with the Hamiltonian $H : D(H) \subset L^2(\mathcal{G}) \to L^2(\mathcal{G})$ defined on the domain

$$D(H) := \Big\{ \Psi \in H^2(\mathcal{G}) \,\Big|\, \sum_{e\prec v} \partial_o \psi_e(v) = \alpha_v \Psi(v) \quad \forall v \in \mathcal{V} \Big\}, \tag{2.4}$$

where $e \prec v$ denotes the set of edges having at least one endpoint identified with the vertex v, and we have denoted by ∂_o the outward derivative from the vertex (more precisely $\partial_o\psi_e(v) = \psi_e'(0)$ when $\psi_e(v)$ is identified by $\psi_e(0)$; $\partial_o\psi_e(v) = -\psi_e'(\ell_e)$ when $\psi_e(v)$ is identified by $\psi_e(\ell_e)$). The action of H is defined by

$$(H\Psi)_e = -\psi_e'' + W_e\psi_e \qquad \forall e \in \mathcal{E}. \tag{2.5}$$

We conclude this section by recalling the concentration-compactness lemma (see Lem. 2.2 below) that will be needed in the proof of Th. 2. For the proof we refer to [**17**] (see also [**5**]).

For any $y \in \mathcal{G}$ and $t > 0$, we denote by $B(y,t) \subset \mathcal{G}$ the open ball of radius t and center y

$$B(y,t) := \{x \in \mathcal{G} \mid d(x,y) < t\},$$

here $d(x,y)$ denotes the distance between two points of the graph, defined as the infimum of the length of the paths connecting x to y.

For any function $\Psi \in L^2(\mathcal{G})$ and $t > 0$ we define the concentration function $\rho(\Psi,t)$ as

$$\rho(\Psi,t) := \sup_{y\in\mathcal{G}} \|\Psi\|^2_{L^2(B(y,t))};$$

and, for any sequence $\{\Psi_n\}_{n\in\mathbb{N}}$, $\Psi_n \in L^2(\mathcal{G})$, the concentrated mass parameter τ as

$$\tau := \lim_{t\to+\infty} \liminf_{n\to\infty} \rho(\Psi_n, t).$$

The parameter τ plays a key role in the concentration-compactness lemma because it distinguishes the occurrence of vanishing, dichotomy or compactness in $H^1(\mathcal{G})$-bounded sequences in a sense precisely defined below.

LEMMA 2.2 (Concentration-compactness). *Let $m > 0$ and $\{\Psi_n\}_{n\in\mathbb{N}}$ be such that: $\Psi_n \in H^1(\mathcal{G})$,*

$$\|\Psi_n\|^2 \to m \quad as \quad n \to \infty\,,$$

$$\sup_{n\in\mathbb{N}} \|\Psi_n'\| < \infty\,.$$

Then there exists a subsequence $\{\Psi_{n_k}\}_{k\in\mathbb{N}}$ such that:

i) *(Compactness) If $\tau = m$, at least one of the two following cases occurs:*
  i_1) *(Convergence) There exists a function $\Psi \in H^1(\mathcal{G})$ such that $\Psi_{n_k} \to \Psi$ in L^p as $k \to \infty$ for all $2 \le p \le \infty$.*
  i_2) *(Runaway) There exists $e^* \in \mathcal{E}^{ext}$, such that for all $t > 0$, and $2 \le p \le \infty$*

$$\lim_{k\to\infty}\left(\sum_{e\neq e^*} \|(\Psi_{n_k})_e\|^p_{L^p(I_e)} + \|(\Psi_{n_k})_{e^*}\|^p_{L^p((0,t))}\right) = 0. \tag{2.6}$$

ii) *(Vanishing) If $\tau = 0$, then $\Psi_{n_k} \to 0$ in L^p as $k \to \infty$ for all $2 < p \le \infty$.*
iii) *(Dichotomy) If $0 < \tau < m$, then there exist two sequences $\{R_k\}_{k\in\mathbb{N}}$ and $\{S_k\}_{k\in\mathbb{N}}$ in $H^1(\mathcal{G})$ such that*

$$\operatorname{supp} R_k \cap \operatorname{supp} S_k = \emptyset \tag{2.7}$$

$$|R_k(x)| + |S_k(x)| \le |\Psi_{n_k}(x)| \qquad \forall x \in \mathcal{G} \tag{2.8}$$

$$\|R_k\|_{H^1(\mathcal{G})} + \|S_k\|_{H^1(\mathcal{G})} \le c\|\Psi_{n_k}\|_{H^1(\mathcal{G})} \tag{2.9}$$

$$\lim_{k\to\infty} \|R_k\|^2 = \tau \qquad\qquad \lim_{k\to\infty} \|S_k\|^2 = m - \tau \tag{2.10}$$

$$\liminf_{k\to\infty} \left(\|\Psi'_{n_k}\|^2 - \|R'_k\|^2 - \|S'_k\|^2\right) \ge 0 \tag{2.11}$$

$$\lim_{k\to\infty} \left(\|\Psi_{n_k}\|_p^p - \|R_k\|_p^p - \|S_k\|_p^p\right) = 0 \qquad 2 \le p < \infty \tag{2.12}$$

$$\lim_{k\to\infty} \left\||\Psi_{n_k}|^2 - |R_k|^2 - |S_k|^2\right\|_\infty = 0. \tag{2.13}$$

3. Proof of Theorem 1

We always consider the functional E on the domain $D(E) = H^1(\mathcal{G})$ and note that

$$E[\Psi] = E^{lin}[\Psi] - \frac{1}{\mu+1}\|\Psi\|^{2\mu+2}_{2\mu+2}.$$

PROOF OF TH. 1. We note the trivial lower bound

$$E[\Psi] \ge \|\Psi'\|^2 - (\Psi, W_-\Psi) - \sum_{v\in\mathcal{V}_-} |\alpha_v||\Psi(v)|^2 - \frac{1}{\mu+1}\|\Psi\|^{2\mu+2}_{2\mu+2}. \tag{3.1}$$

Write $W_- = W_{-,1} + W_{-,\infty}$. By Gagliardo-Nirenberg inequality and setting $\|\Psi\|^2 = m$, one has the following lower bound for $E[\Psi]$,

$$E[\Psi] \ge \|\Psi'\|^2 - a_\mu m^{\frac{2+\mu}{2}}\|\Psi'\|^\mu - b\sqrt{m}\|\Psi'\| - cm, \tag{3.2}$$

with

$$a_\mu = \frac{K_{2\mu+2,2}^{2\mu+2}}{\mu+1}; \quad b = K_{\infty,2}\left(\sum_{v\in\mathcal{V}_-}|\alpha_v| + \|W_{-,1}\|_1\right); \quad c = \|W_{-,\infty}\|_\infty.$$

Next we distinguish two cases. For any $0<\mu<2$ and $m>0$, we note that there exists $\beta \equiv \beta(a_\mu, b, c, m) > 0$, such that $x^2 - a_\mu m^{\frac{2+\mu}{2}} x^\mu - b\sqrt{m}x - cm \geq -\beta$. Hence,

$$E[\Psi] \geq -\beta \qquad 0<\mu<2,\ m>0,$$

so that it must be $\nu_\mu(m) \leq \beta$.

For $\mu = 2$ and $0 < m < m_2^* = 1/\sqrt{a_2}$, we note that $(1-a_2m^2)x^2 - b\sqrt{m}x - cm \geq -b^2m/(4(1-a_2m^2)) - cm$, hence

$$E[\Psi] \geq -b^2m/(4(1-a_2m^2)) - cm \qquad \mu = 2,\ 0<m<m_2^*,$$

so that $\nu_\mu(m) \leq b^2m/(4(1-a_2m^2)) + cm$ under the same conditions on μ and m. □

4. Proof of Theorem 2

In this section we prove the existence of the ground state for small enough mass. Since the problem in the subcritical case $0<\mu<2$ was discussed in [**17**], in the proof we often skip the details whenever the argument used in [**17**] remains unchanged.

Proof of Th. 2. We start by recalling that the bound

$$-\nu_\mu(m) \leq -mE_0 \tag{4.1}$$

follows directly from the inequality $E[\Psi] < E^{lin}[\Psi]$, for all $\Psi \in H^1(\mathcal{G})$. Hence, by Ass. C, it must be $\nu_\mu(m) > 0$.

In the remaining part of the proof we shall show that, both in the subcritical and in the critical case, for $m < m_\mu^*$ minimizing sequences have a convergent subsequence (case i_2 of Lemma 2.2).

We recall that, by Th. 1, we already know that $\nu_\mu(m) < +\infty$ for any $m>0$ if $0<\mu<2$, or for $m<m_2^*$ if $\mu=2$, hence the existence of a lower bound for the infimum in Eq. (1.2) is granted for $m<m_\mu^*$ for any $0<\mu\leq 2$.

Let $\{\Psi_n\}_{n\in\mathbb{N}}$ be a minimizing sequence, i.e., $\Psi_n \in H^1(\mathcal{G})$, $\|\Psi_n\|^2 = m$, and $\lim_{n\to\infty} E[\Psi_n] = -\nu_\mu(m)$. We remark that, when choosing a minimizing sequence, it is enough to assume $\|\Psi_n\|^2 \equiv m_n \to m$ as $n\to\infty$, since in such a case one can define $\widetilde{\Psi}_n = \sqrt{m}\Psi_n/\|\Psi_n\|$ and use the fact that $\lim_{n\to\infty} E[\widetilde{\Psi}_n] = \lim_{n\to\infty} E[\Psi_n]$.

We shall prove that, for any $0<\mu\leq 2$ and m small enough, there exists $\hat{\Psi} \in H^1(\mathcal{G})$ such that $\|\hat{\Psi}\|^2 = m$, $E[\hat{\Psi}] = -\nu_\mu(m)$ and $\Psi_n \to \hat{\Psi}$ in $H^1(\mathcal{G})$.

We claim that, up to taking a subsequence that we still denote by Ψ_n, the following bound holds true

$$\sup_{n\in\mathbb{N}} \|\Psi_n'\| < \infty, \tag{4.2}$$

for any $m>0$ if $0<\mu<2$, and for $0<m<m_2^*$ if $\mu=2$.

To prove the bound (4.2) we start by noticing that, up to taking a subsequence, we can assume that

$$0 < m_n < \sqrt{1+\eta}\, m \quad \text{and} \quad E[\Psi_n] \leq -\nu_\mu(m)/2, \tag{4.3}$$

for all $\eta > 0$. Next we consider first the critical case $\mu = 2$. Fix $0 < m < m_2^* = 1/\sqrt{a_2}$. Then there exists $0 < \eta < 1/2$ such that $m < \sqrt{\frac{1-\eta}{(1+\eta)a_2}}$. In inequality (3.2) we set $\Psi \equiv \Psi_n$ and note that, by (4.3),

$$E[\Psi_n] \geq \eta\|\Psi_n'\|^2 + (1-\eta-(1+\eta)a_2m^2)\|\Psi_n'\|^2 - b(1+\eta)^{1/4}\sqrt{m}\|\Psi_n'\| - c\sqrt{1+\eta}\,m.$$

By the trivial bound $Ax^2 - Bx - C \geq -\frac{B^2}{4A} - C$, for all $A, B, C > 0$, we infer

$$E[\Psi_n] \geq \eta\|\Psi_n'\|^2 - \frac{b^2\sqrt{1+\eta}\,m}{4(1-\eta-(1+\eta)a_2m^2)} - c\sqrt{1+\eta}\,m,$$

for any $m < \sqrt{\frac{1-\eta}{(1+\eta)a_2}}$. The latter bound, together with the fact that $E[\Psi_n] < 0$ by (4.3), implies the claim (4.2).

For $0 < \mu < 2$ we proceed in a similar way. By inequality (3.2), we infer that for all $m > 0$ there exists $\tilde{\beta} \equiv \tilde{\beta}(a_\mu, b, c, m) > 0$, such that $(1-\eta)x^2 - a_\mu m_n^{\frac{2+\mu}{2}} x^\mu - b\sqrt{m_n}x - cm_n \geq -\tilde{\beta}$. Hence,

$$E[\Psi_n] \geq \eta\|\Psi_n'\|^2 - \tilde{\beta} \qquad 0 < \mu < 2,\ m > 0,$$

from which the claim (4.2) follows.

The bound (4.2), together with Gagliardo-Nirenberg inequality and the first bound in (4.3), implies

$$\sup_{n\in\mathbb{N}} \|\Psi_n\|_p < \infty \qquad \forall p \in [2, +\infty].$$

Moreover the following lower bound holds true

$$\frac{1}{\mu+1}\|\Psi_n\|_{2\mu+2}^{2\mu+2} + (\Psi_n, W_-\Psi_n) + \sum_{v\in\mathcal{V}_-} |\alpha_v||\Psi_n(v)|^2 \geq \frac{\nu_\mu(m)}{2}. \tag{4.4}$$

The latter is an immediate consequence of the bounds (3.1) and (4.3).

Next we use Lem. 2.2 and prove that vanishing and dichotomy cannot occur for $\{\Psi_n\}_{n\in\mathbb{N}}$. Set $\tau = \lim_{t\to\infty}\liminf_{n\to\infty}\rho(\Psi_n, t)$.

If $\tau = 0$, then by Lem. 2.2 the l.h.s. in Eq. (4.4) would converge to zero bringing to a contradiction (see [**17**] for the details), hence $\tau > 0$.

Suppose $0 < \tau < m$, then there would exist R_k and S_k satisfying (2.7)-(2.13). It is possible to prove (see [**17**] for the details) that in this case it must be

$$\liminf_{k\to\infty}\left(E[\Psi_{n_k}] - E[R_k] - E[S_k]\right) \geq 0,$$

which implies

$$\limsup_{k\to\infty}\left(E[R_k] + E[S_k]\right) \leq -\nu_\mu(m). \tag{4.5}$$

We use the identity

$$E[\Psi] = \frac{1}{\delta^2}E[\delta\Psi] + \frac{\delta^{2\mu}-1}{\mu+1}\|\Psi\|_{2\mu+2}^{2\mu+2},$$

which holds true for any $\Psi \in H^1(\mathcal{G})$ and $\delta > 0$. Let $\delta_k = \sqrt{m}/\|R_k\|$ and $\gamma_k = \sqrt{m}/\|S_k\|$ so that $\|\delta_k R_k\|^2$, $\|\gamma_k S_k\|^2 = m$. Then, using the above identity and the fact that $E[\delta_k R_k], E[\gamma_k S_k] \geq -\nu_\mu(m)$, one has

$$E[R_k] \geq -\frac{\nu_\mu(m)}{\delta_k^2} + \frac{\delta_k^{2\mu}-1}{\mu+1}\|R_k\|_{2\mu+2}^{2\mu+2}$$

$$E[S_k] \geq -\frac{\nu_\mu(m)}{\gamma_k^2} + \frac{\gamma_k^{2\mu}-1}{\mu+1}\|S_k\|_{2\mu+2}^{2\mu+2}$$

from which

$$E[R_k] + E[S_k] \geq -\nu_\mu(m)\left(\frac{1}{\delta_k^2} + \frac{1}{\gamma_k^2}\right) + \frac{\delta_k^{2\mu}-1}{\mu+1}\|R_k\|_{2\mu+2}^{2\mu+2} + \frac{\gamma_k^{2\mu}-1}{\mu+1}\|S_k\|_{2\mu+2}^{2\mu+2}.$$

Notice that, by (2.10), $\delta_k^2 \to m/\tau$ and $\gamma_k^2 \to 1/(1-\tau/m)$, hence $1/\delta_k^2 + 1/\gamma_k^2 \to 1$. Moreover set $\theta = \min\{(\tau/m)^{-\mu}, (1-\tau/m)^{-\mu}\} > 1$. Then

$$\liminf_{k\to\infty}(E[R_k] + E[S_k]) \geq -\nu_\mu(m) + \frac{\theta-1}{\mu+1}\liminf_{k\to\infty}\|\Psi_{n_k}\|_{2\mu+2}^{2\mu+2} > -\nu_\mu(m), \tag{4.6}$$

where we used the fact that $\liminf_{k\to\infty}\|\Psi_{n_k}\|_{2\mu+2}^{2\mu+2} \neq 0$. The latter claim is proved by noticing that $\liminf_{k\to\infty}\|\Psi_{n_k}\|_{2\mu+2}^{2\mu+2} = 0$ would bring to a contradiction with inequality (4.4). This can be understood by using the inequalities

$$\|\Psi\|_\infty \leq K_{\infty,2\mu+2}\|\Psi'\|^\alpha\|\Psi\|_{2\mu+2}^{1-\alpha}$$

with $\alpha = 1/(\mu+2)$; and

$$(\Psi, W_-\Psi) \leq \|W_-\|_r\|\Psi\|_{2r/(r-1)}^2 \leq K^2_{\frac{2r}{r-1},2\mu+2}\|W_-\|_r\|\Psi'\|^{2\alpha}\|\Psi\|_{2\mu+2}^{2(1-\alpha)} \tag{4.7}$$

which holds true for all $r \in [1, 1+1/\mu]$ and with $\alpha = (1-(\mu+1)(r-1)/r)/(\mu+2)$.

Since inequalities (4.5) and (4.6) cannot be satisfied at the same time we must also exclude the case $0 < \tau < m$. Hence it must be $\tau = m$.

Next we prove that for m small enough the minimizing sequence is not *runaway*. By absurd suppose that $\{\Psi_n\}_{n\in\mathbb{N}}$ is *runaway*, then we have that

$$\lim_{n\to\infty}\Psi_n(v) = 0 \quad \forall v \in \mathcal{V} \qquad \text{and} \qquad \lim_{n\to\infty}(\Psi_n, W_-\Psi_n) = 0. \tag{4.8}$$

The first limit is a direct consequence of Lem. 2.2, Eq. (2.6). The proof of the second one requires a bit more work. Assume that Ψ_n escapes at infinity on the external edge e^* (this can always be done up to taking a subsequence). We note that

$$\lim_{n\to\infty}\int_{I_e}(W_-)_e|(\Psi_n)_e|^2dx = 0 \qquad \forall e \neq e^*,$$

this is a direct consequence of the inequality

$$\int_{I_e}(W_-)_e|(\Psi_n)_e|^2dx \leq \|W_{-,\infty}\|_\infty\|(\Psi_n)_e\|_{L^2(I_e)}^2 + \|W_{-,1}\|_1\|(\Psi_n)_e\|_{L^\infty(I_e)}^2$$

and Lemma 2.2, Eq. (2.6). We are left to prove that

$$\lim_{n\to\infty}\int_0^{+\infty}(W_-)_{e^*}|(\Psi_n)_{e^*}|^2dx = 0. \tag{4.9}$$

We have that for any $\varepsilon > 0$ and $r \geq 1$, there exists $R > 0$ (independent of n) such that

$$\int_R^{+\infty}(W_-)_{e^*}|(\Psi_n)_{e^*}|^2dx \leq \|(W_-)_{e^*}\|_{L^r(R,+\infty)}\|\Psi_n\|_{2r/(r-1)}^2 \leq \varepsilon.$$

For such R, there exists n_0 such that for all $n > n_0$ one has

$$\int_0^R(W_-)_{e^*}|(\Psi_n)_{e^*}|^2dx \leq \|W_-\|_r\|(\Psi_n)_{e^*}\|_{L^{2r/(r-1)}(0,R)}^2 \leq \varepsilon$$

by Eq. (2.6), from which the second limit in (4.8).

Recalling that, by Lem. 2.2 - Eq. (2.6), one has $\lim_{n\to\infty}\|(\Psi_n)_e\|_{L^{2\mu+2}(I_e)}=0$ for all $e\neq e^*$, and by using Eq. (4.8), we infer

$$\lim_{n\to\infty} E[\Psi_n] \geq \lim_{n\to\infty}\left(\|(\Psi_n)'_{e^*}\|^2_{L^2(\mathbb{R}_+)} - \frac{1}{\mu+1}\|(\Psi_n)_{e^*}\|^{2\mu+2}_{L^{2\mu+2}(\mathbb{R}_+)}\right)$$

Let $\chi:\mathbb{R}_+\to[0,1]$ be a function such that $\chi\in C^\infty(\mathbb{R}_+)$, $\chi(0)=0$ and $\chi(x)=1$ for all $x\geq 1$ and define

$$\psi_n^*(x) := \chi(x)(\Psi_n)_{e^*}(x),$$

so that $\psi_n^*(0)=0$, and $\|\psi_n^{*\prime}\|^2_{L^2(\mathbb{R}_+)}\leq c$. We have the following inequalities (we refer to [**17**] for the details)

$$\begin{aligned}-\nu_\mu(m) = \lim_{n\to\infty} E[\Psi_n] &\geq \lim_{n\to\infty}\left(\|\psi_n^{*\prime}\|^2_{L^2(\mathbb{R}_+)} - \frac{1}{\mu+1}\|\psi_n^*\|^{2\mu+2}_{L^{2\mu+2}}\right)\\ &\geq \inf\left\{E_{\mathbb{R}}[\psi]\,\middle|\,\psi\in H^1(\mathbb{R}),\ \|\psi\|^2_{L^2(\mathbb{R})}=m\right\} = -t_\mu(m),\end{aligned} \tag{4.10}$$

see Eqs. (1.6) and (1.9) for the explicit value of $t_\mu(m)$.

By the bounds (4.1) and (4.10) we infer that, if the minimizing sequence Ψ_n is runaway, then it must be

$$0 > -E_0 m \geq -\nu_\mu(m) \geq -t_\mu(m). \tag{4.11}$$

Next we distinguish the subcritical and the critical case. If $0<\mu<2$, it is well known (see, e.g, [**20**]) that the infimum $-t_\mu(m)$ is negative, finite, and that it is indeed attained for any $m>0$, moreover $t_\mu(m)$ is given by Eq. (1.6), with

$$\gamma_\mu = \frac{2-\mu}{2+\mu}\left(2\frac{(\mu+1)^{\frac{1}{\mu}}}{\mu}\int_0^1(1-t^2)^{\frac{1}{\mu}-1}dt\right)^{-\frac{2\mu}{2-\mu}} \qquad 0<\mu<2. \tag{4.12}$$

We conclude that, whenever $-E_0 m < -\gamma_\mu m^{1+\frac{2\mu}{2-\mu}}$ (i.e., $m<(E_0/\gamma_\mu)^{\frac{1}{\mu}-\frac{1}{2}}$) there is a contradiction with the chain of inequalities (4.11), hence the minimizing sequence cannot be runaway and must converge to a certain function $\hat\Psi$.
If $\mu=2$ the infimum $-t_2(m)$ exhibits a critical mass, see Eq. (1.9). Moreover the infimum is attained only at the critical mass $m=\pi\sqrt{3}/2$. Since $-t_\mu(m)=0$, for $m\leq\pi\sqrt{3}/2$, contradicts the chain of inequalities (4.11) we conclude that also in this case the minimizing sequence cannot be runaway and must converge to a certain function $\hat\Psi$. The latter argument, together with the constraint $m<m_2^*$ (needed for the bound (4.2)) and the fact that by [**10**, Prop. 2.3], $(2/\pi)^{1/3}=K_{6,2}(\mathbb{R})\leq K_{6,2}(\mathcal{G})$, tell us that the value of the threshold mass is m_2^*.

By Lem. 2.2 we conclude that for all $0<\mu\leq 2$ and m small enough there exists a state $\hat\Psi\in H^1(\mathcal{G})$ such that minimizing sequences converge, up to taking subsequences, to $\hat\Psi$ in L^p for $p\geq 2$. To establish the convergence of $\Psi_n\to\hat\Psi$ in $H^1(\mathcal{G})$, and conclude the proof of the theorem, one can repeat the general argument used in [**17**]. □

Appendix A. An upper bound for $-\nu_\mu(m)$

In the following proposition we prove that in the subcritical case, if the potential W decays at infinity, the infimum in (1.2) cannot exceed the infimum $-t_\mu(m)$.

PROPOSITION A.1. *Let $0 < \mu < 2$. Let Assumption* A *hold true and assume that $W \in L^r(\mathcal{G})$ for some $r \in [1, +\infty)$. Then*

$$-\nu_\mu(m) \leq -t_\mu(m).$$

PROOF. To prove the claim it is enough to exhibit a sequence Φ_n such that: $\Phi_n \in H^1(\mathcal{G})$, $\|\Phi_n\|^2 \to m$, and

$$E[\Phi_n] \to -t_\mu(m)$$

as $n \to \infty$.

We recall that (see, e.g., [**42**] and [**20**]) for $0 < \mu < 2$ and $m > 0$, the infimum $-t_\mu(m)$ is attained by the function

$$\phi_\omega(x) = [(\mu+1)\omega]^{\frac{1}{2\mu}} \operatorname{sech}^{\frac{1}{\mu}}(\mu\sqrt{\omega}x)$$

with

$$\omega = \left(\frac{m\mu}{2(\mu+1)^{\frac{1}{\mu}} I(\mu)}\right)^{\frac{2\mu}{2-\mu}},$$

where $I(\mu) = \int_0^1 (1-t^2)^{\frac{1}{\mu}-1} dt$. Fix $e^* \in \mathcal{E}^{ex}$ and let Φ_n be defined as

$$(\Phi_n)_e = \begin{cases} \chi\phi_\omega(\cdot - n) & e = e^* \\ 0 & e \neq e^* \end{cases}$$

where χ is a $C^\infty(\mathbb{R}_+)$ function such that $\chi \in C^\infty(\mathbb{R}_+)$, $\chi(0) = 0$ and $\chi(x) = 1$ for all $x \geq 1$. One has

$$\|\Phi_n\|^2 = \|\chi\phi_\omega(\cdot - n)\|^2_{L^2(\mathbb{R}_+)} \to m.$$

Moreover

$$\begin{aligned} E[\Phi_n] =& \|(\chi\phi_\omega(\cdot - n))'\|^2_{L^2(\mathbb{R}_+)} \\ &+ (\chi\phi_\omega(\cdot - n), W_e^* \chi\phi_\omega(\cdot - n))_{L^2(\mathbb{R}_+)} - \frac{\|\chi\phi_\omega(\cdot - n))\|^{2\mu+2}_{L^{2\mu+2}(\mathbb{R}_+)}}{\mu+1} \\ \to& E_{\mathbb{R}}[\phi_\omega] = -t_\mu(m). \end{aligned}$$

The latter claim is an immediate consequence of the the fact that

$$\|\phi'_\omega(\cdot - n)\|^2_{L^2(-\infty,1)} \to 0 \qquad \text{and} \qquad \|\phi_\omega(\cdot - n)\|^p_{L^p(-\infty,R)} \to 0$$

for all $p \in [1, \infty]$ and $R > 0$, as $n \to \infty$ (see also Eq. (4.9)). $\square$

References

[1] Riccardo Adami, *Ground states for NLS on graphs: a subtle interplay of metric and topology*, Math. Model. Nat. Phenom. **11** (2016), no. 2, 20–35, DOI 10.1051/mmnp/201611202. MR3491787

[2] Riccardo Adami, Claudio Cacciapuoti, Domenico Finco, and Diego Noja, *Fast solitons on star graphs*, Rev. Math. Phys. **23** (2011), no. 4, 409–451, DOI 10.1142/S0129055X11004345. MR2804557

[3] Riccardo Adami, Claudio Cacciapuoti, Domenico Finco, and Diego Noja, *On the structure of critical energy levels for the cubic focusing NLS on star graphs*, J. Phys. A **45** (2012), no. 19, 192001, 7, DOI 10.1088/1751-8113/45/19/192001. MR2924493

[4] Riccardo Adami, Claudio Cacciapuoti, Domenico Finco, and Diego Noja, *Stationary states of NLS on star graphs*, Europhys. Lett. EPL **100** (2012), no. 1, 10003.

[5] Riccardo Adami, Claudio Cacciapuoti, Domenico Finco, and Diego Noja, *Constrained energy minimization and orbital stability for the NLS equation on a star graph*, Ann. Inst. H. Poincaré Anal. Non Linéaire **31** (2014), no. 6, 1289–1310, DOI 10.1016/j.anihpc.2013.09.003. MR3280068

[6] Riccardo Adami, Claudio Cacciapuoti, Domenico Finco, and Diego Noja, *Variational properties and orbital stability of standing waves for NLS equation on a star graph*, J. Differential Equations **257** (2014), no. 10, 3738–3777, DOI 10.1016/j.jde.2014.07.008. MR3260240

[7] Riccardo Adami, Claudio Cacciapuoti, Domenico Finco, and Diego Noja, *Stable standing waves for a NLS on star graphs as local minimizers of the constrained energy*, J. Differential Equations **260** (2016), no. 10, 7397–7415, DOI 10.1016/j.jde.2016.01.029. MR3473445

[8] Riccardo Adami, Enrico Serra, and Paolo Tilli, *Lack of ground state for NLSE on bridge-type graphs*, Mathematical technology of networks, Springer Proc. Math. Stat., vol. 128, Springer, Cham, 2015, pp. 1–11, DOI 10.1007/978-3-319-16619-3_1. MR3375151

[9] Riccardo Adami, Enrico Serra, and Paolo Tilli, *NLS ground states on graphs*, Calc. Var. Partial Differential Equations **54** (2015), no. 1, 743–761, DOI 10.1007/s00526-014-0804-z. MR3385179

[10] Riccardo Adami, Enrico Serra, and Paolo Tilli, *Negative energy ground states for the L^2-critical NLSE on metric graphs*, Comm. Math. Phys. **352** (2017), no. 1, 387–406, DOI 10.1007/s00220-016-2797-2. MR3623262

[11] Riccardo Adami, Enrico Serra, and Paolo Tilli, *Threshold phenomena and existence results for NLS ground states on metric graphs*, J. Funct. Anal. **271** (2016), no. 1, 201–223, DOI 10.1016/j.jfa.2016.04.004. MR3494248

[12] Felix Ali Mehmeti, *Nonlinear waves in networks*, Mathematical Research, vol. 80, Akademie-Verlag, Berlin, 1994. MR1287844

[13] Felix Ali Mehmeti, Joachim von Below, and Serge Nicaise (eds.), *Partial differential equations on multistructures*, Lecture Notes in Pure and Applied Mathematics, vol. 219, Marcel Dekker, Inc., New York, 2001. MR1824576

[14] Gregory Berkolaiko and Peter Kuchment, *Introduction to quantum graphs*, Mathematical Surveys and Monographs, vol. 186, American Mathematical Society, Providence, RI, 2013. MR3013208

[15] Jerry L. Bona and Radu C. Cascaval, *Nonlinear dispersive waves on trees*, Can. Appl. Math. Q. **16** (2008), no. 1, 1–18. MR2500096

[16] Claudio Cacciapuoti, Domenico Finco, and Diego Noja, *Topology-induced bifurcations for the nonlinear Schrödinger equation on the tadpole graph*, Phys. Rev. E (3) **91** (2015), no. 1, 013206, 8, DOI 10.1103/PhysRevE.91.013206. MR3416696

[17] Claudio Cacciapuoti, Domenico Finco, and Diego Noja, *Ground state and orbital stability for the NLS equation on a general starlike graph with potentials*, Nonlinearity **30** (2017), no. 8, 3271–3303, DOI 10.1088/1361-6544/aa7cc3. MR3685669

[18] Radu C. Cascaval and C. Travis Hunter, *Linear and nonlinear Schrödinger equations on simple networks*, Libertas Math. **30** (2010), 85–98. MR2675854

[19] Vincent Caudrelier, *On the inverse scattering method for integrable PDEs on a star graph*, Comm. Math. Phys. **338** (2015), no. 2, 893–917, DOI 10.1007/s00220-015-2378-9. MR3351062

[20] Thierry Cazenave, *Semilinear Schrödinger equations*, Courant Lecture Notes in Mathematics, vol. 10, New York University, Courant Institute of Mathematical Sciences, New York; American Mathematical Society, Providence, RI, 2003. MR2002047

[21] Pavel Exner, Jonathan P. Keating, Peter Kuchment, Toshikazu Sunada, and Alexander Teplyaev, *Vladimir A. Geyler, April 29, 1943–April 2, 2007*, Analysis on graphs and its applications, Proc. Sympos. Pure Math., vol. 77, Amer. Math. Soc., Providence, RI, 2008, pp. xiii, DOI 10.1090/pspum/077. MR2459861

[22] Steffen Gilg, Dmitry Pelinovsky, and Guido Schneider, *Validity of the NLS approximation for periodic quantum graphs*, NoDEA Nonlinear Differential Equations Appl. **23** (2016), no. 6, Art. 63, 30, DOI 10.1007/s00030-016-0417-7. MR3571159

[23] Sven Gnutzmann, Uzy Smilansky, and Stanislav Derevyanko, *Stationary scattering from a nonlinear network*, Phys. Rev. A **83** (2011), 033831.

[24] Sven Gnutzmann and Daniel Waltner, *Stationary waves on nonlinear quantum graphs: general framework and canonical perturbation theory*, Phys. Rev. E **93** (2016), no. 3, 032204, 19. MR3652709

[25] Sven Gnutzmann and Daniel Waltner, *Stationary waves on nonlinear quantum graphs: general framework and canonical perturbation theory*, Phys. Rev. E **93** (2016), no. 3, 032204, 19. MR3652709

[26] Sebastian Haeseler, *Heat kernel estimates and related inequalities on metric graphs*, arXiv:1101.3010 [math-ph] (2011), 20pp.

[27] Panagiotis G. Kevrekidis, Dimitrios J. Frantzeskakis, Georgios Theocharis, and Ioannis G. Kevrekidis, *Guidance of matter waves through Y-junctions*, Phys. Lett. A **317** (2003), no. 5, 513–522.

[28] Elliott H. Lieb and Michael Loss, *Analysis*, 2nd ed., Graduate Studies in Mathematics, vol. 14, American Mathematical Society, Providence, RI, 2001. MR1817225

[29] Jeremy L. Marzuola and Dmitry E. Pelinovsky, *Ground state on the dumbbell graph*, Appl. Math. Res. Express. AMRX **1** (2016), 98–145, DOI 10.1093/amrx/abv011. MR3483843

[30] Delio Mugnolo, *Semigroup methods for evolution equations on networks*, Understanding Complex Systems, Springer, Cham, 2014. MR3243602

[31] Diego Noja, *Nonlinear Schrödinger equation on graphs: recent results and open problems*, Philos. Trans. R. Soc. Lond. Ser. A Math. Phys. Eng. Sci. **372** (2014), no. 2007, 20130002, 20, DOI 10.1098/rsta.2013.0002. MR3151078

[32] Diego Noja, Dmitry Pelinovsky, and Gaukhar Shaikhova, *Bifurcations and stability of standing waves in the nonlinear Schrödinger equation on the tadpole graph*, Nonlinearity **28** (2015), no. 7, 2343–2378, DOI 10.1088/0951-7715/28/7/2343. MR3366647

[33] Dmitry Pelinovsky and Guido Schneider, *Bifurcations of standing localized waves on periodic graphs*, Ann. Henri Poincaré **18** (2017), no. 4, 1185–1211, DOI 10.1007/s00023-016-0536-z. MR3626301

[34] Olaf Post, *Spectral analysis on graph-like spaces*, Lecture Notes in Mathematics, vol. 2039, Springer, Heidelberg, 2012. MR2934267

[35] Karimjon K. Sabirov, Zarif A. Sobirov, Doniyor Babajanov, and Davron U. Matrasulov, *Stationary nonlinear Schrödinger equation on simplest graphs*, Phys. Lett. A **377** (2013), no. 12, 860–865, DOI 10.1016/j.physleta.2013.02.011. MR3028326

[36] Enrico Serra and Lorenzo Tentarelli, *Bound states of the NLS equation on metric graphs with localized nonlinearities*, J. Differential Equations **260** (2016), no. 7, 5627–5644, DOI 10.1016/j.jde.2015.12.030. MR3456809

[37] Enrico Serra and Lorenzo Tentarelli, *On the lack of bound states for certain NLS equations on metric graphs*, Nonlinear Anal. **145** (2016), 68–82, DOI 10.1016/j.na.2016.07.008. MR3547674

[38] Zarif A. Sobirov, Davron U. Matrasulov, Karimjon K. Sabirov, Shinichi Sawada, and Katsuhiro Nakamura, *Integrable nonlinear Schrödinger equation on simple networks: connection formula at vertices*, Phys. Rev. E (3) **81** (2010), no. 6, 066602, 10, DOI 10.1103/PhysRevE.81.066602. MR2736292

[39] Zarif A. Sobirov, Doniyor Babajanov, and Davron U. Matrasulov, *Nonlinear standing waves on planar branched systems: Shrinking into metric graph*, Nanosystems: Physics, Chemistry, Mathematics **8** (2017), no. 1, 29–37.

[40] Zarif A. Sobirov, Karimjon K. Sabirov, and Davron U. Matrasulov, *Stationary nonlinear Schrödinger equation on simplest graphs: Boundary conditions and exact solutions*, arXiv:1107.1220 [nlin.SI] (2011), 11pp.

[41] Zarif A. Sobirov, Karimjon K. Sabirov, Davron U. Matrasulov, Abdulla Saidov, and Katsuhiro Nakamura, *Ballistic soliton transport in networks*, NATO Science for Peace and Security Series - C: Environmental Security, pp. 215–231, Springer, 2014.

[42] Catherine Sulem and Pierre-Louis Sulem, *The nonlinear Schrödinger equation*, Applied Mathematical Sciences, vol. 139, Springer-Verlag, New York, 1999. Self-focusing and wave collapse. MR1696311

[43] Lorenzo Tentarelli, *NLS ground states on metric graphs with localized nonlinearities*, J. Math. Anal. Appl. **433** (2016), no. 1, 291–304, DOI 10.1016/j.jmaa.2015.07.065. MR3388792

[44] Hannes Uecker, Daniel Grieser, Zarif A. Sobirov, Doniyor Babajanov, and Davron U. Matrasulov, *Soliton transport in tubular networks: transmission at vertices in the shrinking limit*, Phys. Rev. E (3) **91** (2015), no. 2, 023209, 8, DOI 10.1103/PhysRevE.91.023209. MR3418622

DiSAT, Sezione di Matematica, Università dell'Insubria, via Valleggio 11, I-22100 Como, Italy
Email address: claudio.cacciapuoti@uninsubria.it

Contemporary Mathematics
Volume **717**, 2018
https://doi.org/10.1090/conm/717/14447

Fredholm theory of differential operators on periodic graphs

Vladimir Rabinovich

ABSTRACT. The paper is devoted to the Fredholm theory of differential operators on ifinite metric graphs imbedded in $\mathbb{R}^n$ and periodic with respect to a group $\mathbb{G}$ isomorphic to $\mathbb{Z}^r, 1 \leq r \leq n$. It is assumed that the coefficients of differential operators and vertex conditions are not periodic. Necessary and sufficient conditions for operators under consideration to be Fredholm are obtained.

1. Introduction

During the last two decades, considerable interest has arisen in the study of so-called quantum graphs, that is metric graphs equipped with differential or pseudo-differential operators on the edges. The theory of quantum graphs has a considerable mathematical interest and serves as simplified models to describe important processes in physics, chemistry, and engineering. We refer to the well known reviews, books, and papers on this topic: [**2**],[**3**],[**4**],[**5**],[**6**],[**8**], see also references cited in [**3**].

The main aim of the paper is the investigation of the Fredholm property and essential spectrum of differential operators with variable non-periodic coefficients on infinite periodic graphs Γ with variable non-periodic vertex conditions.

We suppose that Γ is a metric oriented graph embedded in $\mathbb{R}^n$ periodic with respect to a discrete group $\mathbb{G}$ isomorphic to $\mathbb{Z}^r$ where $1 \leq r \leq n$. Let $\mathcal{E}$ be the set of the edges, $\mathcal{V}$ be the set of the vertices of Γ, and $d(v)$ be the valence of the vertex $v \in \mathcal{V}$, that is the number of edges incident to the vertex v. We suppose that $1 \leq d(v) \leq d_0$ where d_0 is independent of $v \in \mathcal{V}$. Let the graph Γ be equipped by a differential operator

$$Au(x) = \sum_{j=0}^{m} a_j(x)u^{(j)}(x), x \in \Gamma \setminus \mathcal{V} \tag{1}$$

with coefficients $a_j \in L^\infty(\Gamma)$ and transmission conditions at every vertex $v \in \mathcal{V}$

$$B_{k,v}u(v) = \sum_{j=0}^{m_k} b_{j,k}(v)u^{(j)}(v) = \varphi_k(v) \in \mathbb{C}^{d(v)}, v \in \mathcal{V}, k = 1, ..., l \tag{2}$$

2010 *Mathematics Subject Classification.* Primary 34B45, 47A10.

Key words and phrases. Quantum graphs, Fredholm theory, Essential spectrum, Localization.

where $l < m$, and $m_k < m - \frac{1}{2}$, $b_{j,k}(v)$ are $d(v) \times d(v)$ complex matrices, $u^{(j)}(v) = (u_1^{(j)}(v), ..., u_{d(v)}^{(j)}(v)) \in \mathbb{C}^{d(v)}$ and $u_i^{(j)}(v), i = 1, ..., d(v)$ are limit values at the vertex $v \in \mathcal{V}$ of the derivatives $u_i^{(j)}(x)$ taken along the edges $e_i \in \mathcal{E}_\nu$ according their orientation. We associate with the differential operator A and transmission conditions (2) an operator $\mathfrak{A}$ acting from the Sobolev space $H^m(\Gamma) = \bigoplus_{e\in\mathcal{E}} H^m(e)$ ($H^m(e)$ is the Sobolev space on the edge $e \in \mathcal{E}$) to the space $L^2(\Gamma)\bigoplus_{v\in\mathcal{V}}\mathbb{C}^{ld(v)}$. We study the Fredholm property of the operator $\mathfrak{A}$. It is well-known that if Γ is a *compact* graph and A is an elliptic operator, that is $\inf_{x\in\Gamma} |a_m(x)| > 0$, then $\mathfrak{A} : H^m(\Gamma) \to L^2(\Gamma)\bigoplus_{v\in\mathcal{V}}\mathbb{C}^{ld(v)}$ is a Fredholm operator for arbitrary transmission conditions (2) (see for instance [**3**], pp. 52-54). Of course this statement is not correct for infinite graphs, and $\mathfrak{A}$ needs some conditions at infinity to be a Fredholm operator.

To formulate these conditions we introduce a compactification of the graph Γ as follows. Let $\tilde{\mathbb{R}}^n$ be a compactification of $\mathbb{R}^n$ by the "sphere" of the infinite distant points homeomorphic to the closed unit ball $\bar{B} = \{x \in \mathbb{R}^n : |x| \leq 1\}$, and let $\tilde{\Gamma}, \tilde{\mathcal{V}}$ be the closures of $\Gamma, \mathcal{V}$, respectively, in the topology of $\tilde{\mathbb{R}}^n$. We set $\Gamma_\infty = \tilde{\Gamma}\setminus\Gamma$, $\mathcal{V}_\infty = \tilde{\mathcal{V}}\setminus\mathcal{V}$. Let $\mathcal{M}(\tilde{\Gamma})$ be a sub-algebras of $L^\infty(\Gamma)$ consisting of functions f that admit an extension $\tilde{f}$ on Γ_∞ such that $\tilde{f} \mid_{\Gamma_\infty} \in C(\Gamma_\infty)$. In the same way the sub-algebra $\mathcal{M}(\tilde{\mathcal{V}})$ of $l^\infty(\mathcal{V})$ is defined. Let $L^\infty_{\mathbb{G}}(\Gamma)$ be a sub-algebra of $L^\infty(\Gamma)$ consisting of functions that are periodic with respect to the group $\mathbb{G}$. Then we denote by $\mathcal{M}_{\mathbb{G}}(\tilde{\Gamma})$ a sub-algebra of $L^\infty(\Gamma)$ generated by $\mathcal{M}(\tilde{\Gamma})$ and $L^\infty_{\mathbb{G}}(\Gamma)$. In the same way the sub-algebra $\mathcal{M}_{\mathbb{G}}(\tilde{\mathcal{V}})$ is defined.

Let the coefficients a_j of A belong to $\mathcal{M}_{\mathbb{G}}(\tilde{\Gamma})$ and the elements of the matrices $b_{j,k}$ belong to $\mathcal{M}_{\mathbb{G}}(\tilde{\mathcal{V}})$. Then we introduce a family of limit operators $\mathfrak{A}^\eta, \eta \in \Gamma_\infty$ defined by the differential operators on the edges

$$A^\eta u(x) = \sum_{j=1}^{m} a_j^\eta(x) u^{(j)}(x), x \in \Gamma \setminus \mathcal{V} \tag{3}$$

where $a_j^\eta(x) = \lim_{g\to\eta\in\Gamma_\infty} a_j(x+g), x \in \Gamma$ belongs to $L^\infty_{\mathbb{G}}(\Gamma)$, and by the vertex operators

$$B_{k,v}^\eta u(v) = \sum_{j=0}^{m_k} b_{j,k}^\eta(v) u^{(j)}(v), v \in \mathcal{V}, k = 1, ..., l, \tag{4}$$

where $b_{j,k}^\eta(x) = \lim_{g\to\eta} b_{j,k}^\eta(x+g), x \in \Gamma$ belongs to $L^\infty_{\mathbb{G}}(\mathcal{V})$.

The main result of the paper is

THEOREM 1. *Let the coefficients of A belong to $\mathcal{M}_{\mathbb{G}}(\tilde{\Gamma})$ and the coefficients of the vertex operators $B_{k,v}$ belong to $\mathcal{M}_{\mathbb{G}}(\tilde{\mathcal{V}})$. Assume that A is an elliptic operator. Then $\mathfrak{A} : H^m(\Gamma) \to L^2(\Gamma)\bigoplus_{v\in\mathcal{V}}\mathbb{C}^{ld(v)}$ is a Fredholm operator if and only if the limit operators $\mathfrak{A}^\eta : H^m(\Gamma) \to L^2(\Gamma)\bigoplus_{v\in\mathcal{V}}\mathbb{C}^{ld(v)}$ are invertible for all $\eta \in \Gamma_\infty$.*

The proof of Theorem 1 is based on a localization of the operator $\mathfrak{A}$ at the infinite distant points of $\tilde{\Gamma}_\infty$. The idea of this localization goes back to the papers [**19**],[**17**], see also [**16**].

It should be noted that all limit operators $\mathfrak{A}^{\eta}, \eta \in \Gamma_{\infty}$ are periodic, that is $\mathfrak{A}^{\eta}$ is defined by the differential operators A^{η} and by the vertex operators $B^{\eta}_{k,v}$ with coefficients which are periodic with respect to $\mathbb{G}$. Hence for the study of the invertibility of these operators can be applied the Floquet theory. Note that the spectral properties of periodic Schrödinger operators on periodic metric graphs have been studied in many papers (see for instance [**3**],Chap.4,[**5**],[**6**],[**7**],[**9**],[**10**], and references cited there). It is well known that their spectra have a band-gap structure obtained by means of the Floquet transform just as it is done in the solid state physics. In particular, the detail spectral analysis of the quantum graphs associated with the carbon nano-structures have been done in [**7**],[**10**].

Note that the essential spectra of discrete operators on periodic combinatorial graphs and the essential spectra of quantum graphs with Kirchhof-Neumann vertex conditions have been studied in [**14**],[**15**], [**18**] by means of the limit operators method (see the book [**13**]).

The paper is organized as follows. In Sec. 2 we give necessary notations and definitions for the graphs under consideration. Sec. 3 is devoted to the proof of Theorem 1. In Sec. 4 we consider applications of Theorem 1 to the Fredholm properties of Schrödinger operators on periodic graphs with non-periodic potentials and non-periodic general vertex conditions. As an application we give the description of the essential spectra of quantum graphs described by the Schrödinger operators with real-valued potentials and general periodic vertex conditions.

2. Notations

1^0. Let X, Y be Banach spaces. We denote by $\mathcal{B}(X,Y)$ the Banach space of all bounded linear operators, acting from X into Y, and by $\mathcal{K}(X,Y)$ we denote the subspace in $\mathcal{B}(X,Y)$ of all compact operators. In the case $X = Y$ we simplify these notations as $\mathcal{B}(X,X) = \mathcal{B}(X)$, $\mathcal{K}(X,X) = \mathcal{K}(X)$.

An operator $A \in \mathcal{B}(X,Y)$ is called a Fredholm operator if $\ker A = \{x \in X : Ax = 0\}$ and *coker* $A = Y/\operatorname{Im} A$ are finite-dimensional spaces. The last condition implies that $\operatorname{Im} A = \{y \in Y : y = Ax, x \in X\}$ is a closed subspace in Y. Note that $A \in \mathcal{B}(X,Y)$ is a Fredholm operator if and only if there exist operators $R, L \in \mathcal{B}(Y,X)$ (left and right regularizators) such that

$$LA = I_X + T_1, AR = I_Y + T_2$$

where $T_1 \in \mathcal{K}(X), T_2 \in \mathcal{K}(Y)$.

2^0. We consider an oriented metric graph Γ embedded in $\mathbb{R}^n$. We suppose that the graph Γ consists of a countably infinite sets of vertices $\mathcal{V}$ and edges $\mathcal{E}$ connecting these vertices. Each edge $e \in \mathcal{E}$ is a line segment connecting vertices, and we suppose that for each pair of vertices $\{\alpha, \beta\}$ there exists not more than one edge connecting this pair. Let $\mathcal{E}_v$ be a set of edges incident to the vertex v (i.e., containing v). We always assume that the valence $d(\nu)$ (the number of the edges in $\mathcal{E}_v$) for any vertex v is finite and positive. Vertices with no incident edges are not allowed. The graph Γ is a metric space with a metric induced by the standard metric of $\mathbb{R}^n$. The topology on Γ is induced by the topology of $\mathbb{R}^n$, and the measure dl on Γ is the line Lebesgue measure on every edge. Moreover we also assume that the graph Γ is a connected set.

Let $\mathbb{G} \subset \mathbb{Z}^n$ be a discrete group

$$\mathbb{G} = \left\{ g \in \mathbb{R}^n : g = \sum_{j=1}^{r} \alpha_j \mathfrak{e}_j, \alpha_j \in \mathbb{Z}, \mathfrak{e}_j \in \mathbb{R}^n \right\},$$

where $\{\mathfrak{e}_1, ..., \mathfrak{e}_r\}, 1 \leq r \leq n$ is a linearly independent system of vectors in $\mathbb{R}^n$. The group $\mathbb{G}$ acts on the graph Γ by shifts $\Gamma \ni x \to x + g \in \Gamma$. We suppose that the action of $\mathbb{G}$ on Γ is co-compact, that is the fundamental domain $\Gamma/\mathbb{G}$ is a compact set in the corresponding quotient topology. We fix a measurable set $\Gamma_0 \subset \Gamma$ with compact closure which contains for every $x \in \Gamma$ exactly one element of the quotient class $x + \mathbb{G} \in \Gamma/\mathbb{G}$. There exists a natural one-to-one mapping $\Gamma_0 \to \Gamma/\mathbb{G}$ which is the composition of the inclusion mapping $\Gamma_0 \subset \Gamma$ and the canonical projection $\Gamma \to \Gamma/\mathbb{G}$. Then

$$\Gamma = \Gamma_0 + \mathbb{G} = \{x \in \mathbb{R} : x = x_0 + g, x_0 \in \Gamma_0, g \in \mathbb{G}\}. \tag{5}$$

Let $\mathcal{V}_0 = \mathcal{V} \cap \Gamma_0$. Then

$$\mathcal{V} = \mathcal{V}_0 + \mathbb{G} = \{x \in \mathbb{R} : x = v_0 + g, v_0 \in \mathcal{V}_0, g \in \mathbb{G}\}. \tag{6}$$

In what follows we assume that $\mathcal{V} \ni v \to d(v) \in \mathbb{N}$ is a periodic function, that is $d(v+g) = d(v)$ for every $v \in \mathcal{V}$ and $g \in \mathbb{G}$. Since Γ is co-compact with respect to the action of the group $\mathbb{G}$ there exists $d_0 \in \mathbb{N}$ such that $1 \leq d(v) \leq d_0$ for every $v \in \mathcal{V}$.

We denote by $L^2(\Gamma)$ the Hilbert space of measurable functions on Γ with the norm

$$\|u\|_{L^2(\Gamma)} = \left(\int_\Gamma |u(x)|^2 \, dx \right)^{1/2} = \left(\sum_{e \in \mathcal{E}} \int_e |u(x)|^2 \, dx \right)^{1/2}.$$

Let $H^s(e), s \in \mathbb{R}$ be the 1-dimensional Sobolev space on the edge e with the standard norm

$$\|u\|_{H^s(e)} = \inf_{lu \in H^s(\mathbb{R})} \|lu\|_{H^s(\mathbb{R})} \tag{7}$$

where we identify the edge e with a segment of $\mathbb{R}$, and the infimum in (7) is taken with respect to all extensions lu of u on $\mathbb{R}$ belonging to $H^s(\mathbb{R})$. We introduce the Sobolev space $H^s(\Gamma)$ on Γ as

$$H^s(\Gamma) = \bigoplus_{e \in \mathcal{E}} H^s(e)$$

with the norm

$$\|u\|_{H^s(\Gamma)} = \left(\sum_{e \in \mathcal{E}} \|u_e\|^2_{H^s(e)} \right)^{1/2}, u_e = u \mid_e .$$

We also use the standard notations:

- If Ω is an open set in $\mathbb{R}^n$ then: $C^\infty(\Omega)$ is the set of infinitely differentiable functions in Ω, $C_0^\infty(\Omega)$ is a subset of $C^\infty(\Omega)$ consisting of functions with compact supports in $\Omega, C_b^\infty(\Omega)$ is a subset of $C^\infty(\Omega)$ consisting of functions bounded in Ω with all derivatives.
- If $\varphi \in C(\mathbb{R}^n)$ we denote by $\hat{\varphi}$ the restriction of φ on the graph Γ, and by $\check{\varphi}$ the restriction of φ on the set of the vertices $\mathcal{V}$.
- If a is a function we denote by aI the operator of multiplication by a acting in an suitable functional space.

- $B_R = \{x \in \mathbb{R}^n : |x| < R\}, B'_R = \mathbb{R}^n \setminus \bar{B}_R$.

3. Fredholm property and essential spectra of differential operators on graphs

1^0. We assume that the graph Γ is oriented, and we fix an orientation on the every edge $e \in \mathcal{E}$. We consider the differential equation on the graph Γ as

$$Au(x) = \sum_{j=0}^{m} a_j(x)u^{(j)}(x) = f(x), x \in \Gamma \setminus \mathcal{V} \tag{8}$$

with coefficients $a_j \in L^\infty(\Gamma)$. We add to equation (8) the vertex (transmission) conditions at $v \in \mathcal{V}$

$$B_{k,v}u(v) = \sum_{j=0}^{m_k} b_{j,k}(v)u^{(j)}_{\mathcal{E}_v}(v) = \varphi_k(v), v \in \mathcal{V}, \varphi_k \in \mathbb{C}^{d(v)}, k = 1, ..., l \tag{9}$$

where $l \leq m - 1$, and $b_{j,k}(v)$ are $d(v) \times d(v)$ complex matrices, $u^{(j)}_{\mathcal{E}_v}(v) = (u^{(j)}_{e_1}(v), ..., u^{(j)}_{e_{d(v)}}(v)) \in \mathbb{C}^{d(v)}$ and $u^{(j)}_{e_i}(v), i = 1, ..., d(v)$ are limit values at the vertex $v \in \mathcal{V}$ of the derivatives $u^{(j)}_{e_i}(x)$ taken along the edges $e_i \in \mathcal{E}_\nu$ according their orientation. We suppose that

$$m_k < m - \frac{1}{2}, \tag{10}$$

for every $k = 1, ..., l$, and

$$\sup_{j=0,...,m_k, v\in\mathcal{V}} \|b_{j,k}(v)\|_{\mathcal{B}(\mathbb{C}^{d(v)})} < \infty. \tag{11}$$

We associate with problem (8),(9) an operator

$$(\mathfrak{A}u)(x, v) = \begin{cases} Au(x), x \in \Gamma \setminus \mathcal{V} \\ B_{k,v}u(v), v \in \mathcal{V}, k = 1, ..., l \end{cases},$$

acting from $H^m(\Gamma)$ into $L^2(\Gamma) \oplus \tilde{l}^2(\mathcal{V})$, where $\tilde{l}^2(\mathcal{V}) = \oplus_{v\in\mathcal{V}}\mathbb{C}^{ld(v)}$. Let $\varphi(v) = (\varphi_1(v), ..., \varphi_l(v)) \in \mathbb{C}^{ld(v)}, v \in \mathcal{V}$ and $\varphi_j(v) \in \mathbb{C}^{d(v)}$. Then we set

$$\|\varphi\|^2_{\tilde{l}^2(\mathcal{V})} = \sum_{v\in\mathcal{V}} \sum_{j=1}^{l} \|\varphi_j(v)\|^2_{\mathbb{C}^{d(v)}}.$$

The well-known estimate $|f_e(\nu)| \leq C_s \|f\|_{H^s(e)}, e \in \mathcal{E}_v, s > 1/2$ implies that

$$\left\|u^{(j)}_{\mathcal{E}_v}(v)\right\|^2_{\mathbb{C}^{d(v)}} = \sum_{i=1}^{d(v)} \left\|u^{(j)}_{e_i}(v)\right\|^2_{\mathbb{C}^{d(v)}} \leq C_j \sum_{i=1}^{d(v)} \|u_{e_i}\|^2_{H^{s-j-1/2}(e_k)}, \text{if } s > j + \frac{1}{2} \tag{12}$$

with a constant $C_j > 0$ independent of $v \in \mathcal{V}$ (see also [**3**], Lemma 1.3.8). Estimates (10) and (12) yield that $\mathfrak{A} : H^m(\Gamma) \to L^2(\Gamma) \oplus \tilde{l}^2(\mathcal{V})$ is a bounded operator.

2^0. We denote by $\tilde{\mathbb{R}}^n$ a compactification of $\mathbb{R}^n$ by the "sphere" $\mathbb{S}^{n-1}_\infty$ of infinite distant points such that $\tilde{\mathbb{R}}^n$ is homeomorphic to the closed unit ball $\bar{B}_1 = \{x \in \mathbb{R}^n : |x| \leq 1\}$. Let $\eta_\omega \in \mathbb{S}^{n-1}_\infty$ be an infinitely distant point associated to the ray l_ω starting from the origin and passing through the point $\omega \in S^{n-1}$. The fundamental system of neighborhoods of the point $\eta_\omega \in \mathbb{S}^{n-1}_\infty$ is formed by the conical sets $U_\omega^R = K_\omega \cap B_R'$ where K_ω is an open cone in $\mathbb{R}^n$ with the top in the origin containing the ray l_ω. Let

$$\mathbb{H}^r = \left\{ x \in \mathbb{R}^n : x = \sum_{j=1}^{r} x_j \mathfrak{e}_j, x_j \in \mathbb{R} \right\}$$

be a subspace in $\mathbb{R}^n$ generated by the system $\{\mathfrak{e}_1, ..., \mathfrak{e}_r\}$ containing the group $\mathbb{G}$. We denote by $\tilde{\mathbb{H}}^r$ the closure of $\mathbb{H}^r$ in the topology of $\tilde{\mathbb{R}}^n$. Let $\tilde{\mathbb{G}}, \tilde{\Gamma}, \tilde{\mathcal{V}}$ be the closures of the sets $\mathbb{G}, \Gamma, \mathcal{V}$ in the topology of $\tilde{\mathbb{R}}^n$ and $\mathbb{H}_\infty^r = \tilde{\mathbb{H}}^r \setminus \mathbb{H}^r, \mathbb{G}_\infty = \tilde{\mathbb{G}} \setminus \mathbb{G}, \Gamma_\infty = \tilde{\Gamma} \setminus \Gamma, \mathcal{V}_\infty = \tilde{\mathcal{V}} \setminus \mathcal{V}$. It follows from (5),(6) that

$$\Gamma_\infty = \mathcal{V}_\infty = \mathbb{G}_\infty = \mathbb{H}_\infty^r. \tag{13}$$

DEFINITION 2. *We denote by $M(\tilde{\Gamma})$ $(M(\tilde{\mathcal{V}}))$ a C^*-sub-algebra of $L^\infty(\Gamma)$ $(l^\infty(\mathcal{V}))$ consisting of functions $a \in L^\infty(\Gamma)$ $(l^\infty(\mathcal{V}))$ such that for every $x \in \Gamma$ $(\mathcal{V})$ there exists a limit*

$$\tilde{a}(\eta_\omega) = \lim_{\mathbb{G} \ni g \to \eta_\omega \in \mathbb{H}_\infty^r} a(x+g)$$

in the topology of $\tilde{\mathbb{R}}^n$ independent of x and the limit function $\tilde{a} \in C(\mathbb{H}_\infty^r)$.

EXAMPLE 3. *Let $a \in L^\infty(\Gamma)$ and $a(x) = \hat{a}_\infty\left(\frac{x}{|x|}\right)$ for $x \in \Gamma \cap B_R'$ where $a_\infty(\omega) \in C(S^{n-1})$. Then*

$$\lim_{\mathbb{G} \ni g \to \eta_\omega \in \mathbb{H}_\infty^r} a(x+g) = \lim_{\mathbb{G} \ni g \to \eta_\omega \in \mathbb{H}_\infty^r} a_\infty\left(\frac{x+g}{|x+g|}\right) = a_\infty(\omega),$$

for every points $x \in \Gamma$ $(x \in \mathcal{V})$. It implies and $\hat{a} \in M(\tilde{\Gamma})$, and $\check{a} \in M(\tilde{\mathcal{V}})$.

DEFINITION 4. *We denote by $L_{\mathbb{G}}^\infty(\Gamma)$ $(l_{\mathbb{G}}^\infty(\mathcal{V}))$ the C^*-sub-algebras of $L^\infty(\Gamma)$ $(l^\infty(\mathcal{V}))$ of periodic with respect to $\mathbb{G}$ functions on Γ $(\mathcal{V})$. We denote by $\mathcal{M}_{\mathbb{G}}(\tilde{\Gamma})$ $(\mathcal{M}_{\mathbb{G}}(\tilde{\mathcal{V}}))$ the C^*-algebras generated by $M(\tilde{\Gamma})$ and $L_{\mathbb{G}}^\infty(\Gamma)$ $(M(\tilde{\mathcal{V}})$ and $l_{\mathbb{G}}^\infty(\mathcal{V}))$, respectively. It means that $\mathcal{M}_{\mathbb{G}}(\Gamma)$ $(\mathcal{M}_{\mathbb{G}}(\mathcal{V}))$ is the closure in $L^\infty(\Gamma)$ $(l^\infty(\mathcal{V}))$ of the set of finite sums $\sum_{j=1}^N a_j b_j$ where $a_j \in M(\tilde{\Gamma})$ $(M(\tilde{\mathcal{V}}))$ and $b_j \in L_{\mathbb{G}}^\infty(\Gamma)$ $(l_{\mathbb{G}}^\infty(\mathcal{V}))$.*

PROPOSITION 5. *Let $a \in \mathcal{M}_{\mathbb{G}}(\tilde{\Gamma})$ $(\mathcal{M}_{\mathbb{G}}(\tilde{\mathcal{V}}))$, $\eta_\omega \in \mathbb{H}_\infty^r$. Then there exist limit*

$$a^{\eta_\omega}(x) = \lim_{\mathbb{G} \ni g \to \eta_\omega \in \mathbb{H}_\infty^r} a(x+g), x \in \Gamma, a^{\eta_\omega} \in L_{\mathbb{G}}^\infty(\Gamma), \tag{14}$$

$$a^{\eta_\omega}(x) = \lim_{\mathbb{G} \ni g \to \eta_\omega \in \mathbb{H}_\infty^r} a(x+g), x \in \mathcal{V}, a^{\eta_\omega} \in L_{\mathbb{G}}^\infty(\mathcal{V}) \tag{15}$$

in the topology $\tilde{\mathbb{R}}^n \cap \Gamma$ $(\tilde{\mathbb{R}}^n \cap \mathcal{V})$, that is for every $\varepsilon > 0$ there exists a neighborhood U_ω^R of the point η_ω such that

$$\sup_{x \in U_\omega^R \cap \Gamma} |a(x) - a^{\eta_\omega}(x)| < \varepsilon, \tag{16}$$

$$\sup_{x \in U_\omega^R \cap \mathcal{V}} |a(x) - a^{\eta_\omega}(x)| < \varepsilon. \tag{17}$$

PROOF. Let $a^N(x) = \sum_{j=1}^N a_j(x)b_j(x)$ where $a_j \in M(\tilde{\Gamma})$ $(a_j \in M(\tilde{\mathcal{V}})), b_j \in L^\infty_{\mathbb{G}}(\Gamma)$ $(b_j \in l^\infty_{\mathbb{G}}(\mathcal{V}))$. Let $x \in \Gamma$ $(\mathcal{V})$. Then

$$\lim_{\mathbb{G}\ni g\to\eta_\omega\in\mathbb{H}^r_\infty} a^N(x+g) = \lim_{\mathbb{G}\ni g\to\eta_\omega\in\mathbb{H}^r_\infty} \sum_{j=1}^N a_j(x+g)b_j(x+g) \tag{18}$$

$$= \sum_{j=1j}^N \tilde{a}(\eta_\omega)b_j(x) \in L^\infty_{\mathbb{G}}(\Gamma) \ (L^\infty_{\mathbb{G}}(\mathcal{V})).$$

For general $a \in \mathcal{M}_{\mathbb{G}}(\tilde{\Gamma})$ $(\mathcal{M}_{\mathbb{G}}(\tilde{\mathcal{V}}))$ formulas (14),(15) are obtained by passing to the limit. Let $\mathbb{G} \ni g_m \to \eta_\omega \in \mathbb{H}^r_\infty$. It follows from (14), (15) that for every $\varepsilon > 0$ there exists a neighborhood U^R_ω of the point η_ω and $m_0 \in \mathbb{N}$ such that

$$\sup_{x\in U^R_\omega\cap\Gamma} |a(x+g_m) - a^{\eta_\omega}(x)| < \varepsilon, a^{\eta_\omega} \in L^\infty_{\mathbb{G}}(\Gamma), \tag{19}$$

$$\sup_{x\in U^R_\omega\cap\mathcal{V}} |a(x+g_m) - a^{\eta_\omega}(x)| < \varepsilon, a^{\eta_\omega} \in L^\infty_{\mathbb{G}}(\mathcal{V}) \tag{20}$$

for every $m > m_0$. Inequalities (19),(20) yield that

$$\sup_{y\in(U^R_\omega+g_m)\cap\Gamma} |a(y) - a^{\eta_\omega}(y)| < \varepsilon, y \in \Gamma,$$

$$\sup_{y\in(U^R_\omega+g_m)\cap\mathcal{V}} |a(y) - a^{\eta_\omega}(y)| < \varepsilon, y \in \mathcal{V}.$$

Since $U^R_\omega + g_m$ is a neighborhood of η_ω we obtain the statement of Proposition 5. $\square$

EXAMPLE 6. *Let $a = a^1 + a^2$, where a^1 is such that in Example 3 and $a^2 \in L^\infty_{\mathbb{G}}(\Gamma)$. Then*

$$\lim_{\mathbb{G}\ni g\to\eta_\omega\in\mathbb{H}^r_\infty} a(x+g) = \lim_{\mathbb{G}\ni g\to\eta_\omega\in\mathbb{H}^r_\infty} a^1_\infty\left(\frac{x+g}{|x+g|}\right) + a^2(x)$$
$$= a^1_\infty(\omega) + a^2(x), x \in \Gamma \ \ (x \in \mathcal{V}).$$

DEFINITION 7. *Let $\varphi \in C_0^\infty(\mathbb{R}^n)$, $\varphi(x) = 1$ for $|x| \le \frac{1}{2}$, $\varphi(x) = 0$ if $|x| \ge 1, 0 \le \varphi(x) \le 1$, and $\varphi_R(x) = \varphi(x/R), \psi_R(x) = 1 - \varphi_R(x)$.*

(i) We say that $\mathfrak{A} : H^m(\Gamma) \to L^2(\Gamma) \oplus \tilde{l}^2(\mathcal{V})$ is a locally Fredholm operator if for every $R > 0$ there exist operators $\mathcal{L}_R, \mathcal{R}_R \in \mathcal{B}(L^2(\Gamma)\oplus\tilde{l}^2(\mathcal{V}), H^m(\Gamma))$ such that

$$\mathcal{L}_R\mathfrak{A}\hat{\varphi}_R I = \hat{\varphi}_R I + T^{(1)}_R, \hat{\varphi}_R A\mathcal{R}_R = \hat{\varphi}_R I + T^{(2)}_R, \tag{21}$$

where $T^{(1)}_R \in \mathcal{K}(H^m(\Gamma)), T^{(2)}_R \in \mathcal{K}(L^2(\Gamma) \oplus \tilde{l}^2(\mathcal{V}))$;

(ii) We say that $\mathfrak{A} : H^m(\Gamma) \to L^2(\Gamma) \oplus \tilde{l}^2(\mathcal{V})$ is a locally invertible at infinity operator if there exists $R_0 > 0$ *and operators $\mathcal{L}^\infty_{R_0}, \mathcal{R}^\infty_{R_0} \in \mathcal{B}(L^2(\Gamma)\oplus\tilde{l}^2(\mathcal{V}), H^m(\Gamma))$ such that*

$$\mathcal{L}^\infty_{R_0}\mathfrak{A}\hat{\psi}_{R^0} I = \hat{\psi}_{R^0} I, \hat{\psi}_{R_0}\mathfrak{A}\mathcal{R}^\infty_{R_0} = \hat{\psi}_{R_0} I. \tag{22}$$

PROPOSITION 8. *Let $\mathfrak{A} : H^m(\Gamma) \to L^2(\Gamma)\oplus\tilde{l}^2(\mathcal{V})$ be a locally Fredholm operator. Then the operator $\mathfrak{A} : H^m(\Gamma) \to L^2(\Gamma) \oplus \tilde{l}^2(\mathcal{V})$ is a Fredholm operator if and only if $\mathfrak{A} : H^m(\Gamma) \to L^2(\Gamma) \oplus \tilde{l}^2(\mathcal{V})$ is a locally invertible at infinity operator.*

PROOF. Let $\mathfrak{A}$ be a locally invertible at infinity operator and $\mathcal{L}_{R_0}^{\infty}, \mathcal{R}_{R_0}^{\infty}$ be the above defined locally inverses operators. We set

$$L_R = \mathcal{L}_R \hat{\varphi}_R I + \mathcal{L}_{R_0}^{\infty} \hat{\psi}_R I,$$

where $R > R_0$ is such that $\hat{\psi}_R \hat{\psi}_{R_0} = \hat{\psi}_R$. Then

$$\begin{aligned} L_R \mathfrak{A} &= \mathcal{L}_R \hat{\varphi}_R \mathfrak{A} + \mathcal{L}_{R_0}^{\infty} \hat{\psi}_R \mathfrak{A} \\ &= \mathcal{L}_R \mathfrak{A} \hat{\varphi}_R I + \mathcal{L}_R [\hat{\varphi}_R, \mathfrak{A}] + \mathcal{L}_{R_0}^{\infty} \mathfrak{A} \hat{\psi}_R + \mathcal{L}_{R_0}^{\infty} \left[\hat{\psi}_R, \mathfrak{A}\right] \\ &= I + K_R + T_R \end{aligned} \tag{23}$$

where $K_R \in \mathcal{K}(H^m(\Gamma)), T_R = T_R' + T_R''$, and

$$T_R' = \mathcal{L}_R [\hat{\varphi}_R, \mathfrak{A}] = \mathcal{L}_R (\hat{\varphi}_R \mathfrak{A} - \mathfrak{A} \hat{\varphi}_R)$$
$$T_R'' = \mathcal{L}_{R_0}^{\infty} [\hat{\varphi}_R, \mathfrak{A}] = \mathcal{L}_{R_0}^{\infty} (\hat{\psi}_R \mathfrak{A} - \mathfrak{A} \hat{\psi}_R).$$

Since

$$\lim_{R\to\infty} \|[\hat{\varphi}_R, \mathfrak{A}]\|_{\mathcal{B}(H^m(\Gamma), L^2(\Gamma)\oplus \tilde{l}^2(\mathcal{V}))} = \lim_{R\to\infty} \left\| \left[\hat{\psi}_R, \mathfrak{A}\right] \right\|_{\mathcal{B}(H^m(\Gamma), L^2(\Gamma)\oplus \tilde{l}^2(\mathcal{V}))} = 0$$

we can find $R > R_0$ such that $\|T_R\|_{\mathcal{B}(H^m(\Gamma))} < 1$. Hence (23) yields that $(I + T_R)^{-1} L_R$ is a left regularizator of $\mathfrak{A}$. In the same way we prove the existence of a right regularizator of $\mathfrak{A}$.

It is easy to prove that if $\mathfrak{A}$ is a Fredholm operator then $\mathfrak{A}$ is locally invertible at infinity. □

THEOREM 9. *Let Γ be a graph periodic with respect to the group $\mathbb{G}$. We assume that: (a) the coefficients a_j of the operator A belong to $\mathcal{M}_{\mathbb{G}}(\tilde{\Gamma})$, (b) the coefficients b_{jk} of the vertex operators $B_{k,v}, k = 1, ..., m$ belong to $\mathcal{M}_{\mathbb{G}}(\tilde{\mathcal{V}})$, (c) A is a uniformly elliptic operator on Γ, that is $\inf_{x\in\Gamma} |a_m(x)| > 0$.*

For every $\eta_\omega \in \mathbb{H}_\infty^r$ we set

$$(\mathfrak{A}^{\eta_\omega} u)(x, v) = \begin{cases} A^{\eta_\omega} u(x), x \in \Gamma \backslash \mathcal{V} \\ B_{k,v}^{\eta_\omega} u(v), v \in \mathcal{V}, k = 1, ..., l \end{cases},$$

where

$$A^{\eta_\omega} u(x) = \sum_{j=0}^{m} a_j^{\eta_\omega}(x) u^{(j)}(x), x \in \Gamma \setminus \mathcal{V},$$

$$B_{k,v}^{\eta_\omega} u(v) = \sum_{j=0}^{m_k} b_{jk}^{\eta_\omega}(v) u_{\mathcal{E}_v}^{(j)}(v), v \in \mathcal{V}, k = 1, ..., l,$$

and $a_j^{\eta_\omega}(x) = \lim_{\mathbb{G}\ni g\to\eta_\omega} a_j(x+g), b_{jk}^{\eta_\omega}(\nu) = \lim_{\mathbb{G}\ni g\to\eta_\omega} b_{jk}(v+g)$.

Then $\mathfrak{A} : H^m(\Gamma) \to L^2(\Gamma) \oplus \tilde{l}^2(\mathcal{V})$ is a Fredholm operator if and only if for every $\eta_\omega \in \mathbb{H}_\infty^r$ the operators $\mathfrak{A}^{\eta_\omega} : H^m(\Gamma) \to L^2(\Gamma) \oplus \tilde{l}^2(\mathcal{V})$ are invertible.

PROOF. (i) Let $\Gamma_R = \Gamma \cap B_R$. We denote by $\mathfrak{A}_R$ the operator acting from $H^m(\Gamma_R) \to L^2(\Gamma_R) \oplus \tilde{l}^2(\mathcal{V}_R)$ which is the restriction of $\mathfrak{A}$ on $H^m(\Gamma_R)$. Since A is an elliptic operator on Γ the operator $\mathfrak{A}_R : H^m(\Gamma_R) \to L^2(\Gamma_R) \oplus \tilde{l}^2(\mathcal{V}_R)$ is a Fredholm operator (see for instance [**3**], chap.2.4). It yields that for *every* $R > 0$ there exist operators $\mathcal{L}_R, \mathcal{R}_R \in \mathcal{B}(L^2(\Gamma) \oplus \tilde{l}^2(\mathcal{V}), H^{2m}(\Gamma))$ such that

$$\mathcal{L}_R \mathfrak{A} \hat{\varphi}_R I = \hat{\varphi}_R I + T_R^{(1)}, \hat{\varphi}_R A \mathcal{R}_R = \hat{\varphi}_R I + T_R^{(2)} \tag{24}$$

where $T_R^{(1)} \in \mathcal{K}(H^m(\Gamma)), T_R^{(2)} \in \mathcal{K}(L^2(\Gamma) \oplus \tilde{l}^2(\mathcal{V}))$. Hence $\mathfrak{A}$ is a locally Fredholm operator.

(ii) According of Proposition 8 for the proof of the Fredholmness of $\mathfrak{A}$ we have to prove that $\mathfrak{A} : H^m(\Gamma) \to L^2(\Gamma) \oplus \tilde{l}^2(\mathcal{V})$ is a locally invertible operator at infinity. Let U_ω^R be a neighborhood of the point η_ω and $\varphi_{R,\omega}$ be a cut-off function of the neighborhood U_ω^R, that is $\varphi_{R,\omega}(x) = \varkappa_R(|x|)\ \phi_\omega(\frac{x}{|x|})$ where:

a) $\varkappa \in C^\infty(\mathbb{R})$, $\varkappa(t) = 0$ for $t \leq 1$ and $\varkappa(t) = 1$ for $t \geq 2$,$0 \leq \varkappa(t) \leq 1, \varkappa_R(x) = \varkappa(\frac{|x|}{R}), R > 0$.

b) $\phi_\omega \in C^\infty(S^{n-1}), \phi_\omega(\theta) = 1, \theta \in \Omega_\omega \subset S^{n-1}$, *supp* $\theta \subset \Omega'_\omega$, where $\Omega_\omega, \Omega'_\omega$ are neighborhoods on S^{n-1} of the point ω such that $\bar{\Omega}_\omega \subset \Omega'_\omega$ and $0 \leq \phi_\omega(x) \leq 1$.

Then because the coefficients a_j of the operator A belong to $\mathcal{M}_{\mathbb{G}}(\tilde{\Gamma})$, and the coefficients b_{jk} of the vertex operators $B_{k,v}$ belong to $\mathcal{M}_{\mathbb{G}}(\tilde{\mathcal{V}})$, for every $\eta_\omega \in \mathbb{H}_\infty^r$ and $\varepsilon > 0$ there exists a neighborhood U_ω^R of η_ω and a cut-off function $\varphi_{\eta_\omega} \in C_b^\infty(U_\omega^R)$ such that

$$\sup_{x\in\Gamma} \left|\hat{\varphi}_{\eta_\omega}(x)(a_j(x) - a_j^{\eta_\omega}(x))\right| < \varepsilon, j = 1, ..., m, \tag{25}$$

and

$$\sup_{v\in\mathcal{V}} \left|\check{\varphi}_{\eta_\omega}(\nu)(b_{jk}(\nu) - b_{jk}^{\eta_\omega}(v))\right| < \varepsilon \tag{26}$$

where $a_j^{\eta_\omega} \in L_{\mathbb{G}}^\infty(\Gamma)$, and $b_{jk}^{\eta_\omega} \in l_{\mathbb{G}}^\infty(\mathcal{V})$. Estimates (25),(26) imply that for every $\epsilon > 0$ there exists a cut-off function φ_{η_ω} such that

$$\|\hat{\varphi}_{\eta_\omega}(\mathfrak{A} - \mathfrak{A}^{\eta_\omega})\| < \epsilon, \|(\mathfrak{A} - \mathfrak{A}^{\eta_\omega})\,\hat{\varphi}_{\eta_\omega} I\| < \epsilon. \tag{27}$$

Estimates (27) and invertibility of the operators $\mathfrak{A}^{\eta_\omega} : H^m(\Gamma) \to L^2(\Gamma) \oplus \tilde{l}^2(\mathcal{V})$, $\eta_\omega \in \mathbb{H}_\infty^r$ yield that for every $\eta_\omega \in \mathbb{H}_\infty^r$ there exist right and left locally inverse operators $R_{\eta_\omega}, L_{\eta_\omega} \in \mathcal{B}(L^2(\Gamma) \oplus \tilde{l}^2(\mathcal{V}), H^{2m}(\Gamma))$ such that

$$L_{\eta_\omega}\mathfrak{A}\hat{\varphi}_{\eta_\omega} I = \hat{\varphi}_{\eta_\omega} I, \hat{\varphi}_{\eta_\omega}\mathfrak{A}R_{\eta_\omega} = \hat{\varphi}_{\eta_\omega} I. \tag{28}$$

If (28) holds we say that the operator $\mathfrak{A} : H^m(\Gamma) \to L^2(\Gamma) \oplus \tilde{l}^2(\mathcal{V})$ is locally invertible at the point $\eta_\omega \in \mathbb{H}_\infty^r$.

The next step is to prove of the local invertibility of the operator $\mathfrak{A}$ at infinity. For this aim we construct an admissible partition of the unity and glue the locally inverse operators at infinity from the locally inverse operators at the points $\eta_\omega \in \mathbb{H}_\infty^r$. Let $S_{\mathbb{H}_\infty^r}^{n-1}$ be a closed set of points on the unit sphere S^{n-1} associated with $\eta_\omega \in \mathbb{H}_\infty^r$. There exists a finite covering $\{\Omega_{\omega_j}\}_{j=1}^N$ of $S_{\mathbb{H}_\infty^r}^{n-1}$ and the above defined cut-off functions $\phi_{\omega_j}(\theta)$ such that *supp* $\phi_{\omega_j} \subset \Omega'_{\omega_j}$ and $\phi_{\omega_j}(\theta) = 1$, for $\theta \in \Omega_{\omega_j}$. We set

$$\psi_{\omega_j}(\omega) = \frac{\phi_{\omega_j}(\theta)}{\sum_{j=1}^N \phi_{\omega_j}(\theta)}, \theta \in S_{\mathbb{H}_\infty^r}^{n-1}. \tag{29}$$

Let $\varphi_{R,\omega_j}(x) = \varkappa_R(|x|)\ \psi_{\omega_j}(\frac{x}{|x|})$. Formula (29) implies that

$$\sum_{j=1}^N \varphi_{R,\omega_j}(x) = \varkappa_R(x), x \in \mathbb{R}^n, R > 0. \tag{30}$$

We showed above that $\mathfrak{A}$ is a locally invertible operator at every point $\eta_{\omega_j} \in \mathbb{H}_\infty^r$. Hence there exists $R_0 > 0$, a system of functions φ_{R_0,ω_j}, and operators

$\mathcal{L}_{\omega_j,R_0}, \mathcal{R}_{\omega_j,R_0}$ such that

$$\mathcal{L}_{\omega_j,R_0}\mathfrak{A}\hat{\varphi}_{R_0,\omega_j}(x)I = \hat{\varphi}_{R_0,\omega_j}I, \hat{\varphi}_{R_0,\omega_j}\mathfrak{A}\mathcal{R}_{\omega_j,R_0} = \hat{\varphi}_{R_0,\omega_j}I. \tag{31}$$

Let

$$\mathcal{L}_R^{\infty} = \sum_{j=1}^{N} \mathcal{L}_{\omega_j,R_0}\hat{\varphi}_{R,\omega_j}I =$$

where $\varkappa_R\varkappa_{R_0} = \varkappa_R$, and $\varphi_{R,\omega_j}\varphi_{R_0,\omega_j} = \varphi_{R,\omega_j}$ for $R\,(> R_0)$ large enough. Then

$$\mathcal{L}_R^{\infty}\mathfrak{A} = \sum_{j=1}^{N} \mathcal{L}_{\omega_j,R_0}\hat{\varphi}_{R,\omega_j}\mathfrak{A} =$$

$$\sum_{j=1}^{N} \mathcal{L}_{\omega_j,R_0}\mathfrak{A}\hat{\varphi}_{R,\omega_j}I + \sum_{j=1}^{N} \mathcal{L}_{\omega_j,R_0}\left[\hat{\varphi}_{R,\omega_j}I, \mathfrak{A}\right] =$$

$$\sum_{j=1}^{N} \hat{\varphi}_{R,\omega_j}I + T_R = \varkappa_R I + T_R$$

where $T_R = \sum_{j=1}^{N} \mathcal{L}_{\omega_j,R_0}\left[\hat{\varphi}_{R,\omega_j}I, \mathfrak{A}\right]$. Since

$$\lim_{R\to\infty} \left\| \left[\hat{\varphi}_{R,\omega_j}I, \mathfrak{A}\right] \right\|_{\mathcal{B}(H^m(\Gamma), L^2(\Gamma)\oplus\tilde{l}^2(\mathcal{V}))} = 0$$

we can choose $R > R_0$ such that $\|T_R\|_{\mathcal{B}(H^m(\Gamma))} < 1$. Let $\varkappa_{R'}\varkappa_R = \varkappa_{R'}$, for $R'\,(> R)$ large enough. Then

$$\mathcal{L}_R^{\infty}\mathfrak{A}\varkappa_{R'}I = (I + T_R)\varkappa_{R'}I.$$

Hence the operator $(I + T_R)^{-1}\mathcal{L}_R^{\infty}$ is a left locally inverse operator at infinity of $\mathfrak{A}$. In a similar way we prove that there exists a right locally inverse at infinity operator for $\mathfrak{A}$. Hence $\mathfrak{A} : H^m(\Gamma) \to L^2(\Gamma) \oplus l^2(\mathcal{V})$ is a locally invertible at infinity operator. This statement with the local Fredholmness of $\mathfrak{A}$ implies that $\mathfrak{A} : H^m(\Gamma) \to L^2(\Gamma)\oplus \tilde{l}^2(\mathcal{V})$ is a Fredholm operator.

(iii) We will prove now the necessity of the conditions of Theorem 9. Let $\mathfrak{A} : H^m(\Gamma) \to L^2(\Gamma) \oplus \tilde{l}^2(\mathcal{V})$ be a Fredholm operator. Then there exists the left and right regularizators of $\mathfrak{A}$, that is the operators $\mathfrak{L}$ and $\mathfrak{R}$ such that

$$\mathfrak{L}\mathfrak{A} = I + \mathfrak{T}_1, \mathfrak{A}\mathfrak{R} = I + \mathfrak{T}_2, \tag{32}$$

where $\mathfrak{T}_1 \in \mathcal{K}(H^m(\Gamma))$ and $\mathfrak{T}_2 \in \mathcal{K}(L^2(\Gamma) \oplus l^2(\mathcal{V}))$. Equality (32) yields the estimates

$$\|u\|_{H^m(\Gamma)} \leq C\,\|\mathfrak{A}u\|_{L^2(\Gamma)\oplus\tilde{l}^2(\mathcal{V})} - \|\mathfrak{T}_1 u\|_{H^m(\Gamma)}, \tag{33}$$

$$\|w\|_{L^2(\Gamma)\oplus\tilde{l}^2(\mathcal{V})} \leq C\,\|\mathfrak{A}^* w\|_{H^{-m}(\Gamma)} - \|\mathfrak{T}_2^* w\|_{L^2(\Gamma)\oplus\tilde{l}^2(\mathcal{V})}, \tag{34}$$

with $C > 0$. Let $V_h u(x) = u(x - h), h \in \mathbb{G}$ be the isometric shift operator in the spaces $H^m(\Gamma), L^2(\Gamma)$ and $\tilde{l}^2(\mathcal{V})$. Let $\mathbb{G} \ni h_k \to \eta_\omega \in \mathbb{H}_\infty^r$. Then for every function $u \in C_0^\infty(\Gamma)$

$$\lim_{m\to\infty} \|V_{-h_k}\mathfrak{A}V_{h_k}u - \mathfrak{A}^{\eta_\omega}u\| = 0. \tag{35}$$

Estimate (33) implies that

$$\|u\|_{H^m(\Gamma)} \leq C\,\|V_{-h_k}\mathfrak{A}V_{h_k}u\|_{L^2(\Gamma)\oplus\tilde{l}^2(\mathcal{V})} - \|\mathfrak{T}_1 V_{h_k}u\|_{L^2(\Gamma)\oplus\tilde{l}^2(\mathcal{V})}, u \in C_0^\infty(\Gamma). \tag{36}$$

Note that the sequence $V_{h_k}u$ weakly tends to 0 in $H^m(\Gamma)$. Hence $\|\mathfrak{T}_1 V_{h_k}u\|_{H^m(\Gamma)} \to 0$ for $m \to \infty$ since $\mathfrak{T}_1$ is a compact operator in $H^m(\Gamma)$. Hence letting $h_k \to \eta_\omega \in \mathbb{H}^r_\infty$ and taking into account (35) we obtain that

$$\|\mathfrak{A}^{\eta_\omega}u\|_{L^2(\Gamma)\oplus\tilde{l}^2(\mathcal{V})} \geq \frac{1}{C}\|u\|_{H^m(\Gamma)}. \tag{37}$$

In the same way estimate (34) yields the estimate

$$\left\|(\mathfrak{A}^{\eta_\omega})^* w\right\|_{H^{-m}(\Gamma)} \geq \frac{1}{C}\|w\|_{L^2(\Gamma)\oplus\tilde{l}^2(\mathcal{V})}. \tag{38}$$

Estimates (37),(38) yield that the operator $\mathfrak{A}^{\eta_\omega}: H^m(\Gamma) \to L^2(\Gamma)\oplus\tilde{l}^2(\mathcal{V})$ is invertible. □

REMARK 10. *Theorem* 9 *reduces the study of the Fredholm property of* $\mathfrak{A}: H^m(\Gamma) \to L^2(\Gamma)\oplus\tilde{l}^2(\mathcal{V})$ *to the study of the invertibility of the operators* $\mathfrak{A}^{\eta_\omega}: H^m(\Gamma) \to L^2(\Gamma)\oplus\tilde{l}^2(\mathcal{V})$ *generated by the differential operators* A^{η_ω} *and the vertex operators* $B^{\eta_\omega}_{k,v}$ *with coefficients periodic with respect to the group* $\mathbb{G}$. *For the investigation of invertibility such operators can be applied the Floquet theory (see for instance* [**3**]*,Chap.4,*[**5**],[**6**],[**7**],[**9**],[**10**]*).*

4. Fredholm property and essential spectrum of Schrödinger operators on Γ

1^0. Let

$$Su(x) = -\frac{d^2u(x)}{dx^2} + q(x)u(x), x \in \Gamma\setminus\mathcal{V} \tag{39}$$

be the Schrödinger operator on the periodic graph Γ with electric potential $q \in L^\infty(\Gamma)$.

We consider the following problem for the operator S on Γ

$$\begin{cases} Su(x) = f(x), x \in \Gamma\backslash\mathcal{V} \\ A(v)u_{\mathcal{E}_v}(v) + B(v)u'_{\mathcal{E}_v}(v) = \varphi(v), v \in \mathcal{V} \end{cases} \tag{40}$$

where $A(v), B(v)$ are $d(v)\times d(v)$ complex matrices, $u_{\mathcal{E}_v}(v) = (u_{e_1}(v), ..., u_{e_{d(v)}}(v)) \in \mathbb{C}^{d(v)}$, $u_{e_i}(v), i = 1, ..., d(v)$ are limit values at the vertex $v \in \mathcal{E}$ of the functions $u_{e_i}(x)$ defined on the edges e_i, the derivatives $\frac{du_{e_i}(v)}{dx}$ taken along the edges $e_i \in \mathcal{E}_\nu$ as outgoing from v. We suppose that

$$\sup_{v\in\mathcal{V}}\|A(v)\|_{\mathcal{B}(\mathbb{C}^{d(v)})} < \infty, \sup_{v\in\mathcal{V}}\|B(v)\|_{\mathcal{B}(\mathbb{C}^{d(v)})} < \infty. \tag{41}$$

We associate with problem (40) the operator

$$\mathfrak{H}u(x,\nu) = \begin{cases} Su(x), x \in \Gamma\backslash\mathcal{V} \\ A(v)u_{\mathcal{E}_v}(v) + B(v)u'_{\mathcal{E}_v}(v), v \in \mathcal{V} \end{cases} \tag{42}$$

acting from $H^2(\Gamma)$ into $L^2(\Gamma)\oplus\tilde{l}^2(\mathcal{V})$.

Theorem 9 yields the following result.

THEOREM 11. *Let* $\Gamma \subset \mathbb{R}^n$ *be an oriented graph periodic with respect to the group* $\mathbb{G}$. *We suppose that: (a) the electric potential* $q \in \mathcal{M}_{\mathbb{G}}(\tilde{\Gamma})$; *(b) the elements of the matrices* $A(v) = (A_{j,k}(v))_{j,k=1}^{d(v)}, B(v) = (B_{j,k}(v))_{j,k=1}^{d(v)}$ *belong to* $\mathcal{M}_{\mathbb{G}}(\tilde{\mathcal{V}})$. *We set*

$$(\mathfrak{H}^{\eta_\omega} u)(x,v) = \begin{cases} S^{\eta_\omega} u(x), x \in \Gamma \backslash \mathcal{V} \\ A^{\eta_\omega}(v) u_{\mathcal{E}_v}(v) + B^{\eta_\omega}(v) u'_{\mathcal{E}_v}(v), v \in \mathcal{V} \end{cases} \tag{43}$$

where

$$S^{\eta_\omega} u(x) = -\frac{d^2 u(x)}{dx^2} + q^{\eta_\omega}(x) u(x), x \in \Gamma \setminus \mathcal{V},$$

$$A^{\eta_\omega}(v) = \left(A_{j,k}^{\eta_\omega}(v)\right)_{j,k=1}^{d(v)}, B^{\eta_\omega}(v) = \left(B_{j,k}^{\eta_\omega}(v)\right)_{j,k=1}^{d(v)}, v \in \mathcal{V}.$$

Then $\mathfrak{H} : H^2(\Gamma) \to L^2(\Gamma) \oplus \tilde{l}^2(\mathcal{V})$ *is a Fredholm operator if and only if for every* $\eta_\omega \in \mathbb{H}_\infty^r$ *the operators* $\mathfrak{H}^{\eta_\omega} : H^2(\Gamma) \to L^2(\Gamma) \oplus \tilde{l}^2(\mathcal{V})$ *are invertible.*

Hence the study of the Fredholm property of $\mathfrak{H} : H^2(\Gamma) \to L^2(\Gamma) \oplus \tilde{l}^2(\mathcal{V})$ is reduced to the investigation of the invertibility of operators $\mathfrak{H}^{\eta_\omega} : H^2(\Gamma) \to L^2(\Gamma) \oplus \tilde{l}^2(\mathcal{V}), \eta_\omega \in \mathbb{H}_\infty^r$ generated by periodic Schrödinger operators on the periodic graph Γ with periodic vertex conditions.

2^0. We assume that the electric potential $q \in L^\infty(\Gamma)$ is a real-valued function and the matrices $A(v)$ and $B(v)$ satisfy the following conditions:

$$d(v) \times 2d(v) \text{ matrix } (A(v), B(v)) \text{ has the maximal rank } d(v); \tag{44}$$

$$\text{the matrix } A(v)B^*(v) \text{ is Hermitian} \tag{45}$$

Then (see [**3**], page 59) the operator (40) admits a self-adjoint extension $\mathcal{H}$ in $L^2(\Gamma)$ with the domain

$$D_{\mathcal{H}} = \left\{u \in H^2(\Gamma) : A(v) u_{\mathcal{E}_v}(v) + B(v) u'_{\mathcal{E}_v}(v) = 0, \text{ for every } v \in \mathcal{V}\right\}.$$

Definition 12. *The essential spectrum* $sp_{ess}\mathcal{H}$ *of* $\mathcal{H}$ *is defined as the set of points* $\lambda \in \mathbb{C}$ *such that the operator* $\mathcal{H} - \lambda I$ *is not Fredholm operator in* $L^2(\Gamma)$ *as unbounded operator (see for instance* [**1**]*).*

We associate with the operator $\mathcal{H}$ the family of self-adjoint operators $\mathcal{H}^{\eta_\omega}$ acting in $L^2(\Gamma)$ generated by the Schrödinger operators S^{η_ω} with domains

$$D_{\mathcal{H}^{\eta_\omega}} = \left\{u \in H^2(\Gamma) : A^{\eta_\omega}(v) u_{\mathcal{E}_v}(v) + B^{\eta_\omega}(v) u'_{\mathcal{E}_v}(v) = 0 \text{ for every } v \in \mathcal{V}\right\}.$$

Then Theorem 11 yields the following result.

Theorem 13. *Let the assumptions of Theorem* 11 *with respect to the potential* q *and the vertex matrices* $A(v), B(v)$ *hold. Moreover matrices* $A(v), B(v)$ *satisfy conditions* (44),(45)*. Then*

$$sp_{ess}\mathcal{H} = \bigcup_{\eta_\omega \in \mathbb{H}_\infty^r} sp\mathcal{H}^{\eta_\omega} \tag{46}$$

where $sp\mathcal{H}^{\eta_\omega}$ *is a spectrum of the unbounded operator* $\mathcal{H}^{\eta_\omega}$ *where* $\mathcal{H}^{\eta_\omega}$ *is defined by the Schrödinger operator* S^{η_ω} *with periodic potentials* q^{η_ω} *and periodic vertex conditions*

$$A^{\eta_\omega}(v) u_{\mathcal{E}_v}(v) + B^{\eta_\omega}(v) u'_{\mathcal{E}_v}(v) = 0. \tag{47}$$

Let for the simplicity $\mathbb{G} = \mathbb{R}^m, 1 \leq m \leq n$. By means of the Floquet transform the operator $\mathcal{H}^{\eta_\omega}$ is represented as $\mathcal{H}^{\eta_\omega} = \oplus_{\theta \in B} \mathcal{H}_\theta^{\eta_\omega}$ where $B = [-\pi, \pi]^m$ is the Brillouin zone, and $\mathcal{H}_\theta^{\eta_\omega}$ are the Bloch Hamiltonians. The operators $\mathcal{H}_\theta^{\eta_\omega}$ are generated by the Schrödinger operator $-\frac{d^2}{dx^2} + q^{\eta_\omega}(x)$ in $L^2(\Gamma_0)$ with domain $\tilde{H}_\theta^2(\Gamma_0)$ consisting of functions $u \in H^2(\Gamma_0)$ satisfying conditions (47) at the every vertex $v \in int\Gamma_0$ and the cyclic conditions $u(v+g) = e^{i\theta \cdot g}u(v), g \in \mathbb{Z}^m$, for the vertices v such that v and $v+g, g \in \mathbb{G}$ belonging to $\overline{\Gamma_0}$. The Bloch Hamiltonian $\mathcal{H}_\theta^{\eta_\omega}$ is a self-adjoint operator with a discrete spectrum $\{\lambda_1^{\eta_\omega}(\theta) \leq \lambda_2^{\eta_\omega}(\theta) \leq ... \leq \lambda_N^{\eta_\omega}(\theta) \leq .....\}$ where $\lambda_j^{\eta_\omega}(\theta), \theta \in B$ are real-valued bounded continuous functions. Then the spectrum of $\mathcal{H}^{\eta_\omega}$ has a band-gap structure $sp\mathcal{H}^{\eta_\omega} = \bigcup_{j=1}^{\infty} \left[\alpha_j^{\eta_\omega}, \beta_j^{\eta_\omega}\right]$ where $\left[\alpha_j^{\eta_\omega}, \beta_j^{\eta_\omega}\right]$ is the range of the mapping $\lambda_j^{\eta_\omega} : B \to \mathbb{R}$.

2^0. Let the potentials q of the Schrödinger operator S is of the form: $q = q_0 + q_1$ where $q_0 \in L_{\mathbb{G}}^\infty(\Gamma), q_1 \in \mathcal{M}(\Gamma)$ are real-valued functions, and $A(v), B(v) \in L_{\mathbb{G}}^\infty(\mathcal{V})$ and satisfy conditions (44),(45). Let $\mathcal{H}_0$ be a periodic operator generated by the Schrödinger operator $-\frac{d^2}{dx^2} + q_0$ with the periodic potential q_0 and the periodic vertex conditions

$$A(v)u_{\mathcal{E}_v}(v) + B(v)u'_{\mathcal{E}_v}(v) = 0, A(v+g) = A(v), B(v+g) = B(\nu). \tag{48}$$

The spectrum of the operator $\mathcal{H}_0$ has the band-gaps structure $sp\mathcal{H}_0 = \bigcup_{j=1}^{\infty} \left[\alpha_j^0, \beta_j^0\right]$.

Note that the operators $\mathcal{H}^{\eta_\omega}$ are generated by the Schrödinger operator $-\frac{d^2}{dx^2} + q_0 + q_1^{\eta_\omega}$ on $\Gamma \smallsetminus \mathcal{V}$ where $q_1^{\eta_\omega} \in \mathbb{R}$ with periodic vertex condition (48). Hence

$$sp\mathcal{H}^{\eta_\omega} = \bigcup_{j=1}^{\infty} \left[\alpha_j^0 + q_1^{\eta_\omega}, \beta_j^0 + q_1^{\eta_\omega}\right].$$

Let dim $\mathbb{G}$ $=m > 1$. Then formula (46) implies that

$$sp_{ess}\mathcal{H} = \bigcup_{j=1}^{\infty} \left[\alpha_j^0 + m(q_1), \beta_j^0 + M(q_1)\right] \tag{49}$$

where

$$m(q_1) = \inf_{\eta_\omega \in \mathbb{H}_\infty^r} \lim_{g \to \eta(\omega)} q_1(x+g), M(q_1) = \sup_{\eta(\omega) \in \eta_\omega \in \mathbb{H}_\infty^1} \lim_{g \to \eta(\omega)} q_1(x+g), x \in \Gamma \tag{50}$$

and the limits in (50) are independent of $x \in \Gamma$. Let dim $\mathbb{G} = 1$. Then $\mathbb{H}_\infty^1 = \{\eta_+, \eta_-\}$ and

$$sp_{ess}\mathcal{H} = \bigcup_{j=1}^{\infty} \left[\alpha_j^0 + q_1^+, \beta_j^0 + q_1^+\right] \bigcup_{j=1}^{\infty} \left[\alpha_j^0 + q_1^-, \beta_j^0 + q_1^-\right]$$

where

$$q_1^+ = \lim_{g \to \eta_+} q_1(x+g), q_1^- = \lim_{g \to \eta_-} q_1(x+g), x \in \Gamma$$

and $q_1^\pm$ are independent of $x \in \Gamma$.

References

[1] M.S.Agranovich,*Elliptic Operators on Closed Manifolds*, Partial Differential Equations VI, Volume 63 of the series Encyclopaedia of Mathematical Sciences, 1-130, (1994), MR1036528 58G15 (35S05)

[2] Gregory Berkolaiko, Robert Carlson, Stephen A. Fulling, and Peter Kuchment (eds.), *Quantum graphs and their applications*, Contemporary Mathematics, vol. 415, American Mathematical Society, Providence, RI, 2006. MR2279143

[3] Gregory Berkolaiko and Peter Kuchment, *Introduction to quantum graphs*, Mathematical Surveys and Monographs, vol. 186, American Mathematical Society, Providence, RI, 2013. MR3013208

[4] P. Exner, J. P. Keating, P. Kuchment, T. Sunada, and A. Teplyaev, *Vladimir A. Geyler, April 29, 1943–April 2, 2007*, Analysis on graphs and its applications, Proc. Sympos. Pure Math., vol. 77, Amer. Math. Soc., Providence, RI, 2008, pp. xiii, DOI 10.1090/pspum/077. MR2459861

[5] Peter Kuchment, *Quantum graphs. I. Some basic structures*, Waves Random Media **14** (2004), no. 1, S107–S128, DOI 10.1088/0959-7174/14/1/014. Special section on quantum graphs. MR2042548

[6] Peter Kuchment, *Quantum graphs. II. Some spectral properties of quantum and combinatorial graphs*, J. Phys. A **38** (2005), no. 22, 4887–4900, DOI 10.1088/0305-4470/38/22/013. MR2148631

[7] Peter Kuchment and Olaf Post, *On the spectra of carbon nano-structures*, Comm. Math. Phys. **275** (2007), no. 3, 805–826, DOI 10.1007/s00220-007-0316-1. MR2336365

[8] Peter Kuchment and Leonid A. Kunyansky, *Spectral properties of high contrast band-gap materials and operators on graphs*, Experiment. Math. **8** (1999), no. 1, 1–28. MR1685034

[9] Peter Kuchment and Boris Vainberg, *On the structure of eigenfunctions corresponding to embedded eigenvalues of locally perturbed periodic graph operators*, Comm. Math. Phys. **268** (2006), no. 3, 673–686, DOI 10.1007/s00220-006-0105-2. MR2259210

[10] Evgeny Korotyaev and Igor Lobanov, *Schrödinger operators on zigzag nanotubes*, Ann. Henri Poincaré **8** (2007), no. 6, 1151–1176, DOI 10.1007/s00023-007-0331-y. MR2355344

[11] E.Korotyaev, N.Saburova,*Scattering on periodic metric graphs*, arXiv:1507.06441v1 [math.SP] 23, (2015).

[12] Konstantin Pankrashkin, *Spectra of Schrödinger operators on equilateral quantum graphs*, Lett. Math. Phys. **77** (2006), no. 2, 139–154, DOI 10.1007/s11005-006-0088-0. MR2251302

[13] V.S.Rabinovich, S.Roch, B.Silbermann,*Limit Operators and its Applications in the Operator Theory*, In ser. Operator Theory: Advances and Applications, vol 150, (2004).

[14] Vladimir S. Rabinovich and Steffen Roch, *The essential spectrum of Schrödinger operators on lattices*, J. Phys. A **39** (2006), no. 26, 8377–8394, DOI 10.1088/0305-4470/39/26/007. MR2238507

[15] Vladimir S. Rabinovich and Steffen Roch, *Essential spectra of difference operators on* $\mathbb{Z}^n$*-periodic graphs*, J. Phys. A **40** (2007), no. 33, 10109–10128, DOI 10.1088/1751-8113/40/33/012. MR2371282

[16] Steffen Roch, Pedro A. Santos, and Bernd Silbermann, *Non-commutative Gelfand theories*, Universitext, Springer-Verlag London, Ltd., London, 2011. A tool-kit for operator theorists and numerical analysts. MR2761883

[17] V.S.Rabinovich,*Pseudodifferential operators on a class of noncompact manifolds*, Math. USSR-Sb., 18:1, 45–59, (1972).

[18] Vladimir Rabinovich, *On the essential spectrum of quantum graphs*, Integral Equations Operator Theory **88** (2017), no. 3, 339–362, DOI 10.1007/s00020-017-2386-6. MR3682194

[19] I.B.Simonenko,*Operators of convolution type in cones*, Math. USSR-Sb., 3:2, 279–293, (1967).

Insituto Politécnico Nacional, México; ESIME Zacatenco
Email address: vladimir.rabinovich@gmail.com

Contemporary Mathematics
Volume **717**, 2018
https://doi.org/10.1090/conm/717/14448

Optimization of the lowest eigenvalue for leaky star graphs

Pavel Exner and Vladimir Lotoreichik

ABSTRACT. We consider the problem of geometric optimization for the lowest eigenvalue of the two-dimensional Schrödinger operator with an attractive δ-interaction of a fixed strength, the support of which is a star graph with finitely many edges of an equal length $L \in (0, \infty]$. Under the constraint of fixed number of the edges and fixed length of them, we prove that the lowest eigenvalue is maximized by the fully symmetric star graph. The proof relies on the Birman-Schwinger principle, properties of the Macdonald function, and on a geometric inequality for polygons circumscribed into the unit circle.

1. Introduction

The topic addressed in this note is the spectral optimization problem for the lowest eigenvalue of the two-dimensional Schrödinger operator with a δ-interaction supported by a star graph. This problem can be regarded as a two-dimensional counterpart of our recent analysis [**EL17**] of δ-interactions supported by conical surfaces in $\mathbb{R}^3$. As in the said paper, in order to obtain the main result we have to combine the Birman-Schwinger principle with a certain geometric inequality. To advocate the use of this principle as a powerful tool of proving spectral optimization results is an additional motivation of our present considerations.

To describe our main result we need to introduce some notations. In what follows we consider a star graph $\Sigma_N = \Sigma_N(L) \subset \mathbb{R}^2$, which has $N \geq 2$ edges of length $L \in (0, \infty]$ each. Being enumerated in the clockwise manner, these edges are, up to an overall rotation, characterized by the angles $\phi = \phi(\Sigma_N) = \{\phi_1, \phi_2, \dots, \phi_N\}$ between the neighboring edges. We also adopt the natural assumptions $\phi_n \in (0, 2\pi)$ for all $n \in \{1, \dots, N\}$ and $\sum_{n=1}^N \phi_n = 2\pi$; when convenient we say that ϕ is the vector of angles for the star graph Σ_N. Furthermore, by Γ_N we denote the star graph with maximum symmetry, having the same number $N \geq 2$ of edges of the same length $L \in (0, \infty]$ each, and whose vector of angles is given by $\phi = \phi(\Gamma_N) = \{\frac{2\pi}{N}, \frac{2\pi}{N}, \dots, \frac{2\pi}{N}\}$. For brevity we refer to Γ_N as to the symmetric star graph. Examples of star graphs Σ_N and Γ_N for $N = 5$ are plotted in Figure 1.

Given a real number $\alpha > 0$, we consider the spectral problem for the self-adjoint operator H_{α,Σ_N} corresponding via the first representation theorem to the closed,

2010 *Mathematics Subject Classification.* Primary 35P15; Secondary 58J50, 81Q37.
Key words and phrases. Singular Schrödinger operator, δ-interaction on a star graph, lowest eigenvalue, spectral optimization problem, Birman-Schwinger principle.

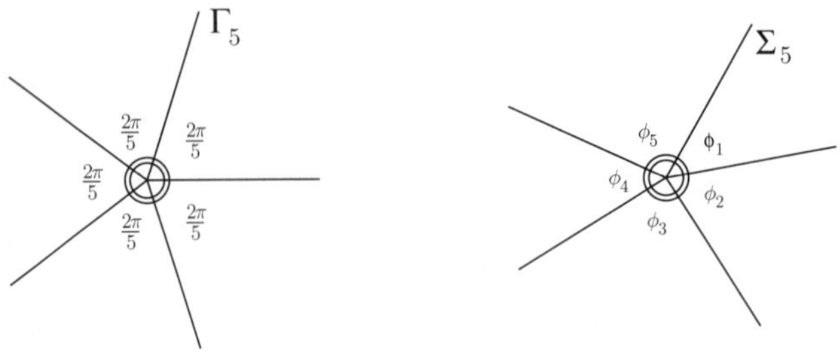

FIGURE 1. The star graphs Γ_5 and Σ_5 with $N = 5$ and $L < \infty$.

densely defined, symmetric, and semi-bounded quadratic form in $L^2(\mathbb{R}^2)$,

$$\mathfrak{h}_{\alpha,\Sigma_N}[u] := \|\nabla u\|^2_{L^2(\mathbb{R}^2;\mathbb{C}^2)} - \alpha \|u|_{\Sigma_N}\|^2_{L^2(\Sigma_N)}, \qquad \mathrm{dom}\,\mathfrak{h}_{\alpha,\Sigma_N} := H^1(\mathbb{R}^2); \tag{1.1}$$

here $u|_{\Sigma_N}$ denotes the usual trace of $u \in H^1(\mathbb{R}^2)$ onto Σ_N; *cf.* [**BEKŠ94**, Sec. 2] and [**BEL14**, Sec. 3.2]. The operator H_{α,Σ_N} is usually called Schrödinger operator with the δ-interaction of strength α supported by Σ_N. In recent years the investigation of Schrödinger operators with singular interactions supported by sets of lower dimensionality became a topic of permanent interest – see, *e.g.*, [**BEL14, BLL13, BEKŠ94, BEW09, DR14, EN03, EV16, LO16, P17**], the monograph [**EK**], and the references therein. The typical physical use of the Hamiltonian H_{α,Σ_N} is to model electron behavior of dilute electron gas in quantum wire networks, that is, very thin structures constructed from semiconductor or other materials, the confinement being realized by potential jump between different material types. With an acceptable idealization we can neglect the 'wire' width and suppose that the interaction is supported by the network skeleton, for instance, by the graph Σ_N. One usually employs the name *'leaky star graph'* for such models [**E08**] to underline that, in contrast to the standard quantum graphs models [**BK**], they do not neglect quantum tunnelling. Another physical use of the Hamiltonian H_{α,Σ_N} can be found in the few-body quantum mechanics with zero-range interactions – see, *e.g.*, [**BK13, BD06, CDR08, HKPC17, LL63**].

Recall that the essential spectrum of H_{α,Σ_N} coincides with the set $[0,\infty)$ if $L < \infty$ and with the set $[-\frac{1}{4}\alpha^2, +\infty)$ if $L = \infty$. The negative discrete spectrum is known to be non-empty and finite unless we have simultaneously $L = \infty$, $N = 2$, and $\phi = \{\pi,\pi\}$; *cf.* Section 2 below for details. By $\lambda_1^\alpha(\Sigma_N)$ we denote the spectral threshold of H_{α,Σ_N} which, except for the mentioned trivial case, is an isolated eigenvalue.

The aim of this paper is to demonstrate that $\lambda_1^\alpha(\Sigma_N)$ is maximized by the symmetric star graph Γ_N. A precise formulation of this claim is the content of the following theorem.

THEOREM 1.1. *For all $\alpha > 0$ we have*

$$\max_{\Sigma_N(L)} \lambda_1^\alpha(\Sigma_N(L)) = \lambda_1^\alpha(\Gamma_N(L)), \tag{1.2}$$

where the maximum is taken over all star graphs with $N \geq 2$ edges of a given length $L \in (0,\infty]$. In the case $L < \infty$ the equality in (1.2) *is achieved if and only if Σ_N and Γ_N are congruent.*

Our method of the proof of Theorem 1.1 relies on the Birman-Schwinger principle for H_{α,Σ_N} and on the trick proposed in [**E05a, EHL06**] and further applied and developed in [**BFK$^+$17, EL17, L18**]. The main geometric ingredient in the proof of Theorem 1.1 is an inequality for the lengths of the diagonals for polygons circumscribed into the unit circle. This inequality can be viewed as a discrete version of the mean chord-length inequality known for smooth arcs [**ACF$^+$03, EHL06, Lu66**]. Finally, we mention that a slightly different discrete version of the mean chord-length inequality is used in [**E05b, E06**] to prove a spectral isoperimetric inequality for Schrödinger operators with point interactions.

Organisation of the paper. In Section 2 we recall the known spectral properties of H_{α,Σ_N} that are needed in this paper. Section 3 is devoted to the Birman-Schwinger principle for H_{α,Σ_N} and its consequences. Theorem 1.1 is proven in Section 4. The paper is concluded by Section 5 containing a discussion of the obtained results and their possible extensions and generalizations.

2. The spectral problem for δ-interactions supported by star graphs

Throughout this section, Σ_N is an arbitrary star graph in $\mathbb{R}^2$ with $N \geq 2$ edges of length $L \in (0,\infty]$ each; *cf.* Figure 1 for the case $N = 5$. The edges of Σ_N are labeled by $\sigma_1, \sigma_2, \ldots, \sigma_N$ being enumerated in the clockwise way. For convenience we extend the sequence of edges in a periodic manner by letting $\sigma_{kN+n} := \sigma_n$ for all $k \in \mathbb{N}$ and $n \in \{1, 2, \ldots, N\}$. By $\phi_n \in (0, 2\pi)$ with $n \in \mathbb{N}$ we denote the magnitude of the angle between the edges σ_n and σ_{n+1}, measured again in the clockwise direction. The vector of angles $\phi = \phi(\Sigma_N) = \{\phi_1, \phi_2, \ldots, \phi_N\}$ satisfies $\phi_n \in (0, 2\pi)$ for all $n \in \{1, \ldots, N\}$ and $\sum_{n=1}^N \phi_n = 2\pi$. At the same time, the (attractive) coupling strength α is a fixed positive number.

We are interested in spectral properties of the self-adjoint operator H_{α,Σ_N} in $L^2(\mathbb{R}^2)$ introduced via the first representation theorem [**K**, Thm. VI 2.1] as associated with the closed, densely defined, symmetric, and semi-bounded quadratic form $\mathfrak{h}_{\alpha,\Sigma_N}$ defined in (1.1); see [**BEKŠ94**, Sec. 2] and also [**BEL14**, Prop. 3.1].

For the sake of completeness, let us add a few words about the operator domain of H_{α,Σ_N}. Let $\widetilde{\Sigma}_N = \Sigma_N(\infty) \subset \mathbb{R}^2$ be the natural extension of Σ_N up to the star graph with semi-infinite edges $\{\widetilde{\sigma}_n\}_{n=1}^N$; in the case $L = \infty$ it is, of course, trivial, we clearly have $\widetilde{\Sigma}_N = \Sigma_N$. The graph $\widetilde{\Sigma}_N$ obviously splits $\mathbb{R}^2$ into N wedge-type domains $\{\Omega_n\}_{n=1}^N$ with the angles $\{\phi_n\}_{n=1}^N$ at their corners. Again for the sake of convenience we extend the sequences of edges for $\widetilde{\Sigma}_N$ and of respective wedges putting $\widetilde{\sigma}_{kN+n} := \widetilde{\sigma}_n$ and $\Omega_{kN+n} := \Omega_n$ for $k \in \mathbb{N}$ and $n \in \{1, \ldots, N\}$. For any $u \in L^2(\mathbb{R}^2)$ we introduce the notation $u_n := u|_{\Omega_n}$, $n \in \mathbb{N}$. Then the operator domain of H_{α,Σ_N} consists of functions $u \in H^1(\mathbb{R}^2)$, which satisfy $\Delta u_n \in L^2(\Omega_n)$, $n \in \mathbb{N}$, in the distributional sense and the δ-type boundary condition on each edge,

$$\partial_{\nu_n} u_n|_{\widetilde{\sigma}_{n+1}} + \partial_{\nu_{n+1}} u_{n+1}|_{\widetilde{\sigma}_{n+1}} = \alpha \chi_{\sigma_n} u|_{\widetilde{\sigma}_{n+1}}, \qquad n \in \mathbb{N},$$

in the sense of traces, where $\chi_{\sigma_n} : \widetilde{\sigma}_n \to \{0, 1\}$, $n \in \mathbb{N}$, is the characteristic function of the set $\sigma_n \subset \widetilde{\sigma}_n$, and where $\partial_{\nu_n} u_n|_{\widetilde{\sigma}_n}$ and $\partial_{\nu_n} u_n|_{\widetilde{\sigma}_{n+1}}$ denote the traces of normal derivatives of u_n onto the edges $\widetilde{\sigma}_n$ and $\widetilde{\sigma}_{n+1}$, respectively, with the normal vector at the boundary of Ω_n pointing outwards. Moreover, for any $u \in \operatorname{dom} H_{\alpha,\Sigma_N}$ we have $H_{\alpha,\Sigma_N} u = -\oplus_{n=1}^N \Delta u_n$. The reader may consult [**BEL14**, Sec. 3.2] for a more precise description of $\operatorname{dom} H_{\alpha,\Sigma_N}$.

It is not surprising that the operator H_{α,Σ_N} has a non-empty essential spectrum. More specifically, we have the following statement.

PROPOSITION 2.1. *Let $\Sigma_N = \Sigma_N(L) \subset \mathbb{R}^2$ be a star graph with $N \geq 2$ edges of length $L \in (0,\infty]$. Then for all $\alpha > 0$ the essential spectrum of H_{α,Σ_N} is characterized as follows:*

(i) $\sigma_{\rm ess}(H_{\alpha,\Sigma_N}) = [0,\infty)$ *if* $L < \infty$;
(ii) $\sigma_{\rm ess}(H_{\alpha,\Sigma_N}) = [-\frac{1}{4}\alpha^2,\infty)$ *if* $L = \infty$.

The claim (i) of this proposition is easy to verify, because the essential spectrum of the Laplacian in the whole space $\mathbb{R}^2$ equals $[0,\infty)$ and introducing a δ-interaction supported by Σ_N with $L < \infty$ leads to a compact perturbation in the sense of resolvent differences; we refer to [**BEKŠ94**, Thm. 3.1] or to [**BEL14**, Thm. 4.2]. For a proof of Proposition 2.1 (ii) see [**EN03**, Prop. 5.4].

As for the discrete spectrum of H_{α,Σ_N}, various properties of it are investigated in or follow from [**BEW09, CDR08, DR14, EI01, EN03, KP18, KL14, P17**]. For our purposes we need the following statement.

PROPOSITION 2.2. *Let $\Sigma_N = \Sigma_N(L) \subset \mathbb{R}^2$ be a star graph with $N \geq 2$ edges of length $L \in (0,\infty]$. If $L = \infty$ and $N = 2$ hold simultaneously, we additionally assume that $\phi(\Sigma_2) \neq \{\pi,\pi\}$. Then $\#\sigma_{\rm dis}(H_{\alpha,\Sigma_N}) \geq 1$ holds for all $\alpha > 0$*[1].

Nontriviality of $\sigma_{\rm dis}(H_{\alpha,\Sigma_N})$ for $L < \infty$ follows from [**KL14**, Thm. 3.1]. It can be alternatively proven via the test function argument as in the proof of [**KL18**, Prop. 2]. The fact that $\sigma_{\rm dis}(H_{\alpha,\Sigma_N})$ is non-void for $L = \infty$ is shown in [**EN03**, Thm. 5.7]; note that by [**KP18**, Thm. E.1] this discrete spectrum is finite.

Summarizing, the essential spectrum of H_{α,Σ_N} coincides with the interval $[0,\infty)$ if $L < \infty$ and with the interval $[-\frac{1}{4}\alpha^2,\infty)$ if $L = \infty$, and there is at least one isolated eigenvalue below the threshold of the essential spectrum, unless simultaneously $L = \infty$, $N = 2$, and $\phi(\Sigma_2) = \{\pi,\pi\}$ hold.

3. Birman-Schwinger principle

In this section we formulate a version of the Birman-Schwinger-type principle for the operator H_{α,Σ_N} and derive a related characterization of its lowest eigenvalue $\lambda_1^\alpha(\Sigma_N)$. A standing assumption throughout this section is $L < \infty$, although most of the results hold or can be reformulated for $L = \infty$ as well.

First of all, we parametrize each edge σ_n of Σ_N by the unit-speed mapping $\sigma_n\colon \mathcal{I} \to \mathbb{R}^2$ with $\mathcal{I} := [0,L]$, *i.e.* $|\sigma_n'(s)| = 1$ holds for all $s \in \mathcal{I}$. We choose this parametrisation in such a way that $\sigma_1(0) = \sigma_2(0) = \cdots = \sigma_N(0)$. Clearly, the Hilbert spaces $L^2(\Sigma_N)$ and $\mathcal{H} := \oplus_{n=1}^N L^2(\mathcal{I})$ can be identified. For any $\psi \in \mathcal{H}$ we denote by $[\psi]_n \in L^2(\mathcal{I})$ (with $n \in \{1,2,\dots,N\}$) its n-th component in the orthogonal decomposition of $\mathcal{H}$.

Furthermore, we define a weakly singular integral operator $Q_{\Sigma_N}(\kappa)\colon \mathcal{H} \to \mathcal{H}$ for $\kappa > 0$ by

$$(3.1) \qquad [Q_{\Sigma_N}(\kappa)\psi]_n(s) := \sum_{m=1}^N \frac{1}{2\pi}\int_0^L K_0\left(\kappa|\sigma_n(s)-\sigma_m(t)|\right)[\psi]_m(t)\mathsf{d}t,$$

[1] Given a self-adjoint operator T we denote by $\#\sigma_{\rm dis}(T)$ the cardinality of the discrete spectrum with the multiplicities of the eigenvalues taken into account.

where $n \in \{1, 2, \ldots, N\}$ and $K_0(\cdot)$ is the modified Bessel function of the second kind having the order $\nu = 0$; *cf.* [**AS64**, §9.6]. In the next proposition we state basic properties of this integral operator.

PROPOSITION 3.1. *Let $\Sigma_N = \Sigma_N(L) \subset \mathbb{R}^2$ be a star graph with $N \geq 2$ edges of length $L \in (0, \infty)$. Then the operator $Q_{\Sigma_N}(\kappa)$ in (3.1) is self-adjoint, compact, and non-negative for all $\kappa > 0$.*

PROOF. Self-adjointness and non-negativity of $Q_{\Sigma_N}(\kappa)$ directly follow from more general results in [**B95**]. Compactness of $Q_{\Sigma_N}(\kappa)$ is a consequence of [**BEKŠ94**, Lem. 3.2]. □

Now we have all the tools to formulate our Birman-Schwinger-type condition for H_{α,Σ_N}.

THEOREM 3.2. *Let $\Sigma_N = \Sigma_N(L) \subset \mathbb{R}^2$ be a star graph with $N \geq 2$ edges of length $L \in (0, \infty)$. Let the self-adjoint operator H_{α,Σ_N} in $L^2(\mathbb{R}^2)$ represent the quadratic form in (1.1) and let the operator-valued function $(0, \infty) \ni \kappa \mapsto Q_{\Sigma_N}(\kappa)$ be as in (3.1). Then the following claims hold:*

(i) $\dim\ker(H_{\alpha,\Sigma_N} + \kappa^2) = \dim\ker(I - \alpha Q_{\Sigma_N}(\kappa))$ *for all* $\kappa > 0$.
(ii) *The mapping* $u \mapsto u|_{\Sigma_N}$ *is a bijection between* $\ker(H_{\alpha,\Sigma_N} + \kappa^2)$ *and* $\ker(I - \alpha Q_{\Sigma_N}(\kappa))$.

PROOF. The claim (i) is a particular case of [**BEKŠ94**, Lem. 2.3 (iv)]. The claim (ii) follows from an abstract statement in [**B95**, Lem. 1]. □

We conclude this section by two corollaries of Theorem 3.2.

COROLLARY 3.3. *Let the assumptions be as in Theorem 3.2, and moreover, let $\kappa > 0$ be such that $\lambda_1^\alpha(\Sigma_N) = -\kappa^2$. Then the following claims hold:*

(i) $\ker(I - \alpha Q_{\Sigma_N}(\kappa)) = \operatorname{span}\{\psi_\star\}$ *where* $\psi_\star \in \mathcal{H}$ *is a positive function.*
(ii) *For the symmetric star graph Γ_N the corresponding function $\psi_\star$ depends only on the distance from the vertex of Γ_N; i.e.* $[\psi_\star]_1 = [\psi_\star]_2 = \cdots = [\psi_\star]_N$.

PROOF. The proof of (i) is completely analogous to the proofs of [**EL17**, Prop. 2.3] and of [**L18**, Cor. 3.3], hence we skip it to avoid self-repetition.

In order to prove (ii) we define the subspace

$$\mathcal{H}_{\rm sym} = \{\psi \oplus \psi \oplus \cdots \oplus \psi \colon \psi \in L^2(\mathcal{I})\}$$

of the Hilbert space $\mathcal{H}$. It is not difficult to check that the operator $Q_{\Gamma_N}(\kappa)$ can be rewritten as an orthogonal sum $Q^{\|}_{\Gamma_N}(\kappa) \oplus Q^{\perp}_{\Gamma_N}(\kappa)$ with respect to the orthogonal decomposition $\mathcal{H} = \mathcal{H}_{\rm sym} \oplus \mathcal{H}_{\rm sym}^{\perp}$. Thus, the function $\psi_\star$ in (i) is an eigenfunction of at least one of the operators $Q^{\|}_{\Gamma_N}(\kappa)$ and $Q^{\perp}_{\Gamma_N}(\kappa)$.

It is clear that $\psi_\star$, being a positive function, can not belong to the space $\mathcal{H}_{\rm sym}^{\perp}$. This allows us to conclude $\psi_\star \in \mathcal{H}_{\rm sym}$ by which the claim of (ii) is proven. □

Next we provide the second consequence of Theorem 3.2. The proof of this corollary is analogous to the one of [**EL17**, Prop. 2.3] or of [**L18**, Cor. 3.4] which is why we can skip it.

COROLLARY 3.4. *Let the assumptions be as in Theorem 3.2. Let $\kappa > 0$ be fixed. Then the following claims hold:*

(i) $\sup \sigma(\alpha Q_{\Sigma_N}(\kappa)) \geq 1$ *if and only if* $\lambda_1^\alpha(\Sigma_N) \leq -\kappa^2$.
(ii) $\sup \sigma(\alpha Q_{\Sigma_N}(\kappa)) = 1$ *if and only if* $\lambda_1^\alpha(\Sigma_N) = -\kappa^2$.

4. Proof of Theorem 1.1

Now we are finally in position to establish Theorem 1.1. Throughout this section, we assume that $\Sigma_N = \Sigma_N(L) \subset \mathbb{R}^2$ is a star graph with $N \geq 2$ edges of length $L \in (0, \infty]$.

First of all, we suppose that the star graph has edges of finite length $L < \infty$, and that they are parametrized as in Section 3 via the mappings $\{\sigma_n\}_{n=1}^N$. Let $\Gamma_N \subset \mathbb{R}^2$ be the symmetric star graph having the same length $L \in (0, \infty)$ of the edges, which are also parametrized by the unit-speed mappings $\{\gamma_n\}_{n=1}^N$ in the same way. In addition, we assume that Σ_N is not congruent to Γ_N, in other words, the corresponding vectors of angles $\phi(\Sigma_N)$ and $\phi(\Gamma_N)$ do not coincide. Without loss of generality we may assume that the vertices of Σ_N and of Γ_N are both located at the origin.

Recall that $\lambda_1^\alpha(\Sigma_N)$ and $\lambda_1^\alpha(\Gamma_N)$ denote the lowest (negative) eigenvalues of H_{α,Σ_N} and of H_{α,Γ_N}, respectively. Furthermore, we fix $\kappa > 0$ such that $\lambda_1^\alpha(\Gamma_N) = -\kappa^2$. By Corollary 3.3 (i) we have $\ker(I - \alpha Q_{\Gamma_N}(\kappa)) = \operatorname{span}\{\psi_\star\}$, where $\psi_\star \in \mathcal{H} \cong L^2(\Gamma_N)$ is a positive function, which according to Corollary 3.3 (ii) depends on the distance from the origin only. We normalise $\psi_\star$ so that $\|\psi_\star\|_{L^2(\Gamma_N)} = 1$. Note that by Corollary 3.4 (ii) we also have $\sup \sigma(\alpha Q_{\Gamma_N}(\kappa)) = 1$.

Let $\mathbb{S}^1 \subset \mathbb{R}^2$ be the unit circle centred at the origin. In what follows we fix the points on $\mathbb{S}^1$ by $y_n := L^{-1}\sigma_n(L)$ and $x_n := L^{-1}\gamma_n(L)$ for all $n \in \{1, \ldots, N\}$. The sequences of points $\{y_n\}_{n=1}^N$ and $\{x_n\}_{n=1}^N$ can be interpreted as vertices of polygons circumscribed into $\mathbb{S}^1$. Next, we extend these sequences to all $n \in \mathbb{N}$ in a natural periodic manner with the period N. We also adopt the notation $\angle\sigma_n\sigma_m \in (0, 2\pi)$ for the magnitude of the angle between the edges σ_m and σ_n, measured from σ_m to σ_n in the clockwise direction. Using the cosine law for triangles and Jensen's inequality applied for the concave function $(0, \pi) \ni x \mapsto \sin^2(x)$ we find

$$
\begin{aligned}
\sum_{n=1}^N |y_{n+m} - y_n|^2 &= 4\sum_{n=1}^N \sin^2\left(\frac{\angle\sigma_{n+m}\sigma_n}{2}\right) \leq 4N\sin^2\left(\sum_{n=1}^N \frac{\angle\sigma_{n+m}\sigma_n}{2N}\right) \\
&= 4N\sin^2\left(\sum_{k=0}^{m-1}\sum_{n=1}^N \frac{\angle\sigma_{n+k+1}\sigma_{n+k}}{2N}\right) \qquad (4.1) \\
&= 4N\sin^2\left(\frac{\pi m}{N}\right) = \sum_{n=1}^N |x_{n+m} - x_n|^2,
\end{aligned}
$$

for any $m \in \mathbb{N}_0$. Note that the sequence $\{\angle\sigma_{n+1}\sigma_n\}_{n=1}^N$ does not consist of equal numbers, since the star graph Σ_N is not congruent to the symmetric star graph Γ_N. Thus, the inequality in (4.1) is strict at least for $m = 1$, because the function $(0, \pi) \ni x \mapsto \sin^2(x)$ is strictly concave. Furthermore, the cosine law for triangles implies that for any $s, t \in \mathcal{I}$ and $n, m \in \mathbb{N}$ one has the identities

$$
\begin{aligned}
|\sigma_n(s) - \sigma_m(t)|^2 &= (s - t)^2 + st|y_n - y_m|^2, \\
|\gamma_n(s) - \gamma_m(t)|^2 &= (s - t)^2 + st|x_n - x_m|^2.
\end{aligned}
$$

Next we introduce an auxiliary function by

$$\mathbb{R}_+ \ni x \mapsto F_{s,t}(x) := K_0\left(\kappa\sqrt{(s-t)^2 + stx}\right), \qquad \text{for } s,t \in \mathcal{I},\ \kappa > 0.$$

Differentiating it with respect to x twice, we get

$$F'_{s,t}(x) = -\frac{\kappa st K_1(\kappa\sqrt{(s-t)^2 + stx})}{2\sqrt{(s-t)^2 + stx}} < 0,$$

$$F''_{s,t}(x) = \frac{\kappa s^2t^2 K_1(\kappa\sqrt{(s-t)^2 + stx})}{4((s-t)^2 + stx)^{3/2}} - \frac{\kappa^2 s^2t^2 K_1'(\kappa\sqrt{(s-t)^2 + stx})}{4((s-t)^2 + stx)} > 0,$$

where we have employed that the Bessel function $K_1(\cdot)$ is monotonously decreasing and positive [**AS64**, §9.6]. Thus, we conclude that $F_{s,t}$ is a strictly decreasing and strictly convex function. Using the min-max principle we can further infer that

$$\begin{aligned}
2\pi \sup \sigma(Q_{\Sigma_N}(\kappa)) &\geq \int_0^L \int_0^L \left(\sum_{n=1}^N \sum_{m=1}^N K_0\left(\kappa|\sigma_n(s) - \sigma_m(t)|\right)\right) \psi_\star(s)\psi_\star(t)\mathsf{d}s\mathsf{d}t \\
&= \sum_{m=0}^{N-1} \int_0^L \int_0^L \left(\sum_{n=1}^N F_{s,t}\left(|y_{n+m} - y_n|^2\right)\right) \psi_\star(s)\psi_\star(t)\mathsf{d}s\mathsf{d}t.
\end{aligned}$$

Plugging now the geometric inequality (4.1) into the inequality above and using the positivity of $\psi_\star$, properties of the function $F_{s,t}(\cdot)$, and Jensen's inequality, we obtain

$$\begin{aligned}
2\pi \sup \sigma(Q_{\Sigma_N}(\kappa)) &\geq N \sum_{m=0}^{N-1} \int_0^L \int_0^L \left(F_{s,t}\left(\sum_{n=1}^N \frac{|y_{n+m} - y_n|^2}{N}\right)\right) \psi_\star(s)\psi_\star(t)\mathsf{d}s\mathsf{d}t \\
&> \sum_{m=0}^{N-1} \int_0^L \int_0^L \left(\sum_{n=1}^N F_{s,t}\left(|x_{n+m} - x_n|^2\right)\right) \psi_\star(s)\psi_\star(t)\mathsf{d}s\mathsf{d}t \\
&= \int_0^L \int_0^L \left(\sum_{n=1}^N \sum_{m=1}^N K_0\left(\kappa|\gamma_n(s) - \gamma_m(t)|\right)\right) \psi_\star(s)\psi_\star(t)\mathsf{d}s\mathsf{d}t \\
&= 2\pi \sup \sigma(Q_{\Gamma_N}(\kappa)) = \frac{2\pi}{\alpha}.
\end{aligned}$$

This means that $\sup \sigma(\alpha Q_{\Sigma_N}(\kappa)) > 1$ and using Corollary 3.4 we get the inequality

$$\lambda_1^\alpha(\Sigma_N) < -\kappa^2 = \lambda_1^\alpha(\Gamma_N)$$

which concludes the proof for $L < \infty$.

In the case of semi-infinite edges we first show that $H_{\alpha,\Sigma_N(L)}$ and $H_{\alpha,\Gamma_N(L)}$ converge to $H_{\alpha,\Sigma_N(\infty)}$ and $H_{\alpha,\Gamma_N(\infty)}$ in the strong resolvent sense as $L \to \infty$. Indeed, the sequences of quadratic forms $\mathfrak{h}_{\alpha,\Sigma_N(L)}$ and $\mathfrak{h}_{\alpha,\Gamma_N(L)}$ are decreasing in L in the sense of form ordering and for any $u \in H^1(\mathbb{R}^2)$ the dominated convergence theorem implies $\lim_{L\to\infty} \mathfrak{h}_{\alpha,\Sigma_N(L)}[u] = \mathfrak{h}_{\alpha,\Sigma_N(\infty)}[u]$ and $\lim_{L\to\infty} \mathfrak{h}_{\alpha,\Gamma_N(L)}[u] = \mathfrak{h}_{\alpha,\Gamma_N(\infty)}[u]$. Hence, the desired strong resolvent convergence follows from [**RS-I**, Thm. S.16]. By [**W**, Satz 9.26 (b)] we then get

$$\lim_{L\to\infty} \lambda_1^\alpha(\Sigma_N(L)) = \lambda_1^\alpha(\Sigma_N(\infty)) \quad \text{and} \quad \lim_{L\to\infty} \lambda_1^\alpha(\Gamma_N(L)) = \lambda_1^\alpha(\Gamma_N(\infty)).$$

Now the claim for infinite star graphs follows by passing to the limit $L \to \infty$ in the inequality $\lambda_1^\alpha(\Sigma_N(L)) \leq \lambda_1^\alpha(\Gamma_N(L))$ demonstrated above. □

REMARK 4.1. Note that for $L = \infty$ the equality in Theorem 1.1 is apparently achieved *if and only if* $\Sigma_N(\infty) \cong \Gamma_N(\infty)$, however, the used method of the proof through the spectral convergence is not refined enough to make this conclusion. We do not pursue this question further, instead we review below several less trivial extensions of the above considerations.

5. Discussion

The present analysis adds one more item to the long list of various 'isoperimetric' results having in mathematical physics a tradition almost a century old starting from the papers [**F23**, **K24**]. The symmetry embedded in the result comes naturally from the symmetry of the interaction. This would change when the latter is violated, for instance, by choosing the edges of the graph having different lengths, or alternatively by adding a potential bias in the spirit of [**EV16**] to the Schrödinger operator in question. There is no need to stress that the problem then becomes more complicated. Moreover, the solution may not be obvious also in situations where the support does exhibit a symmetry but the interaction is modified; as examples one can mention star graphs of the type considered here supporting a δ'-interaction as in [**BEL14**], or a three-dimensional 'star' with edges supporting a δ-interaction of co-dimension two as in [**BFK$^+$17**, **EK02**, **EK08**, **EK16**]. In both these cases one naturally expects that the extremal configuration(s) could depend on the number N of the star rays.

Acknowledgements

The authors acknowledge the support by the grant No. 17-01706S of the Czech Science Foundation (GAČR).

References

[AS64] *Handbook of mathematical functions, with formulas, graphs, and mathematical tables*, Edited by Milton Abramowitz and Irene A. Stegun, Dover Publications, Inc., New York, 1966. MR0208797

[ACF$^+$03] Aaron Abrams, Jason Cantarella, Joseph H. G. Fu, Mohammad Ghomi, and Ralph Howard, *Circles minimize most knot energies*, Topology **42** (2003), no. 2, 381–394, DOI 10.1016/S0040-9383(02)00016-2. MR1941441

[BEL14] Jussi Behrndt, Pavel Exner, and Vladimir Lotoreichik, *Schrödinger operators with δ- and δ'-interactions on Lipschitz surfaces and chromatic numbers of associated partitions*, Rev. Math. Phys. **26** (2014), no. 8, 1450015, 43, DOI 10.1142/S0129055X14500159. MR3256859

[BK13] Jens Bolte and Joachim Kerner, *Quantum graphs with singular two-particle interactions*, J. Phys. A **46** (2013), no. 4, 045206, 25, DOI 10.1088/1751-8113/46/4/045206. MR3029862

[BD06] Raymond Brummelhuis and Pierre Duclos, *Effective Hamiltonians for atoms in very strong magnetic fields*, J. Math. Phys. **47** (2006), no. 3, 032103, 41, DOI 10.1063/1.2178155. MR2219780

[BFK$^+$17] Jussi Behrndt, Rupert L. Frank, Christian Kühn, Vladimir Lotoreichik, and Jonathan Rohleder, *Spectral theory for Schrödinger operators with δ-interactions supported on curves in $\mathbb{R}^3$*, Ann. Henri Poincaré **18** (2017), no. 4, 1305–1347, DOI 10.1007/s00023-016-0532-3. MR3626305

[BLL13] Jussi Behrndt, Matthias Langer, and Vladimir Lotoreichik, *Schrödinger operators with δ and δ'-potentials supported on hypersurfaces*, Ann. Henri Poincaré **14** (2013), no. 2, 385–423, DOI 10.1007/s00023-012-0189-5. MR3028043

[BK] Gregory Berkolaiko and Peter Kuchment, *Introduction to quantum graphs*, Mathematical Surveys and Monographs, vol. 186, American Mathematical Society, Providence, RI, 2013. MR3013208

[B95] Johannes F. Brasche, *On the spectral properties of singular perturbed operators*, Dirichlet forms and stochastic processes (Beijing, 1993), de Gruyter, Berlin, 1995, pp. 65–72. MR1366424

[BEKŠ94] J. F. Brasche, P. Exner, Yu. A. Kuperin, and P. Šeba, *Schrödinger operators with singular interactions*, J. Math. Anal. Appl. **184** (1994), no. 1, 112–139, DOI 10.1006/jmaa.1994.1188. MR1275948

[BEW09] B. M. Brown, M. S. P. Eastham, and I. G. Wood, *Estimates for the lowest eigenvalue of a star graph*, J. Math. Anal. Appl. **354** (2009), no. 1, 24–30, DOI 10.1016/j.jmaa.2008.12.014. MR2510414

[CDR08] H. D. Cornean, P. Duclos, and B. Ricaud, *On the skeleton method and an application to a quantum scissor*, Analysis on graphs and its applications, Proc. Sympos. Pure Math., vol. 77, Amer. Math. Soc., Providence, RI, 2008, pp. 657–672, DOI 10.1090/pspum/077/2459895. MR2459895

[DR14] V. Duchêne and N. Raymond, *Spectral asymptotics of a broken δ-interaction*, J. Phys. A **47** (2014), no. 15, 155203, 19, DOI 10.1088/1751-8113/47/15/155203. MR3191674

[E05a] Pavel Exner, *An isoperimetric problem for leaky loops and related mean-chord inequalities*, J. Math. Phys. **46** (2005), no. 6, 062105, 10, DOI 10.1063/1.1914728. MR2149809

[E05b] Pavel Exner, *An isoperimetric problem for point interactions*, J. Phys. A **38** (2005), no. 22, 4795–4802, DOI 10.1088/0305-4470/38/22/004. MR2148622

[E06] Pavel Exner, *Point interaction polygons: an isoperimetric problem*, Mathematical physics of quantum mechanics, Lecture Notes in Phys., vol. 690, Springer, Berlin, 2006, pp. 55–64, DOI 10.1007/3-540-34273-7_7. MR2234903

[E08] Pavel Exner, *Leaky quantum graphs: a review*, Analysis on graphs and its applications, Proc. Sympos. Pure Math., vol. 77, Amer. Math. Soc., Providence, RI, 2008, pp. 523–564, DOI 10.1090/pspum/077/2459890. MR2459890

[EHL06] Pavel Exner, Evans M. Harrell, and Michael Loss, *Inequalities for means of chords, with application to isoperimetric problems*, Lett. Math. Phys. **75** (2006), no. 3, 225–233, DOI 10.1007/s11005-006-0053-y. MR2211029

[EI01] P. Exner and T. Ichinose, *Geometrically induced spectrum in curved leaky wires*, J. Phys. A **34** (2001), no. 7, 1439–1450, DOI 10.1088/0305-4470/34/7/315. MR1819942

[EK02] P. Exner and S. Kondej, *Curvature-induced bound states for a δ interaction supported by a curve in $\mathbb{R}^3$*, Ann. Henri Poincaré **3** (2002), no. 5, 967–981, DOI 10.1007/s00023-002-8644-3. MR1937610

[EK08] P. Exner and S. Kondej, *Hiatus perturbation for a singular Schrödinger operator with an interaction supported by a curve in $\mathbb{R}^3$*, J. Math. Phys. **49** (2008), no. 3, 032111, 19, DOI 10.1063/1.2845419. MR2406779

[EK16] Pavel Exner and Sylwia Kondej, *Strong coupling asymptotics for Schrödinger operators with an interaction supported by an open arc in three dimensions*, Rep. Math. Phys. **77** (2016), no. 1, 1–17, DOI 10.1016/S0034-4877(16)00005-7. MR3461794

[EK] Pavel Exner and Hynek Kovařík, *Quantum waveguides*, Theoretical and Mathematical Physics, Springer, Cham, 2015. MR3362506

[EL17] Pavel Exner and Vladimir Lotoreichik, *A spectral isoperimetric inequality for cones*, Lett. Math. Phys. **107** (2017), no. 4, 717–732, DOI 10.1007/s11005-016-0917-8. MR3623278

[EN03] P. Exner and K. Němcová, *Leaky quantum graphs: approximations by point-interaction Hamiltonians*, J. Phys. A **36** (2003), no. 40, 10173–10193, DOI 10.1088/0305-4470/36/40/004. MR2023303

[EV16] Pavel Exner and Semjon Vugalter, *On the existence of bound states in asymmetric leaky wires*, J. Math. Phys. **57** (2016), no. 2, 022104, 15, DOI 10.1063/1.4941139. MR3456722

[F23] G. Faber, Beweis dass unter allen homogenen Membranen von gleicher Fläche und gleicher Spannung die kreisförmige den tiefsten Grundton gibt, *Sitz. bayer. Akad. Wiss.* (1923), 169–172.

[HKPC17] Jonas Have, Hynek Kovařík, Thomas G. Pedersen, and Horia D. Cornean, *On the existence of impurity bound excitons in one-dimensional systems with zero range interactions*, J. Math. Phys. **58** (2017), no. 5, 052106, 16, DOI 10.1063/1.4983921. MR3654567

[K] Tosio Kato, *Perturbation theory for linear operators*, Classics in Mathematics, Springer-Verlag, Berlin, 1995. Reprint of the 1980 edition. MR1335452

[KP18] Magda Khalile and Konstantin Pankrashkin, *Eigenvalues of Robin Laplacians in infinite sectors*, Math. Nachr. **291** (2018), no. 5-6, 928–965, DOI 10.1002/mana.201600314. MR3795565

[KL14] Sylwia Kondej and Vladimir Lotoreichik, *Weakly coupled bound state of 2-D Schrödinger operator with potential-measure*, J. Math. Anal. Appl. **420** (2014), no. 2, 1416–1438, DOI 10.1016/j.jmaa.2014.06.053. MR3240086

[K24] E. Krahn, *Über eine von Rayleigh formulierte Minimaleigenschaft des Kreises* (German), Math. Ann. **94** (1925), no. 1, 97–100, DOI 10.1007/BF01208645. MR1512244

[KL18] David Krejčiřík and Vladimir Lotoreichik, *Optimisation of the lowest Robin eigenvalue in the exterior of a compact set*, J. Convex Anal. **25** (2018), no. 1, 319–337. MR3756939

[LL63] Elliott H. Lieb and Werner Liniger, *Exact analysis of an interacting Bose gas. I. The general solution and the ground state*, Phys. Rev. (2) **130** (1963), 1605–1616. MR0156630

[L18] V. Lotoreichik, Spectral isoperimetric inequalities for singular interactions on open arcs, *to appear in Appl. Anal.*, `arXiv:1609.07598`.

[LO16] Vladimir Lotoreichik and Thomas Ourmières-Bonafos, *On the bound states of Schrödinger operators with δ-interactions on conical surfaces*, Comm. Partial Differential Equations **41** (2016), no. 6, 999–1028, DOI 10.1080/03605302.2016.1168843. MR3509521

[Lu66] Lükő Gábor, *On the mean length of the chords of a closed curve*, Israel J. Math. **4** (1966), 23–32, DOI 10.1007/BF02760067. MR0200795

[P17] Konstantin Pankrashkin, *Variational proof of the existence of eigenvalues for star graphs*, Functional analysis and operator theory for quantum physics, EMS Ser. Congr. Rep., Eur. Math. Soc., Zürich, 2017, pp. 447–458. MR3677034

[RS-I] Michael Reed and Barry Simon, *Methods of modern mathematical physics. I*, 2nd ed., Academic Press, Inc. [Harcourt Brace Jovanovich, Publishers], New York, 1980. Functional analysis. MR751959

[W] Joachim Weidmann, *Lineare Operatoren in Hilberträumen. Teil 1* (German), Mathematische Leitfäden. [Mathematical Textbooks], B. G. Teubner, Stuttgart, 2000. Grundlagen. [Foundations]. MR1887367

Department of Theoretical Physics, Nuclear Physics Institute, Czech Academy of Sciences, 25068 Řež near Prague, Czechia—and—Doppler Institute for Mathematical Physics and Applied Mathematics, Czech Technical University, Břehová 7, 11519 Prague, Czechia

Email address: `exner@ujf.cas.cz`

Department of Theoretical Physics, Nuclear Physics Institute, Czech Academy of Sciences, 25068 Řež near Prague, Czechia

Email address: `lotoreichik@ujf.cas.cz`

Contemporary Mathematics
Volume **717**, 2018
https://doi.org/10.1090/conm/717/14449

Local eigenvalue asymptotics of the perturbed Krein Laplacian

Vincent Bruneau and Georgi Raikov

Abstract. We consider the Krein Laplacian on a regular bounded domain $\Omega \subset \mathbb{R}^d$, $d \geq 2$, perturbed by a real-valued multiplier V vanishing on the boundary. Assuming that V has a definite sign, we investigate the asymptotics of the functions counting the eigenvalues of $K+V$ which converge to the origin from below or from above. We show that the effective Hamiltonian that governs the main asymptotic term of these functions is the harmonic Toeplitz operator T_V with symbol V, unitarily equivalent to a pseudodifferential operator on the boundary. In the cases where V admits a power-like decay at $\partial\Omega$, or V is compactly supported in Ω, and Ω and $\operatorname{supp} V$ are radially symmetric, we obtain the main asymptotic term of the eigenvalue counting functions.

1. Introduction

In this article we study the spectral properties of the perturbed Krein Laplacian $K+V$ in a bounded domain $\Omega \subset \mathbb{R}^d$, $d \geq 2$, with boundary $\partial\Omega \in C^\infty$. The Krein Laplacian $K := -\Delta$ is defined on sufficiently regular functions $u : \Omega \to \mathbb{C}$ which satisfy the boundary condition

$$\frac{\partial u}{\partial \nu} = \mathcal{D}u \quad \text{on} \quad \partial\Omega,$$

where ν is the unit outer normal vector at $\partial\Omega$, and $\mathcal{D}$ is the Dirichlet-to-Neumann operator, a first-order elliptic pseudodifferential operator (ΨDO), self-adjoint in $L^2(\partial\Omega)$. Then $K \geq 0$ is self-adjoint in $L^2(\Omega)$, and one of its remarkable properties is that its essential spectrum is not empty. Namely, $\sigma_{\rm ess}(K) = \{0\}$, and the zero is an isolated eigenvalue of K of infinite multiplicity. Further, we assume that the perturbation of K is the multiplier by the function $V \in C(\overline{\Omega}; \mathbb{R})$. Then, evidently the operator $K+V$, on the domain of K, is self-adjoint in $L^2(\Omega)$. Moreover,

$$\sigma_{\rm ess}(K+V) = V(\partial\Omega)$$

(see Theorem 2.4 below). Assuming that V vanishes identically on $\partial\Omega$, we get

$$\sigma_{\rm ess}(K+V) = \sigma_{\rm ess}(K) = \{0\}.$$

2010 *Mathematics Subject Classification.* Primary 47B35, 35J25, 35P15, 35P20.

Key words and phrases. Harmonic Toeplitz operators, Krein Laplacian, eigenvalue asymptotics, effective Hamiltonian.

The second author was partially supported by the Chilean Scientific Foundation *Fondecyt* under Grant 1170816 and by the Visiting Scholars Program of IDEX Bordeaux.

However, in contrast to the unperturbed operator K, the zero is an accumulation point of the discrete eigenvalues of the perturbed operator $K+V$. We suppose in addition that V has a definite (negative or positive) sign and investigate the asymptotic distribution of the eigenvalues of $K+V$ adjoining the origin. First, in Theorem 2.5 we show that the effective Hamiltonian governing the eigenvalue counting functions for $K+V$ is the Toeplitz operator $T_V := PV_{|\mathrm{Ker}\,K}$, where P is the orthogonal projection onto $\mathrm{Ker}\,K$. That is why, in Section 3 we discuss the general spectral properties of T_V. Further, in Section 4 we assume that V admits a power-like decay at $\partial\Omega$, and examine the eigenvalue asymptotics for the compact operator T_V, unitarily equivalent to a classical ΨDO on the boundary. We obtain the main asymptotic term and a sharp remainder estimate of the eigenvalue counting function for T_V. Finally, in Section 5, we analyze the case where V is compactly supported in Ω. More precisely, we suppose that Ω is the unit ball in $\mathbb{R}^d$ while $\mathrm{supp}\,V$ is the concentric ball of radius $c \in (0,\infty)$, and obtain the main asymptotic term of the eigenvalue counting function for T_V.

This article is an extended version of the talk given by the second author at the conference QMath13 *Mathematical Results in Quantum Physics* held at Georgia Tech, Atlanta, USA, in October 2016. He is grateful to the organizers of the conference and those of the Session *Quantum Mechanics on Graphs and Similar Structures* for the opportunity to present this work.

A more detailed exposition of some of the results of this paper can be found in [**13**].

2. The Krein Laplacian and its perturbations

Let $\Omega \subset \mathbb{R}^d$, $d \geq 2$, be a bounded domain with smooth boundary $\partial\Omega$.
For $s \in \mathbb{R}$ denote by $H^s(\Omega)$ and $H^s(\partial\Omega)$ the Sobolev spaces on Ω and $\partial\Omega$ respectively. Moreover, as usual, we denote by $H^s_0(\Omega)$, $s > 1/2$, the closure of $C_0^\infty(\Omega)$ in $H^s(\Omega)$. Set also $H^2_D(\Omega) := H^2(\Omega) \cap H^1_0(\Omega)$. Define the minimal Laplacian

$$\Delta_{\min} := \Delta, \quad \mathrm{Dom}\,\Delta_{\min} = H^2_0(\Omega).$$

The operator $\Delta_{\min}$ is symmetric but not self-adjoint in $L^2(\Omega)$, since we have

$$\Delta^*_{\min} =: \Delta_{\max} = \Delta, \quad \mathrm{Dom}\,\Delta_{\max} = \left\{u \in L^2(\Omega) \,|\, \Delta\, u \in L^2(\Omega)\right\}, \tag{2.1}$$

Δu being the distributional Laplacian of $u \in L^2(\Omega)$. Note that

$$\mathrm{Ker}\,\Delta_{\max} = \mathcal{H}(\Omega) := \left\{u \in L^2(\Omega) \,|\, \Delta u = 0 \text{ in } \Omega\right\}. \tag{2.2}$$

It is well known that $\mathcal{H}(\Omega)$ is a closed subspace of $L^2(\Omega)$ (see e.g. [**24**]). The Laplace equation $\Delta u = 0$ in (2.2) is understood *a priori* in the distributional sense. However, by the Weyl lemma, if u belongs to $\mathcal{D}'(\Omega)$, the class of distributions over $C_0^\infty(\Omega)$, and $\Delta u = 0$, then $u \in C^\infty(\Omega)$ (see the original work [**31**] for a proof in the case $u \in L^1_{\mathrm{loc}}(\Omega)$, and the monograph [**17**] whose Chapter 10 contains an extension to general $u \in \mathcal{D}'(\Omega)$).

Lemma 2.1. *The domain* $\mathrm{Dom}\,\Delta_{\max}$ *admits the direct-sum decomposition*

$$\mathrm{Dom}\,\Delta_{\max} = \mathcal{H}(\Omega) \dotplus H^2_D(\Omega). \tag{2.3}$$

PROOF. Let us first show that the sum at the r.h.s. of (2.3) is direct. Assume that $u_1 \in \mathcal{H}(\Omega)$, $u_2 \in H_D^2(\Omega)$, and $u_1 + u_2 = 0$. Then u_2 satisfies the homogeneous boundary-value problem

$$\begin{cases} \Delta u_2 = 0 & \text{in} \quad \Omega, \\ u_2 = 0 & \text{on} \quad \partial\Omega. \end{cases}$$

Hence, $u_2 = 0$, and $u_1 = 0$. Evidently, if $u_1 \in \mathcal{H}(\Omega)$, $u_2 \in H_D^2(\Omega)$, then $u_1 + u_2 \in \operatorname{Dom} \Delta_{\max}$. Pick now $u \in \operatorname{Dom} \Delta_{\max}$, and define the Dirichlet Laplacian

$$\Delta_D := \Delta, \quad \operatorname{Dom} \Delta_D := H_D^2(\Omega).$$

Then u_1 and u_2 defined by $u_2 := \Delta_D^{-1} \Delta u$, $u_1 := u - u_2$ clearly satisfy

$$u_1 \in \mathcal{H}(\Omega), \quad u_2 \in H_D^2(\Omega), \quad u = u_1 + u_2.$$

□

Introduce the Krein Laplacian

$$K := -\Delta, \quad \operatorname{Dom} K = \mathcal{H}(\Omega) \dot{+} H_0^2(\Omega).$$

The operator $K \geq 0$, self-adjoint in $L^2(\Omega)$, is the von Neumann - Krein "soft" extension of $-\Delta_{\min}$, remarkable for the fact that any other self-adjoint extension $S \geq 0$ of $-\Delta_{\min}$ satisfies

$$(S + I)^{-1} \leq (K + I)^{-1}$$

(see [**25, 30**]). Evidently, $\operatorname{Ker} K = \mathcal{H}(\Omega)$. The domain $\operatorname{Dom} K$ admits a more explicit description in the terms of the Dirichlet-to-Neumann operator $\mathcal{D}$. For $f \in C^\infty(\partial\Omega)$, $\mathcal{D}f$ is defined by

$$\mathcal{D}f = \frac{\partial u}{\partial \nu}_{|\partial\Omega},$$

where u is the solution of the boundary-value problem

$$\begin{cases} \Delta u = 0 & \text{in} \quad \Omega, \\ u = f & \text{on} \quad \partial\Omega. \end{cases}$$

The operator $\mathcal{D}$ is a first-order elliptic operator; by the elliptic regularity, it extends to a bounded operator form $H^s(\partial\Omega)$ into $H^{s-1}(\partial\Omega)$, $s \in \mathbb{R}$. Then we have

$$\operatorname{Dom} K = \left\{ u \in \operatorname{Dom} \Delta_{\max} \;\middle|\; \frac{\partial u}{\partial \nu}_{|\partial\Omega} = \mathcal{D}\left(u_{|\partial\Omega}\right) \right\}$$

(see [**18**, Theorem III.1.2]).

Denote by L the restriction of K onto $\operatorname{Dom} K \cap \mathcal{H}(\Omega)^\perp$ where $\mathcal{H}(\Omega)^\perp := L^2(\Omega) \ominus \mathcal{H}(\Omega)$. Then, L is self-adjoint in the Hilbert space $\mathcal{H}(\Omega)^\perp$.

PROPOSITION 2.2. ([**25**], [**4**, Theorem 5.1]) *The spectrum of L is purely discrete and positive, and, hence, L^{-1} is a compact operator on $\mathcal{H}(\Omega)^\perp$. As a consequence, $\sigma_{\rm ess}(K) = \{0\}$, and the zero is an isolated eigenvalue of K of infinite multiplicity.*

The Krein Laplacian K arises naturally in the so called *abstract buckling problem*

$$\begin{cases} \Delta^2 u = -\lambda \Delta u, \\ u_{|\partial\Omega} = \frac{\partial u}{\partial \nu}_{|\partial\Omega} = 0, \\ u \in \operatorname{Dom} \Delta_{\max}. \end{cases}$$

Then the admissible values of $\lambda \neq 0$ for which the buckling problem has a non-trivial solution, coincide with the non-zero eigenvalues of K (see e.g. [**5**, **19**]).

Let $V \in C(\overline{\Omega}; \mathbb{R})$. Then the operator $K+V$ with domain $\mathrm{Dom}\, K$ is self-adjoint in $L^2(\Omega)$. In this article, we investigate the spectral properties of $K+V$.

Remark: The perturbations K_V of the Krein Laplacian K discussed in [**6**] are of different nature than the perturbations $K+V$ considered here. Namely, the authors of [**6**] assume that $V \geq 0$, define the maximal operator $K_{V,\max}$ as

$$K_{V,\max} := -\Delta + V, \quad \mathrm{Dom}\, K_{V,\max} := \mathrm{Dom}\, \Delta_{\max},$$

and set

$$K_V := -\Delta + V, \quad \mathrm{Dom}\, K_V := \mathrm{Ker}\, K_{V,\max} \dotplus H_0^2(\Omega).$$

Thus, if $V \neq 0$, then the operators K_V and $K_0 = K$ are self-adjoint on different domains, while the operators $K+V$ introduced here are self-adjoint on the same domain $\mathrm{Dom}\, K$. It is shown in [**6**] that for any $0 \leq V \in L^\infty(\Omega)$ we have $K_V \geq 0$, $\sigma_{\rm ess}(K_V) = \{0\}$, and the zero is an isolated eigenvalue of K_V of infinite multiplicity. As we will see in what follows, the spectral properties of $K+V$ could be quite different.

In Theorem 2.4 below we locate the essential spectrum of the operator $K+V$. For its proof we need some additional notations and definitions. Let $P : L^2(\Omega) \to L^2(\Omega)$ be the orthogonal projection onto $\mathcal{H}(\Omega)$. Assume that $V \in C(\overline{\Omega})$, and introduce *the harmonic Toeplitz operator* $T_V := PV : \mathcal{H}(\Omega) \to \mathcal{H}(\Omega)$. Evidently, T_V is bounded, and if V is real-valued, then T_V is self-adjoint. Note that T_V could be well defined as a bounded and even compact operator for a much wider class of symbols V which are locally integrable in Ω and satisfy certain regularity properties near $\partial\Omega$.

We start our analysis with the location of the essential spectrum $\sigma_{\rm ess}(T_V)$ of T_V, and a criterion for the compactness of T_V in the case $V \in C(\overline{\Omega})$.

PROPOSITION 2.3. *Let $\Omega \subset \mathbb{R}^d$, $d \geq 2$, be a bounded domain with boundary $\partial\Omega \in C^\infty$. Let $V \in C(\overline{\Omega})$.*
(i) [**14**, Theorem 4.5] *We have $\sigma_{\rm ess}(T_V) = V(\partial\Omega)$.*
(ii) [**14**, Corollary 4.7] *The operator T_V is compact in $\mathcal{H}(\Omega)$ if and only if $V = 0$ on $\partial\Omega$.*

Now we are in position to locate $\sigma_{\rm ess}(K+V)$.

THEOREM 2.4. *Let $V \in C(\overline{\Omega}; \mathbb{R})$. Then we have*

$$\sigma_{\rm ess}(K+V) = V(\partial\Omega). \tag{2.4}$$

In particular, $\sigma_{\rm ess}(K+V) = \{0\}$ if and only if $V_{|\partial\Omega} = 0$.

PROOF. First, we will show that the operator

$$(K+V-i)^{-1} - (K+PVP-i)^{-1} \tag{2.5}$$

is compact. Set $Q := I - P$. Then

$$(K+V-i)^{-1} - (K+PVP-i)^{-1} = \tag{2.6}$$
$$-(K+V-i)^{-1}(K-i)(K-i)^{-1}(QV+PVQ)(K-i)^{-1}(K-i)(K+PVP-i)^{-1}.$$

Evidently, the operators $(K+V-i)^{-1}(K-i)$, $(K-i)(K+PVP-i)^{-1}$, P, and V, are bounded in $L^2(\Omega)$. Moreover, using the orthogonal decomposition $L^2(\Omega) = \mathcal{H}(\Omega) \oplus \mathcal{H}(\Omega)^\perp$, and bearing in mind Proposition 2.2, we find that the operators $Q(K-i)^{-1}$ and $(K-i)^{-1}Q$ are compact in $L^2(\Omega)$. Now the compactness of the operator defined in (2.5) follows from (2.6). Therefore,

$$\sigma_{\text{ess}}(K+V) = \sigma_{\text{ess}}(K+PVP). \tag{2.7}$$

Further, we have $K+PVP = T_V \oplus L$ in $L^2(\Omega) = \mathcal{H}(\Omega) \oplus \mathcal{H}(\Omega)^\perp$, and, hence,

$$\sigma_{\text{ess}}(K+PVP) = \sigma_{\text{ess}}(T_V) \cup \sigma_{\text{ess}}(L). \tag{2.8}$$

By Proposition 2.3 (i), we have $\sigma_{\text{ess}}(T_V) = V(\partial\Omega)$, and by Proposition 2.2, $\sigma_{\text{ess}}(L) = \emptyset$. Thus, (2.7) and (2.8) imply (2.4). □

In the rest of the section we assume that $0 \leq V \in C(\overline{\Omega})$ with $V_{|\partial\Omega} = 0$, and investigate the asymptotic distribution of the discrete spectrum of the operators $K \pm V$, adjoining the origin. In particular, we show that the harmonic Toeplitz operator T_V is the effective Hamiltonian governing the main asymptotic term of the corresponding eigenvalue counting functions (see (2.11) – (2.12) below).

Let $\lambda_0 := \inf \sigma(L)$. By Proposition 2.2, we have $\lambda_0 > 0$. Introduce the eigenvalue counting functions

$$\mathcal{N}_-(\lambda) := \operatorname{Tr} \mathbb{1}_{(-\infty,-\lambda)}(K-V), \quad \lambda > 0,$$
$$\mathcal{N}_+(\lambda) := \operatorname{Tr} \mathbb{1}_{(\lambda,\lambda_0)}(K+V), \quad \lambda \in (0,\lambda_0).$$

Here and in the sequel $\mathbb{1}_S$ denotes the characteristic function of the set S; thus $\mathbb{1}_{\mathcal{I}}(T)$ is the spectral projection of the operator $T = T^*$ corresponding to the interval $\mathcal{I} \subset \mathbb{R}$. Let $T = T^*$ be a compact operator in a Hilbert space, and $s > 0$. Set

$$n_\pm(s;T) := \operatorname{Tr} \mathbb{1}_{(s,\infty)}(\pm T). \tag{2.9}$$

Thus, $n_+(s;T)$ (resp., $n_-(s;T)$) is just the number of the eigenvalues of T larger than s (resp., smaller than $-s$), counted with their multiplicities.

If $T_j = T_j^*$, $j = 1,2$, are two compact operators, then the Weyl inequalities

$$n_\pm(s_1+s_2; T_1+T_2) \leq n_\pm(s_1;T_1) + n_\pm(s_2;T_2) \tag{2.10}$$

hold for $s_j > 0$, $j = 1,2$, (see e.g. [**10**, Theorem 9, Section 9.2]).

THEOREM 2.5. *Assume that* $0 \leq V \in C(\overline{\Omega})$ *and* $V_{|\partial\Omega} = 0$. *Then for any* $\varepsilon \in (0,1)$ *we have*

$$n_+(\lambda;T_V) \leq \mathcal{N}_-(\lambda) \leq n_+((1-\varepsilon)\lambda;T_V) + O(1), \tag{2.11}$$

$$n_+((1+\varepsilon)\lambda;T_V) + O(1) \leq \mathcal{N}_+(\lambda) \leq n_+((1-\varepsilon)\lambda;T_V) + O(1), \tag{2.12}$$

as $\lambda \downarrow 0$.

PROOF. By the Birman-Schwinger principle [**7**, Lemma 1.1], we have

$$\mathcal{N}_-(\lambda) = n_+(1;(K+\lambda)^{-1/2}V(K+\lambda)^{-1/2}) = n_+(1;V^{1/2}(K+\lambda)^{-1}V^{1/2}), \ \lambda > 0. \tag{2.13}$$

It follows from the mini-max principle that

$$n_+(1;(K+\lambda)^{-1/2}V(K+\lambda)^{-1/2}) \geq$$
$$n_+(1;P(K+\lambda)^{-1/2}V(K+\lambda)^{-1/2}P) = n_+(\lambda;PVP) = n_+(\lambda;T_V),$$

which, combined with the first equality in (2.13), implies the lower bound in (2.11). Further, by the Weyl inequalities (2.10), we have

$$n_+(1; V^{1/2}(K+\lambda)^{-1}V^{1/2}) \leq$$

$$n_+((1-\varepsilon)\lambda; V^{1/2}PV^{1/2}) + n_+(\varepsilon; V^{1/2}Q(K+\lambda)^{-1}V^{1/2}),\ \lambda > 0,\ \varepsilon \in (0,1), \tag{2.14}$$

where, as above, $Q = I - P$. Evidently,

$$n_+(s; V^{1/2}PV^{1/2}) = n_+(s; PVP) = n_+(s; T_V), \quad s > 0, \tag{2.15}$$

while Proposition 2.2 easily implies that for any $\varepsilon > 0$ we have

$$n_\pm(\varepsilon; V^{1/2}Q(K+\lambda)^{-1}V^{1/2}) = O(1), \quad \lambda \to 0. \tag{2.16}$$

Putting together (2.13) and (2.14) – (2.16), we obtain the upper bound in (2.11). In order to prove (2.12), we recall that the generalized Birman-Schwinger principle (see e.g. [**2**, Theorem 1.3]), easily entails

$$\mathcal{N}_+(\lambda) = n_-(1; V^{1/2}(K-\lambda)^{-1}V^{1/2}) + O(1), \quad \lambda \downarrow 0. \tag{2.17}$$

By the Weyl inequalities, the estimates

$$n_+((1+\varepsilon)\lambda; V^{1/2}PV^{1/2}) - n_+(\varepsilon; V^{1/2}Q(K-\lambda)^{-1}V^{1/2}) \leq$$

$$n_-(1; V^{1/2}(K-\lambda)^{-1}V^{1/2}) \leq \tag{2.18}$$

$$n_+((1-\varepsilon)\lambda; V^{1/2}PV^{1/2}) + n_-(\varepsilon; V^{1/2}Q(K-\lambda)^{-1}V^{1/2})$$

hold true for every $\varepsilon \in (0,1)$. Now (2.17), (2.18), (2.15), and (2.16), imply (2.12). □

3. General properties of harmonic Toeplitz operators

In this section we establish sufficient conditions which guarantee $T_V \in S_p$, the pth Schatten-von Neumann class, or $T_V \in S_{p,\mathrm{w}}$, the weak counterpart of S_p.

We first introduce the notations we need. Let X and Y be separable Hilbert spaces. We denote by $\mathcal{L}(X,Y)$ (resp., $S_\infty(X,Y)$) the class of linear bounded (resp., compact) operators $T : X \to Y$. Next, $S_p(X,Y)$, $p \in (0,\infty)$, is the class of compact operators $T : X \to Y$ for which the functional

$$\|T\|_p := \left(\mathrm{Tr}\ (T^*T)^{p/2}\right)^{1/p}$$

is finite.

Let $T \in S_\infty(X,Y)$. Pick $s > 0$, and bearing in mind notation (2.9), set

$$n_*(s;T) := n_+(s^2; T^*T). \tag{3.1}$$

Thus, $n_*(s;T)$ is the number of the singular values of the operator T, larger than s, and counted with their multiplicities. Evidently, if $T = T^* \in S_\infty(X,X)$, then

$$n_\pm(s;T) \leq n_*(s;T), \quad s > 0. \tag{3.2}$$

Further, $S_{p,\mathrm{w}}(X,Y)$, $p \in (0,\infty)$, is the class of operators $T \in S_\infty(X,Y)$ for which the functional

$$\|T\|_{p,\mathrm{w}} := \sup_{s>0} s\, n_*(s;T)^{1/p} \tag{3.3}$$

is finite.

If $X = Y$, we write $\mathcal{L}(X)$, $S_p(X)$, and $S_{p,\mathrm{w}}(X)$, instead of $\mathcal{L}(X,X)$, $S_p(X,X)$, and $S_{p,\mathrm{w}}(X,X)$, respectively. Moreover, whenever appropriate, we omit X and Y in the notations $\mathcal{L}$, S_p, and $S_{p,\mathrm{w}}$.

Let us now turn to the study of the spectral properties of the harmonic Toeplitz operators $T_V = PV$. It is well known that the projection P onto $\mathcal{H}(\Omega)$ (see (2.2)) admits an integral kernel $\mathcal{R} \in C^\infty(\Omega \times \Omega)$, called *the reproducing kernel* of P (see e.g. [**14**, **24**]). Thus

$$(Pu)(x) = \int_\Omega \mathcal{R}(x,y)u(y)dy, \quad x \in \Omega, \quad u \in L^2(\Omega).$$

Let $\{\varphi_j\}_{j\in\mathbb{N}}$ be an orthogonal basis in $\mathcal{H}(\Omega)$. Then

$$\mathcal{R}(x,y) = \sum_{j\in\mathbb{N}} \varphi_j(x)\overline{\varphi_j(y)}, \quad x,y \in \Omega, \tag{3.4}$$

the series being locally uniformly convergent in $\Omega \times \Omega$. Evidently, $\mathcal{R}(x,y)$ is independent of the choice of the basis $\{\varphi_j\}_{j\in\mathbb{N}}$. Moreover, the kernel $\mathcal{R}$ is real-valued and symmetric. For $x \in \Omega$ put

$$\varrho(x) := \mathcal{R}(x,x).$$

Then, (3.4) implies that

$$|\mathcal{R}(x,y)| \le \varrho(x)^{1/2}\,\varrho(y)^{1/2}, \quad x,y \in \Omega.$$

For $x, y \in \Omega$, set

$$r(x) := \operatorname{dist}(x, \partial\Omega), \quad \delta(x,y) := |x-y| + r(x) + r(y). \tag{3.5}$$

LEMMA 3.1. [**24**, Theorem 1.1] *For any multiindices $\alpha, \beta \in \mathbb{Z}_+^d$ there exists a constant $C_{\alpha,\beta} \in (0,\infty)$ such that*

$$\left|D_x^\alpha D_y^\beta \mathcal{R}(x,y)\right| \le \frac{C_{\alpha,\beta}}{\delta(x,y)^{d+|\alpha|+|\beta|}}, \quad x,y \in \Omega. \tag{3.6}$$

Moreover, there exists a constant $C \in (0,\infty)$ such that

$$\varrho(x) \ge Cr(x)^{-d}, \quad x \in \Omega. \tag{3.7}$$

For a Borel set $\mathcal{A} \subset \Omega$ set $\rho(\mathcal{A}) := \int_{\mathcal{A}} \varrho(x)dx$. By (3.6) with $\alpha = \beta = 0$, and (3.7), ρ is an infinite σ-finite measure on Ω which is absolutely continuous with respect to the Lebesgue measure.

Our next goal is to establish conditions which guarantee $T_V \in S_p(\mathcal{H}(\Omega))$, $p \in [1,\infty)$, or $T_V \in S_{p,\mathrm{w}}(\mathcal{H}(\Omega))$, $p \in (1,\infty)$. For $p \in (0,\infty)$ define $L^p_{\mathrm{w}}(\Omega; d\rho)$ as the class of ρ-measurable functions $u : \Omega \to \mathbb{C}$ for which the quasinorm

$$\|u\|_{L^p_{\mathrm{w}}(\Omega;d\rho)} := \sup_{t>0} t\rho\left(\{x \in \Omega \mid |u(x)| > t\}\right)^{1/p}$$

is finite.

PROPOSITION 3.2. *Let $\Omega \subset \mathbb{R}^d$, $d \ge 2$ be a bounded domain with boundary $\partial\Omega \in C^\infty$.*

(i) *Assume $V \in L^p(\Omega; d\rho)$, $p \in [1,\infty)$. Then $T_V \in S_p(\mathcal{H}(\Omega))$ and*

$$\|T_V\|_p \le \|V\|_{L^p(\Omega;d\rho)}. \tag{3.8}$$

(ii) *Assume $V \in L^p_{\mathrm{w}}(\Omega; d\rho)$, $p \in (1,\infty)$. Then $T_V \in S_{p,\mathrm{w}}(\mathcal{H}(\Omega))$ and*

$$\|T_V\|_{p,\mathrm{w}} \le \|V\|_{L^p_{\mathrm{w}}(\Omega;d\rho)}. \tag{3.9}$$

PROOF. Let us consider the operator PVP as defined on $L^2(\Omega)$. Evidently,

$$\|T_V\|_p = \|PVP\|_p, \quad \|T_V\|_{p,\mathrm{w}} = \|PVP\|_{p,\mathrm{w}}, \quad p \in (0,\infty). \tag{3.10}$$

We have $PVP = F^* e^{i\arg V} F$ where $F : L^2(\Omega) \to L^2(\Omega)$ is the operator with integral kernel

$$|V(x)|^{1/2}\mathcal{R}(x,y), \quad x,y \in \Omega.$$

Assume $V \in L^1(\Omega; d\rho)$. Then

$$\|PVP\|_1 \le \|F^*\|_2 \|e^{i\arg V}\| \|F\|_2 = \|F\|_2^2 = \|V\|_{L^1(\Omega;d\rho)}. \tag{3.11}$$

Assume now $V \in L^\infty(\Omega; d\rho)$. Since $\|P\| = 1$ and $d\rho$ is absolutely continuous with respect to the Lebesgue measure,

$$\|PVP\| \le \|V\|_{L^\infty(\Omega)} = \|V\|_{L^\infty(\Omega;d\rho)}. \tag{3.12}$$

Interpolating between (3.11) and (3.12), and applying [**8**, Theorem 3.1], we find that

$$\|PVP\|_p \le \|V\|_{L^p(\Omega;d\rho)}, \quad p \in [1,\infty),$$
$$\|PVP\|_{p,\mathrm{w}} \le \|V\|_{L^p_{\mathrm{w}}(\Omega;d\rho)}, \quad p \in (1,\infty),$$

which combined with (3.10), implies (3.8) and (3.9). □

Remark: Let $\mu \ge 0$ be a finite Borel measure on Ω. In this case, the harmonic Toeplitz operator T_μ is defined by

$$(T_\mu u)(x) := \int_\Omega \mathcal{R}(x,y)u(y)d\mu(y), \quad u \in \mathcal{H}(\Omega), \quad x \in \Omega.$$

If $d\mu(x) = V(x)dx$ with $0 \le V \in L^1(\Omega)$, then, of course, $T_\mu = T_V$. Criteria on μ which guarantee the boundedness of T_μ, the compactness of T_μ, or the inclusion $T_\mu \in S_p(\mathcal{H}(\Omega))$, $p \in [1,\infty)$, are contained in [**14**]. These criteria are formulated in terms of the Berezin transform $\tilde{\mu}$ of the measure μ, defined by

$$\tilde{\mu}(x) := \varrho(x)^{-1} \int_\Omega \mathcal{R}(x,y)^2 d\mu(y), \quad x \in \Omega.$$

The combination of Theorem 2.5 with Proposition 3.2 entails the following

COROLLARY 3.3. *Let Ω be a bounded domain with boundary $\partial\Omega \in C^\infty$. Assume that $V \in C(\overline{\Omega};\mathbb{R}) \cap L^p_{\mathrm{w}}(\Omega; d\rho)$, $p \in (1,\infty)$. Then for any $\varepsilon \in (0,1)$ we have*

$$\mathcal{N}_\pm(\lambda) \le (1-\varepsilon)^{-p}\lambda^{-p}\|V\|^p_{L^p_{\mathrm{w}}(\Omega;d\rho)} + O(1), \tag{3.13}$$

for sufficiently small $\lambda > 0$.

PROOF. Estimate (3.13) follows immediately from (2.11) - (2.12), combined with (3.2) and (3.3). □

At the end of this section we show that if the symbol V is compactly supported in Ω, then $T_V \in S_p$ for any $p \in (0,\infty)$, even if the behavior of V is quite irregular. In fact, we will replace in this case V by $\phi \in \mathcal{E}'(\Omega)$, the class of distributions over $\mathcal{E}(\Omega) := C^\infty(\Omega)$. We recall that $\phi \in \mathcal{D}'(\Omega)$ is in $\mathcal{E}'(\Omega)$, if and only if $\operatorname{supp}\phi$ is compact in Ω. If $\phi \in \mathcal{E}'(\Omega)$, we define $T_\phi : \mathcal{H}(\Omega) \to \mathcal{H}(\Omega)$ as the operator with integral kernel

$$K_\phi(x,y) := (\phi, \mathcal{R}(x,\cdot)\mathcal{R}(\cdot,y)), \quad x,y \in \Omega,$$

where $(\cdot,\cdot)$ denotes the pairing between $\mathcal{E}'(\Omega)$ and $\mathcal{E}(\Omega)$. Of course, if $\phi = \mu$ and $\mu \ge 0$ is a finite Borel measure such that $\operatorname{supp}\mu$ is compact in Ω, then $T_\phi = T_\mu$.

PROPOSITION 3.4. *Let $\Omega \subset \mathbb{R}^d$, $d \geq 2$, be a bounded domain with boundary $\partial\Omega \in C^\infty$. Assume that $\phi \in \mathcal{E}'(\Omega)$. Then we have $T_\phi \in S_p(\mathcal{H}(\Omega))$ for any $p \in (0, \infty)$, and, hence,*

$$n_*(\lambda; T_\phi) = O(\lambda^{-\alpha}), \quad \lambda \downarrow 0, \tag{3.14}$$

for any $\alpha \in (0, \infty)$.

PROOF. Since $\operatorname{supp} \phi$ is compact in Ω, we have $K_\phi \in C^\infty(\overline{\Omega} \times \overline{\Omega})$. Therefore, (3.14) follows immediately from, say, [**8**, Proposition 2.1]. □

Remarks: (i) In Section 5 we will show that if Ω is the unit ball in $\mathbb{R}^d$, and $V \geq 0$ is compactly supported, and possesses a partial radial symmetry, then the eigenvalues of T_V decay exponentially fast.
(ii) Harmonic Toeplitz operators T_ϕ with $\phi \in \mathcal{E}'(\Omega)$ were considered in [**3**] where, in particular, it was proved that $\operatorname{rank} T_\phi < \infty$, if and only if $\operatorname{supp} \phi$ is finite.

4. Spectral asymptotics of T_V for V of power-like decay at $\partial\Omega$

4.1. Statement of the main results. In this section we assume that V is smooth and positive near the boundary, and admits a power-like decay at $\partial\Omega$, and investigate the asymptotic behavior of the discrete spectrum of T_V near the origin. We obtain the main asymptotic term of $n_+(\lambda; T_V)$ as $\lambda \downarrow 0$, and give a sharp estimate of the remainder (see Theorems 4.1 and 4.2 below).

In what follows we consider $\partial\Omega$ as a compact $(d-1)$-dimensional Riemannian manifold with metric tensor $g(y) := \{g_{jk}(y)\}_{j,k=1}^{d-1}$, $y \in \partial\Omega$, generated by the Euclidean metrics in $\mathbb{R}^d$, and denote by dS the measure induced by g on $\partial\Omega$.

Let $a, \tau \in C^\infty(\overline{\Omega})$ satisfy $a > 0$ on $\overline{\Omega}$, $\tau > 0$ on Ω, and $\tau = r := \operatorname{dist}(\cdot, \partial\Omega)$ (see (3.5)) in a neighborhood of $\partial\Omega$. Assume that

$$V(x) = \tau(x)^\gamma a(x), \quad \gamma \geq 0, \quad x \in \Omega. \tag{4.1}$$

Set $a_0 := a_{|\partial\Omega}$.

THEOREM 4.1. *Assume that V satisfies (4.1) with $\gamma > 0$. Then we have*

$$n_+(\lambda; T_V) = \mathcal{C}\, \lambda^{-\frac{d-1}{\gamma}} \left(1 + O(\lambda^{\frac{1}{\gamma}})\right), \quad \lambda \downarrow 0, \tag{4.2}$$

where

$$\mathcal{C} := \omega_{d-1} \left(\frac{\Gamma(\gamma+1)^{\frac{1}{\gamma}}}{4\pi} \right)^{d-1} \int_{\partial\Omega} a_0(y)^{\frac{d-1}{\gamma}} \, dS(y), \tag{4.3}$$

and $\omega_n = \pi^{n/2}/\Gamma(1+n/2)$ is the Lebesgue measure of the unit ball in $\mathbb{R}^n$, $n \geq 1$.

The proof of Theorem 4.1 can be found in Subsection 4.2.

The assumption of Theorem 4.1 that V satisfies (4.1) *in the whole domain* Ω is rather restrictive. In the following theorem we consider Toeplitz operators with symbols which satisfy (4.1) with $\gamma > 0$ only in a neighborhood of $\partial\Omega$ while they can have quite an irregular behavior on a compact subset of Ω; in particular, away from the boundary, these symbols are not obliged to be smooth and positive.

THEOREM 4.2. *Let V satisfy the assumptions of Theorem 4.1, and $\phi \in \mathcal{E}'(\Omega;\mathbb{R})$. Then we have*

$$n_+(\lambda; T_{V+\phi}) = \mathcal{C}\,\lambda^{-\frac{d-1}{\gamma}}\left(1 + O(\lambda^{\frac{\varepsilon}{\gamma}})\right), \quad \lambda \downarrow 0, \tag{4.4}$$

where $T_{V+\phi} := T_V + T_\phi$, $\mathcal{C}$ is the constant defined in (4.3), *$\varepsilon = 1$ if $d \geq 3$, and $\varepsilon < 1$ is arbitrary if $d = 2$.*

The proof of Theorem 4.2 is contained in Subsection 4.3.

Remark: Let $d \geq 3$. Then Theorem 4.2 implies that (4.2) remains true if we replace T_V by $T_{V+\phi}$ with $\phi \in \mathcal{E}'(\Omega;\mathbb{R})$. In particular, (4.2) is valid if $V \in L^1_{\rm loc}(\Omega;\mathbb{R})$ satisfies (4.1) with $\gamma > 0$ only in a neighborhood of $\partial\Omega$. If $d = 2$, the prize we have to pay for the substitution of T_V by $T_{V+\phi}$ is that the remainder estimate in (4.2) is better than in (4.4). However, as mentioned in the remark at the end of Subsection 4.3, if $d = 2$ and the distribution $\phi \in \mathcal{E}'(\Omega)$ is non-negative, then (4.4) holds true also with $\varepsilon = 1$.

Combining Theorems 2.4 and 4.2, we obtain the following

COROLLARY 4.3. *Assume that $0 \leq V \in C(\overline{\Omega})$, and* (4.1) *with $\gamma > 0$ holds true in a neighborhood of $\partial\Omega$. Then we have*

$$\lim_{\lambda \downarrow 0} \lambda^{\frac{d-1}{\gamma}} \mathcal{N}_\pm(\lambda) = \mathcal{C}, \tag{4.5}$$

$\mathcal{C}$ being the constant defined in (4.3).

Remark: Assume the hypotheses of Corollary 4.3. Then, recalling (3.6) with $\alpha = \beta = 0$, and (3.7), we find that $V \in L^p_{\rm w}(\Omega; d\rho)$ if and only if $p = \frac{d-1}{\gamma}$. Thus, if $\gamma < d - 1$, then (4.5) implies that the order of our estimate (3.13) is sharp.

4.2. Proof of Theorem 4.1. Assume that $f \in H^s(\partial\Omega)$, $s \in \mathbb{R}$. Then the boundary-value problem

$$\begin{cases} \Delta u = 0 & \text{in} \quad \Omega, \\ u = f & \text{on} \quad \partial\Omega, \end{cases} \tag{4.6}$$

admits a unique solution $u \in H^{s+1/2}(\Omega)$, we have

$$\|u\|_{H^{s+1/2}(\Omega)} \asymp \|f\|_{H^s(\partial\Omega)}, \tag{4.7}$$

and, therefore, the mapping $f \mapsto u$ defines an isomorphism between $H^s(\partial\Omega)$ and $H^{s+1/2}(\Omega)$ (see [**26**, Sections 5, 6, 7, Chapter 2]).

If $s = 0$, we set

$$u = Gf. \tag{4.8}$$

By (4.7) with $s = 0$, and the compactness of the embedding of $H^{1/2}(\Omega)$ into $L^2(\Omega)$, we find that the operator $G : L^2(\partial\Omega) \to L^2(\Omega)$ is compact. By [**16**, Theorem 12, Section 2.2], we have

$$u(x) = \int_{\partial\Omega} \mathcal{K}(x,y) f(y) dS(y), \quad x \in \Omega, \tag{4.9}$$

where

$$\mathcal{K}(x,y) := -\frac{\partial \mathcal{G}}{\partial \nu_y}(x,y), \quad x \in \Omega, \quad y \in \partial\Omega, \tag{4.10}$$

$\mathcal{G}$ is the Dirichlet Green function associated with Ω. Note that

$$\mathcal{K} \in C^\infty(\Omega \times \partial\Omega). \tag{4.11}$$

LEMMA 4.4. *We have*

$$\operatorname{Ker} G = \{0\}, \tag{4.12}$$

$$\overline{\operatorname{Ran} G} = \mathcal{H}(\Omega). \tag{4.13}$$

PROOF. Relation (4.12) follows from (4.7) with $s = 0$. Let us check (4.13). Pick $u \in \mathcal{H}(\Omega)$. Then, by (4.6) with $s = -1/2$, we have $f := u_{|\partial\Omega} \in H^{-1/2}(\partial\Omega)$. Let $f_n \in L^2(\partial\Omega)$, $n \in \mathbb{N}$, and

$$\lim_{n\to\infty} \|f_n - f\|_{H^{-1/2}(\partial\Omega)} = 0. \tag{4.14}$$

Set $u_n := Gf_n$. Then $u_n \in \operatorname{Ran} G$, $n \in \mathbb{N}$, and by (4.7) with $s = -1/2$, and (4.14), we have $\lim_{n\to\infty} \|u_n - u\|_{L^2(\Omega)} = 0$ which implies (4.13). □

Set $J := G^*G$. Then the operator $J = J^* \geq 0$ is compact in $L^2(\partial\Omega)$. Due to (4.12), we have $\operatorname{Ker} J = \{0\}$. Let $\{\lambda_j\}_{j\in\mathbb{N}}$ be the non-increasing sequence of the eigenvalues $\lambda_j > 0$ of J, and let $\{\phi_j\}_{j\in\mathbb{N}}$ be the corresponding orthonormal eigenbasis in $L^2(\partial\Omega)$ with $J\phi_j = \lambda_j\phi_j$, $j \in \mathbb{N}$. Define the operator J^{-1}, self-adjoint in $L^2(\partial\Omega)$, by

(4.15)

$$J^{-1}u := \sum_{j\in\mathbb{N}} \lambda_j^{-1}\langle u, \phi_j\rangle\phi_j, \quad \operatorname{Dom} J^{-1} := \left\{ u \in L^2(\partial\Omega) \mid \sum_{j\in\mathbb{N}} \lambda_j^{-2} |\langle u, \phi_j\rangle|^2 < \infty \right\},$$

$\langle \cdot, \cdot \rangle$ being the scalar product in $L^2(\partial\Omega)$. Evidently, $\overline{JJ^{-1}} = J^{-1}J = I$.

Further, introduce the polar decomposition of the operator $G = U|G| = UJ^{1/2}$ where $U : L^2(\partial\Omega) \to L^2(\Omega)$ is an isometric operator. By Lemma 4.4, we have $\operatorname{Ker} U = \{0\}$ and $\operatorname{Ran} U = \mathcal{H}(\Omega)$. Thus, we obtain the following

PROPOSITION 4.5. *The orthogonal projection P onto $\mathcal{H}(\Omega)$ satisfies*

$$P = GJ^{-1}G^* = UU^*. \tag{4.16}$$

Assume that V satisfies (4.1) with $\gamma \geq 0$, and set $J_V := G^*VG$; from this point of view, we have $J = J_1$.

PROPOSITION 4.6. *Let V satisfy* (4.1) *with $\gamma \geq 0$. Then the operator T_V is unitarily equivalent to (the closure of) the operator $J^{-1/2}J_VJ^{-1/2}$.*

PROOF. By (4.16), we have

$$PVP = UJ^{-1/2}G^*VGJ^{-1/2}U^* = UJ^{-1/2}J_VJ^{-1/2}U^*,$$

and the operator U maps unitarily $L^2(\partial\Omega)$ onto $\mathcal{H}(\Omega)$. □

Let $y \in \partial\Omega$ and $\eta \in T_y^*\partial\Omega$. Set

$$|\eta| = |\eta|_y := \left(\sum_{j,k=1}^{d-1} g^{jk}(y)\eta_j\eta_k\right)^{1/2},$$

where $\{g^{jk}(y)\}_{j,k=1}^{d-1}$ is the matrix inverse to $g(y)$.

PROPOSITION 4.7. *Under the assumptions of Proposition* 4.6, *the operator* $J^{-1/2}J_V J^{-1/2}$ *is a* ΨDO *with principal symbol*

$$2^{-\gamma}\Gamma(\gamma+1)|\eta|^{-\gamma}a_0(y), \quad (y,\eta) \in T^*\partial\Omega. \tag{4.17}$$

PROOF. Using the pseudo-differential calculus due to L. Boutet de Monvel (see [**11**, **12**]), M. Engliš showed recently in [**15**, Sections 6, 7] that if V satisfies (4.1) with $\gamma \geq 0$, then the operator J_V is a ΨDO with principal symbol

$$2^{-\gamma-1}\Gamma(\gamma+1)|\eta|^{-\gamma-1}a_0(y), \quad (y,\eta) \in T^*\partial\Omega.$$

In particular, $J = J_1$ is a ΨDO with principal symbol $2^{-1}|\eta|^{-1}$. Then the pseudo-differential calculus (see e.g. [**29**, Chapters I, II]) easily implies that $J^{-1/2}$ is a ΨDO with principal symbol $2^{1/2}|\eta|^{1/2}$, and $J^{-1/2}J_V J^{-1/2}$ is a ΨDO with principal symbol defined in (4.17). □

Now we are in position to prove Theorem 4.1. It is easy to see that under its assumptions we have $\operatorname{Ker} J^{-1/2}J_V J^{-1/2} = \{0\}$. Using the spectral theorem, define the operator

$$A := \left(J^{-1/2}J_V J^{-1/2}\right)^{-1/\gamma}$$

(cf. (4.15)). Then, by the pseudo-differential calculus, A is a ΨDO with principal symbol

$$2\Gamma(\gamma+1)^{-1/\gamma}|\eta|a_0(y)^{-1/\gamma}, \quad (y,\eta) \in T^*\partial\Omega.$$

By Proposition 4.6 and the spectral theorem, we have

$$n_+(\lambda; T_V) = n_+(\lambda; J^{-1/2}J_V J^{-1/2}) = \operatorname{Tr} \mathbb{1}_{(-\infty,\lambda^{-1/\gamma})}(A), \quad \lambda > 0. \tag{4.18}$$

A classical result of L. Hörmander [**22**] easily implies that

$$\operatorname{Tr} \mathbb{1}_{(-\infty,E)}(A) =$$

$$(2\pi)^{-d+1}\left|\left\{(y,\eta) \in T^*\partial\Omega \,\middle|\, 2\Gamma(\gamma+1)^{-1/\gamma}|\eta|a_0(y)^{-1/\gamma} < E\right\}\right| + O(E^{-(d-2)}) =$$

$$\mathcal{C}E^{d-1}(1+O(E^{-1})), \quad E \to \infty, \tag{4.19}$$

where $|\cdot|$ is the Lebesgue measure on $T^*\partial\Omega$, and $\mathcal{C}$ is the constant defined in (4.3). Combining (4.18) and (4.19), we arrive at (4.2).

Remark: The natural idea to parametrize the functions $u \in \mathcal{H}(\Omega)$ by their restrictions on $\partial\Omega$ has been used in the theory of harmonic Toeplitz operators and related areas by various authors; it could be traced back at least to the classical work [**11**], and has been recently applied in [**15**] in order to obtain a suitable representation of the operator J_V. We would like also to mention here the article [**9**] where the authors consider the operator generated by the ratio of two quadratic differential forms defined on the solutions of a homogeneous elliptic equation. The order of the numerator is lower than the order of the denominator, and, since the

domain considered is supposed to be bounded and to have a regular boundary, the operator generated by the ratio is compact.
The harmonic Toeplitz operator T_V could be interpreted as the operator generated by the quadratic-form ratio

$$\frac{\int_\Omega V|u|^2\,dx}{\int_\Omega |u|^2\,dx}, \quad u \in \mathcal{H}(\Omega). \tag{4.20}$$

Note that both the numerator and the denominator in (4.20) are of zeroth order, and the compactness of T_V is now due to the fact that V vanishes at $\partial\Omega$.
In spite of the differences between the operators considered in [**9**], and the harmonic Toeplitz operators studied here, the unitary equivalence of T_V and $J^{-1/2}J_VJ^{-1/2}$ established in our Proposition 4.6 has much in common with the reduction to a ΨDO on $\partial\Omega$, performed in [**9**].

4.3. Proof of Theorem 4.2. The Weyl inequalities (2.10) imply

$$\begin{gathered} n_+(\lambda(1+\lambda^\theta);T_V) - n_-(\lambda^{1+\theta};T_\phi) \le \\ n_+(\lambda;T_{V+\phi}) \le \\ n_+(\lambda(1-\lambda^\theta);T_V) + n_+(\lambda^{1+\theta};T_\phi), \end{gathered} \tag{4.21}$$

for $\lambda \in (0,1)$ and $\theta > 0$. By (4.2),

$$\begin{gathered} n_+(\lambda(1\pm\lambda^\theta);T_V) = \\ \mathcal{C}\left(\lambda(1\pm\lambda^\theta)\right)^{-\frac{d-1}{\gamma}} + O\left(\lambda^{-\frac{d-2}{\gamma}}\right) = \mathcal{C}\lambda^{-\frac{d-1}{\gamma}} + O\left(\lambda^{-\frac{d-2}{\gamma}}\right), \quad \lambda \in (0,1), \end{gathered} \tag{4.22}$$

provided that $\theta > 1/\gamma$. Next, by estimate (3.14), we have

$$n_\pm(\lambda^{1+\theta};T_\phi) = O(\lambda^{-\alpha(1+\theta)}), \lambda > 0, \tag{4.23}$$

for any $\alpha \in (0,\infty)$. Assume $d \ge 3$ and choose $\alpha \in \left(0, \frac{d-2}{\gamma(1+\theta)}\right)$. Then (4.4) follows from (4.21) - (4.23). If $d = 2$, then we can pick any $\varepsilon \in (0,1)$ and choose $\alpha \in \left(0, \frac{1-\varepsilon}{\gamma(1+\theta)}\right)$, in order to check that in this case (4.21) – (4.23) again imply (4.4).

Remark: Arguing as in the proof of Theorem 4.1 (see Propositions 4.6 and 4.7), we can show that $T_{V+\phi}$ with $\phi \in \mathcal{E}'(\Omega;\mathbb{R})$ is unitarily equivalent to a self-adjoint ΨDO with principal symbol defined in (4.17). The only problem to extend in a straightforward manner our proof of Theorem 4.1 to $T_{V+\phi}$ is that this operator may have a non trivial kernel unless, for example, $\phi \ge 0$. In particular, if $d = 2$ and $\phi \in \mathcal{E}'(\Omega;\mathbb{R})$ satisfies $\phi \ge 0$, then (4.4) holds also for $\varepsilon = 1$. Finally, if we apply the results stated without proof in [**23**, Appendix H] (see in particular Theorem H.1 and the remark after it), we could simplify slightly the proofs of our Theorems 4.1 and 4.2, and could show that Theorem 4.2 holds true with $\varepsilon = 1$ for every $d \ge 2$ without any additional assumptions.

5. Spectral asymptotics of T_V for compactly supported V

In this section we consider the asymptotics of $n_+(\lambda; T_V)$ as $\lambda \downarrow 0$ in the case where V is compactly supported in Ω, i.e. when V vanishes identically in a neighborhood of $\partial\Omega$. In this case T_V admits an integral kernel which is in the class $C^\infty(\overline{\Omega} \times \overline{\Omega})$, and T_V can be considered as a ΨDO of order $-\infty$.

Set

$$B_R := \{x \in \mathbb{R}^d \,|\, |x| < R\}, \quad d \geq 2, \quad R \in (0, \infty).$$

Since we are still unable to handle the case of general bounded Ω and compactly supported V, we suppose that Ω is the unit ball B_1 in $\mathbb{R}^d$ while $\operatorname{supp} V$ coincides with B_c with $c \in (0,1)$. Using the known fact that if V is proportional to $\mathbb{1}_{B_c}$, then the eigenvalues of T_V can be calculated explicitly, we obtain the main asymptotic term of $n_+(\lambda; T_V)$ as $\lambda \downarrow 0$, for generic T_V such that $\operatorname{supp} V = B_c$.

Let $\Omega = B_1$. Thus, $\partial\Omega = \mathbb{S}^{d-1} := \{x \in \mathbb{R}^d \,|\, |x| = 1\}$. The space $\mathcal{H}(B_1)$ admits an explicit orthonormal eigenbasis which we are now going to describe. Recall that $k(k+d-2)$, $k \in \mathbb{Z}_+$, are the eigenvalues of the Beltrami-Laplace operator $-\Delta_{\mathbb{S}^{d-1}}$, self-adjoint in $L^2(\mathbb{S}^{d-1})$ (see e.g. [**29**, Section 22]). Moreover,

$$\dim \operatorname{Ker}\,(-\Delta_{\mathbb{S}^{d-1}} - k(k+d-2)I) =: m_k = \binom{d+k-1}{d-1} - \binom{d+k-3}{d-1}$$

where $\binom{m}{n} = \frac{m!}{(m-n)!\,n!}$ if $m \geq n$, and $\binom{m}{n} = 0$ if $m < n$ (see e.g. [**29**, Theorem 22.1]). Set

$$M_k := \binom{d+k-1}{d-1} + \binom{d+k-2}{d-1}, \quad k \in \mathbb{Z}.$$

Evidently,

$$M_k = \frac{2k^{d-1}}{(d-1)!}\left(1 + O\left(k^{-1}\right)\right), \quad k \to \infty, \tag{5.1}$$

(see e.g. [**1**, Eq. 6.1.47]). By induction, we easily find that

$$\sum_{j=0}^{k} m_j = M_k, \quad k \in \mathbb{Z}_+. \tag{5.2}$$

Let $\psi_{k,\ell}$, $\ell = 1, \ldots, m_k$, be an orthonormal basis in $\operatorname{Ker}\,(-\Delta_{\mathbb{S}^{d-1}} - k(k+d-2)I)$, $k \in \mathbb{Z}_+$. It is well known that $\psi_{k,\ell}$ are restrictions on $\mathbb{S}^{d-1}$ of homogeneous polynomials of degree k, harmonic in $\mathbb{R}^d$ (see e.g [**29**, Section 22]). Then the functions

$$\phi_{k,\ell}(x) := \sqrt{2k+d}\,|x|^k \psi_{k,\ell}(x/|x|), \quad x \in B_1, \quad \ell = 1, \ldots, m_k, \quad k \in \mathbb{Z}_+,$$

form an orthonormal basis in $\mathcal{H}(B_1)$. Let $\mathcal{H}_k(B_1)$, $k \in \mathbb{Z}_+$, be the subspace of $\mathcal{H}(B_1)$ generated by $\phi_{k,\ell}$, $\ell = 1, \ldots, m_k$.

Further, let $V(x) = v(|x|)$, $x \in B_1$, and let $v : [0,1) \to \mathbb{R}$ satisfy $\lim_{r\uparrow 1} v(r) = 0$, $v \in L^1((0,1); r^{d-1}dr)$. Then T_V is self-adjoint and compact in $\mathcal{H}(B_1)$, and

$$T_V u = \mu_k u, \quad u \in \mathcal{H}_k(B_1), \tag{5.3}$$

where

$$\mu_k(v) := (2k+d)\int_0^1 v(r) r^{2k+d-1} dr, \quad k \in \mathbb{Z}_+. \tag{5.4}$$

Set

$$\xi(s;v) = \#\{k \in \mathbb{Z}_+ \mid \mu_k(v) > s\}, \quad s > 0.$$

Let us calculate the eigenvalues of T_V in the simple model situation where $v(r) = b\,\mathbb{1}_{[0,c]}(r)$, $r \in [0,1)$, with $b > 0$ and $c \in (0,1)$. Then (5.4) implies

$$\mu_k(v) = b\,c^{2k+d}, \quad k \in \mathbb{Z}_+. \tag{5.5}$$

Evidently, the sequence $\{\mu_k(v)\}_{k\in\mathbb{Z}_+}$ is decreasing. Setting $V(x) := v(|x|)$, $x \in \mathbb{R}^d$, we get

$$n_+(\lambda; T_V) = M_{\xi(\lambda;v)-1}, \quad \lambda > 0. \tag{5.6}$$

Let us discuss the asymptotics of $n_+(\lambda; T_V)$ as $\lambda \downarrow 0$. By (5.5),

$$\xi(\lambda; v) = \frac{1}{2}\frac{|\ln\lambda|}{|\ln c|} + O(1), \quad \lambda \downarrow 0. \tag{5.7}$$

By (5.6), (5.1), and (5.7), we get

$$n_+(\lambda; T_V) = \frac{2^{-d+2}}{(d-1)!|\ln c|^{d-1}}|\ln\lambda|^{d-1} + O\left(|\ln\lambda|^{-d+2}\right), \quad \lambda \downarrow 0. \tag{5.8}$$

Remark: The fact that the basis $\{\phi_{k,\ell}\}$ diagonalizes the operator T_V with radially symmetric symbol V, acting in $\mathcal{H}(B_1)$, was noted in [**28**, Part 2.3.2], and was used there, in particular, to obtain asymptotic relations of type (5.8). The fact that the Toeplitz operators with radially symmetric symbols, acting in the holomorphic *Fock-Segal-Bargmann* space, are diagonalized in a certain canonic basis, was utilized already in [**21**, **27**]. A similar result concerning Toeplitz operators with radially symmetric symbols, acting in the holomorphic *Bergman* space, can be found in [**20**].

Next, we use (5.8) in order to study the spectral asymptotics for Toeplitz operators with symbols V which possess partial radial symmetry.

THEOREM 5.1. *Let $\Omega = B_1$. Assume that $V : B_1 \to [0,\infty)$ satisfies $V \in L^\infty(B_1)$ and* $\operatorname{supp} V = \overline{B_c}$ *for some $c \in (0,1)$. Suppose moreover that for any $\delta \in (0,c)$ we have* $\operatorname{ess\,inf}_{x\in B_\delta} V(x) > 0$. *Then*

$$\lim_{\lambda\downarrow 0} |\ln\lambda|^{-d+1}\, n_+(\lambda; T_V) = \frac{2^{-d+2}}{(d-1)!|\ln c|^{d-1}}. \tag{5.9}$$

PROOF. Pick $\delta \in (0,c)$. Then for almost every $x \in B_1$ we have

$$b_-\mathbb{1}_{B_\delta}(x) \le V(x) \le b_+\mathbb{1}_{B_c}(x),$$

where

$$b_- := \operatorname{ess\,inf}_{x\in B_\delta} V(x), \quad b_+ := \operatorname{ess\,sup}_{x\in B_1} V(x).$$

Then the mini-max principle and (5.8) imply

$$\frac{2^{-d+2}}{(d-1)!|\ln\delta|^{d-1}} \le$$

$$\liminf_{\lambda\downarrow 0} |\ln\lambda|^{-d+1}\, n_+(\lambda; T_V) \le \limsup_{\lambda\downarrow 0} |\ln\lambda|^{-d+1}\, n_+(\lambda; T_V) \le$$

$$\frac{2^{-d+2}}{(d-1)!|\ln c|^{d-1}}.$$

Letting $\delta \uparrow c$, we obtain (5.9). □

Remark: Hopefully, in a future work we will extend the result of Theorem 5.1 to more general domains Ω, and more general compactly supported V.

Putting together Theorems 2.5 and 5.1, we obtain the following

COROLLARY 5.2. *Let* $\Omega = B_1 \subset \mathbb{R}^d$, $d \geq 2$, $0 \leq V \in C(\overline{B_1})$. *Assume that* $\operatorname{supp} V = \overline{B_c}$ *for some* $c \in (0,1)$, *and that for any* $\delta \in (0,c)$ *we have* $\inf_{x \in B_\delta} V(x) > 0$. *Then*

$$\lim_{\lambda \downarrow 0} |\ln \lambda|^{-d+1} \mathcal{N}_\pm(\lambda) = \frac{2^{-d+2}}{(d-1)!|\ln c|^{d-1}}.$$

References

[1] Milton Abramowitz and Irene A. Stegun, *Handbook of mathematical functions with formulas, graphs, and mathematical tables*, National Bureau of Standards Applied Mathematics Series, vol. 55, For sale by the Superintendent of Documents, U.S. Government Printing Office, Washington, D.C., 1964. MR0167642

[2] Stanley Alama, Percy A. Deift, and Rainer Hempel, *Eigenvalue branches of the Schrödinger operator $H - \lambda W$ in a gap of $\sigma(H)$*, Comm. Math. Phys. **121** (1989), no. 2, 291–321. MR985401

[3] Alexey Alexandrov and Grigori Rozenblum, *Finite rank Toeplitz operators: some extensions of D. Luecking's theorem*, J. Funct. Anal. **256** (2009), no. 7, 2291–2303, DOI 10.1016/j.jfa.2008.11.006. MR2498766

[4] Alberto Alonso and Barry Simon, *The Birman-Kreĭn-Vishik theory of selfadjoint extensions of semibounded operators*, J. Operator Theory **4** (1980), no. 2, 251–270. MR595414

[5] Mark S. Ashbaugh, Fritz Gesztesy, Marius Mitrea, Roman Shterenberg, and Gerald Teschl, *The Krein-von Neumann extension and its connection to an abstract buckling problem*, Math. Nachr. **283** (2010), no. 2, 165–179, DOI 10.1002/mana.200910067. MR2604115

[6] Mark S. Ashbaugh, Fritz Gesztesy, Marius Mitrea, and Gerald Teschl, *Spectral theory for perturbed Krein Laplacians in nonsmooth domains*, Adv. Math. **223** (2010), no. 4, 1372–1467, DOI 10.1016/j.aim.2009.10.006. MR2581375

[7] M. Š. Birman, *On the spectrum of singular boundary-value problems* (Russian), Mat. Sb. (N.S.) **55 (97)** (1961), 125–174. MR0142896

[8] M. Š. Birman and M. Z. Solomjak, *Estimates for the singular numbers of integral operators* (Russian), Uspehi Mat. Nauk **32** (1977), no. 1(193), 17–84, 271. MR0438186

[9] M. Š. Birman and M. Z. Solomjak, *Asymptotic behavior of the spectrum of variational problems on solutions of elliptic equations* (Russian), Sibirsk. Mat. Zh. **20** (1979), no. 1, 3–22, 204. MR523133

[10] M. Sh. Birman and M. Z. Solomjak, *Spectral theory of selfadjoint operators in Hilbert space*, Mathematics and its Applications (Soviet Series), D. Reidel Publishing Co., Dordrecht, 1987. Translated from the 1980 Russian original by S. Khrushchëv and V. Peller. MR1192782

[11] Louis Boutet de Monvel, *Boundary problems for pseudo-differential operators*, Acta Math. **126** (1971), no. 1-2, 11–51, DOI 10.1007/BF02392024. MR0407904

[12] L. Boutet de Monvel and V. Guillemin, *The spectral theory of Toeplitz operators*, Annals of Mathematics Studies, vol. 99, Princeton University Press, Princeton, NJ; University of Tokyo Press, Tokyo, 1981. MR620794

[13] V. Bruneau, G. Raikov, *Spectral properties of harmonic Toeplitz operators and applications to the perturbed Krein Laplacian*, to appear in Asymp. Anal. (2018).

[14] Boo Rim Choe, Young Joo Lee, and Kyunguk Na, *Toeplitz operators on harmonic Bergman spaces*, Nagoya Math. J. **174** (2004), 165–186, DOI 10.1017/S0027763000008837. MR2066107

[15] Miroslav Engliš, *Boundary singularity of Poisson and harmonic Bergman kernels*, J. Math. Anal. Appl. **429** (2015), no. 1, 233–272, DOI 10.1016/j.jmaa.2015.03.081. MR3339073

[16] Lawrence C. Evans, *Partial differential equations*, 2nd ed., Graduate Studies in Mathematics, vol. 19, American Mathematical Society, Providence, RI, 2010. MR2597943

[17] Lars Gårding, *Some points of analysis and their history*, University Lecture Series, vol. 11, American Mathematical Society, Providence, RI; Higher Education Press, Beijing, 1997. MR1469493

[18] Gerd Grubb, *A characterization of the non-local boundary value problems associated with an elliptic operator*, Ann. Scuola Norm. Sup. Pisa (3) **22** (1968), 425–513. MR0239269

[19] Gerd Grubb, *Spectral asymptotics for the "soft" selfadjoint extension of a symmetric elliptic differential operator*, J. Operator Theory **10** (1983), no. 1, 9–20. MR715550

[20] S. Grudsky, A. Karapetyants, and N. Vasilevski, *Toeplitz operators on the unit ball in* $\mathbb{C}^n$ *with radial symbols*, J. Operator Theory **49** (2003), no. 2, 325–346. MR1991742

[21] S. M. Grudsky and N. L. Vasilevski, *Toeplitz operators on the Fock space: radial component effects*, Integral Equations Operator Theory **44** (2002), no. 1, 10–37, DOI 10.1007/BF01197858. MR1913421

[22] Lars Hörmander, *The spectral function of an elliptic operator*, Acta Math. **121** (1968), 193–218, DOI 10.1007/BF02391913. MR0609014

[23] Victor Ivrii, *Microlocal analysis and precise spectral asymptotics*, Springer Monographs in Mathematics, Springer-Verlag, Berlin, 1998. MR1631419

[24] Hyeonbae Kang and Hyungwoon Koo, *Estimates of the harmonic Bergman kernel on smooth domains*, J. Funct. Anal. **185** (2001), no. 1, 220–239, DOI 10.1006/jfan.2001.3761. MR1853757

[25] M. Krein, *The theory of self-adjoint extensions of semi-bounded Hermitian transformations and its applications. I* (Russian, with English summary), Rec. Math. [Mat. Sbornik] N.S. **20(62)** (1947), 431–495. MR0024574

[26] J.-L. Lions and E. Magenes, *Non-homogeneous boundary value problems and applications. Vol. I*, Springer-Verlag, New York-Heidelberg, 1972. Translated from the French by P. Kenneth; Die Grundlehren der mathematischen Wissenschaften, Band 181. MR0350177

[27] Georgi D. Raikov and Simone Warzel, *Quasi-classical versus non-classical spectral asymptotics for magnetic Schrödinger operators with decreasing electric potentials*, Rev. Math. Phys. **14** (2002), no. 10, 1051–1072, DOI 10.1142/S0129055X02001491. MR1939760

[28] Grigori Rozenblum, *On lower eigenvalue bounds for Toeplitz operators with radial symbols in Bergman spaces*, J. Spectr. Theory **1** (2011), no. 3, 299–325, DOI 10.4171/JST/13. MR2831755

[29] M. A. Shubin, *Pseudodifferential operators and spectral theory*, 2nd ed., Springer-Verlag, Berlin, 2001. Translated from the 1978 Russian original by Stig I. Andersson. MR1852334

[30] J. v. Neumann, *Allgemeine Eigenwerttheorie Hermitescher Funktionaloperatoren* (German), Math. Ann. **102** (1930), no. 1, 49–131, DOI 10.1007/BF01782338. MR1512569

[31] Hermann Weyl, *The method of orthogonal projection in potential theory*, Duke Math. J. **7** (1940), 411–444. MR0003331

UMR CNRS 5251, Institut de Mathématiques de Bordeaux, Université de Bordeaux, 351, Cours de la Libération, 33405 Talence, France

Email address: `vbruneau@math.u-bordeaux.fr`

Facultad de Matemáticas, Pontificia Universidad Católica de Chile, Vicuña Mackenna 4860, Santiago de Chile

Email address: `graikov@mat.uc.cl`

Contemporary Mathematics
Volume **717**, 2018
https://doi.org/10.1090/conm/717/14450

Singular spectrum and recent results on hierarchical operators

Per von Soosten and Simone Warzel

ABSTRACT. We use trace class scattering theory to exclude the possibility of absolutely continuous spectrum in a large class of self-adjoint operators with an underlying hierarchical structure and provide applications to certain random hierarchical operators and matrices. We proceed to contrast the localizing effect of the hierarchical structure in the deterministic setting with previous results and conjectures in the random setting. Furthermore, we survey stronger localization statements truly exploiting the disorder for the hierarchical Anderson model and report recent results concerning the spectral statistics of the ultrametric random matrix ensemble.

1. Hierarchical operators

Random Hamiltonians with hierarchical structure provide highly simplified and analytically tractable toy models for the Anderson localization-delocalization transition, whose full version is beyond the reach of the currently available mathematical machinery. In this article, we define a notion of abstract hierarchical operator and use trace class scattering theory to show that the hierarchical structure forbids the formation of absolutely continuous spectrum even in the deterministic setting. For deterministic operators, such a result is optimal regarding the spectral type since changing even a single potential value can produce singular continuous spectrum [**3**]. The secondary aim of this article is to survey recent results concerning localization in two concrete disordered hierarchical models, the hierarchical Anderson model introduced by Bovier in [**2**] and the ultrametric random matrix ensemble of Fyodorov, Ossipov and Rodriguez [**12**]. This is done in Sections 2 and 3, where we also show how to capture these models in the abstract framework and apply our main result, thereby proving the absence of absolutely continuous spectrum for the infinite-volume Hamiltonians of these models.

The idea of exploiting the stability of the absolutely continuous spectrum to rule out its presence altogether goes back at least to the work of Dombrowski [**4–6**] and was rediscovered and popularized in Simon and Spencer's treatment of one-dimensional Schrödinger operators with potential barriers [**24**]. These arguments are based on comparing the operator in question with a direct sum of finite-dimensional matrices, which has only pure-point spectrum. If the resolvents agree

2010 *Mathematics Subject Classification.* Primary 47-06; Secondary 47B80.
Key words and phrases. Localization, absolutely continuous spectrum, hierarchical operators.
This work was supported by the DFG (WA 1699/2-1).

up to a trace class perturbation, the Kato-Rosenblum theorem [**23**] asserts that the absolutely continuous spectra of the two operators are equal, which is to say, both are empty. We follow the same approach here, but the hierarchical structure allows the direct comparison of the two operators without passing to the resolvents first.

Let us now turn to a detailed definition of hierarchical operators and the statement and proof of our main result. Inspired by [**15**], we define a hierarchy in some (necessarily countable) set X to be a sequence $\{\mathcal{P}_r\}_{r=0}^{\infty}$ of partitions of X such that:

- the members of $\mathcal{P}_0$ are finite sets,
- each member of $\mathcal{P}_r$ is a finite union of members of $\mathcal{P}_{r-1}$, and
- for any $x, y \in X$ there exists $r \geq 0$ such that x and y lie in a common member of $\mathcal{P}_r$.

Thus, if $x \in X$ and $r \geq 0$, there exists a unique member of $\mathcal{P}_r$ containing x, which we will denote by $B_r(x)$. We say that an operator H on $\ell^2(X)$ is hierarchical if it is of the form

$$H = \sum_{r\geq 0} \sum_{B\in\mathcal{P}_r} H(B) \tag{1.1}$$

where $H(B)$ is a self-adjoint operator on $\ell^2(B)$ for every $B \in \mathcal{P}_r$ and

$$\sum_{r\geq 1} \|H(B_r(x))\|_1 < \infty \tag{1.2}$$

for all $x \in X$. In (1.2), $\|\cdot\|_p$ is the Schatten p-norm, i.e., the ℓ^p-norm of the singular values. The requirement (1.2) is not strictly necessary for (1.1) to make sense as an operator on $\ell^2(X)$, but we will need it to classify the spectrum, and there is no real loss of generality in assuming it from the beginning. The action of H on the function

$$\delta_x(y) = \begin{cases} 1 & \text{if } y = x \\ 0 & \text{otherwise} \end{cases}$$

depends only on those summands of (1.1) with $x \in B$, that is, $H\delta_x = S_x\delta_x$ with

$$S_x = \sum_{r\geq 1} H(B_r(x)).$$

By (1.2), the sum defining $S_x\delta_x$ converges for all $x \in X$ so the domain of H contains at least the functions with finite support. In the main new result of this article we prove that H defines a self-adjoint operator without any absolutely continuous spectrum if the couplings $H(B_r(x))$ become sufficiently weak as $r \to \infty$, by showing that the removal of S_x completely disconnects the system.

THEOREM 1.1. *Every hierarchical operator H is self-adjoint with $\sigma_{ac}(H) = \emptyset$.*

PROOF. Setting $X_0 = B_0(x)$, there exist $\{x_r\}$ in X such that $X_r = B_{r-1}(x_r)$ is a disjoint sequence of finite sets satisfying

$$B_r(x) = B_{r-1}(x) \cup X_r, \quad B_{r-1}(x) \cap X_r = \emptyset$$

for all $r \geq 1$. This means that

$$X = \bigcup_{r\geq 0} B_r(x) = \bigcup_{r\geq 0} X_r$$

and hence

$$\ell^2(X) = \bigoplus_{r\geq 0} \ell^2(X_r)$$

is a direct sum of finite-dimensional subspaces. If $y \in X_r$, then

$$H\delta_y = \sum_{r\geq 0} H(B_r(y))\delta_y$$

and therefore

$$F\delta_y := (H - S_x)\delta_y = \sum_{s=0}^{r-1} H(B_s(y))\delta_y$$

since $X_r \subset B_r(x)$. Thus $\ell^2(X_r)$ is an invariant subspace for F because it is an invariant subspace of $H(B_s(y))$ for every $s \leq r-1$, and this proves that F is a direct sum of finite-dimensional matrices. Since S_x is trace class, $H = F + S_x$ is self-adjoint on the domain of F and

$$\sigma_{ac}(H) = \sigma_{ac}(F) = \emptyset$$

by the Kato-Rosenblum theorem. □

2. Hierarchical Schrödinger operators

We now turn to the definition introduced in [**2**] of a hierarchical analogue of the finite-difference Laplacian on $\mathbb{Z}^d$. To capture the effect of nearest-neighbor hopping, or local averaging, it is natural to choose

$$H(B) = p_r|\varphi_B\rangle\langle\varphi_B|,$$

if $r \geq 1$ and $B \in \mathcal{P}_r$, where

$$\varphi_B = |B|^{-1/2} 1_B$$

is the normalized maximally extended state in $\ell^2(B)$. Moreover, the decay in the interaction strength over long distances may be encoded in the requirement that the coefficients p_r satisfy

$$\sum_{r=1}^{\infty} |p_r| < \infty.$$

We leave the choice of $H(B)$ with $B \in \mathcal{P}_0$ open for now by setting $H(B) = 0$ in this case, and define the hierarchical Laplacian as Δ as the hierarchical operator H furnished by (1.1) with this specific choice of $H(B)$.

An important quantity associated with a hierarchical Laplacian is the spectral dimension, whose definition is motivated by matching the decay of the density of states near the band edge to the lattice case. The precise definition,

$$d_s = \lim_{\lambda\downarrow 0} \frac{\ln \langle \delta_k, 1_{[\lambda_\infty - \lambda, \lambda_\infty]}(\Delta)\delta_k\rangle}{\ln \sqrt{\lambda}}$$

with $\lambda_\infty = \sup \sigma(\Delta)$, is equal to the spatial dimension d if Δ is replaced by the d-dimensional lattice Laplacian. The spectral dimension can be computed explicitly for a large class of hierarchies $\{\mathcal{P}_r\}$ and coefficients p_r (see [**15**]).

A hierarchical Schrödinger operator is the sum of a hierarchical Laplacian and a potential,

$$H = \Delta + V,$$

with

$$V = \sum_{x \in X} V_x |\delta_x\rangle\langle\delta_x|$$

for $V_x \in \mathbb{R}$. We note that this H is still hierarchical in the sense of (1.1) if we choose $\mathcal{P}_0$ to be the trivial partition, whose members are singletons, and set

$$H(\{x\}) = V_x.$$

Since $\{p_r\}$ was assumed to be a summable sequence, it is clear that $\|S_x\|_1 < \infty$ for any $x \in X$, and thus Theorem 1.1 answers the most immediate question concerning the localization properties of H.

COROLLARY 2.1. *Every hierarchical Schrödinger operator satisfies* $\sigma_{ac}(H) = \emptyset$.

To obtain the stronger localization results that follow, the presence of true disorder is crucial, and we shall henceforth require that the $\{V_x\}$ are drawn independently from a common density $\rho \in L^\infty$, thereby obtaining the usual hierarchical Anderson model. Furthermore, it is often necessary to have some fixed decay rate of p_r in terms of the partitions $\mathcal{P}_r$, so we focus on an explicit model, whose configuration space is $X = \mathbb{N}_0$, and whose hierarchy is given in terms of the partitions $\{\mathcal{P}_r\}$ defined by

$$\mathbb{N}_0 = \{0, ..., 2^r - 1\} \cup \{2^r, ..., 2 \cdot 2^r - 1\} \cup ... \tag{2.1}$$

It is now easiest to take $p_r = \epsilon\, 2^{-cr}$ for some $\epsilon, c \in (0, \infty)$, as we shall do here, and one may check that the spectral dimension is then

$$d_s = \frac{2}{c}.$$

In particular, the assumption $c > 0$ amounts to the requirement that the spectral dimension be finite.

A more complete characterization of $\sigma(H)$ was first given by Molchanov [**21**] and Kritchevski [**14, 15**], who proved that H almost surely has only pure point spectrum, provided that $d_s < 4$ is small enough (i.e. $c > 1/2$ is large enough) or that ρ is a mixture of Cauchy distributions with barycenters strictly separated from the real axis. Their argument, which is based on approximation by the truncations

$$H_s = \sum_{r=0}^{s} \sum_{B \in \mathcal{P}_r} H(B)$$

and the Simon-Wolff criterion [**25**], was extended to the full parameter range $c > 0$ and $\rho \in L^\infty$ in [**28**]. Moreover, it is possible to show that the spatial decay of the eigenfunctions of H with respect to the metric

$$d(j,k) = \min \{r \geq 0 \,|\, j \text{ and } k \text{ belong to a common member of } \mathcal{P}_r\} \tag{2.2}$$

is essentially controlled by the decay of p_r.

THEOREM 2.2 ([**28**]). *The spectrum of H is almost surely of pure-point type with eigenfunctions satisfying*

$$\sum_{k \in \mathbb{N}_0} 2^{\frac{c}{4} d(0,k)} |\psi_E(k)|^2 < \infty \tag{2.3}$$

for any $E \in \sigma(H)$.

Via the RAGE theorem [**23**], Theorem 2.2 has the dynamical implication that, for almost every realization of the disorder, the quantum probability of a particle ever leaving a finite radius R vanishes asymptotically as $R \to \infty$. However, the complete lack of control of the implied amplitude of the wave function in (2.3) makes this an assertion of a fundamentally qualitative nature and does not completely rule out the appearance of more subtle delocalization effects in finite volumes. Indeed, prior to the work [**28**], but after the appearance of [**14, 15, 21**], Metz et. al. [**19**] analyzed the hierarchical Anderson model numerically in large finite volumes by relying on renormalization group equations (see also [**22**]) whose effect on the parameters is

$$\mathcal{R}(\{p_r\}_{r\geq 1}, \rho) = ((p_{r+1})_{r\geq 1}, T_{p_1}\rho),$$

where $T_p\rho$ is the probability density of

$$\left(\frac{1}{2V} + \frac{1}{2V'}\right)^{-1} + p,$$

and V and V' are drawn independently from ρ. If $c < 1$ and ρ is the density of a sufficiently weak Gaussian random variable, the authors of [**19**] claimed delocalization at the special energy $E = \sum p_r$ in the sense of the inverse participation ratios satisfying

$$\lim_{n\to\infty} \frac{1}{|B_n|\nu(E)} \mathbb{E}\left[\sum_{\lambda\in\sigma(H_n)} \|\psi_\lambda\|_4 \tilde{\delta}_{E-\lambda}\right] = 0 \tag{2.4}$$

($\tilde{\delta}_{E-\lambda}$ being a suitably regularized Dirac delta). In (2.4), ν denotes the infinite-volume density of states defined by

$$\mathbb{E}\,\langle\delta_0, f(H)\delta_0\rangle = \int f(E)\nu(E)\,dE$$

and $\{\psi_\lambda\}$ are the normalized eigenfunctions of the finite-volume Hamiltonian

$$H_n = 1_{B_n(0)} H 1_{B_n(0)}.$$

Nevertheless, the renormalization dynamics $\mathcal{R}$ appear to effectively drive the Hamiltonian into a regime of high disorder, where strong finite-volume localization bounds apply generically. In fact, for any interval $I \subset \mathbb{R}$ one can prove that there exists $\delta > 0$ such that

$$\sup_{E\in I} \|T_{p_r}...T_{p_1}\varrho(\cdot + E)\|_\infty = \mathcal{O}(2^{(c-\delta)r}) \tag{2.5}$$

whenever:

- $d_s < 2$,
- V is Gaussian and $d_s < 4$, or
- V has a Cauchy component and $d_s < \infty$.

This observation shows that the strength of the random potential does not change significantly in comparison to the decrease in the strength of the kinetic term,

$$(\mathcal{R}p)_r = p_{r+1} = 2^{-c}p_r,$$

and can be used to prove localization in terms of the eigenfunction correlator

$$Q_n(j, k; I) = \sup\, |\langle\delta_k, f(H_n)\delta_j\rangle|,$$

where the supremum ranges over those $f \in C_0$ with supp$f \subset I$ and $\|f\|_\infty \leq 1$.

THEOREM 2.3 ([**28**]). *If (2.5) is satisfied in a bounded interval $I \subset \mathbb{R}$, then there exist $C, \mu \in (0, \infty)$ such that*

$$\sup_{n\in\mathbb{N}_0}\sup_{j\in\mathbb{N}_0}\sum_{k\in\mathbb{N}_0} 2^{\mu d(j,k)}\mathbb{E}\left[Q_n(j,k;I)\right] \le C|I|.$$

It follows that

$$\sum_{k:d(j,k)\ge R} \mathbb{E}\,|\langle \delta_k, 1_I(H)e^{itH}\delta_j\rangle|^2 \le C\,2^{-\mu R},$$

which gives a quantitative averaged bound on the quantum probability that a particle, which was started at $j \in \mathbb{N}_0$ and subsequently filtered by energy, ever leaves $B_R(j)$. Theorem 2.3 is not compatible with the claims of [**19**] since it implies lower bounds for the inverse participation ratios of the eigenfunctions; the details can be found in [**28**].

A further canonical method of probing localization properties in finite systems consists of examining the statistics of the energy levels, i.e. the local behavior of the eigenvalues on the microscopic scale. This is best quantified by the rescaled eigenvalue point process

$$\mu_n(f) = \sum_{\lambda\in\sigma(H_n)} f(2^n(\lambda - E)). \tag{2.6}$$

In terms of this random measure, the renormalization group approach lets one prove the following theorem, which was previously proved by Kritchevski [**16**] for $d_s < 1$.

THEOREM 2.4 ([**28**]). *Suppose (2.5) is satisfied in an open set $I \subset \mathbb{R}$ and $E \in I$ is a Lebesgue point of ν. Then, μ_n converges in distribution to a Poisson point process with intensity $\nu(E)$ as $n \to \infty$.*

It remains an open problem to prove (2.5) for more general densities ρ, and we hope to inspire further work in this direction. Still, even without a general theorem of this kind, our results leave no doubt that the hierarchical approximation in finite dimension is too crude to capture the Anderson transition on the lattice.

3. Ultrametric ensemble

In this section we continue to consider the hierarchy (2.1) and retain the notation d for the restriction of the hierarchical metric (2.2) to the finite volume $B_n = \{1, 2, ..., 2^n\}$. The second model we wish to study in this article, the ultrametric ensemble, consists of hierarchical matrices on $\ell^2(B_n)$ whose off-diagonal entries are also random variables with variances satisfying power-law decay in $d(k,\ell)$. We are thus led to consider

$$H_n = \sum_{r=0}^{n} 2^{-\frac{(1+c)}{2}r}\Phi_{n,r}, \tag{3.1}$$

where $c \in \mathbb{R}$ is a real parameter and the entries of $\Phi_{n,r}$ are independent centered real Gaussian random variables with variances

$$\mathbb{E}\,|\langle \delta_\ell, \Phi_{n,r}\delta_k\rangle|^2 = 2^{-r}\begin{cases} 2 & \text{if } d(k,\ell) = 0 \\ 1 & \text{if } 1 \le d(k,\ell) \le r \\ 0 & \text{otherwise.}\end{cases}$$

With this convention, each $\Phi_{n,r}$ is a direct sum of 2^{n-r} random matrices drawn independently from the Gaussian Orthogonal Ensemble (GOE), rather than the Gaussian Unitary Ensemble used in [**12**], but our results are not based on any additional symmetries and are valid in both cases.

The definition (3.1) shows that H_n has independent entries whose typical magnitudes are approximately

$$\left(\mathbb{E}\,|\langle \delta_\ell, H_n\delta_k\rangle|^2\right)^{1/2} \approx 2^{-\frac{2+c}{2}d(\ell,k)},$$

and hence the ultrametric ensemble interpolates between a perfectly localized random potential and a completely delocalized Wigner matrix as the parameter c varies in $\mathbb{R}$. Indeed, this ensemble is a hierarchical analogue of the Power-Law Random Band Matrices [**20**] in the sense which was first introduced to statistical mechanical models by Dyson [**8**]. The authors of [**12**] argue for a localization-delocalization transition in the eigenfunctions of H_n as the parameter varies from $c > 0$ to $c < 0$ with a theoretical physics level of rigor and support this claim with numerical simulations.

An infinite-volume version of the ultrametric ensemble with parameter $c \in \mathbb{R}$ may be defined formally using the prescription (1.1), by letting each $H(B)$ be an independent and appropriately scaled GOE, that is,

$$\mathbb{E}\,|\langle \delta_x, H(B)\delta_y\rangle|^2 = |B|^{-(1+c)}\frac{1+\delta_{xy}}{|B|}.$$

Of course, this H can only be defined as an operator on $\ell^2(X)$ if $c > -1$, and in this case one may easily check that $H\delta_x = S_x\delta_x$ converges almost surely to an element of $\ell^2(X)$. As with the hierarchical Anderson model, we may apply Theorem 1.1 to exclude absolutely continuous spectrum when S_x is trace class.

Corollary 3.1. *If $c > 1$, the infinite-volume ultrametric operator H satisfies $\sigma_{ac}(H) = \emptyset$ almost surely.*

Proof. By applying Jensen's inequality twice we get the inequality

$$\begin{aligned}\mathbb{E}\,\|H(B_r(x))\|_1 &\le |B_r(x)|^{1/2}\mathbb{E}\,\|H(B_r(x))\|_2\\ &\le |B_r(x)|^{1/2}\left(\sum_{y,y'\in B_r(x)}\mathbb{E}\,|\langle \delta_{y'}, H(B_r(x))\delta_y\rangle|^2\right)^{1/2}\\ &= |B_r(x)|^{\frac{1-c}{2}}\end{aligned}$$

for the expected trace norm of $H(B_r(x))$. Because each member of $\mathcal{P}_{r+1}$ is the union of two members of $\mathcal{P}_r$, we have $|B_r(x)| \ge 2^r$ so Markov's inequality shows that

$$\sum_{r\ge 0}\mathbb{P}\left(\|H(B_r(x))\|_1 \ge |B_r(x)|^{\frac{1-c}{4}}\right) < \infty,$$

provided $c > 1$. Therefore, the Borel-Cantelli lemma implies that there almost surely exists some constant $C < \infty$ such that

$$\|H(B_r(x))\|_1 \le C|B_r(x)|^{\frac{1-c}{4}} \le C\,2^{\frac{1-c}{4}r}$$

for all $r \ge 0$. Hence

$$\|S_x\|_1 \le \sum_{r\ge 0}\|H(B_r(x))\|_1 < \infty$$

and Theorem 1.1 implies the result. □

We remark that Theorem 1.1 fails to cover the entirety of the regime in which the infinite volume operator is well-defined, and it is an open question whether the formation of continuous spectrum is possible in the remaining parameter range.

To obtain more detailed results in finite volumes, we now focus on the eigenvalue statistics of H_n. The effect of a GOE perturbation Φ on the spectrum of an arbitrary symmetric $N \times N$ matrix H_0 can be described dynamically by thinking of $H_0 + \Phi$ as $H_0 + \Phi(t)$ where $\Phi(t)$ is the symmetric stochastic matrix flow

$$\langle \delta_\ell, \Phi(t)\delta_k \rangle = \sqrt{\frac{1+\delta_{k\ell}}{N}} B_{k\ell}(t)$$

and $\{B_{k\ell}\}$ is a symmetric array of independent standard Brownian motions. The observation of Dyson [**7**] was that the eigenvalues $\lambda_1(t) \leq ... \leq \lambda_N(t)$ then undergo Dyson Brownian motion, that is, evolve according to the stochastic differential equation

$$d\lambda_j(t) = \sqrt{\frac{2}{N}} dB_j(t) + \frac{1}{N} \sum_{i \neq j} \frac{dt}{\lambda_j(t) - \lambda_i(t)}.$$

The relevance of this discussion to the ultrametric ensemble can be seen from the relation

$$H_n = \sum_{r=0}^{n-1} 2^{-\frac{(1+c)}{2} r} \Phi_{n,r} + 2^{-\frac{(1+c)}{2} r} \Phi_{n,n} = H_{n-1} \oplus H'_{n-1} + \Phi(t),$$

where the system size N is now 2^n, $t = 2^{-(1+c)n}$, and H'_{n-1} is an independent copy of H_{n-1}. One may thus construct the spectrum of H_n by initializing $\sigma(H_0)$ to consist of a single standard normal random variable and following the recursion

(1) Sample an independent copy $\sigma(H'_{k-1})$ of $\sigma(H_{k-1})$
(2) Let $\sigma(H_k)$ be the evolution of $\sigma(H_{k-1}) \cup \sigma(H'_{k-1})$ under Dyson Brownian motion with duration $t = 2^{-(1+c)k}$.

In his original paper, Dyson motivated the conjecture that the Dyson Brownian motion of N particles locally equilibrates at times greater than $t = N^{-1}$, that is, the particle configuration $\{\lambda_j(t)\}$ becomes extremely rigid locally and the k-point correlation functions are very close to those of the GOE. Thus, one does not expect the addition of independent randomness in step (1) to significantly affect the spectrum of H_k on small enough scales when $c < 0$ and the level statistics should asymptotically agree with the GOE. On the other hand, if $c > 0$, the Dyson Brownian motion in step (2) is not running for a long enough time to compensate the additional fluctuations of step (1) and the resulting level statistics are only a slight perturbation of a point process with many independent components. This second point is the basic idea behind the proof of the following theorem of [**27**]. As before, ν denotes the density of states in the infinite volume and μ_n is the random measure (2.6).

THEOREM 3.2 ([**27**]). *Suppose $c > 0$ and $E \in \mathbb{R}$ is a Lebesgue point of ν. Then, μ_n converges in distribution to a Poisson point process with intensity $\nu(E)$ as $n \to \infty$.*

Similar considerations for the off-diagonal spectral measures also allow one to prove localization estimates for the eigenfunctions of H_n.

THEOREM 3.3 (Eigenfunction localization). *Suppose $c > 0$ and let $E \in \mathbb{R}$. Then, there exist $w, \mu, \kappa > 0$, $C < \infty$, and a sequence m_n with $n - m_n \to \infty$ such that for every $x \in B_n$ the ℓ^2-normalized eigenfunctions satisfy*

$$\mathbb{P}\left(\sum_{y \in B_n \setminus B_{m_n}(x)} Q_n(x, y; W) > 2^{-\mu n}\right) \leq C\, 2^{-\kappa n}$$

with

$$W = \left[E_0 - 2^{-(1-w)n}, E_0 + 2^{-(1-w)n}\right].$$

For times $t \gg N^{-1}$, Landon, Sosoe and Yau [**17**] have recently proved the fixed energy universality (and more) of the Dyson Brownian motion for very general regularly spaced initial conditions. A key feature of this result is that as t approaches $\mathcal{O}(1)$, the scale of the required local law may also be relaxed towards $\mathcal{O}(1)$. If $c < -1$, then the ℓ^2-norm of $H_n \delta_0$,

$$Z_{n,c} := \left(\sum_{\ell \in B_n} \mathbb{E}\left|\langle \delta_\ell, \left(\sum_{r=0}^{n} 2^{-\frac{(1+c)}{2} r} \Phi_{n,r}\right) \delta_0 \rangle\right|^2\right)^{1/2},$$

is approximately $2^{-\frac{1+c}{2} n}$. Therefore, the rescaled Hamiltonian $Z_{n,c}^{-1} H_n$, whose spectrum lies in $[-2, 2]$ with high probability, has spread

$$M := \left(\max_{k, \ell \in B_n} \mathbb{E}\left|\langle \delta_\ell, H_n \delta_k \rangle\right|^2\right)^{-1} \approx Z_{n,c}^2,$$

which grows like a positive power of the system size. Hence, the results of [**10**] show that the semicircle law is valid up to scales of order M^{-1}. This is consistent with the idea that the Dyson Brownian motion equilibrates globally for times of order $\mathcal{O}(1)$. The analysis of [**17**] can now be applied to prove universality of the k-point correlation function, i.e. the k-th marginal of the symmetrized eigenvalue density ρ_{H_n}:

$$\rho_{H_n}^{(k)}(\lambda_1, ..\lambda_k) = \int_{\mathbb{R}^{2^n - k}} \rho_{H_n}(\lambda_1, ..., \lambda_{2^n})\, d\lambda_{k+1} ... d\lambda_{2^n}.$$

The precise theorem of [**17**] is formulated in terms of

$$\begin{aligned}\Psi_{n,E}^{(k)}(\alpha_1, ..., \alpha_k) = {}& \rho_{H_n}^{(k)}\left(E + 2^{-n}\frac{\alpha_1}{\rho_{sc}(E)}, ..., E + 2^{-n}\frac{\alpha_k}{\rho_{sc}(E)}\right) \\ & - \rho_{GOE}^{(k)}\left(E + 2^{-n}\frac{\alpha_1}{\rho_{sc}(E)}, ..., E + 2^{-n}\frac{\alpha_k}{\rho_{sc}(E)}\right),\end{aligned}$$

where $\rho_{GOE}^{(k)}$ denotes the k-point correlation function of the $2^n \times 2^n$ GOE and ρ_{sc} is the density of the semicircle law.

THEOREM 3.4 (cf. [**10**] and [**17**]). *Suppose $c < -1$, $E \in (-2, 2)$ and $k \geq 1$. Then*

$$\lim_{n \to \infty} \int_{\mathbb{R}^k} O(\alpha) \Psi_{n,E}^{(k)}(\alpha)\, d\alpha = 0 \tag{3.2}$$

for every $O \in C_c^\infty(\mathbb{R}^k)$.

A corollary of the existence of a local law for the local resolvent on scale M^{-1} is the complete delocalization of the eigenfunctions, in the sense that

$$\|\psi_{E,n}\|_\infty = \mathcal{O}(M^{-1/2})$$

with overwhelming probability in mesoscopic spectral windows (see [**9**, Thm. 2.21]).

In case $c \in [-1, 0)$, the density of states is no longer given by the semicircle law but one still expects the local statistics to exhibit random matrix universality. A technical difficulty in applying the results of [**17**] is the lack of a lower bound on the non-averaged local density of states. This gap in the understanding of the local statistics in the ultrametric ensemble can be closed for the simpler Rosenzweig-Porter model, for which the techniques used in the proof of Theorem 3.2 yield Poisson statistics if $t \ll N^{-1}$, and the results of [**17**] can be applied to obtain random matrix universality for all $t \gg N^{-1}$.

THEOREM 3.5 ([**17**] and [**27**]). *Suppose $V_1, ...V_N$ are independent random variables with density $\rho \in L^\infty$ and $t = N^{-(1+c)}$ for some $c \in \mathbb{R}$. Then, as $N \to \infty$,*

$$H_N = diag(V_1, ..., V_N) + \Phi(t)$$

satisfies the following statements:

(1) *If $c > 0$, the level statistics μ_N near a Lebesgue point E of ρ converge in distribution to a Poisson point process with intensity $\rho(E)$.*
(2) *If $c < 0$, the k-point correlation functions $\rho_N^{(k)}$ are universal in the sense of (3.2).*

This theorem partially confirms the picture sketched in [**11**, **13**], but leaves open the conjectured existence of non-ergodic delocalized states in the regime $N^{-1} \ll t \ll 1$. The existence of such a phase was recently proved in the independent works [**1**] and [**26**], complementing the eigenfunction bounds in [**18**] and [**27**].

References

[1] L. Benigni, *Eigenvectors distribution and quantum unique ergodicity for deformed Wigner matrices*, Preprint available at arXiv:1711.07103, 2017.

[2] Anton Bovier, *The density of states in the Anderson model at weak disorder: a renormalization group analysis of the hierarchical model*, J. Statist. Phys. **59** (1990), no. 3-4, 745–779, DOI 10.1007/BF01025849. MR1063180

[3] R. Del Rio, N. Makarov, and B. Simon, *Operators with singular continuous spectrum. II. Rank one operators*, Comm. Math. Phys. **165** (1994), no. 1, 59–67. MR1298942

[4] Joanne Dombrowski, *Tridiagonal matrix representations of cyclic selfadjoint operators*, Pacific J. Math. **114** (1984), no. 2, 325–334. MR757504

[5] J. Dombrowski, *Tridiagonal matrix representations of cyclic selfadjoint operators. II*, Pacific J. Math. **120** (1985), no. 1, 47–53. MR808928

[6] J. Dombrowski, *Cyclic operators, commutators, and absolutely continuous measures*, Proc. Amer. Math. Soc. **100** (1987), no. 3, 457–463, DOI 10.2307/2046429. MR891145

[7] Freeman J. Dyson, *A Brownian-motion model for the eigenvalues of a random matrix*, J. Mathematical Phys. **3** (1962), 1191–1198, DOI 10.1063/1.1703862. MR0148397

[8] Freeman J. Dyson, *Existence of a phase-transition in a one-dimensional Ising ferromagnet*, Comm. Math. Phys. **12** (1969), no. 2, 91–107. MR0436850

[9] L. Erdős, *Universality of Wigner random matrices: a survey of recent results*, Russian Mathematical Surveys **66** (2011), no. 3, 507.

[10] László Erdős, Antti Knowles, Horng-Tzer Yau, and Jun Yin, *The local semicircle law for a general class of random matrices*, Electron. J. Probab. **18** (2013), no. 59, 58, DOI 10.1214/EJP.v18-2473. MR3068390

[11] D. Facoetti, P. Vivo, and G. Biroli, *From non-ergodic eigenvectors to local resolvent statistics and back: A random matrix perspective*, EPL (Europhysics Letters) **115** (2016), no. 4, 47003.

[12] Y. V. Fyodorov, A. Ossipov, and A. Rodriguez, *The Anderson localization transition and eigenfunction multifractality in an ensemble of ultrametric random matrices*, Journal of Statistical Mechanics: Theory and Experiment **2009** (2009), no. 12, L12001.

[13] V. E. Kravtsov, I. M. Khaymovich, E. Cuevas, and M. Amini, *A random matrix model with localization and ergodic transitions*, New Journal of Physics **17** (2015), no. 12, 122002.

[14] Evgenij Kritchevski, *Hierarchical Anderson model*, Probability and mathematical physics, CRM Proc. Lecture Notes, vol. 42, Amer. Math. Soc., Providence, RI, 2007, pp. 309–322. MR2352276

[15] Evgenij Kritchevski, *Spectral localization in the hierarchical Anderson model*, Proc. Amer. Math. Soc. **135** (2007), no. 5, 1431–1440, DOI 10.1090/S0002-9939-06-08614-X. MR2276652

[16] Evgenij Kritchevski, *Poisson statistics of eigenvalues in the hierarchical Anderson model*, Ann. Henri Poincaré **9** (2008), no. 4, 685–709, DOI 10.1007/s00023-008-0369-5. MR2413200

[17] B. Landon, P. Sosoe, and H.-T. Yau, *Fixed energy universality for Dyson Brownian motion*, Preprint available at arXiv:1609.09011, 2016.

[18] Ji Oon Lee and Kevin Schnelli, *Local deformed semicircle law and complete delocalization for Wigner matrices with random potential*, J. Math. Phys. **54** (2013), no. 10, 103504, 62, DOI 10.1063/1.4823718. MR3134604

[19] F. L. Metz, L. Leuzzi, G. Parisi, and V. Sacksteder, *Transition between localized and extended states in the hierarchical Anderson model*, Phys. Rev. B **88** (2013), 045103.

[20] A. D. Mirlin, Y. V. Fyodorov, F.-M. Dittes, J. Quezada, and T. H. Seligman, *Transition from localized to extended eigenstates in the ensemble of power-law random banded matrices*, Phys. Rev. E **54** (1996), 3221–3230.

[21] S. Molchanov, *Hierarchical random matrices and operators. Application to Anderson model*, Multidimensional statistical analysis and theory of random matrices (Bowling Green, OH, 1996), VSP, Utrecht, 1996, pp. 179–194. MR1463464

[22] Cécile Monthus and Thomas Garel, *A critical Dyson hierarchical model for the Anderson localization transition*, J. Stat. Mech. Theory Exp. **5** (2011), P05005, 27. MR2972048

[23] Michael Reed and Barry Simon, *Methods of modern mathematical physics. III*, Academic Press [Harcourt Brace Jovanovich, Publishers], New York-London, 1979. Scattering theory. MR529429

[24] Barry Simon and Thomas Spencer, *Trace class perturbations and the absence of absolutely continuous spectra*, Comm. Math. Phys. **125** (1989), no. 1, 113–125. MR1017742

[25] Barry Simon and Tom Wolff, *Singular continuous spectrum under rank one perturbations and localization for random Hamiltonians*, Comm. Pure Appl. Math. **39** (1986), no. 1, 75–90, DOI 10.1002/cpa.3160390105. MR820340

[26] P. von Soosten and S. Warzel, *Non-ergodic delocalization in the Rosenzweig-Porter model*, Preprint available at arXiv:1709.10313, 2017.

[27] ———, *The phase transition in the ultrametric ensemble and local stability of Dyson Brownian motion*, Preprint available at arXiv:1705.00923, 2017.

[28] Per von Soosten and Simone Warzel, *Renormalization group analysis of the hierarchical Anderson model*, Ann. Henri Poincaré **18** (2017), no. 6, 1919–1947, DOI 10.1007/s00023-016-0549-7. MR3649447

Zentrum Mathematik, TU München, Boltzmannstr. 3, 85747 Garching, Germany
Email address: vonsoost@ma.tum.de

Zentrum Mathematik, TU München, Boltzmannstr. 3, 85747 Garching, Germany
Email address: warzel@ma.tum.de

Contemporary Mathematics
Volume **717**, 2018
https://doi.org/10.1090/conm/717/14451

On Parseval frames of exponentially decaying composite Wannier functions

David Auckly and Peter Kuchment

ABSTRACT. Let L be a periodic self-adjoint linear elliptic operator in $\mathbb{R}^n$ with coefficients periodic with respect to a lattice Γ, e.g. Schrödinger operator $(i^{-1}\partial/\partial_x - A(x))^2 + V(x)$ with periodic magnetic and electric potentials A, V, or a Maxwell operator $\nabla \times \varepsilon(x)^{-1}\nabla\times$ in a periodic medium. Let also S be a finite part of its spectrum separated by gaps from the rest of the spectrum. We address here the question of existence of a finite set of exponentially decaying Wannier functions $w_j(x)$ such that their Γ-shifts $w_{j,\gamma}(x) = w_j(x-\gamma)$ for $\gamma \in \Gamma$ span the whole spectral subspace corresponding to S. It was shown by D. Thouless in 1984 that a topological obstruction sometimes exists to finding exponentially decaying $w_{j,\gamma}$ that form an orthonormal (or any) basis of the spectral subspace. This obstruction has the form of non-triviality of certain finite dimensional (with the dimension equal to the number of spectral bands in S) analytic vector bundle (Bloch bundle), which we denote Λ_S. It was shown by G. Nenciu in 1983 that in the presence of time reversal symmetry (which implies absence of magnetic fields), and if S is a single band, the bundle is trivial and thus the desired Wannier functions do exist. In 2007, G. Panati proved that in dimensions $n \leq 3$, even if S consists of several spectral bands, the time reversal symmetry removes the obstruction as well. If the bundle is non-trivial, it was shown in 2009 by one of the authors that it is always possible to find a finite number l (estimated there as $m \leq l \leq 2^n m$) of exponentially decaying Wannier functions w_j such that their Γ-shifts form a tight (Parseval) frame in the spectral subspace. A Parseval frame is the next best thing after an orthonormal basis (unavailable in the presence of the topological obstacle). This appears to be the best one can do when the topological obstruction is present. Here we significantly improve the estimate on the number of extra Wannier functions needed, showing that in physical dimensions the number l can be chosen equal to $m+1$, i.e. only one extra family of Wannier functions is required. This is the lowest number possible in the presence of the topological obstacle. The result for dimension four is also stated (without a proof), in which case $m+2$ functions are needed.

2010 *Mathematics Subject Classification.* Primary 35Q40, 35P10, 47F05, 81V55; Secondary 65N25.

Key words and phrases. Spectral theory, periodic operator, condensed matter, Bloch bundle, Wannier function.

The first author was partially supported by NSF grant DMS - 0604994. The second author was supported in part by NSF grants DMS-0406022 and DMS-1517938.

1. Introduction

Wannier functions, along with Bloch waves, play an important role in various areas of physics and material science, condensed matter theory, photonic crystal theory (see, e.g., [**17**, **22**] for general discussion of photonic crystals). They give a very useful tool for description of electronic properties of solids, theory of polarization, photonic crystals, numerical analysis using tight-binding approximation, etc. (see, e.g. [**1**, **3**, **5**, **19**, **23**, **24**, **26**–**28**, **32**, **33**, **37**, **41**, **41**–**43**, **45**] and references therein for Wannier functions and their applications). As it is formulated in [**27**], strongly localized Wannier functions "are the solid-state equivalent of "localized molecular orbitals"..., and thus provide an insightful picture of the nature of chemical bonding." It is crucial to have the Wannier functions decaying as fast as possible. Thus the problem of choosing a finite number of exponentially decaying Wannier functions whose lattice shifts form an orthonormal basis in the spectral subspace corresponding an isolated part S of the spectrum has been intensively considered in physics literature since the paper by W. Kohn [**19**], who showed that this was possible in $1D$. The problem becomes non-trivial in $2D$ and higher dimensions. Indeed, a topological obstruction, not present in $1D$, might arise, as shown by D. Thouless [**40**]. Existence of such a basis is known to be equivalent to triviality of certain analytic vector fiber bundle (**Bloch bundle**), which we will denote Λ_S (see (13)). G. Nenciu [**32**] showed in 1983 (see also [**15**]) in presence of time reversal symmetry, and if S is a single spectral band, the bundle is trivial and thus the obstacle disappears. G. Panati [**33**] proved in 2007, that in dimensions $n \leq 3$, even if S consists of several spectral bands, the time reversal symmetry still removes the obstruction. In this case one resorts to the so called **composite**, or **generalized Wannier functions** that correspond to a finite family of bands, rather than to a single band (see Section 2 below). The activity in this direction is still high and even increasing [**27**, **42**].

What can one do if the topological obstacle is present and thus a family of Wannier functions with the described basis properties cannot exist? One can try to relax the exponential decay condition, but there is not much one can do without stumbling upon the topological obstruction (e.g., a slow decay such as summability of L^2 norms over the shifted copies of the Wigner-Seitz cell is already impossible [**23**]).

In the positive direction, the following result was proven in [**23**]:

Theorem *Let L be a self-adjoint elliptic Γ-periodic operator in $\mathbb{R}^n, n \geq 1$ and $S \subset \mathbb{R}$ be the union of m spectral bands of L. Suppose that S is separated from the rest of the spectrum by gaps. Then there exists a finite number l of exponentially decaying composite Wannier functions $w_j(x)$ such that their shifts $w_{j,\gamma} := w_j(x - \gamma), \gamma \in \Gamma$ form a tight (Parseval) frame in the spectral subspace H_S of the operator L. This means that for any $f(x) \in H_S$, the equality holds*

$$\int_{\mathbb{R}^n} |f(x)|^2 dx = \sum_{j,\gamma} |\int_{\mathbb{R}^n} f(x)\overline{w_{j,\gamma}(x)}dx|^2. \tag{1}$$

Here the number $l \in [m, 2^n m]$ is equal to the smallest dimension of a trivial bundle containing an equivalent copy of Λ_S. In particular, $l = m$ if and only if Λ_S is trivial, in which case an orthonormal basis of exponentially decaying composite Wannier functions exists.

Here a **frame** means an overdetermined (rather than basis) system of functions, and thus the orthonormal property is not achievable. The tight (Parseval) property is the best analog of orthonormality one can get in this case (see, e.g., [**14**]). For instance, it allows the control of the L^2 norms in terms of the projections onto the Wannier system. Moreover, the overdeterminancy makes the computation stabler.

The significant deficiencies of the above result of [**23**] are that, first, the upper bound $l \leq 2^n m$ is ridiculously high; second, even for a manageable value of $l > m$ it is not clear how to practically create the overdetermined system of l Wannier functions, since the proof in [**23**] is not constructive. Even in the case of a trivial Bloch bundle, it is not that clear how one can actually find the needed holomorphic basis of Bloch functions, and this is still under investigation [**7, 8, 10, 35**].

We address the first of these issues in this text. In Section 2 we survey the main notions and results of the theory that will be needed for the rest of the text. In order to make the text independent, this section is rather long. An expert can skip through it or go immediately to Section 3, where it is shown that in the "physical dimension" $n = 3$ only one extra Wannier functions is needed, i.e. $l = m + 1$ (announced in [**24**, Section 6.5]). Section 4 contains (without a proof) an analog of this result, when $n = 4$. Here $m + 2$ Wannier functions might be required. We discuss how one can try to find the "missing" $(m + 1)$st function in Section 5. In Section 6 final remarks and conclusions are provided. The texts ends with the Acknowledgments section.

2. Main notions and auxiliary results

Let $L(x, D)$ be a bounded from below self-adjoint elliptic operator in $\mathbb{R}^n$. The specific nature of the operator will be irrelevant (e.g., matrix operators, such as Dirac or Maxwell can be allowed). One can think, without loss of generality, of one's favorite periodic operator, e.g. the Schrödinger operator $(\frac{1}{i}\frac{\partial}{\partial x} - A(x))^2 + V(x)$ with real periodic magnetic and electric potentials A, V. Some conditions need to be imposed on the potentials to define a self-adjoint operator L in $L^2(\mathbb{R}^n)$ (e.g., [**9, 36**]). However, for what follows most of these details are not germane to the main issue, so we assume "sufficiently nice" (e.g., smooth) coefficients of L.

Let Γ be a (**Bravais** [**1**]) **lattice** in $\mathbb{R}^n$, i.e. the set of integer linear combinations of vectors of a basis $a_1, \dots, a_n$ in $\mathbb{R}^n$ [1]. The coefficients of L are assumed to be periodic with respect to the shifts by vectors $\gamma \in \Gamma$. We fix a **fundamental domain W** of Γ, i.e. such that its Γ-shifts cover the whole $\mathbb{R}^n$ with only boundary overlap.

Let Γ^* the **reciprocal** (or **dual**) lattice [**1**] to Γ. It lives in the dual space $(\mathbb{R}^n)^*$, but if an inner product $(\cdot, \cdot)$ in $\mathbb{R}^n$ is fixed, Γ^* can be realized in $\mathbb{R}^n$ as the set of all vectors κ such that $(\kappa, \gamma) \in 2\pi\mathbb{Z}$ for all $\gamma \in \Gamma$ (if $\Gamma = \mathbb{Z}^n$, then $\Gamma^* = 2\pi\mathbb{Z}^n$). We fix a fundamental domain **B** for Γ^* (e.g., the **first Brillouin zone** [**1**]). Then $\mathbb{T} := \mathbb{R}^n/\Gamma$ ($\mathbb{T}^* := \mathbb{R}^n/\Gamma^*$) is a torus and Γ-(Γ^*-) periodic functions on $\mathbb{R}^n$ are naturally identified with functions on $\mathbb{T}$ ($\mathbb{T}^*$).

The periodicity of the spectral problem $Lu = \lambda u$ with respect to Γ suggests to use the well known **Bloch-Floquet transform** (the name varies from a source to

[1] No generality will be lost if the reader assumes that Γ is the integer lattice $\mathbb{Z}^n$.

a source) [**1**, **20**–**22**, **36**] :

$$f(x) \mapsto \hat{f}(k, x) := \sum_{\gamma \in \Gamma} f(x+\gamma)e^{-ik\cdot\gamma}. \tag{2}$$

Here k is a real (or complex) n-dimensional vector, which is called **quasi-momentum** (or **crystal momentum** and **Bloch momentum**). Assuming that f decays sufficiently fast, there is no convergence problem. It is easy to check that for any (even complex) quasi-momentum k, the function $\hat{f}(k, x)$ is Γ^*-periodic with respect to k and is of the **Bloch** (also called **Floquet**) **form** with respect to x, i.e.

$$\hat{f}(k, x) = e^{ik\cdot x} v_k(x), \tag{3}$$

where $v_k(x)$ is Γ-periodic. The values $x \in \mathbf{W}$ and $k \in \mathbf{B}$ are sufficient for determining the whole function $\hat{f}(k, x)$. One can consider $\hat{f}(k, x)$ as a function $\hat{f}(k, \cdot)$ on $\mathbf{B}$ (or better, on the torus $\mathbb{T}^*$) with values in a space of functions on $\mathbf{W}$. Considering the torus $\mathbb{T}^*$, it is more natural to consider $\hat{f}(k, \cdot)$ as a function of the **Floquet multiplier** $z := e^{ik} := (e^{ik\cdot a_1}, \ldots, e^{ik\cdot a_n})$, rather than of the quasi-momentum k. Then the torus $\mathbb{T}^*$ becomes the unit torus $\{z||z_j| = 1, j = 1, \ldots, n\}$ in $\mathbb{C}^n$. We will also need some complex neighborhoods of the space of real quasi-momenta and of the torus $\mathbb{T}^*$, defined for a given $\alpha > 0$:

$$\mathcal{D}_\alpha = \{k \in \mathbb{C}^n | \, |\mathrm{Im}\ k \cdot a_j| < \alpha, j = 1, ..., n\}, \tag{4}$$

and its image under the transform $k \mapsto z$

$$\Omega_\alpha = \{z = (z_1, ..., z_n) \in \Omega \,|\, e^{-\alpha} < |z_j| < e^\alpha,\ j = 1, ..., n\}. \tag{5}$$

Here $\{a_j\}$ is the basis of Γ mentioned before.

DEFINITION 2.1.

- *If Ω is an n-dimensional complex domain and H is a Hilbert space, We will denote by $A(\Omega, H)$ the space of all H-valued analytic functions on Ω, equipped with the topology of uniform convergence on compacta.*
- *The space $L^2_\alpha(\mathbb{R}^n)$ consists of all functions $f \in L^2_{loc}(\mathbb{R}^n)$ such that for any $0 < b < \alpha$ the following expression is finite:*

$$\psi_b(f) := sup_{\gamma\in\Gamma} \|f\|_{L^2(\mathbf{W}+\gamma)} e^{b|\gamma|} < \infty. \tag{6}$$

This space is equipped with the natural topology defined by the semi-norms ψ_b.

We now quote some standard results about the Bloch-Floquet transform (analogs of the standard Plancherel, Paley-Wiener, and inversion theorems for the Fourier series), see e.g., [**21**, **24**]:

THEOREM 2.2.

(1) *If $f \in L^2(\mathbb{R}^n)$ and $K \subset \mathbb{R}^n$ is a compact, then the series (2) converges in the space $L^2(\mathbb{T}^*, L^2(K))$. Moreover, the following equality (Plancherel theorem) holds:*

$$\|f\|^2_{L^2(\mathbb{R}^n)} = \int_{\mathbf{B}} \|\hat{f}(k, \cdot)\|^2_{L^2(\mathbf{W})} d'k = \int_{\mathbb{T}^*} \|\hat{f}(z, \cdot)\|^2_{L^2(\mathbf{W})} d'z, \tag{7}$$

where $d'k$ is the normalized to total measure 1 Lebesque measure on $\mathbf{B}$, and $d'z$ is the normalized Haar measure on $\mathbb{T}^*$.

(2) *For any* $\alpha \in (0, \infty]$, *Bloch-Floquet transform*

$$f \mapsto \hat{f}$$

is a topological isomorphism of the space $L^2_\alpha(\mathbb{R}^n)$ *onto* $A(\Omega_\alpha, L^2(\mathbf{W}))$.

(3) *For any* $f \in L^2(\mathbb{R}^n)$ *the following inversion formula holds:*

$$f(x) = \int_{\mathbb{T}^*} \hat{f}(k, x) d'k, \; x \in \mathbb{R}^n. \tag{8}$$

The first statement claims that the Bloch-Floquet transform is an isometry between the natural Hilbert spaces, the second shows that exponential decay transforms into analyticity in a neighborhood of the torus $\mathbb{T}^*$ (a Paley-Wiener type theorem), and the third one provides an inversion of the transform.

REMARK 2.3. *In the definition (5) of the domain* Ω_α *one can allow a vector* $\alpha = (\alpha_1, \ldots, \alpha_n)$, *modifying accordingly (5) to*

$$\Omega_\alpha = \{z = (z_1, ..., z_n) \in \Omega \,|\, e^{-\alpha_j} < |z_j| < e^{\alpha_j}, \, j = 1, ..., n\}. \tag{9}$$

Then the second statement of Theorem 2.2 still holds, if one modifies accordingly the definition (6) of the space L^2_α, *requiring anisotropic decay.*

The Bloch-Floquet transform block diagonalizes the periodic operator L, from which one obtains the well known [**1**, **20**, **21**, **24**, **36**] **band-gap spectral structure** of the operator L. Namely,

THEOREM 2.4. [**21**, **44**]
Let $L(k)$ *be the operator* L *acting on the Bloch functions (3) with a fixed quasi-momentum* k. *If we label (for real* k*) the eigenvalues of* L *in nondecreasing order as*

$$\lambda_1(k) \leq \lambda_2(k) \leq \cdots \mapsto \infty, \tag{10}$$

then

(1) ***band functions*** $\lambda_j(k)$ *are continuous,* Γ^**-periodic, and piece-wise analytic in* k;

(2) *if* I_j *is the range of* $\lambda_j(k)$ *(the* j**th band***), then the spectrum of* L *is*

$$\sigma(L) = \bigcup_j I_j.$$

(3) *The corresponding* ***Bloch eigenfunctions*** $\phi_j(k, x)$ *(i.e.,* $L\phi_j = \lambda_j \phi_j$*) can be chosen* $\phi_j(k, \cdot)$ *as a piece-wise analytic* L^2*-functions on* $\mathbf{B}$ *with values in* $L^2(\mathbf{W})$, *whose norm in* $L^2(\mathbf{W})$ *is almost everywhere constant and can be chosen equal to 1. One can also assume that* $\phi(k, \cdot)$ *is* Γ^**-periodic with respect to* k.

2.1. Bloch and Wannier functions. There are two bases of functions (distributions) extremely useful for studying constant coefficients linear PDEs in $\mathbb{R}^n$. These are the basis of delta functions (well localized in the space) and the basis of plane waves (well localized in the dual space). They are dual to each other under the Fourier transform. One would want to have similar bases in the periodic case. Analogs of plane waves is clear - the Bloch functions. One suspects that an analog of delta functions could be obtained from these by the Bloch-Floquet transform. This indeed works (to some degree) and leads to the so called Wannier functions.

Let $\phi(k,x)$ be a Bloch function with the quasi-momentum k, depending "sufficiently nicely" on k. E.g., in our applications we can always assume that $\phi \in L^2(\mathbb{T}^*, L^2(\mathbf{W}))$.

DEFINITION 2.5. *The* Wannier function $w(x)$ *corresponding to the Bloch function* $\phi(k,x)$ *is*

$$w(x) = \int\limits_{\mathbb{T}^*} \phi(k,x)d'k, \quad x \in \mathbb{R}^n. \tag{11}$$

Comparing this definition with (8), one sees that the Wannier function $w(x)$ is just the inverse Bloch-Floquet transform of $\phi(k,x)$, and vice-versa, $\phi(k,x)$ is the Bloch-Floquet transform of $w(x)$.

It is often useful to have a Wannier function $w(x)$ normalized and having mutually orthogonal lattice shifts $w(x-\gamma)$. The following auxiliary result [**23**] is useful:

LEMMA 2.6.

(1) *The Wannier function* $w(x)$ *belongs to* $L^2(\mathbb{R}^n)$ *and*

$$\int\limits_{\mathbb{R}^n} |w(x)|^2 dx = \int\limits_{\mathbf{B}} \|\phi(k,\cdot)\|^2_{L^2(\mathbf{W})} d'k. \tag{12}$$

(2) *Functions* $w_\gamma(x) := w(x-\gamma)$ *are mutually orthogonal for* $\gamma \in \Gamma$ *iff the function* $\phi(k,x)$ *in (11) has a* k*-independent norm in* $L^2(\mathbf{W})$.

The most important property of Wannier functions, due to which one wants to use them for numerical calculations in many problems of physics is their decay. An exponential decay is often desired.

A direct consequence of Theorem 2.2) is that smoothness with respect to k of the Bloch function $\phi_j(k,x)$ corresponds to decay of $w(x)$ [2]:

LEMMA 2.7. [**21**, Section 2.2]

(1) *If* $\sum\limits_{\gamma\in\Gamma} \|w_j\|_{L^2(\mathbf{W}+\gamma)} < \infty$, *then* $\phi_j(k,\cdot)$ *is a continuous* $L^2(\mathbf{W})$*-valued function on* $\mathbb{T}^*$.
(2) *Infinite differentiability of* $\phi_j(k,\cdot)$ *as a function on* $\mathbb{T}^*$ *with values in* $L^2(\mathbf{W})$ *is equivalent to the decay of* $\|w_j\|_{L^2(\mathbf{W}+\gamma)}$ *faster than any power of* $|\gamma|$ *for* $\gamma \in \Gamma$.
(3) *Analyticity of* $\phi_j(k,\cdot)$ *as a function on* $\mathbb{T}^*$ *with values in* $L^2(\mathbf{W})$ *is equivalent to the exponential decay of* $\|w_j\|_{L^2(\mathbf{W}+\gamma)}$.

2.2. Wannier bases for spectral subspaces. Suppose that a compact set $S \subset \mathbb{R}$ is the union of m ($m \geq 1$) bands of the spectrum of L (with overlaps allowed) and is separated by spectral gaps at the bottom and the top from the rest of the spectrum. An example is the part of the spectrum from its bottom till the first gap. We call such subset S a **composite band** (versus just a **band** when $m = 1$).

Surrounding S by a closed contour $\Sigma \in \mathbb{C}$, separating S from the rest of the spectrum, one can define in the space $L_2(\mathbb{R}^n)$ the Riesz projector [**36**] P onto the

[2]Smoothness also means matching the values and all derivatives across the boundaries of Brillouin zones.

spectral subspace H_S that corresponds to S. Restricting one's attention only to this subspace, one wants to have a convenient basis for numerical computations. Orthonormal bases of exponentially decaying Wannier functions are known to be the excellent candidates [**27–29, 33**]. So, let us see how and whether one can find such bases.

Applying the above spectral projector to the direct integral decomposition of the operator L (keeping the same contour Σ for all $k \in B$), one gets an analytic with respect to k family $P(k)$ of m-dimensional spectral projectors for the operators $L(k)$. It thus produces the m-dimensional **Bloch bundle** Λ_S over the torus $\mathbb{T}^*$

$$\Lambda_S = \bigcup_{\mathbb{T}^*} Ran(P(k)), \tag{13}$$

which corresponds to the spectral subset S. Then elements of the space H_S correspond under Bloch-Floquet transform to L_2-sections of this bundle.

The discussion above (see more details in [**23**]) shows that in order to have an orthonormal basis of exponentially decaying Wannier functions in H_S, one need to find an *analytic with respect to k* basis $\phi_j(k,x)|_{j=1}^m$ of Bloch functions (sections of Λ_S) and take its (inverse) Bloch-Floquet transform. One immediately realises that (even if we were looking for a basis of continuous sections rather than analytic ones) this requires the bundle to be trivial. As it is shown in [**23**], the so called Oka's principle (implemented by Grauert [**11–13**]) assures that there is no additional obstructions to getting an analytic, rather than just a continuous basis. Thus, topological non-triviality of this bundle is the only possible obstacle to the existence of desirable Wannier bases. It was shown by D. Thouless [**40**], that such obstacle can indeed occur. (For topological insulators this is not an obstacle, but rather a blessing [**2**].)

There are cases when the triviality is known. E.g., G. Nenciu [**32**] showed in 1983 (see also [**15**]) that when $m = 1$ (single band), presence of time reversal symmetry guarantees triviality of the bundle and thus the obstacle disappears. G. Panati [**33**] proved in 2007, that in dimensions $n \leq 3$, even if $m > 1$, the time reversal symmetry still removes the obstruction. However, what can one do if the obstruction exists? In [**23**] one of the authors showed that if one allows some overdeterminancy of Wannier functions (rather than basis property), the obstacle disappears. In order to describe the result, we need to introduce some (rather popular nowadays) notions concerning frames.

2.3. Parseval frames. We recall here the notion of the so called **tight (or Parseval) frame** of vectors (e.g., [**14**]) that replaces orthonormality in the overdetermined case.

DEFINITION 2.8.

- *A set of vectors $v_j, j = 1, 2, \ldots$ in a Hilbert space H is said to be a* **frame**, *if for some constants $A, B > 0$ and any vector $u \in H$ the following inequality holds:*

$$A \sum_j |(u, v_j)|^2 \leq \|u\|^2 \leq B \sum_j |(u, v_j)|^2.$$

- *This set is said to be a* **tight (or Parseval) frame** *if* $A, B = 1$, *i.e. for any vector* $u \in H$ *the equality holds*

$$\sum_j |(u, v_j)|^2 = \|u\|^2.$$

The name "Parseval frame" is justified by the fact that the formulas for expansion and synthesis for Parseval frames are exactly the same as for orthonormal bases. This follows from the simple observation [**14**] that tight frames are exactly the orthogonal projections of orthonormal bases from larger Hilbert spaces. At the same time, the overdetermined nature of the collection of vectors provides improved stability with respect to errors.

2.4. Wannier Parseval frames. The following result was proven in [**23**]:

THEOREM 2.9. *Let L be a self-adjoint elliptic Γ-periodic operator in $\mathbb{R}^n, n \geq 1$ and $S \subset \mathbb{R}$ be the union of m spectral bands of L. Suppose that S is separated from the rest of the spectrum. Then there exists a finite number l of exponentially decaying composite Wannier functions $w_j(x)$ such that their shifts $w_{j,\gamma} := w_j(x - \gamma), \gamma \in \Gamma$ form a tight (Parseval) frame in the spectral subspace H_S of the operator L. This means that for any $f(x) \in H_S$, the equality holds*

$$\int\limits_{\mathbb{R}^n} |f(x)|^2 dx = \sum_{j,\gamma} |\int\limits_{\mathbb{R}^n} f(x)\overline{w_{j,\gamma}(x)} dx|^2. \tag{14}$$

Here the number $l \in [m, 2^n m]$ is equal to the smallest dimension of a trivial bundle containing an equivalent copy of Λ_S. In particular, $l = m$ if and only if Λ_S is trivial, in which case an orthonormal basis of exponentially decaying composite Wannier functions exists.

Our goal here is to prove the result announced in [**24**, section 6.5], which in physical dimensions drastically reduces the number of extra Wannier functions needed.

3. In dimension three, only one extra Wannier function is needed

As the techniques and results of [**23**] surveyed in the section 2 show, the main issue is finding the smallest rank of a trivial bundle containing a copy of the Bloch bundle in question. Moreover, due to the Grauert theorem [**11–13**] (an instance of the Oka's principle), only topological triviality is needed[3].

When the dimension of the base does not exceed 3, the answer is given by the following, probably known, theorem, a proof of which we provide for completeness. All complexes below are assumed to be compact.

THEOREM 3.1. *Any complex vector bundle of rank m over any 3-dimensional CW complex X embeds into a trivial bundle of rank $m + 1$.*

Let us prove first the following auxiliary statement:

LEMMA 3.2. *Any complex vector bundle of rank larger than 1 over a 3-dimensional CW complex X admits a non-vanishing section.*

[3]For basic notions involved in this and the next sections, one can refer to [**16, 30, 38**].

PROOF. Let us introduce a fiber metric on the bundle. Clearly one can find a non-zero (in fact unit length) section over the 0-skeleton of X. Let us proceed by induction with respect to the dimension of the skeleton. Assuming there is a section over the k-skeleton, consider extending to the $(k+1)$-skeleton. By induction hypothesis, we have a section over the boundary of each (k + 1)-cell. Such may be viewed as a map from the boundary of the cell to the unit sphere in the fiber, i.e. $S^k \mapsto S^{2m-1}$. However, for $m > 1$, the fundamental group and the second homotopy group of S^{2m-1} are trivial, so the section can be extended. □

We can now return to proving Theorem 3.1.

PROOF. Lemma 3.2 shows that any complex vector bundle E of rank greater than 1 over a 3-complex contains a trivial sub-bundle of rank 1. I.e., E is the direct sum of a trivial bundle and its orthogonal complement. By induction, any complex vector m-bundle over a 3-complex is isomorphic to $L \bigoplus \mathbb{C}^{m-1}$, where L is a linear bundle.

Computing the first Chern class, we to obtain $c_1(L \bigoplus \mathbb{C}^{m-1}) = c_1(L)$, and the first Chern class of the line bundle L is determined by the first Chern class of the bundle E.

Complex line bundles are classified by the first Chern class, i.e. are isomorphic if and only if their first Chern classes coincide. Moreover, any second cohomology class is the 1rst Chern class of some complex line bundle.

Now one checks that $L^* \bigoplus E = L^* \bigoplus L \bigoplus \mathbb{C}^{m-1}$, where L^* is the dual bundle to L, has trivial 1st Chern class, and so is trivial. □

REMARK 3.3.

(1) *Under conditions of Theorem* 3.1, *one sees from its proof that any m-bundle has a trivial $m-1$ subbundle.*
(2) *One can show that in dimension 4, the number $m+1$ should be replaced by $m+2$.*

We can now establish the main result:

THEOREM 3.4. *Let L be a self-adjoint elliptic Γ-periodic operator in $\mathbb{R}^n, n \leq 3$ and $S \subset \mathbb{R}$ be the union of m spectral bands of L, separated by gaps from the rest of the spectrum. Then there exist $m+1$ exponentially decaying composite Wannier functions $w_j(x)$, such that their shifts $w_{j,\gamma} := w_j(x-\gamma), \gamma \in \Gamma$ form a tight (Parseval) frame in the spectral subspace H_S of the operator L. This means that for any $f(x) \in H_S$, the equality holds*

$$\int_{\mathbb{R}^n} |f(x)|^2 dx = \sum_{j,\gamma} |\int_{\mathbb{R}^n} f(x)\overline{w_{j,\gamma}(x)} dx|^2. \tag{15}$$

A choice of m such functions is possible if and only if Λ_S is trivial, in which case an orthonormal basis of exponentially decaying composite Wannier functions exists.

We provide here just an outline of the proof, with missing details that can be found in [**23**]

PROOF. Consider the m-dimensional Bloch bundle Λ_S, which is analytic on a neighborhood Ω_α (see (5)) of the torus $\mathbb{T}^* \subset \mathbb{C}^d$, with $d \leq 3$. As it is noted in [**23**], this neighborhood is a Stein domain, which deformation retracts to the torus, and thus analytic and topological classifications of vector bundles here coincide, due to the Grauert Theorem [**11–13**]. According to Theorem 3.1, there is an analytic line bundle L such that $\Lambda_S \bigoplus L$ is trivial.

Consider the infinite dimensional bundle

$$\Sigma_S := \bigcup_{\Omega_a} Ran(I - P(k)), \tag{16}$$

complementary to Λ_S.

Due to Kuiper's theorem [**25**], Σ_S is topologically, and due to Bungart's version [**4**] of the Grauert Theorem, analytically trivial as well. As shown in [**46**] (see the details in [**23**]), an isomorphic copy of L (which we denote with the same letter L) can be found in Σ_S. Then the $(m+1)$-subbundle $\Lambda_S \bigoplus L$ in the trivial bundle $\Omega_\alpha \times L_2(W)$ is trivializable. Thus, the construction of [**23**] (picking a holomorphic basis in $\Lambda_S \bigoplus L$ and then projecting it into Λ_S) provides the required Parseval frame and thus finishes the proof. □

4. Dimension four

The four-dimensional situation is more complex, but less physically relevant, so we formulate the corresponding topological result here without the proof.

The answer to the question of trivialization of vector bundles over a 4-complex is given in the following proposition.

THEOREM 4.1. *Let E be a rank m complex vector bundle E over a 4-complex X.*

1. *E is trivial if and only if the total Chern class $c(E)$ is equal to 1.*
2. *E embeds in a trivial rank $m+1$ bundle if and only if $c_2(E) = c_1(E)^2$.*
3. *Any rank m complex vector bundle over a 4-complex embeds into a trivial rank $m+2$ bundle.*

Exactly as in the previous section, using Theorem 4.1 and following the scheme of the proof in [**23**], one obtains

THEOREM 4.2. *Let L be a self-adjoint elliptic Γ-periodic operator in $\mathbb{R}^4$ and $S \subset \mathbb{R}$ be the union of m spectral bands of L, separated by gaps from the rest of the spectrum. Then there exist $m+2$ exponentially decaying composite Wannier functions $w_j(x)$, such that their shifts $w_{j,\gamma} := w_j(x-\gamma), \gamma \in \Gamma$ form a tight (Parseval) frame in the spectral subspace H_S of the operator L. This means that for any $f(x) \in H_S$, the equality holds*

$$\int_{\mathbb{R}^4} |f(x)|^2 dx = \sum_{j,\gamma} |\int_{\mathbb{R}^4} f(x)\overline{w_{j,\gamma}(x)}dx|^2. \tag{17}$$

5. Generic choice of a "missing" function

Now the question arises of how one can find these $m+1$ Wannier functions. The question is non-trivial even in the absence of the topological obstacle. There are recent strong works approaching this issue (see, e.g., [**6–8, 10, 31, 35**]).

Without answering this question, we make a comment here that might turn out to be useful.

Let us assume that we know how to find the Wannier functions in absence of the topological obstruction (i.e., how to find a holomorphic basis of a trivial Bloch bundle Λ_S). Then, if the obstacle is present, the proofs of Theorems 2.9 and 3.4

show that one finds a linear bundle L such that $\Lambda_S \bigoplus L$ is trivial, embeds it into Σ_S, and uses a holomorphic basis of $\Lambda_S \bigoplus L$ to produce the desired frame of Wannier functions. The question is: assuming that such L is found, how can one practically imbed it into Σ_S? The idea is that a transversality theorem (e.g., [**18**]) implies that essentially "almost any" bundle morphism from L into Σ_S will do.

To simplify the situation, consider a linear analytic bundle M, such that the 2-bundle $N := L \bigoplus M$ is trivial. Then, an imbedding of N into Σ_S would generate an imbedding of L.

It is easy to find a non-trivial morphism of M into Σ_S. Indeed, consider for instance a two-dimensional linear subspace Q in $Ran\,(I - P(k))$ for some $k = k_0$ and consider the trivialization map from M to $\Omega_\alpha \times Q$, which is an isomorphism by construction. However, $\Omega_\alpha \times Q$ does not sit inside of Σ_S, except of the fiber at the point k_0. Multiplying by the projector-function $(I - P(k))$, we fix this difficulty and obtain a non-trivial bundle morphism $\Phi : M \mapsto \Sigma_S$, which might fail to be imbedding, except in a neighborhood of the base point k_0.

THEOREM 5.1. *A generic perturbation of any morphism $\Phi : M \mapsto \Sigma_S$ produces an imbedding of M into Σ_S.*

In fact, the statement holds also if one replaces Σ_S (which is infinite dimensional and trivial [**23, 46**]) by any holomorphic vector bundle R of a sufficiently large rank r. Indeed, in this case one deals with a $r \times 2$-matrix function on a three-dimensional manifold. One needs to make sure that the rank of all matrices is maximal. For a sufficiently high r, the sub-variety of the matrices of smaller rank has co-dimension larger than the dimension of the set of parameters Ω_α. Then application of a transversality theorem [**18**] finishes the proof.

So, a "random" perturbation of the non-trivial morphism described above should produce an imbedding and thus, as it is described above, a Parseval frame of exponentially decaying Wannier functions.

6. Conclusion and remarks

(1) We have shown that in the spatial dimension up to three, for any isolated m-band S of the spectrum of an elliptic self-adjoint operator periodic with respect to a lattice Γ, there exists a family of $m+1$ exponentially decaying composite (generalized) Wannier functions such that their Γ-shifts form a Parseval frame in the spectral subspace H_S that corresponds to S. This result was announced in [**24**].

In dimension 4, one needs $m + 2$ such functions.

(2) Theorem 2.2 and Lemma 2.7 show that the best rate of exponential decay one can expect from Wannier functions is dictated by the value α in the domain Ω_α of analyticity of the spectral projectors $P(k)$ (the larger α, the faster is the decay). The proof of the main result of this paper shows that the Parseval frames allow for the same best rate of decay.

One can improve on the decay rate by allowing anisotropic exponential decay, and thus using a multi-index $\alpha = (\alpha_1, \dots, \alpha_n)$ and corresponding domains Ω_α (see Remark 2.3).

(3) The proof of Theorem 3.1 shows (see Remark 3.3) that when the spatial dimension does not exceed 3, $m - 1$ holomorphic linearly independent sections can be found in Λ_S, even if the bundle is non-trivial. So, it is just

one family of Wannier functions that is missing and has to be replaced by two.

(4) The original restriction made on the operator is not essential. The results also apply in an abstract situation of a periodic elliptic self-adjoint operator on an abelian covering of a compact manifold, graph, or quantum graph, as long as the torsion free rank of the abelian deck group (rather than the spatial dimension) does not exceed 4. Neither formulations, nor proofs require any modifications in this case.

(5) The results of this article are of pure existence nature and do not provide any answer to the question of how one can actually construct the required number of the Wannier functions. However, a remark on generic nature of the required functions is made.

(6) One should mention that a powerful numerical machinery for creating bases and frames of decaying Wannier functions has been developed in [**3**, **27**–**29**], and studied analytically in [**10**, **31**, **33**, **34**, **45**]. Recent progress in direction of actual construction of Wannier bases and frames has also been achieved [**6**–**8**, **10**]. We do not consider his issue, except showing in Section 5 that a "generic" choice of the missing $(m+1)$st function should work.

(7) The main result of the paper was announced without proof in [**24**, Section 6.5].

Acknowledgments

The two authors discussed and partially resolved this problem at the MSRI program on Inverse Problems and Applications in the Fall of 2010. At the time the first author was Associate Director at the institute and his research was partially supported by NSF grant DMS - 0604994. The second author expresses his gratitude to the NSF grants DMS-0406022 and DMS-1517938 for partial support of this work and to J. Corbin, A. Levitt, D. Monaco, G. Panati, and M. Zaidenberg for useful discussions.

References

[1] N.W. Ashcroft and N.D.Mermin, *Solid State Physics*, Holt, Rinehart and Winston, New York-London, 1976.

[2] B. Andrei Bernevig, *Topological insulators and topological superconductors*, Princeton University Press, Princeton, NJ, 2013. With Taylor L. Hughes. MR3185492

[3] Ch. Brouder, G. Panati, M. Calandra, Ch. Mourougane, and N. Marzari, Exponential localization of Wannier functions in insulators, Phys. Rev. Lett. **98** (2007), 046402.

[4] Lutz Bungart, *On analytic fiber bundles. I. Holomorphic fiber bundles with infinite dimensional fibers*, Topology **7** (1967), 55–68, DOI 10.1016/0040-9383(86)90015-7. MR0222338

[5] K. Busch, S. F. Mingaleev, A. Garcia-Martin, M. Schillinger, and Daniel Hermann, The Wannier function approach to photonic crystal circuits, J. Phys.: Condens. Matter **15** (2003), R1233–R1256.

[6] Cancès, E.; Levitt, A.; Panati, G.; Stoltz, G.: Robust determination of maximally-localized Wannier functions, Phys. Rev. B **95** (2017), 075114.

[7] Horia D. Cornean, Ira Herbst, and Gheorghe Nenciu, *On the construction of composite Wannier functions*, Ann. Henri Poincaré **17** (2016), no. 12, 3361–3398, DOI 10.1007/s00023-016-0489-2. MR3568019

[8] H. D. Cornean and D. Monaco, Parseval frames of localized Wannier functions, arXiv: 1704.00932

[9] H. L. Cycon, R. G. Froese, W. Kirsch, and B. Simon, *Schrödinger operators with application to quantum mechanics and global geometry*, Springer Study Edition, Texts and Monographs in Physics, Springer-Verlag, Berlin, 1987. MR883643

[10] Domenico Fiorenza, Domenico Monaco, and Gianluca Panati, *Construction of real-valued localized composite Wannier functions for insulators*, Ann. Henri Poincaré **17** (2016), no. 1, 63–97, DOI 10.1007/s00023-015-0400-6. MR3437825

[11] Hans Grauert, *Approximationssätze für holomorphe Funktionen mit Werten in komplexen Rämen* (German), Math. Ann. **133** (1957), 139–159, DOI 10.1007/BF01343296. MR0098197

[12] Hans Grauert, *Holomorphe Funktionen mit Werten in komplexen Lieschen Gruppen* (German), Math. Ann. **133** (1957), 450–472, DOI 10.1007/BF01343758. MR0098198

[13] Hans Grauert, *Analytische Faserungen über holomorph-vollständigen Räumen* (German), Math. Ann. **135** (1958), 263–273, DOI 10.1007/BF01351803. MR0098199

[14] Deguang Han, Keri Kornelson, David Larson, and Eric Weber, *Frames for undergraduates*, Student Mathematical Library, vol. 40, American Mathematical Society, Providence, RI, 2007. MR2367342

[15] B. Helffer and J. Sjöstrand, *Équation de Schrödinger avec champ magnétique et équation de Harper* (French), Schrödinger operators (Sønderborg, 1988), Lecture Notes in Phys., vol. 345, Springer, Berlin, 1989, pp. 118–197, DOI 10.1007/3-540-51783-9_19. MR1037319

[16] Dale Husemoller, *Fibre bundles*, McGraw-Hill Book Co., New York-London-Sydney, 1966. MR0229247

[17] J. D. Joannopoulos, S. G. Johnson, J. N. Winn, and R. D. Meade, *Photonic Crystals: Molding the Flow of Light*, Princeton University Press 2008.

[18] Sh. Kaliman and M. Zaĭdenberg, *A tranversality theorem for holomorphic mappings and stability of Eisenman-Kobayashi measures*, Trans. Amer. Math. Soc. **348** (1996), no. 2, 661–672, DOI 10.1090/S0002-9947-96-01482-1. MR1321580

[19] W. Kohn, *Analytic properties of Bloch waves and Wannier functions*, Phys. Rev. (2) **115** (1959), 809–821. MR0108284

[20] P. A. Kuchment, *Floquet theory for partial differential equations* (Russian), Uspekhi Mat. Nauk **37** (1982), no. 4(226), 3–52, 240. MR667973

[21] Peter Kuchment, *Floquet theory for partial differential equations*, Operator Theory: Advances and Applications, vol. 60, Birkhäuser Verlag, Basel, 1993. MR1232660

[22] Peter Kuchment, *The mathematics of photonic crystals*, Mathematical modeling in optical science, Frontiers Appl. Math., vol. 22, SIAM, Philadelphia, PA, 2001, pp. 207–272. MR1831334

[23] Peter Kuchment, *Tight frames of exponentially decaying Wannier functions*, J. Phys. A **42** (2009), no. 2, 025203, 16, DOI 10.1088/1751-8113/42/2/025203. MR2525287

[24] Peter Kuchment, *An overview of periodic elliptic operators*, Bull. Amer. Math. Soc. (N.S.) **53** (2016), no. 3, 343–414, DOI 10.1090/bull/1528. MR3501794

[25] Nicolaas H. Kuiper, *The homotopy type of the unitary group of Hilbert space*, Topology **3** (1965), 19–30, DOI 10.1016/0040-9383(65)90067-4. MR0179792

[26] J. M. Luttinger and W. Kohn, Motion of electrons and holes in a perturbed periodic field, Phys. Rev. **97** (1955), 869–883.

[27] N. Marzari, I. Souza, and D. Vanderbilt, An introduction to maximally-localized Wannier functions, Highlight of the Month, Psi-K Newsletter **57** (2003), 129–168. http://psi-k.dl.ac.uk/psi-k/newsletters.html.

[28] N. Marzari and D. Vanderbilt, Maximally localized generalized Wannier functions for composite energy bands, Phys. Rev. B **56** (1997), no. 20, 12847–12865.

[29] Marzari, N.; Mostofi A.A.; Yates J.R.; Souza I.; Vanderbilt D. : Maximally localized Wannier functions: Theory and applications. Rev. Mod. Phys. 84 (2012), 1419.

[30] John W. Milnor and James D. Stasheff, *Characteristic classes*, Princeton University Press, Princeton, N. J.; University of Tokyo Press, Tokyo, 1974. Annals of Mathematics Studies, No. 76. MR0440554

[31] Domenico Monaco, Gianluca Panati, Adriano Pisante, and Stefan Teufel, *Optimal Decay of Wannier functions in Chern and Quantum Hall Insulators*, Comm. Math. Phys. **359** (2018), no. 1, 61–100, DOI 10.1007/s00220-017-3067-7. MR3781446

[32] G. Nenciu, *Existence of the exponentially localised Wannier functions*, Comm. Math. Phys. **91** (1983), no. 1, 81–85. MR719812

[33] Gianluca Panati, *Triviality of Bloch and Bloch-Dirac bundles*, Ann. Henri Poincaré **8** (2007), no. 5, 995–1011, DOI 10.1007/s00023-007-0326-8. MR2342883

[34] Gianluca Panati and Adriano Pisante, *Bloch bundles, Marzari-Vanderbilt functional and maximally localized Wannier functions*, Comm. Math. Phys. **322** (2013), no. 3, 835–875, DOI 10.1007/s00220-013-1741-y. MR3079333

[35] Read, N. : Compactly-supported Wannier functions and algebraic K-theory Wannier functions, Phys. Rev. B **95** (2017), 115309.

[36] Michael Reed and Barry Simon, *Methods of modern mathematical physics. IV. Analysis of operators*, Academic Press [Harcourt Brace Jovanovich, Publishers], New York-London, 1978. MR0493421

[37] M. Schillinger, S. Mingaleev, D. Hermann, and K. Busch, Highly localized Wannier functions for the efficient modeling of Photonic Crystal Circuits, SPIE, Photonic crystal materials and devices, Conference No3, San Jose CA , ETATS-UNIS (24/01/2005) **5773** (2005), 324–335.

[38] Norman Steenrod, *The topology of fibre bundles*, Princeton Landmarks in Mathematics, Princeton University Press, Princeton, NJ, 1999. Reprint of the 1957 edition; Princeton Paperbacks. MR1688579

[39] Elias M. Stein and Guido Weiss, *Introduction to Fourier analysis on Euclidean spaces*, Princeton University Press, Princeton, N.J., 1971. Princeton Mathematical Series, No. 32. MR0304972

[40] D. J. Thouless, Wannier functions for magnetic sub-bands, J. Phys. C: Solid State Phys. **17** (184), L325-7.

[41] G. H. Wannier, The Structure of electronic excitation levels in insulating crystals, Phys. Rev. **52** (1937), 191–197.

[42] http://www.wannier.org/

[43] D. M. Whittaker and M. P. Croucher, Maximally localized Wannier functions for photonic lattices, Phys. Rev. B **67** (2003), 085204.

[44] Calvin H. Wilcox, *Theory of Bloch waves*, J. Analyse Math. **33** (1978), 146–167, DOI 10.1007/BF02790171. MR516045

[45] Jiao Yang, S. Mingaleev, M. Schillinger, D. A. B. Miller, Shanhui Fan, and K. Busch, Wannier basis design and optimization of a photonic crystal waveguide crossing, IEEE photonics technology letters **17** (2005) no. 9, 1875–1877.

[46] M. G. Zaĭdenberg, S. G. Kreĭn, P. A. Kučment, and A. A. Pankov, *Banach bundles and linear operators* (Russian), Uspehi Mat. Nauk **30** (1975), no. 5(185), 101–157. MR0415661

Mathematics Department, Kansas State University, Manhattan, Kansas
Email address: dav@math.ksu.edu

Mathematics Department, Texas A&M University, College Station, Texas 77843-3368
Email address: kuchment@math.tamu.edu

Contemporary Mathematics
Volume **717**, 2018
https://doi.org/10.1090/conm/717/14452

Non-demolition measurements of observables with general spectra

M. Ballesteros, N. Crawford, M. Fraas, J. Fröhlich, and B. Schubnel

ABSTRACT. It has recently been established that, in a non-demolition measurement of an observable $\mathcal{N}$ with a finite point spectrum, the density matrix of the system approaches an eigenstate of $\mathcal{N}$, i.e., it "purifies" over the spectrum of $\mathcal{N}$. We extend this result to observables with general spectra. It is shown that the spectral density of the state of the system converges to a delta function exponentially fast, in an appropriate sense. Furthermore, for observables with absolutely continuous spectra, we show that the spectral density approaches a Gaussian distribution over the spectrum of $\mathcal{N}$. Our methods highlight the connection between the theory of non-demolition measurements and classical estimation theory.

1. Introduction

In an indirect measurement, information about a quantum system S is obtained by performing a sequence of standard von Neumann measurements on probes that have previously interacted with S. A theory of indirect measurements has been proposed by Kraus [**1**]. Upon tracing out the degrees of freedom of the probes, the effective time evolution of the system is described by jump operators, V_ξ, indexed by probe measurement outcomes ξ, which act on the Hilbert space of pure state vectors of the system S. These operators encode the statistics of measurement results and the conditional evolution of the system. If a result ξ is recorded in a probe measurement, the state, $|\psi\rangle$, of the system changes according to the rule

$$|\psi\rangle \quad \to \quad \frac{V_\xi|\psi\rangle}{||V_\xi|\psi\rangle||}. \tag{1}$$

In order to describe the probabilities of different measurement outcomes, ξ, the set, $\mathcal{X}$, of all such outcomes must be equipped with a probability measure. In particular, we need to specify an a-priori measure, μ, "counting" the different measurement results. The probability measure on $\mathcal{X}$ is then chosen to be

$$\langle\psi|V_\xi^* V_\xi|\psi\rangle d\mu(\xi). \tag{2}$$

2010 *Mathematics Subject Classification.* Primary 81P15, 81P16; Secondary 62F12.
Key words and phrases. Quantum non-demolition measurement, Large sample theory.
M. Ballesteros was supported in part by grants PAPIIT (UNAM) IN 103918 and SEP-CONACYT 254062.

Consistency imposes a normalisation condition on the jump operators, namely

$$\int_{\mathcal{X}} V_\xi^* V_\xi \mathrm{d}\mu(\xi) = 1.$$

Apart from this condition, the operators V_ξ can be chosen arbitrarily. Eqs. (1, 2) are consequences of the *Born rule*. The precise form of the jump operators V_ξ can be derived from the joint Hamiltonian evolution of the system and the probes and the Born rule applied to the probe measurements. We do not repeat this derivation here; but see, e.g., [**2**]. In our analysis we will make a fixed choice of jump operators.

If the system interacts with a sequence of independent probes, Eqs. (1, 2) can be iterated so as to obtain the probability of recording a sequence, $\underline{\xi} = (\xi_1, \xi_2, \dots)$, of measurement results, along with the corresponding changes of the state of the system. A fundamental problem in the theory of indirect measurements is to understand the asymptotic behaviour of the probability distribution on the space of sequences $\underline{\xi}$ of measurement outcomes and of the state of the system. Various aspects of this problem have been studied by different authors: Conditions for asymptotic purification have been given in [**3**], entropy production has been studied in [**4**], and conditions for uniqueness of the invariant measure have been derived in [**5**]; a general approach has been outlined in [**6**].

We consider a special case of indirect measurements – so called non-demolition measurements – with the feature that all jump operators V_ξ are functions of the observable $\mathcal{N}$ of S that one wants to measure. The motivation to study this case comes from experiments carried out in the group of Haroche and Raimond [**7**] whose theoretical description fits into the framework developed in our paper. In these experiments it is observed that the state of the system gradually approaches an eigenstate of a certain observable $\mathcal{N}$ (the number of photons stored in a cavity). A theoretical description of this phenomenon has been proposed in [**8**], see also [**3**], and studied more fully in a series of papers; see [**6**, **9**, **10**], where further details are provided. In the present paper, we continue this line of research by relaxing the conditions on the spectrum of the observable $\mathcal{N}$; (in all previous works only observables with discrete spectra have been considered).

We will exhibit the phenomenon of "purification" over the spectrum of $\mathcal{N}$ for the example of an observable with a general spectrum: Let $|\psi_k\rangle$ denote the state of the system S after the kth probe measurement. We show that, for all continuous functions f, $\langle\psi_k|f(\mathcal{N})|\psi_k\rangle$ approaches the value $f(\nu)$, for some point ν in the spectrum of $\mathcal{N}$, as $k \to \infty$, and that the frequency of occurrence of a specific point ν is given by Born's rule applied to the initial state, $|\psi_0\rangle$, of S. We also determine the rate of approach to the limit, as $k \to \infty$. (For precise statements of assumptions and results see Theorem 2.2.)

If the spectrum of $\mathcal{N}$ is non-degenerate and discrete the convergence of the spectral measure of $|\psi_k\rangle$ implies that $|\psi_k\rangle$ approaches an eigenstate of $\mathcal{N}$, as $k \to \infty$. If the observable $\mathcal{N}$ has continuous spectrum this conclusion does not hold, because there are no normalisable eigenstates associated with points ν in the continuous spectrum of $\mathcal{N}$. If, however, the spectrum of $\mathcal{N}$ is purely absolutely continuous then states of S can be represented as functions, $\psi(\nu)$, on the spectrum of $\mathcal{N}$ that are square-integrable with respect to some measure absolutely continuous with respect to Lebesgue measure. Under suitable hypotheses, the wave function, $\psi_k(\nu)$, of the state $|\psi_k\rangle$ then turns out to approach a Gaussian function whose width shrinks to

0, as $k \to \infty$. The exact description of this convergence result and the assumptions implying that the states $|\psi_k\rangle$ converge, as $k \to \infty$, are given in Theorem 2.7.

To arrive at this generalisation of known results has required a shift in perspective: It turns out to be useful to map the quantum-mechanical problem (or, at least, parts of it) onto a problem of classical parameter estimation. The phenomenon of purification over the spectrum of an observable then turns into the one of consistency of the maximum likelihood estimator. The hypotheses required for consistency are well known. For our readers' convenience, we will present detailed proofs using quantum mechanical language. Appropriate references for the results underlying our analysis will be provided as well.

In the next section we describe our setting and summarise our main results. We divide these results into law-of-large-number type results valid for arbitrary spectra of the observable, provided some regularity conditions hold, and central-limit type results for which absolute continuity of the spectrum of the observable is needed. In accordance with this division, proofs are presented in Section 3 and Section 4, respectively.

2. Setup and Main Results

General mixed states of a physical system S are density matrices, ρ, acting on a separable Hilbert space $\mathcal{H}$. Let μ be a σ-finite (counting) measure on a measure space $(\mathcal{X}, \Sigma_P)$ of probe measurement outcomes. We consider a family of measurable bounded operators V_ξ acting on $\mathcal{H}$ required to satisfy the normalisation condition

$$\int_{\mathcal{X}} V_\xi^* V_\xi \, \mathrm{d}\mu(\xi) = 1. \tag{3}$$

Furthermore, we introduce the space, $\Xi \equiv \mathcal{X}^{\mathbb{N}}$, of infinite sequences $\underline{\xi} \equiv (\xi_1, \xi_2, \dots)$ of probe measurement outcomes equipped with the standard sigma algebra, Σ, generated by cylinder sets. The measure space (Ξ, Σ) has a natural filtration $(\Sigma_k)_{k=1,2,\dots}$, where Σ_k consists of sets determined by the first k measurement results. Without danger of confusion we will identify a cylinder set $E \in \Sigma_k$ with its base $E \in \Sigma_P^{\times k}$. We define an operator-valued stochastic process, V_k, adapted to the above filtration by setting

$$V_k(\underline{\xi}) = V_{\xi_k} \dots V_{\xi_1}.$$

With every density matrix ρ on $\mathcal{H}$ and every $k = 1, 2, \dots$, we associate a probability measure, $\mathbb{P}_\rho^{(k)}$, on (Ξ, Σ_k) by setting

$$\mathbb{P}_\rho^{(k)}(E) := \int_E \mathrm{tr}(V_k(\underline{\xi})\rho V_k^*(\underline{\xi}))\mathrm{d}\mu(\xi_1)\dots\mathrm{d}\mu(\xi_k), \quad E \in \Sigma_k. \tag{4}$$

By a well known lemma due to Kolmogorov, these measures determine a unique measure, $\mathbb{P}_\rho$, on the space (Ξ, Σ). We define a "posterior state" by

$$\rho_k(\underline{\xi}) := \frac{V_k(\underline{\xi})\rho V_k^*(\underline{\xi})}{\mathrm{tr}(V_k(\underline{\xi})\rho V_k^*(\underline{\xi}))}. \tag{5}$$

If the initial state ρ is a rank-one projection, $|\psi_0\rangle\langle\psi_0|$, then the the posterior state is a rank-one projection, $|\psi_k(\underline{\xi})\rangle\langle\psi_k(\underline{\xi})|$, with $|\psi_k(\underline{\xi})\rangle = V_k(\underline{\xi})|\psi_0\rangle / ||V_k(\underline{\xi})|\psi_0\rangle||$.

In the case of non-demolition measurements considered in this paper, the operators V_ξ are functions of a self-adjoint bounded operator $\mathcal{N}$, where $\mathcal{N}$ is the "observable" to be measured. By functional calculus, the operators $V_\xi \equiv V_\xi(\mathcal{N})$

are then determined by a measurable family, $V_\xi : \nu \in \mathbb{R} \mapsto V_\xi(\nu) \in \mathbb{C}$, of bounded complex-valued functions satisfying a normalisation condition

$$\int_{\mathcal{X}} |V_\xi(\nu)|^2 \mathrm{d}\mu(\xi) = 1, \quad \text{for all} \quad \nu \in \mathbb{R}.$$

This normalisation condition implies Eq. (3). Let $\mathrm{d}\lambda_\rho(\nu)$ denote the spectral measure of $\mathcal{N}$ with respect to a state ρ, (i.e., $\mathrm{tr}(g(\mathcal{N})\rho) = \int_{\sigma(\mathcal{N})} g(\nu)\mathrm{d}\lambda_\rho(\nu)$, for an arbitrary Borel-measurable, integrable, function g). For a cylinder set E of the form $E_1 \times \cdots \times E_k$, with $E_j \in \Sigma_P, j = 1, \ldots, k$, the measure $\mathbb{P}_\rho$ introduced in (4) is given by

$$\mathbb{P}_\rho(E) = \int_{\sigma(\mathcal{N})} \mathrm{d}\lambda_\rho(\nu)\mu_\nu(E_1)\ldots\mu_\nu(E_k). \qquad \mu_\nu(E_j) = \int_{E_j} \mathrm{d}\mu(\xi) f(\xi|\nu), \tag{6}$$

where

$$f(\xi|\nu) := |V_\xi(\nu)|^2 \tag{7}$$

acquires the meaning of a conditional probability distribution. We denote by $\mathbb{E}_\nu[\cdot]$ the expectation with respect to the measure μ_ν. Eq. (6) is the *de Finetti decomposition* [**11**] of the measure $\mathbb{P}_\rho$.

Let

$$l(\nu|\xi) := \log f(\xi|\nu),$$

be the log-likelihood function, and define by

$$l_k(\nu) \equiv l_k(\nu|\underline{\xi}) := \frac{1}{k}\sum_{j=1}^{k} l(\nu|\xi_j) \tag{8}$$

the log-likelihood function of a sequence of k measurements. The maximum-likelihood estimator of the value of $\mathcal{N}$ is then given by

$$\hat{\mathcal{N}}_k := \mathrm{argmax}_{\nu\in\sigma(\mathcal{N})} l_k(\nu). \tag{9}$$

For any given realization of $\underline{\xi}$, there may be more than one ν for which the RHS is maximized. If such an ambiguity arises we choose the value of $\hat{\mathcal{N}}_k$ from the set of maximas according to some predetermined rule. In [**12**], Theorem 18.19, it is proved that this can be done in a measurable fashion under the hypotheses adopted in our paper.

Following the notational convention introduced in the previous paragraph, we mostly forgo the $\underline{\xi}$- dependence of functions. We write $\rho_k \equiv \rho_k(\underline{\xi}), V_k \equiv V_k(\underline{\xi})$, etc. The $\underline{\xi}$- dependence is re-introduced at points where we feel that it will be helpful in following our arguments.

2.1. Law of Large Numbers. We prove convergence results for $\hat{\mathcal{N}}_k$ and for the states ρ_k introduced in (5) under minimal hypotheses corresponding to assumptions required for the consistency of a maximum likelihood estimator.

Throughout this article, we assume that $\mathcal{N}$ is a bounded operator. Hence $\sigma(\mathcal{N})$ is a compact subset of $\mathbb{R}$ equipped with the induced metric.

ASSUMPTION 2.1. *For each $\nu \in \sigma(\mathcal{N})$, the function $f(\xi|\nu)$ belongs to $L^1(\mathcal{X}, \mathrm{d}\mu)$ and has the following further properties.*

(1) **Identifiability.** *The map*
$$\nu \in \sigma(\mathcal{N}) \quad \mapsto \quad f(\xi|\nu) \in L^1(\mathcal{X}, \mathrm{d}\mu)$$
is injective, i.e., for $\nu \neq \nu'$, the functions $f(\xi|\nu)$ and $f(\xi|\nu')$ are not identical.

(2) **Continuity.** *For every $\xi \in \mathcal{X}$, the function $f(\xi|\nu)$ is continuous on the spectrum $\sigma(\mathcal{N})$ of the observable $\mathcal{N}$.*

(3) **Dominance.** *The log-likelihood function $l(\nu|\xi) = \log f(\xi|\nu)$ is dominated in the sense that*
$$\sup_{\nu' \in \sigma(\mathcal{N})} |l(\nu'|\xi)| \in L^1(\mathcal{X}, f(\xi|\nu)\mathrm{d}\mu(\xi)),$$
for all $\nu \in \sigma(\mathcal{N})$.

The last part of this assumption guarantees that the *relative entropy,*

$$S(\nu|N) = \min_{\nu' \in N} \mathbb{E}_\nu[l(\nu|\xi) - l(\nu'|\xi)], \tag{10}$$

is well defined for any point ν and any closed subset N of the spectrum $\sigma(\mathcal{N})$. Jensen's inequality implies that the relative entropy is non-negative, $S(\nu|N) \geq 0$, and Assumption 2.1.1 implies that $S(\nu|N) = 0$ if and only if $\nu \in N$. Moreover, from [**12**], Theorem 18.19, we infer that $S(\cdot|N)$ is measurable.

For a Borel set $N \in \sigma(\mathcal{N})$, we let $\Pi(N)$ denote the spectral projection of $\mathcal{N}$ associated with N.

THEOREM 2.2. *Given Assumption 2.1, the maximum likelihood estimator $\hat{\mathcal{N}}_k$ converges almost surely to a random variable $\hat{\mathcal{N}}_\infty$, and, for any Borel set $N \subset \sigma(\mathcal{N})$,*

$$\mathbb{P}_\rho(\underline{\xi} : \hat{\mathcal{N}}_\infty \in N) = \mathrm{tr}(\Pi(N)\rho). \tag{11}$$

Furthermore, if N is the closure of an arbitrary open subset O of the spectrum $\sigma(\mathcal{N})$ of the operator $\mathcal{N}$ contained in the support of the measure λ_ρ, then

$$-\lim_{k\to\infty} \frac{1}{k} \log \mathrm{tr}(\Pi(N)\rho_k) = S(\hat{\mathcal{N}}_\infty|N), \quad \mathbb{P}_\rho - \textit{almost surely.} \tag{12}$$

The first part of the theorem says that the density matrices $\rho_k, k = 1, 2, \ldots,$ purify over the spectrum of $\sigma(\mathcal{N})$. As $k \to \infty$, the spectral measure of $\mathcal{N}$ associated with ρ_k concentrates on the point $\hat{\mathcal{N}}_\infty$. The speed of convergence of this concentration is described in the second part of the theorem. It is quantified by a large deviation principle with a rate expressed in terms of the relative entropy (10). For observables with discrete spectrum, various versions of this statement have been established previously, see [**6, 9, 10**].

2.2. Central Limit Theorem. To describe the asymptotic behavior of $\hat{\mathcal{N}}_k$ corresponding to the central limit theorem we require additional regularity assumptions. It is convenient to assume that the probability distributions $f(\xi|\nu)$ are defined for every $\nu \in \mathbb{R}$ in such manner that $f(\xi|\cdot)$ is continuous for all ξ, and that $f(\xi|\nu) = 1$ for all ν outside a compact interval containing $\sigma(\mathcal{N})$.

ASSUMPTION 2.3. *For all $\nu \in \mathbb{R}$, the following conditions hold:*

(1) **Positivity.** *For all $\nu \in \sigma(\mathcal{N})$, the probability distribution function $f(\xi|\nu)$, see Eq. (7), is strictly positive.*

(2) **Continuity.** *The probability distribution* $f(\xi|\nu)$ *is twice continuously differentiable in* ν.

(3) **Integrability.** *There exists a function* $g(\xi) \in L^1(\mathcal{X}, \mathrm{d}\mu_\nu(\xi))$ *such that*

$$|\partial_\nu^j f(\xi|\nu')| < g(\xi), \quad j = 1, 2,$$

for all $\nu' \in \mathbb{R}$, *and the following differentiation under the integral sign holds true,*

$$\int \partial_\nu^j f(\xi|\nu)\mathrm{d}\mu(\xi) = \partial_\nu^j \int f(\xi|\nu)\mathrm{d}\mu(\xi) \ (= 0), \qquad j \in \{1, 2\}. \tag{13}$$

The log-likelihood function is square-integrable, $\partial_\nu l(\nu|\xi) \in L^2(\mathcal{X}, \mathrm{d}\mu_\nu(\xi))$, *and* $\mathbb{E}_\nu[-\partial_\nu^2 l(\nu|\xi))]$ *is strictly positive, for all* $\nu \in \sigma(\mathcal{N})$.

The quantity $F(\nu) := \mathbb{E}_\nu[(\partial_\nu l(\nu|\xi))^2)]$ is known as the "Fisher Information" of the family of distributions $f(\xi|\nu)$. Since $f(\xi|\nu)$ is a probability distribution, the right hand side of Eq. (13) vanishes. This then implies that the Fisher information is also given by $F(\nu) = -\mathbb{E}_\nu[\partial_\nu^2 l(\nu|\xi)]$. The following theorem is a version of the central limit theorem, adapted to our setting.

THEOREM 2.4. *Suppose Assumptions* 2.1 *and* 2.3 *are valid. Then convergence in distribution,*

$$\sqrt{k}[\hat{\mathcal{N}}_k - \hat{\mathcal{N}}_\infty] \quad \xrightarrow{d} \quad \mathscr{N}(0, F^{-1}(\hat{\mathcal{N}}_\infty)), \tag{14}$$

holds, where $\mathscr{N}(0, \sigma^2)$ *is the normal distribution with zero mean and variance* σ^2.

In a more restricted setting, such a result has been proven in [**6**]. Note that, in Eq. (14), the variance of the limit-distribution is itself a random variable. In more precise terms, the following holds: For every $\nu \in \sigma(\mathcal{N})$, let X_ν be a random variable with distribution $\mathscr{N}(0, F^{-1}(\nu))$. Pick some $a \in \mathbb{R}$, and define $A_k := \left(\sqrt{k}(\hat{\mathcal{N}}_k - \hat{\mathcal{N}}_\infty)\right)^{-1}((-\infty, a])$. Then (see Eq. (6))

$$\lim_{k\to\infty} \mathbb{P}_\rho(A_k) = \lim_{k\to\infty} \int_{\sigma(\mathcal{N})} \mathrm{d}\lambda_\rho(\nu)\mu_\nu^{\otimes\mathbb{N}}(A_k) = \int_{\sigma(\mathcal{N})} \mathrm{d}\lambda_\rho(\nu)\mathbb{P}(X_\nu \in (-\infty, a]). \tag{15}$$

In view of Lemma 3.2 below, it is claimed that $\hat{\mathcal{N}}_\infty = \nu$, $\mu_\nu^{\mathbb{N}}$-almost everywhere. Here $\mathbb{P}(X_\nu \in (-\infty, a])$ denotes the probability that $X_\nu \in (-\infty, a]$.

In order to conclude asymptotic normality of the posterior states ρ_k, we require further assumptions on $\mathcal{N}$.

ASSUMPTION 2.5.

(1) **Uniform multiplicity.** *We suppose that* $\mathcal{N}$ *is of uniform multiplicity* n, $\mathcal{H} = L^2(\mathbb{R}, \mathbb{C}^n, \mathrm{d}\Theta)$ *for a regular Borel measure* Θ, *and*

$$(\mathcal{N}f)(\nu) = \nu f(\nu). \tag{16}$$

(2) **Absolute continuity.** Θ *is absolutely continuous with respect to the Lebesgue measure* λ. *We denote its Radon-Nikodym derivative by* h,

$$\mathrm{d}\Theta(\nu) = h(\nu)\mathrm{d}\lambda(\nu).$$

(3) **Regularity.** *The function* h *is supported on the spectrum* $\sigma(\mathcal{N})$ *of* $\mathcal{N}$, *and it is continuous and strictly positive on* $\sigma(\mathcal{N})$. *The boundary of* $\sigma(\mathcal{N})$ *is assumed to have Lebesgue measure* 0.

The notion of uniform multiplicity is discussed, e.g., in [**13**, Section VII.2]. The first part of the third assumption is redundant, as the support of Θ equals $\sigma(\mathcal{N})$. In Remark 4.2 we explain how to weaken this assumption.

For a Hilbert space $\mathcal{K}$, we denote by $\mathcal{B}_1(\mathcal{K})$ the space of trace class operators. Under Assumption 2.5, an operator $\tau \in \mathcal{B}_1(\mathcal{H})$ is an integral operator with a matrix-valued integral kernel $\tau(\nu, \nu')$, where $\nu, \nu' \in \mathbb{R}$:

$$(\tau f)(\nu) = \int_{\mathbb{R}} \tau(\nu, \nu') f(\nu') d\Theta(\nu').$$

Theorem 2.12 in [**14**] states that if the kernel of τ is continuous then the trace of τ is given by

$$\mathrm{tr}(\tau) = \int \mathrm{tr}_{\mathbb{C}^n}\Big(\tau(\nu, \nu)\Big) d\Theta(\nu), \tag{17}$$

where $\mathrm{tr}(\cdot)$ denotes the trace with respect to $L^2(\mathbb{R}, \mathbb{C}^n, d\Theta)$ and $\mathrm{tr}_{\mathbb{C}^n}(\cdot)$ is the trace on $n \times n$ complex matrices, $\mathbb{M}_n(\mathbb{C})$. Eq. (17) remains valid also for an integral kernel $\tau(\nu, \nu')$ that is continuous for ν, ν' in a bounded measurable set N and vanishes for $\nu \notin N$ or $\nu' \notin N$.

For a real valued function $\theta(\nu|\xi)$, the transformation

$$V_\xi(\nu) \quad \mapsto \quad \exp(i\theta(\nu|\xi)) V_\xi(\nu) \tag{18}$$

transforms the density matrix by a random gauge transformation

$$\rho_k(\nu, \nu') \quad \mapsto \quad e^{ik\theta_k(\nu)} \rho_k(\nu, \nu') e^{-ik\theta_k(\nu)},$$

where

$$\theta_k(\nu) := \frac{1}{k} \sum_{j=1}^{k} \theta(\nu|\xi_k).$$

We address the question of convergence of such transformation as k goes to infinity in Remark 2.8. In the main text we fix a convenient gauge.

ASSUMPTION 2.6. *The function $V_\xi(\nu)$ is real and positive.*

The time evolved density matrix then takes the form

$$\rho_k(\nu, \nu') = \frac{e^{\frac{1}{2}kl_k(\nu)} \rho(\nu, \nu') e^{\frac{1}{2}kl_k(\nu')}}{\int_{\sigma(\mathcal{N})} e^{kl_k(\nu)} \mathrm{tr}(\rho(\nu, \nu)) \mathrm{d}\Theta(\nu)}. \tag{19}$$

We recall that $l_k(\nu)$ was defined in Eq (8). The following theorem shows that the density matrix ρ_k is close to a Gaussian state as k tends to infinity. We define the following normalized Gaussian kernel

$$\boldsymbol{G}_F(\nu, \nu') = \frac{1}{\int \mathrm{e}^{-\frac{F}{2}x^2}\, dx} \mathrm{e}^{-\frac{F}{4}(\nu^2 + \nu'^2)}.$$

Moreover, for a density matrix ρ with a continuous kernel $\rho(\nu, \nu')$ we set

$$c_\rho(\nu) := \frac{\rho(\nu, \nu)}{\mathrm{tr}_{\mathbb{C}^n} \rho(\nu, \nu)}. \tag{20}$$

When the numerator above vanishes, we set c_ρ to be zero.

For a fixed sequence $\underline{\xi}$ we set

$$\mathcal{B}_1^{(k)} := \mathcal{B}_1\Big(L^2\big[\mathbb{R}, \mathbb{C}^n, h(\hat{\mathcal{N}}_k + \frac{\nu}{\sqrt{k}}) \mathrm{d}\lambda(\nu)\big]\Big).$$

We recall the definition of Fisher information at a point ν, $F(\nu) = \mathbb{E}_\nu[(\partial_\nu l(\nu|\xi))^2)]$.

THEOREM 2.7. *We require Assumptions* 2.1, 2.3, 2.5 *and* 2.6. *Moreover, we assume that that* $\rho(\nu,\nu')$, *restricted to* $\sigma(\mathcal{N})\times\sigma(\mathcal{N})$, *is continuous. In addition, we require that for every* ν

$$\lim_{k\to\infty} e^{-kl_k(\hat{\mathcal{N}}_k)}\int_{\mathbb{R}} e^{kl_k(\hat{\mathcal{N}}_k+\frac{\nu'}{\sqrt{k}})}\mathrm{d}\lambda(\nu') = \int_{\mathbb{R}} \mathrm{e}^{-\frac{F(\nu)}{2}\nu'^2}\mathrm{d}\lambda(\nu') \tag{21}$$

holds $\mu_\nu^{\otimes\mathbb{N}}$*- almost surely. Then,*

$$\lim_{k\to\infty}\left\|\frac{1}{\sqrt{k}}\rho_k(\hat{\mathcal{N}}_k+\frac{\nu}{\sqrt{k}},\hat{\mathcal{N}}_k+\frac{\nu'}{\sqrt{k}}) - \frac{c_\rho(\hat{\mathcal{N}}_\infty)}{h(\hat{\mathcal{N}}_\infty)}\boldsymbol{G}_{F(\hat{\mathcal{N}}_\infty)}(\nu,\nu')\right\|_{\mathcal{B}_1^{(k)}} = 0,$$

almost surely with respect to the measure $\mathbb{P}_\rho$.

Notice that Theorem 2.2 implies that $\hat{\mathcal{N}}_\infty\in\sigma(\mathcal{N})$ almost surely. For this reason the denominator $h(\hat{\mathcal{N}}_\infty)$ is almost surely strictly positive. Assumption (21) is natural in the context of the Bernstein - von Mises Theorem [**15**, Theorem 21].

REMARK 2.8. *Suppose that Assumption* 2.6 *is not satisfied and consider the polar decomposition of* $V_\xi(\nu)$*:*

$$V_\xi(\nu) = \exp(-i\theta(\nu|\xi))\exp(\frac{1}{2}l(\nu|\xi)), \tag{22}$$

where $\theta(\nu|\xi)$ *is a phase. Assume that the function* $\theta(\nu|\xi)$ *is twice continuously differentiable in* ν*, for almost all* ξ*, and that for every* $\nu'\in\mathbb{R}$ *there exists a function* $g\in L^1(\mathcal{X},d\mu_{\nu'}(\xi))$ *such that* $|\partial_\nu^2\theta(\nu|\xi)|<g(\xi)$*, for all* $\nu\in\mathbb{R}$*. Then the conclusion of Theorem* 2.7 *would be*

$$\frac{1}{\sqrt{k}}e^{i\sqrt{k}\partial_\nu\theta_k(\hat{\mathcal{N}}_k)(\nu-\nu')}\rho_k(\hat{\mathcal{N}}_k+\frac{\nu}{\sqrt{k}},\hat{\mathcal{N}}_k+\frac{\nu'}{\sqrt{k}})$$
$$\to \frac{c_\rho(\hat{\mathcal{N}}_\infty)}{h(\hat{\mathcal{N}}_\infty)}\boldsymbol{G}_{F(\hat{\mathcal{N}}_\infty)}(\nu,\nu')\mathrm{e}^{-\frac{i}{2}\mathbb{E}_{\hat{\mathcal{N}}_\infty}[\partial_\nu^2\theta(\hat{\mathcal{N}}_\infty)|\xi)]^2(\nu^2-\nu'^2)}$$

in the same topology as specified in the theorem.

3. Law of Large Numbers

The purpose of this section is to prove the convergence results for $\hat{\mathcal{N}}_k$ and ρ_k, as $k\to\infty$, formulated in Theorem 2.2. We split the statements of this theorem into two parts. The main tool employed in the proofs of both parts will be the uniform law of large numbers, which we now recall, see [**15**, Theorem 16].

THEOREM 3.1. *Let* M *be a compact metric space, and let* $X_j(a)\overset{d}{=}X(a), a\in M$*, be a sequence of i.i.d. random variables. Suppose that* $X(a)$ *is almost surely continuous in* a*, and assume that there is a positive random variable* g *with a finite first moment such that* $|X(a)|<g$ *holds almost surely, for all* $a\in M$*. Then*

$$\sup_{a\in M}|\frac{1}{N}\sum_{j=1}^N X_j(a) - \mathbb{E}[X(a)]| \quad \xrightarrow{a.s.} \quad 0,$$

as N *tends to infinity.*

The next lemma claims a convergence result for $\hat{\mathcal{N}}_k$ with respect to the measure $\mu_\nu^{\otimes\mathbb{N}}$. This is a classical result first proven in [**16**]. We follow a proof given in [**15**].

LEMMA 3.2. *Suppose that Assumption* 2.1 *holds true. Then*

$$\lim_{k\to\infty} \hat{\mathcal{N}}_k = \nu,$$

almost surely with respect to $\mu_\nu^{\otimes \mathbb{N}}$.

PROOF. We fix $\nu \in \sigma(\mathcal{N})$ and consider the random variable $l_k(\nu')$ – see Eq. (8) – on the measure space $(\Xi, \mu_\nu^{\otimes \mathbb{N}})$. Given Assumption 2.1, all conditions needed to apply the uniform law of large numbers for $l_k(\nu)$ are satisfied, and we have that

$$\sup_{\nu' \in \sigma(\mathcal{N})} |l_k(\nu'|\underline{\xi}) - \mathbb{E}_\nu[l(\nu'|\xi)]| \quad \to \quad 0\,, \quad \mu_\nu^{\otimes \mathbb{N}} - \text{almost surely.}$$

In particular, for any closed subset N of $\sigma(\mathcal{N})$,

$$S_k(\nu|N) = \min_{\nu' \in N} (l_k(\nu) - l_k(\nu'))$$

converges almost surely to $S(\nu|N)$.
Let U be an open neighborhood of ν and let U^c its complement in $\sigma(\mathcal{N})$. Then, by Assumption 2.1.1, $S(\nu|U^c) > 0$, and we conclude that $\mu_\nu^{\otimes \mathbb{N}}$-almost surely there exists $k_0 \equiv k_0(\underline{\xi})$ such that $S_k(\nu|U^c) > 0$, for all $k > k_0$. By definition,

$$0 = S_k(\nu|\sigma(\mathcal{N})) = S_k(\nu|\hat{\mathcal{N}}_k),$$

whence $\hat{\mathcal{N}}_k$ belongs to U almost surely. It follows that $\hat{\mathcal{N}}_k$ converges to ν almost surely. □

The convergence result for $\hat{\mathcal{N}}_k$ (Eq. (11) in Theorem 2.2) is a direct consequence of the lemma.

PROPOSITION 3.3. *Given Assumption* 2.1 *the maximum likelihood estimator* $\hat{\mathcal{N}}_k$ *converges almost surely, as* $k \to \infty$, *and, for any Borel set* $N \subset \sigma(\mathcal{N})$,

$$\mathbb{P}_\rho(\underline{\xi} : \lim_{k\to\infty} \hat{\mathcal{N}}_k \in N) = \mathrm{tr}(\Pi(N)\rho). \tag{23}$$

PROOF. By $\mathcal{A}$ we denote the set of points $\underline{\xi}$ for which $\lim_{k\to\infty} \hat{\mathcal{N}}_k(\underline{\xi})$ exists. The limit is denoted by $\hat{\mathcal{N}}_\infty(\underline{\xi})$. Since $\limsup_k \hat{\mathcal{N}}_k$ and $\liminf_k \hat{\mathcal{N}}_k$ are measurable functions, and since $\mathcal{A}$ is the set of points on which these two functions coincide, the set $\mathcal{A}$ is measurable. Lemma 3.2 implies that $\mu_\nu^{\otimes \mathbb{N}}(\mathcal{A}) = 1$, for every ν. Almost sure convergence with respect to $\mathbb{P}_\rho$, i.e. $\mathbb{P}_\rho(\mathcal{A}) = 1$, then follows from Eq. (6):

$$\mathbb{P}_\rho(\mathcal{A}) = \int_{\sigma(\mathcal{N})} \mu_\nu^{\otimes \mathbb{N}}(\mathcal{A}) \mathrm{d}\lambda_\rho(\nu) = \int_{\sigma(\mathcal{N})} \mathrm{d}\lambda_\rho(\nu) = 1.$$

Next, we prove Eq. (23) using Eq. (6):

$$\mathbb{P}_\rho\Big(\hat{\mathcal{N}}_\infty^{-1}(N)\Big) = \int_{\sigma(\mathcal{N})} \mu_\nu^{\otimes \mathbb{N}}\Big(\hat{\mathcal{N}}_\infty^{-1}(N)\Big) \mathrm{d}\lambda_\rho(\nu) = \int_N \mathrm{d}\lambda_\rho(\nu) = \mathrm{tr}(\Pi(N)\rho),$$

where we use that $\hat{\mathcal{N}}_\infty = \nu$, $\mu_\nu^{\otimes \mathbb{N}}$-almost surely; (see Lemma 3.2). □

The second part of Theorem 2.2, concerning the speed of concentration of ρ_k around $\hat{\mathcal{N}}_\infty$, is the content of the following proposition.

PROPOSITION 3.4. *We require Assumption 2.1. Let N be the closure of an arbitrary open subset O of the spectrum $\sigma(\mathcal{N})$ of the operator $\mathcal{N}$. Suppose that N is contained in the support of the measure λ_ρ. Then*

$$-\lim_{k\to\infty}\frac{1}{k}\log\operatorname{tr}(\Pi(N)\rho_k) = S(\hat{\mathcal{N}}_\infty|N), \quad \mathbb{P}_\rho - \textit{almost surely.}$$

PROOF. We prove the Proposition in two steps.
Step 1: We prove that, for every ν' in the support of the measure λ_ρ,

$$-\lim_{k\to\infty}\frac{1}{k}\log\operatorname{tr}(\Pi(N)\rho_k) = S(\nu'|N), \quad \mu_{\nu'}^{\otimes\mathbb{N}} - \text{almost surely.} \tag{24}$$

The quantity $\operatorname{tr}(\Pi(N)\rho_k)$ can be expressed in terms of the likelihood function as

$$\operatorname{tr}(\Pi(N)\rho_k) = \frac{\int_N \mathrm{d}\lambda_\rho(\nu)\mathrm{e}^{kl_k(\nu)}}{\int_{\sigma(\mathcal{N})}\mathrm{d}\lambda_\rho(\nu)\mathrm{e}^{kl_k(\nu)}}.$$

Let $I := \max_{\nu\in N}\mathbb{E}_{\nu'}[l(\nu|\xi)]$. We note that Eq. (24) follows from

$$\lim_{k\to\infty}\frac{1}{k}\log\int_N \mathrm{d}\lambda_\rho(\nu)\mathrm{e}^{kl_k(\nu)} = I, \quad \mu_{\nu'}^{\otimes\mathbb{N}} - \text{almost surely,}$$

since $\mathbb{E}_{\nu'}[l(\nu'|\xi)] = \max_{\nu\in\sigma(\mathcal{N})}\mathbb{E}_{\nu'}[l(\nu|\xi)]$. We fix some $\varepsilon > 0$ and show that, for large enough k,

$$\left|\frac{1}{k}\log\int_N \mathrm{d}\lambda_\rho(\nu)\mathrm{e}^{kl_k(\nu)}\mathrm{e}^{-Ik}\right| \le \varepsilon,$$

$\mu_{\nu'}^{\otimes\mathbb{N}}$-almost surely.
Bound from above: Since $l_k(\nu|\underline{\xi})$ converges uniformly to $\mathbb{E}_{\nu'}[l(\nu|\xi)]$, by the uniform law of large numbers, we have that $\mu_{\nu'}^{\otimes\mathbb{N}}$-almost surely $l_k(\nu|\underline{\xi}) - I \le \varepsilon$, for large $k \equiv k(\underline{\xi})$ and all $\nu \in N$. Then

$$\frac{1}{k}\log\int_N \mathrm{d}\lambda_\rho(\nu)\mathrm{e}^{kl_k(\nu|\underline{\xi})}\mathrm{e}^{-Ik} \le \frac{1}{k}\log\int_N \mathrm{d}\lambda_\rho(\nu)\mathrm{e}^{k\varepsilon} \le \varepsilon.$$

Bound from below: Consider the open set

$$U(\varepsilon) := \{\nu\in\sigma(\mathcal{N}) : I - \mathbb{E}_{\nu'}[l(\nu|\xi)] < \varepsilon\}.$$

Since I is the maximum over $\nu \in N$ of $\mathbb{E}_{\nu'}[l(\nu|\xi)]$, we have that $N\cap U(\varepsilon) \neq \emptyset$. Since $N = \overline{O}$, the interior of $N\cap U(\varepsilon)$ is non-empty. Thus, since N is contained in the support of λ_ρ, it follows that $\lambda_\rho(N\cap U(\varepsilon)) > 0$, for every $\varepsilon > 0$. By the uniform law of large numbers, $\left|\mathbb{E}_{\nu'}[l(\nu|\xi)] - l_k(\nu|\underline{\xi})\right| < \varepsilon/4$, for all ν and for sufficiently large $k \equiv k(\underline{\xi})$, $\mu_{\nu'}^{\otimes\mathbb{N}}$-almost surely. It follows that $I - l_k(\nu|\underline{\xi}) < \varepsilon/2$, $\mu_{\nu'}^{\otimes\mathbb{N}}$-almost surely, for every $\nu \in U(\varepsilon/4)$ and for sufficiently large $k \equiv k(\underline{\xi})$. Hence

$$\begin{aligned}\frac{1}{k}\log\int_N \mathrm{d}\lambda_\rho(\nu)\mathrm{e}^{kl_k(\nu|\underline{\xi})}\mathrm{e}^{-Ik} &\ge \frac{1}{k}\log\int_{N\cap U(\varepsilon/4)}\mathrm{d}\lambda_\rho(\nu)\mathrm{e}^{-k\varepsilon/2}\\ &= -\varepsilon/2 + \frac{1}{k}\log\lambda_\rho(U(\varepsilon/4)) > -\varepsilon,\end{aligned}$$

if we choose $k > 2\varepsilon^{-1}|\log\lambda_\rho(U(\varepsilon/4))|$; (note that $N\cap U(\varepsilon/2)$ has a strictly positive measure).
Step 2: We now prove the proposition. We set

$$\mathcal{B} := \left\{\underline{\xi}\,\middle|\, -\lim_{k\to\infty}\frac{1}{k}\log\operatorname{tr}(\Pi(N)\rho_k(\underline{\xi})) = S(\hat{\mathcal{N}}_\infty(\underline{\xi})|N)\right\}. \tag{25}$$

Because the set where $\lim_{k\to\infty}\frac{1}{k}\log\operatorname{tr}(\Pi(N)\rho_k(\underline{\xi}))$ exists is measurable, and the function $S(\hat{\mathcal{N}}_\infty(\underline{\xi})|N)$ is measurable, the set $\mathcal{B}$ is measurable. Moreover, Lemma 3.2 implies that $\hat{\mathcal{N}}_\infty = \nu'$, $\mu_{\nu'}^{\otimes\mathbb{N}}$-almost surely, which when combined with Step 1 proves that

$$\lim_{k\to\infty} -\frac{1}{k}\log\operatorname{tr}(\Pi(N)\rho_k) = S(\nu'|N) = S(\hat{\mathcal{N}}_\infty|N), \quad \mu_{\nu'}^{\otimes\mathbb{N}} - \text{almost surely.} \tag{26}$$

We conclude that $\mu_{\nu'}^{\otimes\mathbb{N}}(\mathcal{B}) = 1$, for every ν' in the support of λ_ρ. Finally, Eq (6) implies the desired result, $\mathbb{P}_\rho(\mathcal{B}) = 1$. □

4. Central Limit Theorem

In this section we prove Theorems 2.4 and 2.7 and conclude by discussing some extensions of our results.

4.1. Proof of Theorem 2.4. The proof is an adaptation, to our setting, of the proof of a theorem due to Cramér: see [**15**, Theorem 18].

PROOF OF THEOREM 2.4. We first fix $\upsilon \in \sigma(\mathcal{N})$. To understand the relevance of the Fisher Information, one notes that, since $\partial_\nu l_k(\hat{\mathcal{N}}_k) = 0$, the mean value theorem implies

$$\partial_\nu l_k(\upsilon) = \partial_\nu l_k(\upsilon) - \partial_\nu l_k(\hat{\mathcal{N}}_k) = -\partial_\nu^2 l_k(\nu')[\hat{\mathcal{N}}_k - \upsilon],$$

for some ν' in the interval between υ and $\hat{\mathcal{N}}_k$. As a consequence of the central limit theorem, the assumption that the Fisher information is finite implies that $\sqrt{k}\partial_\nu l_k(\upsilon)$ converges in distribution (with respect to $\mu_\upsilon^{\otimes\mathbb{N}}$) to a Gaussian random variable with mean 0 and variance $\mathbb{E}_\upsilon[(\partial_\nu l(\upsilon|\xi))^2)]$; (recall that $\mathbb{E}_\upsilon(\partial_\nu l(\upsilon|\xi)) = 0$).

By the uniform law of large numbers and the fact that $\hat{\mathcal{N}}_k$ converges to υ (almost surely with respect to $\mu_\upsilon^{\otimes\mathbb{N}}$), $\partial_\nu^2 l_k(\nu')$ converges to $\mathbb{E}_\upsilon[\partial_\nu^2 l(\upsilon|\xi)]$. Furthermore, one may check, using Assumption 2.3.2, that, for any point ν, $\mathbb{E}_\nu[\partial_\nu^2 l(\nu|\xi)] = -\mathbb{E}_\nu[(\partial_\nu l(\nu|\xi))^2)]$. Combining this with the last statement of the previous paragraph we arrive at the following convergence result:

$$\sqrt{k}[\hat{\mathcal{N}}_k - \upsilon] \quad \xrightarrow{d} \quad \mathscr{N}(0, F^{-1}(\upsilon)), \tag{27}$$

where the above convergence is in distribution with respect to the measure $\mu_\upsilon^{\otimes\mathbb{N}}$. Take $a \in \mathbb{R}$ and set $A_k = \left(\sqrt{k}(\hat{\mathcal{N}}_k - \hat{\mathcal{N}}_\infty)\right)^{-1}((-\infty, a])$, and

$$g_k(\upsilon) := \mu_\upsilon^{\otimes\mathbb{N}}\Big(\left(\sqrt{k}(\hat{\mathcal{N}}_k - \upsilon)\right)^{-1}((-\infty, a])\Big).$$

Eq. (6) implies that

$$\mathbb{P}_\rho(A_k) = \int_{\sigma(\mathcal{N})} d\lambda_\rho(\upsilon)\mu_\upsilon^{\otimes\mathbb{N}}(A_k) = \int_{\sigma(\mathcal{N})} d\lambda_\rho(\upsilon) g_k(\upsilon), \tag{28}$$

where we used Lemma 3.2. Finally, the desired result follows from the Lebesgue dominated convergence theorem and (27), (28).

□

4.2. Proof of Theorem 2.7. We consider the isometry

$$\iota : L^2(\mathbb{R}, \mathbb{C}^n, d\Theta) \to L^2(\mathbb{R}, \mathbb{C}^n, d\lambda), \qquad \phi \mapsto \iota(\phi) := \sqrt{h}\phi, \tag{29}$$

and denote by ι^{-1} the inverse on its range extended by zero on the orthogonal subspace. It follows that ι naturally defines an isometry between $\mathcal{B}_1(L^2(\mathbb{R}, \mathbb{C}^n, d\Theta))$ and $\mathcal{B}_1(L^2(\mathbb{R}, \mathbb{C}^n, d\lambda))$ given by

$$\begin{aligned}\mathcal{B}_1(L^2(\mathbb{R}, \mathbb{C}^n, d\Theta)) \ni \tau \mapsto \boldsymbol{\tau} &:= \iota\tau\iota^{-1} \\ &= \sqrt{h(\nu)}\tau(\nu,\nu')\sqrt{h(\nu')} \in \mathcal{B}_1(L^2(\mathbb{R}, \mathbb{C}^n, d\lambda)).\end{aligned} \tag{30}$$

We repeatedly use the above transformation to change ρ_k into $\boldsymbol{\rho}_k$.

PROPOSITION 4.1. *We require Assumptions* 2.1, 2.3, 2.5 *and* 2.6, *and we assume that $\rho(\nu, \nu')$ is continuous in $(\nu,\nu') \in \sigma(\mathcal{N}) \times \sigma(\mathcal{N})$. Suppose, moreover, that, for every ν,*

$$\lim_{k\to\infty} e^{-kl_k(\hat{\mathcal{N}}_k)} \int_{\mathbb{R}} e^{kl_k(\hat{\mathcal{N}}_k + \frac{\nu'}{\sqrt{k}})} \mathrm{d}\lambda(\nu') = \int_{\mathbb{R}} \mathrm{e}^{-\frac{F(\nu)}{2}\nu'^2} \mathrm{d}\lambda(\nu') \tag{31}$$

holds μ_ν- almost surely. Then, in the standard topology of trace class operators on the Hilbert space $L^2(\mathbb{R}, \mathbb{C}^n, d\lambda)$,

$$\lim_{k\to\infty} \frac{1}{\sqrt{k}} \boldsymbol{\rho}_k(\hat{\mathcal{N}}_k + \frac{\nu}{\sqrt{k}}, \hat{\mathcal{N}}_k + \frac{\nu'}{\sqrt{k}}) = c_\rho(\hat{\mathcal{N}}_\infty)\boldsymbol{G}_{F(\hat{\mathcal{N}}_\infty)}(\nu,\nu'), \tag{32}$$

almost surely with respect to the measure $\mathbb{P}_\rho$.

PROOF. Our proof is divided into two steps.

Step 1: Let υ belong to the interior of $\sigma(\mathcal{N})$. It follows that $\boldsymbol{\rho}(\nu,\nu')$ is continuous at the point (υ,υ), viewed as a function in $\mathbb{R}^2$. We prove that, almost surely with respect to $\mu_\upsilon^{\otimes\mathbb{N}}$,

$$\lim_{k\to\infty} \frac{1}{\sqrt{k}} \boldsymbol{\rho}_k\Big(\hat{\mathcal{N}}_k + \frac{\nu}{\sqrt{k}}, \hat{\mathcal{N}}_k + \frac{\nu'}{\sqrt{k}}\Big) = c_\rho(\upsilon)\boldsymbol{G}_{F(\upsilon)}(\nu,\nu') \tag{33}$$

in the standard topology of trace class operators on the Hilbert space $L^2(\mathbb{R}, \mathbb{C}^n, \mathrm{d}\lambda)$. Convergence with respect to the trace norm is, in general, not obvious because there are not many explicit formulas to compute this norm (unlike what happens with the Hilbert-Schmidt norm). To prove (33) we will use Theorem 2.19 in [**14**] that asserts that a sequence $(\tau_k)_{k\in\mathbb{N}}$ converges to τ with respect to the trace-norm if τ_k and τ_k^* converge strongly to τ and τ^* (respectively) and the trace-norm of τ_k converges to the trace-norm of τ. Notice that (17) alone does not help because even if τ_k and τ are positive, for every k, $\tau_k - \tau$ is not necessarily positive and, therefore, formula (17) cannot be used to estimate its trace-norm.

In the remaining of this proof we will use a couple of times the following result of measure theory that is a direct consequence of Theorem 1.21 and Lemma 1.32 in [**17**]: Suppose that f_k are positive functions on L^p, $1 \le p < \infty$, that converge point-wise (a.e.) to a positive function $f \in L^p$. Moreover, suppose that $(\eta_k)_{k\in\mathbb{N}}$ is a sequence of bounded measurable functions that is uniformly bounded (i.e., $\sup_{x,k} |\eta_k(x)| < \infty$) and converges point-wise to a function η (a.e.). Assume in addition that $\int f_k^p \to \int f^p$. Then

$$\lim_{k\to\infty} f_k\eta_k = f\eta, \qquad \text{with respect to the } L^p\text{-norm}. \tag{34}$$

We use this observation together with Eq. (17) to prove convergence with respect to the trace norm.

Set

$$\tau_k := \frac{1}{\sqrt{k}} \boldsymbol{\rho}_k \Big(\hat{\mathcal{N}}_k + \frac{\nu}{\sqrt{k}}, \hat{\mathcal{N}}_k + \frac{\nu'}{\sqrt{k}} \Big).$$

We define

$$C_k := \Big(\int e^{kl_k(\nu)} \mathrm{tr}(\rho(\nu,\nu)) h(\nu) \mathrm{d}\lambda(\nu) \Big) \mathrm{e}^{-kl_k(\hat{\mathcal{N}}_k)} \sqrt{k}$$

and

$$\widetilde{\tau}_k := C_k \tau_k. \tag{35}$$

Now, we will compute the limit, when k tends to infinity, of C_k. Since the trace of τ_k equals 1, we have that

$$\lim_{k\to\infty} C_k = \lim_{k\to\infty} \mathrm{tr}(\widetilde{\tau}_k), \tag{36}$$

whenever one of these limits exists. Eq. (19) implies that

$$\widetilde{\tau}_k(\nu, \nu') = \mathrm{e}^{-kl_k(\hat{\mathcal{N}}_k)} \mathrm{e}^{\frac{1}{2}kl_k(\hat{\mathcal{N}}_k + \frac{\nu}{\sqrt{k}})} \boldsymbol{\rho}(\hat{\mathcal{N}}_k + \frac{\nu}{\sqrt{k}}, \hat{\mathcal{N}}_k + \frac{\nu'}{\sqrt{k}}) \mathrm{e}^{\frac{1}{2}kl_k(\hat{\mathcal{N}}_k + \frac{\nu'}{\sqrt{k}})}. \tag{37}$$

We define the random function

$$X_k(\nu) := \mathrm{e}^{-\frac{1}{2}kl_k(\hat{\mathcal{N}}_k)} \mathrm{e}^{\frac{1}{2}kl_k(\hat{\mathcal{N}}_k + \frac{\nu}{\sqrt{k}})}.$$

By Taylor's formula – note that $\partial_\nu l_k(\hat{\mathcal{N}}_k) = 0$, because $\hat{\mathcal{N}}_k$ is where the maximum occurs – we have that

$$l_k(\hat{\mathcal{N}}_k + \frac{\nu}{\sqrt{k}}) - l_k(\hat{\mathcal{N}}_k) = \frac{\nu^2}{2k} \partial_\nu^2 l_k(\nu_1)$$

for some $\nu_1 \in (\hat{\mathcal{N}}_k, \hat{\mathcal{N}}_k + \frac{\nu}{\sqrt{k}})$. Hence the random function X_k has the form

$$X_k(\nu) = e^{\frac{\nu^2}{4} \partial_\nu^2 l_k(\nu_1)}.$$

By the uniform law of large numbers (cf. the proof of Theorem 2.4) $-\partial_\nu^2 l_k(\nu)$ converges uniformly to $\mathbb{E}_\upsilon\big(-\partial_\nu^2 l_k(\nu)\big)$, which is strictly positive at $\nu = \upsilon$ by Item 3. in Assumption 2.3. We define

$$\Gamma(\nu) := e^{-\frac{\nu^2}{4} F(\upsilon)}.$$

Using Lemma 3.2 and the uniform law of large numbers we have that (almost surely with respect to $\mu_\upsilon^{\otimes \mathbb{N}}$)

(38)

$$X_k(\nu) \to \Gamma(\nu), \quad \boldsymbol{\rho}(\hat{\mathcal{N}}_k + \frac{\nu}{\sqrt{k}}, \hat{\mathcal{N}}_k + \frac{\nu}{\sqrt{k}}) \to \boldsymbol{\rho}(\upsilon,\upsilon), \quad \widetilde{\tau}_k(\nu,\nu') \to \Gamma(\nu)\boldsymbol{\rho}(\upsilon,\upsilon)\Gamma(\nu'),$$

almost surely with respect to $\mu_\upsilon^{\otimes \mathbb{N}}$. Next we use (34) with $f_k(\nu) = X_k(\nu)^2$, $f = \Gamma(\nu)^2$, $\eta_k(\nu) = \mathrm{tr}_{\mathbb{C}^n} \boldsymbol{\rho}(\hat{\mathcal{N}}_k + \frac{\nu}{\sqrt{k}}, \hat{\mathcal{N}}_k + \frac{\nu}{\sqrt{k}})$ and $\eta(\nu) = \mathrm{tr}_{\mathbb{C}^n} \boldsymbol{\rho}(\upsilon,\upsilon)$ (Eq. (38) together with Assumption (31) are the requirements for (34)). We obtain (see Eq. (17)):

$$\begin{aligned} \lim_{k\to\infty} \mathrm{tr}(\widetilde{\tau}_k) &= \lim_{k\to\infty} \int X_k(\nu)^2 \mathrm{tr}_{\mathbb{C}^n} \boldsymbol{\rho}(\hat{\mathcal{N}}_k + \frac{\nu}{\sqrt{k}}, \hat{\mathcal{N}}_k + \frac{\nu}{\sqrt{k}}) d\lambda(\nu) \\ &= \int \Gamma(\nu)^2 \mathrm{tr}_{\mathbb{C}^n} \boldsymbol{\rho}(\upsilon,\upsilon) d\lambda(\nu) = \mathrm{tr}(\Gamma(\nu)\boldsymbol{\rho}(\upsilon,\upsilon)\Gamma(\nu')). \end{aligned} \tag{39}$$

Then Eq. (36) imply that (almost surely with respect to $\mu_{\upsilon}^{\otimes \mathbb{N}}$)

$$\lim_{k\to\infty} C_k = \mathrm{tr}(\Gamma(\nu)\boldsymbol{\rho}(\upsilon,\upsilon)\Gamma(\nu')) = \int \Gamma(\nu)\mathrm{tr}_{\mathbb{C}^n}\boldsymbol{\rho}(\upsilon,\upsilon)\Gamma(\nu)d\lambda(\nu). \tag{40}$$

Next we will prove that

$$s - \lim_{k\to\infty} \widetilde{\tau}_k = \Gamma(\nu)\boldsymbol{\rho}(\upsilon,\upsilon)\Gamma(\nu') \tag{41}$$

(here $s-\lim$ represents the strong limit), which together with (40) and Theorem 2.19 in [**14**] implies that the limit in (41) holds true also with respect to the trace norm. We will actually prove a stronger result, namely that the limit in (41) is valid with respect to the Hilbert-Schmidt norm. We recall that for a Hilbert-Schmidt operator $O \equiv O(\nu,\nu')$ acting on $L^2(\mathbb{R},\mathbb{C}^n,d\lambda)$, its Hilbert-Schmidt norm is given by

$$\|O\|_{\mathrm{HS}}^2 = \int \|O(\nu,\nu')\|_{\mathrm{HS}(\mathbb{C}^n)}^2 d\lambda(\nu)d\lambda(\nu'), \tag{42}$$

where $\|\cdot\|_{\mathrm{HS}(\mathbb{C}^n)}^2$ is the Hilbert-Schmidt norm for operators in $\mathbb{C}^n$. We use again (34), but now we take $p=2$. Moreover, we set $f_k(\nu,\nu') = X_k(\nu)X_k(\nu')$, $f(\nu,\nu') = \Gamma(\nu)\Gamma(\nu')$, $\eta_k(\nu,\nu') = \boldsymbol{\rho}(\hat{\mathcal{N}}_k + \frac{\nu}{\sqrt{k}}, \hat{\mathcal{N}}_k + \frac{\nu'}{\sqrt{k}})$ and $\eta(\nu,\nu') = \boldsymbol{\rho}(\upsilon,\upsilon)$ (Eq. (38) together with Assumption (31) are the requirements for (34)). We obtain:

$$\begin{aligned}\lim_{k\to\infty} &\|\widetilde{\tau}_k - \Gamma(\nu)\boldsymbol{\rho}(\upsilon,\upsilon)\Gamma(\nu')\|_{\mathrm{Hs}}^2 \\ &= \lim_{k\to\infty} \int \left\|\widetilde{\tau}_k(\nu,\nu') - \Gamma(\nu)\boldsymbol{\rho}(\upsilon,\upsilon)\Gamma(\nu')\right\|_{\mathrm{Hs}(\mathbb{C}^n)}^2 d\lambda(\nu)d\lambda(\nu') = 0.\end{aligned} \tag{43}$$

Eqs. (39) and (43) then lead to

$$\lim_{k\to\infty} \widetilde{\tau}_k(\nu,\nu') = \Gamma(\nu)\boldsymbol{\rho}(\upsilon,\upsilon)\Gamma(\nu'), \tag{44}$$

with respect to the trace norm (see Theorem 2.19 in [**14**] and (40), (43)). Then, (35), (40) and (44) imply that

$$\lim_{k\to\infty} \tau_k(\nu,\nu') = \frac{1}{\mathrm{tr}\Big(\Gamma(\nu)\boldsymbol{\rho}(\upsilon,\upsilon)\Gamma(\nu')\Big)}\Gamma(\nu)\boldsymbol{\rho}(\upsilon,\upsilon)\Gamma(\nu'), \tag{45}$$

which directly implies Eq. (33).

Step 2: We prove Eq. (32). Set $\mathcal{C}$ be the set of points $\underline{\xi}$ such that (32) holds. As we have argued above (see the proof of Proposition 3.3), the set $\mathcal{C}$ is measurable. Since $\hat{\mathcal{N}}_\infty = \upsilon$, almost surely with respect to $\mu_{\upsilon}^{\otimes \mathbb{N}}$, (see Lemma 3.2), Step 1 implies that $\mu_{\upsilon}^{\otimes \mathbb{N}}(\mathcal{C}) = 1$, for every υ in the interior of $\sigma(\mathcal{N})$, that by assumption has λ_ρ measure 1. Finally, Eq (6) implies the desired result, $\mathbb{P}_\rho(\mathcal{C}) = 1$. □

We are ready to prove Theorem 2.7

Proof of Theorem 2.7. We only prove that for every υ in the interior of $\sigma(\mathcal{N})$ (we abbreviate $F \equiv F(\upsilon)$)

$$\lim_{k\to\infty} \left\| \frac{1}{\sqrt{k}}\rho_k(\hat{\mathcal{N}}_k + \frac{\nu}{\sqrt{k}}, \hat{\mathcal{N}}_k + \frac{\nu'}{\sqrt{k}}) - \frac{c_\rho(\upsilon)}{h(\upsilon)}\boldsymbol{G}_F(\nu,\nu')\right\|_{\mathcal{B}_1^{(k)}} = 0, \tag{46}$$

almost surely with respect to $\mu_{\upsilon}^{\otimes \mathbb{N}}$. The rest of the proof follows as in the proof of Step 2 in Proposition 4.1.

For every trace class operator $\tau \equiv \tau(\nu,\nu')$ acting on $L^2(\mathbb{R},\mathbb{C}^n,hd\lambda)$, we set

$$\tau^{(k)}(\nu,\nu') = \tau(\hat{\mathcal{N}}_k + \nu/\sqrt{k}, \hat{\mathcal{N}}_k + \nu'/\sqrt{k}),$$

acting on $L^2(\mathbb{R},\mathbb{C}^n,h(\hat{\mathcal{N}}_k + \nu/\sqrt{k})d\lambda)$. Next we set

$$\boldsymbol{\tau}^{(k)}(\nu,\nu') = \sqrt{h^{(k)}(\nu)}\tau^{(k)}(\nu,\nu')\sqrt{h^{(k)}(\nu')},$$

as an operator in $L^2(\mathbb{R},\mathbb{C}^n,d\lambda)$. As we argued in Eq. (30), $\boldsymbol{\tau}^{(k)}$ and $\tau^{(k)}$ have the same norm, in their respective spaces. In the proof of Proposition 4.1 we prove that

$$\begin{aligned}(47)\quad \lim_{k\to\infty}\left\|\frac{1}{\sqrt{k}}\boldsymbol{\rho}_k^{(k)} - c_\rho(v)\boldsymbol{G}_F(\nu,\nu')\right\|_{\mathcal{B}_1(L^2(\mathbb{R},\mathbb{C}^n,d\lambda))} \\ = \lim_{k\to\infty}\left\|\frac{1}{\sqrt{k}}\rho_k^{(k)} - c_\rho(v)G_F(\nu,\nu')\right\|_{\mathcal{B}_1^{(k)}} = 0,\end{aligned}$$

almost surely with respect to $\mu_v^{\otimes \mathbb{N}}$, where $G_F(\nu,\nu') := \frac{\boldsymbol{G}_F(\nu,\nu')}{\sqrt{h^{(k)}(\nu)}\sqrt{h^{(k)}(\nu')}}$. Moreover,

(48)

$$\begin{aligned}&\lim_{k\to\infty}\left\|\frac{c_\rho(v)}{h(v)}G_F(\nu,\nu') - c_\rho(v)G_F(\nu,\nu')\right\|_{\mathcal{B}_1^{(k)}} \\ &= \lim_{k\to\infty}\left\|\frac{c_\rho(v)\sqrt{h^{(k)}(\nu)}\sqrt{h^{(k)}(\nu')}}{h(v)}\boldsymbol{G}_F(\nu,\nu') - c_\rho(v)\boldsymbol{G}_F(\nu,\nu')\right\|_{\mathcal{B}_1(L^2(\mathbb{R},\mathbb{C}^n,d\lambda))} = 0.\end{aligned}$$

The proof of the last statement, which is left to the reader, can be made either by a direct computation or by the same procedure that we applied repeatedly to prove a trace convergence of operators. Finally (47) and (48) imply Eq. (46). □

REMARK 4.2. *A) Assumptions* 2.1, 2.3 *require the validity of various conditions for all points in the spectrum $\sigma(\mathcal{N})$ of $\mathcal{N}$. If these conditions only hold true in some open interval $N \subset \sigma(\mathcal{N})$ then our conclusions hold when conditioned on $\hat{\mathcal{N}}_\infty \in N$.*

B) By a direct integral version of the spectral decomposition of $\mathcal{N}$, there exists a Hilbert space bundle $\mathcal{H}_\nu$ over a measure space $(\sigma(\mathcal{N}),\lambda)$ such that

$$(49)\qquad \mathcal{H} \simeq \int_{\sigma(\mathcal{N})}^{\oplus} \mathcal{H}_\nu$$

and, under this isometry

$$(50)\qquad \mathcal{N} \simeq \int_{\sigma(\mathcal{N})}^{\oplus} \nu$$

where ν is an abbreviation of $\nu \times 1_\nu$ acting on $\mathcal{H}_\nu$. A Hilbert space bundle is called trivial if all the spaces $\mathcal{H}_\nu$ are isomorphic to a fixed space $\mathcal{H}^{(0)}$, so that $\mathcal{H}$ is isomorphic to the space of square-integrable $\mathcal{H}^{(0)}$-valued functions on $\sigma(\mathcal{N})$, i.e., $\mathcal{H} \simeq L^2(\sigma(\mathcal{N}),\mathcal{H}^{(0)},\lambda)$. Any Hilbert space bundle can be decomposed into a countable sum of trivial bundles. Theorem 2.7 *can then be applied separately within each trivial bundle.*

References

[1] Karl Kraus, *States, effects, and operations*, Lecture Notes in Physics, vol. 190, Springer-Verlag, Berlin, 1983. Fundamental notions of quantum theory; Lecture notes edited by A. Böhm, J. D. Dollard and W. H. Wootters. MR725167

[2] Alexander S. Holevo, *Statistical structure of quantum theory*, Lecture Notes in Physics. Monographs, vol. 67, Springer-Verlag, Berlin, 2001. MR1889193

[3] Hans Maassen and Burkhard Kümmerer, *Purification of quantum trajectories*, Dynamics & stochastics, IMS Lecture Notes Monogr. Ser., vol. 48, Inst. Math. Statist., Beachwood, OH, 2006, pp. 252–261, DOI 10.1214/lnms/1196285826. MR2306206

[4] T. Benoist, V. Jakšić, Y. Pautrat, and C.-A. Pillet, *On entropy production of repeated quantum measurements I. General theory*, Comm. Math. Phys. **357** (2018), no. 1, 77–123, DOI 10.1007/s00220-017-2947-1. MR3764564

[5] T. Benoist, M. Fraas, Y. Pautrat, and C. Pellegrinni. Invariant measure for quantum trajectories. *arXiv preprint arXiv:1703.10773*, 2017.

[6] M. Ballesteros, M. Fraas, J. Fröhlich, and B. Schubnel, *Indirect acquisition of information in quantum mechanics*, J. Stat. Phys. **162** (2016), no. 4, 924–958, DOI 10.1007/s10955-015-1410-y. MR3456983

[7] C. Guerlin, J. Bernu, S. Deleglise, C. Sayrin, S. Gleyzes, S. Kuhr, M. Brune, J.M. Raimond, and S. Haroche. Progressive field-state collapse and quantum non-demolition photon counting. *Nature*, 448(7156):889-893, 2007.

[8] M. Bauer and D. Bernard. Convergence of repeated quantum nondemolition measurements and wave-function collapse. *Phys. Rev. A*, 84(4):044103, 2011.

[9] Michel Bauer, Denis Bernard, and Tristan Benoist, *Iterated stochastic measurements*, J. Phys. A **45** (2012), no. 49, 494020, 17, DOI 10.1088/1751-8113/45/49/494020. MR2999738

[10] Michel Bauer, Tristan Benoist, and Denis Bernard, *Repeated quantum non-demolition measurements: convergence and continuous time limit*, Ann. Henri Poincaré **14** (2013), no. 4, 639–679, DOI 10.1007/s00023-012-0204-x. MR3046451

[11] Bruno de Finetti, *La prévision : ses lois logiques, ses sources subjectives* (French), Ann. Inst. H. Poincaré **7** (1937), no. 1, 1–68. MR1508036

[12] Charalambos D. Aliprantis and Kim C. Border, *Infinite dimensional analysis*, 3rd ed., Springer, Berlin, 2006. A hitchhiker's guide. MR2378491

[13] Michael Reed and Barry Simon, *Methods of modern mathematical physics. I*, 2nd ed., Academic Press, Inc. [Harcourt Brace Jovanovich, Publishers], New York, 1980. Functional analysis. MR751959

[14] Barry Simon, *Trace ideals and their applications*, 2nd ed., Mathematical Surveys and Monographs, vol. 120, American Mathematical Society, Providence, RI, 2005. MR2154153

[15] Thomas S. Ferguson, *A course in large sample theory*, Texts in Statistical Science Series, Chapman & Hall, London, 1996. MR1699953

[16] Abraham Wald, *Note on the consistency of the maximum likelihood estimate*, Ann. Math. Statistics **20** (1949), 595–601. MR0032169

[17] Olav Kallenberg, *Foundations of modern probability*, Probability and its Applications (New York), Springer-Verlag, New York, 1997. MR1464694

Department of Mathematical Physics, Applied Mathematics and Systems Research Institute (IIMAS), National Autonomous University of Mexico (UNAM), Mexico

Department of Mathematics, Technion, Israel

Department of Mathematics, Virginia Tech, United States

Institut für Theoretische Physik, ETH Zurich, Switzerland

Swiss Federal Railways (SBB), Switzerland

Contemporary Mathematics
Volume **717**, 2018
https://doi.org/10.1090/conm/717/14453

Subfactors and quantum information theory

Pieter Naaijkens

ABSTRACT. We consider quantum information tasks in an operator algebraic setting, where we consider normal states on von Neumann algebras. In particular, we consider subfactors $\mathfrak{N} \subset \mathfrak{M}$, that is, unital inclusions of von Neumann algebras with trivial center. One can ask the following question: given a normal state ω on $\mathfrak{M}$, how much can one learn by only doing measurements from $\mathfrak{N}$? We argue how the Jones index $[\mathfrak{M} : \mathfrak{N}]$ can be used to give a quantitative answer to this, showing how the rich theory of subfactors can be used in a quantum information context. As an example we discuss how the Jones index can be used in the context of wiretap channels.

Subfactors also occur naturally in physics. Here we discuss two examples: rational conformal field theories and Kitaev's toric code on the plane, a prototypical example of a topologically ordered model. There we can directly relate aspects of the general setting to physical properties such as the quantum dimension of the excitations. In the example of the toric code we also show how we can calculate the index via an approximation with finite dimensional systems. This explicit construction sheds more light on the connection between topological order and the Jones index.

1. Introduction

Quantum information can be defined as the study of information processing in a quantum setting, and is the quantum analog of classical (Shannon) information theory. In a typical scenario, Alice encodes classical information in a quantum state ρ, and sends it in some way to Bob. This transmission is rarely perfect, so Bob receives a state $\mathcal{E}(\rho)$, where $\mathcal{E}$ is a completely positive map (a *quantum channel*) describing how the state is affected by the transmission. Bob's task is then to recover the classical information that Alice wanted to send, and a natural question is to ask what the rate of information is that can be send in this way (which depends on the channel $\mathcal{E}$). By now there are multiple textbooks which provide an introduction to the field, see for example [**20**, **38**, **51**].

Usually only finite dimensional systems are considered (or infinitely many copies of a finite systems, in case one is interested in asymptotics). Comparatively little work has been done on quantum systems with *infinitely* many degrees of freedom. Nevertheless, there are some examples, such as a study of the Holevo capacity [**43**] and a discussion of error correction in infinite Hilbert spaces [**3**] (this list is by

This project has received funding from the European Union's Horizon 2020 research and innovation programme under the Marie Sklodowska-Curie grant agreement No 657004. I would also like to thank Leander Fiedler for interesting discussions.

no means complete). See also [**39**] for an overview of some technical properties of quantum channels in the operator algebraic picture.

One of the main goals of this paper is to show that the theory of operator algebras gives us new tools to study quantum information tasks in infinite systems. It is our hope that the exposition is accessible to both quantum information theorists as well as to operator algebraists. We do this by relating more technical operator algebraic aspects to more familiar tasks in quantum information. Our running example will be subfactors, unital inclusions $\mathfrak{N} \subset \mathfrak{M}$ of von Neumann algebras with trivial center (see the next section for a brief introduction to von Neumann algebras). Subfactors are extensively studied in the operator algebra community, and have found applications in other fields as well (see for example [**23, 32, 36**]).

Rather than focussing on the states of a system, in our infinite dimensional setting we find it easier and more transparent to work in the Heisenberg picture, with a focus on the observables of the theory. The duality between the state (or Schrödinger) picture and the Heisenberg picture is of course well-known in quantum mechanics. So for us a quantum channel will be a normal unital completely positive map $\mathcal{E} : \mathfrak{M} \to \mathfrak{N}$. A particular example is a *conditional expectation*, in which case $\mathcal{E}$ is a "projection" from $\mathfrak{M}$ onto a subalgebra $\mathfrak{N}$. We will give a precise definition in Section 2. It can be thought of as a generalisation of the partial trace (which indeed is an example of a conditional expectation) that does not require a bipartite structure.

The physical interpretation is as follows. We can think of $\mathfrak{M}$ modelling the observables in some quantum system. The subalgebra $\mathfrak{N}$ then describes a subsystem, in the sense that only a subset of the observables can be measured. In particular, there may be states on $\mathfrak{M}$ that can be distinguished with measurements in $\mathfrak{M}$, but not when one is restricted to $\mathfrak{N}$. We can then think of a conditional expectation $\mathcal{E} : \mathfrak{M} \to \mathfrak{N}$ as describing the quantum operation of restricting observables in $\mathfrak{M}$ to the subsystem, or alternatively describing how much of their ability to distinguish states they lose. In the dual picture, $\mathcal{E}$ can be used to extend states on $\mathfrak{N}$ to the bigger system $\mathfrak{M}$ in a canonical way. In fact, $\omega \circ \mathcal{E}$ is the unique extension of a state ω on $\mathfrak{N}$ that minimizes the relative entropy with respect to any invariant state $\varphi \circ \mathcal{E} = \varphi$ of $\mathfrak{M}$.

The interesting thing about subfactors is that there typically is a natural or "preferred" conditional expectation. To each such conditional expectation one can define an "index". There is a unique conditional expectation $\mathcal{E}$ minimizing this index, and the corresponding index is written as $[\mathfrak{M} : \mathfrak{N}]$ [**22, 28, 32**].[1] This is called the *Jones index*. We recall the definition in Section 4. The upshot is that this index can be interpreted as an information quantity related to question discussed above. This is done through expressing the index in terms of Holevo's χ-quantity [**25**], which gives a bound on the amount of classical information one can recover from an ensemble of states. Although such normal conditional expectations are important in the theory of subfactors, applications to quantum information have received little attention. Here we try to fill this gap, and show how subfactors give rise to examples of quantum channels, which have a natural operational interpretation.

We also discuss two examples of physical systems where a subfactor $\mathfrak{N} \subset \mathfrak{M}$ arises naturally: rational conformal field theory [**23**] and the toric code on the

[1]For Type II_1 factors one should choose the trace-invariant conditional expectation. In general this is not necessarily the one that minimizes the index.

plane [**36**]. In these systems, the index is related to the superselection sectors (or "charges") of the model, providing a physical mechanism with which additional states can be distinguished. Some of these ideas have been discussed earlier in [**11**, **13**]. Here we provide a more mathematical account of the main ingredients.

The definition of the index is rather technical and in many cases not amenable to a direct calculation. In particular for models such as the toric code it would be helpful to have an approximation procedure that gives us the index from (a sequence of) finite dimensional models. Here we present and work out a new way to do precisely that in the toric code. This also clarifies the role of the superselection sectors of the model in the quantum information task described above. It provides a way to relate a method introduced by Haah [**16**] to define the charges in a model, to the definition of superselection sectors that is appropriate for infinite systems.

The paper is organized as follows. We first recall the mathematical setting of quantum information in infinite systems. Then, in Section 3, we recall the definition of relative entropy and discuss how it can be used to say something on how well two states can be distinguished. Section 4 gives a recap of the main definitions in subfactor theory, and shows how the Jones index is related to the relative entropy. This is applied in Section 5 to a wiretap scenario. Finally, in the last two sections we give examples of how such subfactors occur in physical systems, and show how in the toric code one can use approximations by finite dimensional systems to calculate the index.

2. Quantum channels and operator algebras

Often in quantum information theory one considers finite dimensional systems. That is, systems given by (tensor products of) finite dimensional Hilbert spaces. Let $\mathcal{H}$ be such a finite dimensional Hilbert space. The physical operations on $\mathcal{H}$ are described by $\mathfrak{B}(\mathcal{H})$, with self-adjoint elements corresponding to measurable observables, and states are given by density operators on $\mathcal{H}$ (that is, positive operators on $\mathcal{H}$ with unit trace). A *quantum channel* Φ is a linear map that sends states on a system $\mathcal{H}_A$ to states on a possibly different system $\mathcal{H}_B$. Mathematically, it can be modeled as a trace-preserving completely positive linear map between the bounded linear operators of each of these systems. Positivity makes sure that positive operators are sent to positive operators, and hence because of the trace-preserving condition, density operators (or states) to density operators. Complete positivity implies that we can consider the composite system $\mathcal{H}_A \otimes \mathbb{C}^n$, and $\Phi \otimes \mathrm{Id}_n$ will still be a quantum channel.

Physically a quantum channel describes what happens with a state after some action. Examples range from time evolution over a finite amount of time, restricting a state to a smaller set of observables, or describing what happens if we sent a quantum state from one place to another subject to noise induced by interaction with the environment (a "noisy channel"). Typical questions are then, for example, what the maximum rate of (classical or quantum) information is that can be sent through this channel, and if we can improve this by using error correction schemes. We refer to [**51**] for an in-depth treatment of many applications and results.

Here we will be interested in infinite quantum systems, which are no longer described by finite dimensional Hilbert spaces. In addition, the algebra generated by the observables of the system is no longer $\mathfrak{B}(\mathcal{H})$, but rather some subalgebra $\mathfrak{M} \subset \mathfrak{B}(\mathcal{H})$, in particular those subalgebras that are von Neumann algebras. We will recall some of the basic definitions and facts of the theory of such operator algebras, although some familiarity with the theory of von Neumann algebras will be helpful. An introduction to von Neumann algebras and their role in physics can be found in [**5**]. See also [**24**] for a brief discussion in the context of quantum information.

For concreteness, suppose that $\mathfrak{M} \subset \mathfrak{B}(\mathcal{H})$ is a von Neumann algebra. That is, $\mathfrak{M}$ is closed under the $*$-operation (often called the Hermitean conjugate in physics), and satisfies $\mathfrak{M} = \mathfrak{M}''$. Here the prime denotes the commutant,

$$\mathfrak{M}' = \{X \in \mathfrak{B}(\mathcal{H}) : [A, X] = 0 \, \forall A \in \mathfrak{M}\}.$$

Hence $\mathfrak{M}$ is equal to its double commutant. There are different topologies one can consider on a von Neumann algebra. It is most natural to consider *normal* maps. A linear functional $\varphi : \mathfrak{M} \to \mathbb{C}$ is normal if there are sequences $\xi_n, \psi_n \in \mathcal{H}$ with $\sum_n \|\xi_n\|^2 < \infty$ (and similarly for ψ_n), such that $\varphi(X) = \sum \langle \xi_n, X\psi_n \rangle$ for all $X \in \mathfrak{M}$. A linear map $\mathcal{E} : \mathfrak{M} \to \mathfrak{N}$ is normal if and only if $\omega \circ \mathcal{E}$ is a normal state on $\mathfrak{M}$ for any normal state ω of $\mathfrak{N}$. Equivalently, $\sup_\lambda \mathcal{E}(X_\lambda) = \mathcal{E}(\sup X_\lambda)$ for any increasing and bounded (in norm) net X_λ of positive elements in $\mathfrak{M}$. We will only consider normal maps.

A von Neumann algebra is called a *factor* if $\mathfrak{M} \cap \mathfrak{M}' = \mathbb{C}I$, i.e. its center is trivial. Every von Neumann algebra can be decomposed as a direct sum (or, more generally, a direct integral) of factors. As was already known to von Neumann, factors can be classified into three types, denoted I, II and III (with further subdivisions possible). Factors of Type I are the most familiar: those are precisely those von Neumann algebras that are isomorphic to $\mathfrak{B}(\mathcal{H})$ for some Hilbert space $\mathcal{H}$. They are completely classified by the dimension of the Hilbert space. For example, a factor of Type I_n is isomorphic to $M_n(\mathbb{C})$. Every finite dimensional von Neumann algebra is a direct sum of Type I factors.

We will be most concerned with infinite dimensional systems. In particular, in the language of von Neumann algebras, we are primarily interested in systems where the observables are modeled by a von Neumann algebra that is not of Type I. This already makes the situation very different from the finite dimensional setting. Consider for example the case where the observables are given by a Type III factor $\mathfrak{M}$. In such a factor, every projection P is Murray-von Neumann equivalent to I, that is, there is an isometry $W \in \mathfrak{M}$ such that $W^*W = I$ and $WW^* = P$. A consequence is that there a *no* normal pure states.[2] Let ω be a normal state implemented by a vector Ω (this can always be achieved in a suitable representation). If $\mathfrak{M}$ is Type III, then also $\mathfrak{M}'$ is of Type III, and by choosing a non-trivial projection P in $\mathfrak{M}'$ (which always exists), we can find isometries V, W such that $VV^* + WW^* = I$. Then, with $A \in \mathfrak{M}$,

$$\omega(A) = \langle V^*\Omega, AV^*\Omega \rangle + \langle W^*\Omega, AW^*\Omega \rangle.$$

By a suitable choice of V and W the states on the right hand side can be made distinct and hence ω is not a pure state.

This is not just a mathematical curiosity, but also of physical relevance. For example, under quite natural assumptions one can show that the local observables in relativistic quantum field theory are Type III factors. See [**53**] for a discussion of some of the consequences this has, for example on the entanglement properties

[2] Since $\mathfrak{M}$ is in particular a C^*-algebra, it *does* have pure states, but these states cannot be normal.

of the vacuum. It is therefore indeed useful to consider the more general operator algebraic setting.

As mentioned an important role is played by quantum channels. In the operator algebraic picture it is more natural to consider the Heisenberg picture. That is, rather than viewing a channel as a map of states, we view it as a map between observable algebras: in the Heisenberg picture a *quantum channel* is a normal unital completely positive map $\mathcal{E} : \mathfrak{M} \to \mathfrak{N}$ between von Neumann algebras. Of course these two notions of a channel are dual, and there is a unique "adjoint" map $\mathcal{E}_* : \mathfrak{N}_* \to \mathfrak{M}_*$, which sends a normal state ω on $\mathfrak{N}$ to a normal state $\omega \circ \mathcal{E}$ on $\mathfrak{N}$. Conversely, if $\mathcal{E}_*$ is completely positive, it uniquely defines a normal unital completely positive map $\mathcal{E} : \mathfrak{M} \to \mathfrak{N}$.

A fundamental result of Stinespring says that any completely positive map $\mathcal{E} : \mathfrak{M} \to \mathfrak{B}(\mathcal{H})$ is of the form $\mathcal{E}(X) = V^*\pi(X)V$, where $\pi : \mathfrak{M} \to \mathfrak{B}(\mathcal{K})$ is a representation of $\mathfrak{M}$ on some Hilbert space $\mathcal{K}$, and bounded linear map $V : \mathcal{H} \to \mathcal{K}$ [**45**]. If $\mathcal{E}$ is unital, V is an isometry. If $\mathcal{E}$ is normal, then π can also be chosen to be normal. If all Hilbert spaces are finite dimensional, a consequence of Stinespring's theorem is that $\mathcal{E}(X) = \sum_{i=1}^N V_i X V_i^*$, where the V_i are called the *Kraus operators*. For infinite systems this is more subtle, but see for example [**21**].

A particular example of a quantum channel is a *conditional expectation* $\mathcal{E} : \mathfrak{M} \to \mathfrak{N}$ from a von Neumann algebra $\mathfrak{M}$ onto a subalgebra $\mathfrak{N}$ (more generally, one could also consider inclusions of C^*-algebras). Conditional expectations play a fundamental role in subfactor theory. One can find different (but equivalent) definitions in the literature. We will use the following one:

DEFINITION 2.1. A *conditional expectation* from a von Neumann algebra (or, more generally, a C^*-algebra) $\mathfrak{M}$ onto $\mathfrak{N}$ is a positive linear map $\mathcal{E} : \mathfrak{M} \to \mathfrak{N}$ such that $\mathcal{E}(ABC) = A\mathcal{E}(B)C$ and $\mathcal{E}(A) = A$ for all $A, C \in \mathfrak{N}$ and $B \in \mathfrak{M}$.

We will only consider unital conditional expectations between von Neumann algebras, and assume that they are normal. Note that if $\mathcal{E}$ is unital, the condition that $\mathcal{E}(A) = A$ for $A \in \mathfrak{N}$ already follows from $\mathcal{E}(ABC) = A\mathcal{E}(B)C$.

Conditional expectations were introduced by Umegaki [**48**] and Dixmier [**10**] as a non-commutative generalization of conditional expectations in probability theory. An important result by Tomiyama says that every projection of norm one, that is, a linear map $\mathcal{E}$ from a C^*-algebra $\mathfrak{M}$ onto a C^*-subalgebra of $\mathfrak{M}$ such that $\mathcal{E} \circ \mathcal{E} = \mathcal{E}$ and $\|\mathcal{E}\| = 1$, is in fact a conditional expectation [**47**]. The converse is also true: every conditional expectation is a projection of norm one. Finally, it can be shown that $\mathcal{E}$ is in fact *completely* positive. In particular, $\mathcal{E}$ is a quantum channel. Proofs of all these statements can be found in [**46**, §9] or section II.6.10 of [**4**].

EXAMPLE 2.2. Consider a bipartite system $\mathcal{H} = \mathcal{H}_1 \otimes \mathcal{H}_2 = \mathbb{C}^m \otimes \mathbb{C}^n$. Let τ be the normalized trace on $\mathcal{H}_2$. Then we can define a map Tr_B by demanding $\mathrm{Tr}_2(A \otimes B) = A\tau(B)$ for all $A \in \mathfrak{B}(\mathcal{H}_1)$ and $B \in \mathfrak{B}(\mathcal{H}_2)$. By identifying A with $A \otimes I$, this defines a linear map $\mathrm{Tr}_2 : \mathfrak{M} \to \mathfrak{N}$, with $\mathfrak{M} = \mathfrak{B}(\mathcal{H}_1 \otimes \mathcal{H}_2)$ and $\mathfrak{N} = \mathfrak{B}(\mathcal{H}_1) \otimes \mathcal{H}_2$. It is easy to check that Tr_2 is a conditional expectation. Note that up to a normalization factor, this is the partial trace of the second system.

The existence of a conditional expectation is not guaranteed and implies some conditions on $\mathfrak{N}$. For example, if $\mathcal{E} : \mathfrak{M} \to \mathfrak{N}$ is a normal conditional expectation, and $\mathfrak{M}$ is of Type I, then so is $\mathfrak{N}$. Similarly, if $\mathfrak{M}$ is semi-finite, so is $\mathfrak{N}$ (cf. [**46**, §10]).

3. Distinguishing states and relative entropy

Entropies and relative entropies play an essential role in quantum (and indeed, also classical) information theory. Many tasks, such as determining the amount of information one can send through a channel, ultimately boil down to calculation of certain entropies. Another such task that will be relevant for us is distinguishing states. Before we recall how entropies play a role here, we recall the definition of the relative entropy in the context of von Neumann algebras. This was first introduced by Araki [**1, 2**]. A more modern treatment as well as an overview of subsequent results can be found in [**39**].

Let $\mathfrak{M}$ be a von Neumann algebra and suppose that ω, φ are two positive normal functionals on $\mathfrak{M}$. Moreover, suppose that ω is implemented by a vector ξ, which can always be realized by switching to the Haagerup standard form if necessary [**15**]. The vector ξ induces a positive functional ω'_ξ on the commutant $\mathfrak{M}'$, and it is possible to define the *spatial derivative* $\Delta(\varphi/\omega'_\xi)$ [**8**], which generally is an unbounded operator. The *relative entropy* is then defined as[3]

$$S(\omega, \varphi) := \begin{cases} -\langle \xi, \log \Delta(\varphi/\omega'_\xi)\xi\rangle & \text{if } \operatorname{supp}\omega \leq \operatorname{supp}\varphi \\ +\infty & \text{otherwise} \end{cases}. \tag{3.1}$$

Here $\operatorname{supp}\omega$ is the support projection of ω, that is, the smallest projection P such that $\omega(P) = \omega(I)$. Note that we do not restrict to states, but consider *all* positive normal linear functionals.

For the remainder of this paper the precise technical details of this definition are not important. We will however frequently use that for finite dimensional systems, the definition reduces to the following equation, which will be more familiar to the quantum information community:

$$S(\rho, \sigma) = \operatorname{Tr}(\rho \log(\rho) - \rho \log(\sigma)) \tag{3.2}$$

if $\operatorname{supp}(\rho) \leq \operatorname{supp}(\sigma)$, and $+\infty$ otherwise. Here we identify the positive linear functionals ρ and σ with the corresponding positive matrices, i.e. $\rho(A) = \operatorname{Tr}(\rho A)$. This notion of relative entropy was first studied by Umegaki [**49**].

A fundamental result is that for normal states $S(\omega, \varphi) \geq 0$, with equality if and only if $\omega = \varphi$. So even though $S(\omega, \varphi)$ is not a metric, it tells us something on how distinct the two states are. It can sometimes tell us more than just being distinct. Suppose that we have a normal state φ. Then we can consider a finite probability distribution $\{p_x\}$ and a corresponding set of normal states $\{\varphi_x\}$ such that $\varphi = \sum_x p_x \varphi_x$. Physically this corresponds to a procedure where we prepare a state by choosing one of the states φ_x with probability p_x. Suppose now that after preparing the state we give it to Bob, who is allowed to know from which ensemble of states the state is selected, but does not know the probability distribution $\{p_x\}$. Bob's task is to recover the probability distribution. If the states φ_x have orthogonal support, Bob can recover the distribution arbitrarily well, given enough copies of the state φ.

This is no longer true if the states do not have orthogonal support. Nevertheless, it is still possible to give a bound on the amount of information that can be obtained.

[3] The reader should be warned that in the literature sometimes the order of the arguments is reversed in the definition.

To this end, define the *Holevo χ-quantity* for such a decomposition as [**25**]

$$\chi(\{p_x\}, \{\varphi_x\}) := \sum_x p_x S(\varphi_x, \varphi). \tag{3.3}$$

It can also be written in terms of entropies (rather than *relative* entropies), but for infinite systems the relative entropy formulation is preferred (since entropies are often infinite in that case). The χ-quantity gives a bound on the amount of classical information that can be recovered from an ensemble, and appears in many different places in quantum information. For example, a fundamental result is the *coding theorem*, which says that $\max_{\{p_x, \varphi_x\}} \chi(\{p_x\}, \{\varphi_x\})$ is equal to the *classical* capacity of the classical-quantum channel $x \mapsto \varphi_x$ [**19, 41**]. That is, the rate of classical information that can be transmitted without error in the limit of asymptotically many uses of the channel.

For infinite systems, however, the quantity (3.3) is not that useful in itself: for faithful normal states φ of von Neumann algebras of Type II or Type III, we can make this quantity as large as we want by choosing an appropriate decomposition of φ (this follows from [**39**, Lemma 6.10]). Hence while for a given decomposition this is still a meaningful quantity, one has to take care when optimizing over all possible decompositions.

Here we will consider a slightly different scenario. Suppose we have a unital inclusion of von Neumann algebras $\mathfrak{N} \subset \mathfrak{M}$. We will think of $\mathfrak{M}$ as describing "full" set of observables, while $\mathfrak{N}$ is a more limited set of observables, for example describing the situation where a third party has only access to a limited part of the whole quantum system. Then if ω is a normal state on $\mathfrak{M}$, we can again consider a decomposition $\omega = \sum p_x \omega_x$ as before. Since $\mathfrak{M}$ contains more operations than $\mathfrak{N}$, one would expect that in general an observer who has access to all operations in $\mathfrak{M}$ would do a better job of recovering $\{p_x\}$ than an observer who only has access to operations in $\mathfrak{N}$. That is, an observer who has to work with the *restricted* states $\omega_x|\mathfrak{N}$. From the discussion above it is reasonable to stipulate that the advantage that $\mathfrak{M}$ has over $\mathfrak{N}$ can be quantified by

$$S_{\mathfrak{M}|\mathfrak{N}}(\{p_x\}, \{\varphi_x\}) := \chi(\{p_x\}, \{\varphi_x\}) - \chi(\{p_x\}, \{\varphi_x|\mathfrak{N}\}). \tag{3.4}$$

This quantity is also called the *entropic disturbance* of the quantum channel that restricts normal states on $\mathfrak{M}$ to $\mathfrak{N}$ and will play a central role in the remainder of this article. Some properties and applications of this quantity in infinite dimensional systems are discussed in [**44**]. For other applications, see Section 5 below.

4. Subfactors, conditional expectations, and Jones index

We now restrict to a particularly interesting class of examples of inclusions $\mathfrak{N} \subset \mathfrak{M}$: the subfactors. There are many aspects to such subfactors, but for our purposes one of the most important features is that one can define an *index* $[\mathfrak{M} : \mathfrak{N}]$ which, heuristically speaking, gives a measure of how much bigger $\mathfrak{M}$ is compared to $\mathfrak{N}$. It can be seen as the generalization of the index of a subgroup in a group. Subfactors and the index were first studied by Jones in the Type II_1 case [**22**]. Many of the results were later extended to general factors by Kosaki [**28**] and Longo [**32, 33**], among others.

We first recall the definition of a subfactor.

DEFINITION 4.1. A *subfactor* is a unital inclusion of factorial von Neumann algebras $\mathfrak{N} \subset \mathfrak{M}$. It is called *irreducible* if $\mathfrak{N}' \cap \mathfrak{M} = \mathbb{C}I$.

The theory of subfactors is very rich and examples come up in many different contexts, some of which we will discuss in Section 6.

EXAMPLE 4.2. Let $\mathfrak{M} = \mathfrak{B}(\mathcal{H}_1 \otimes \mathcal{H}_2)$ and $\mathfrak{N} = \mathfrak{B}(\mathcal{H}_1) \otimes I$. Then $\mathfrak{N} \subset \mathfrak{M}$ is a subfactor. It is not irreducible (unless $\dim \mathcal{H}_2 = 1$), since $\mathfrak{N}' \cap \mathfrak{M} = (\mathfrak{B}(\mathcal{H}_1) \otimes I)' = I \otimes \mathfrak{B}(\mathcal{H}_2)$.

A key role in the theory is played by the conditional expectations introduced in Section 2. We will mainly consider irreducible subfactors. In that case, if there is a faithful conditional expectation $\mathcal{E} : \mathfrak{M} \to \mathfrak{N}$, it is unique [**32**, Sect. 5]. We are now in a position to define the *index* of a subfactor $\mathfrak{N} \subset \mathfrak{M}$. This is the quantity that will allow us to connect an inclusion of von Neumann algebras to a quantum information quantity.

DEFINITION 4.3. Let $\mathfrak{N} \subset \mathfrak{M}$ be an irreducible subfactor with both algebras not of Type I. Let $\mathcal{E} : \mathfrak{M} \to \mathfrak{N}$ be the unique faithful conditional expectation (if it exists). Then we define the Jones (or Jones-Kosaki-Longo) index by

$$[\mathfrak{M} : \mathfrak{N}] := (\sup\{\lambda > 0 : \mathcal{E}(x) \geq \lambda x \text{ for all } x \in \mathfrak{M}_+\})^{-1}.$$

If such a conditional expectation does not exist, we set $[\mathfrak{M} : \mathfrak{N}] = \infty$.

The index measures how much bigger $\mathfrak{M}$ is compared to $\mathfrak{N}$. One can show that $[\mathfrak{M} : \mathfrak{N}] \geq 1$, with equality if and only if $\mathfrak{M} = \mathfrak{N}$. It also gives a bound on the dimension of $\mathfrak{M}$ seen as a module over $\mathfrak{N}$ (see for example [**29**, Sect. 3.4]).

This is not the original definition of Jones, but coincides with his if $\mathcal{E}$ is the trace-preserving conditional expectation of a Type II_1 subfactor. The equivalence to the definition given here is due to Pimsner and Popa in the Type II_1 case [**40**]. For general subactors, one can define $\text{Ind}(\mathcal{E})$ for a conditional expectation $\mathcal{E} : \mathfrak{M} \to \mathfrak{N}$, either through a Pimsner-Popa type inequality or using modular theory (the definition there is quite technical, so we will not repeat it here). It can be shown that there is a unique conditional expectation $\mathcal{E}_0$ minimizing the index, and one can define $[\mathfrak{M} : \mathfrak{N}]$ using this expectation. It should be noted, however, that for Type II_1 factors this minimal conditional expectation need not coincide with the trace-preserving conditional expectation that Jones uses for his index. Since for irreducible subfactors the faithful conditional expectation is unique (if it exists), this is not an issue. We refer to [**29**] for details on the index theory.

A simple argument relates the index to the entropic quantity $S_{\mathfrak{M}|\mathfrak{N}}$ (c.f. [**40**, Cor. 4.1] or [**37**, Prop. 10.2.2][4]).

LEMMA 4.4. *Let $\mathfrak{N} \subset \mathfrak{M}$ be a finite index irreducible subfactor and $\mathcal{E}$ the corresponding conditional expectation. Then for any normal state $\varphi = \varphi \circ \mathcal{E}$ of $\mathfrak{M}$ we have*

$$S_{\mathfrak{M}|\mathfrak{N}}(\{p_x\}, \{\varphi_x\}) \leq \log [\mathfrak{M} : \mathfrak{N}],$$

where $\varphi = \sum_x p_x \varphi_x$.

PROOF. Since $\mathcal{E}$ is a faithful conditional expectation, by Theorem 5.15 of [**39**] $S(\omega, \varphi \circ \mathcal{E}) = S(\omega|\mathfrak{N}, \varphi|\mathfrak{N}) + S(\omega, \omega \circ \mathcal{E})$ for any normal state ω on $\mathfrak{M}$. Using this we can rewrite equation (3.4) as

$$S_{\mathfrak{M}|\mathfrak{N}}(\{p_x\}, \{\varphi_x\}) = \sum_x p_x S(\varphi_x, \varphi_x \circ \mathcal{E}).$$

[4]The author would like to thank Ben Hayes for pointing out this reference.

From Definition 4.3 it follows that $\omega \circ \mathcal{E} \geq [\mathfrak{M} : \mathfrak{N}]^{-1}\omega$ for any normal state ω. This implies that

$$\sum_x p_x S(\varphi_x, \varphi_x \circ \mathcal{E}) \leq \sum_x p_x S(\varphi_x, [\mathfrak{M} : \mathfrak{N}]^{-1}\varphi_x) = \sum p_x \log[\mathfrak{M} : \mathfrak{N}],$$

where in the first step we used Corollary 5.12 of [**39**], and in the second step the scaling properties of the relative entropy. □

The following key result, due to Pimsner and Popa for Type II_1 factors and Hiai for the general case show that the bound can actually be attained. It is in fact sufficient to consider only *faithful* states $\varphi \circ \mathcal{E} = \varphi$, but since for the applications we are interested in this restriction is not very natural, we consider the general case.

THEOREM 4.5 (Pimsner-Popa [**40**], Hiai [**17**,**18**]). *Let $\mathfrak{N} \subset \mathfrak{M}$ be an irreducible subfactor with finite index. Then we have*

$$\log[\mathfrak{M} : \mathfrak{N}] = \sup_{\varphi : \varphi \circ \mathcal{E} = \varphi} \sup_{\{p_x\}, \{\varphi_x\}} S_{\mathfrak{M}|\mathfrak{N}}(\{p_x\}, \{\varphi_x\}).$$

Here $\mathcal{E} : \mathfrak{M} \to \mathfrak{N}$ is the conditional expectation associated to the subfactor. The first supremum is over all normal states on $\mathfrak{M}$ that leave $\mathcal{E}$ invariant, and the second is over all finite decompositions of such a φ.

The condition that the subfactor is irreducible is not essential, but simplifies the statement (and is enough for our purposes). The reason is that in this case there is a unique conditional expectation $\mathcal{E} : \mathfrak{M} \to \mathfrak{N}$. In the general case, one has to consider the *minimal* conditional expectation.

The use of entropies in subfactor theory was motivated by work of Connes and Størmer, who were interested in a non-commutative generalization of entropies as they are used in classical dynamical systems (see [**37**] for a review). Their relative entropy of two operator algebras provides the connection between the algebraic definition of the index in Definition 4.3 to the formula in terms of relative entropies of states given in the theorem above. Our motivation is different: we are mainly interested in the application of these results to quantum information theory, in particular to channel capacities, as discussed in the next section. This concludes our brief discussion of subfactors.

5. Wiretap channels and private classical capacity

There are various quantum information tasks that can be described in the subfactor setting. For example, an application to secret sharing was discussed in [**13**]. Here we outline how we can interpret the subfactor in the context of wiretapping channels. This provides a new application of entropies in subfactors, and the first in quantum information. Classical wiretap channels were introduced by Wyner [**52**]. Their quantum counterparts were first studied by Schumacher and Westmoreland [**42**].

Let us first consider, following [**20**, Sect. 10.4], the typical setup for a quantum wiretap channel. Consider finite dimensional Hilbert spaces $\mathcal{H}_A$ (controlled by Alice), $\mathcal{H}_B$ (controlled by Bob) and $\mathcal{H}_E$ (controlled by an eavesdropper Eve). Alice uses an isometry $V : \mathcal{H}_A \to \mathcal{H}_B \otimes \mathcal{H}_E$ to encode states ρ_A in the target Hilbert space consisting of Bob's and Eve's parts. Since Bob and Eve can only control their respective parts, they will have access to the states $\rho_B := \mathrm{Tr}_E(V\rho_A V^*)$ and

$\rho_E := \mathrm{Tr}_B(V\rho_A V^*)$, respectively. Note that we can describe this by two quantum channels Φ_B and Φ_E.

Bob can obtain some information on the initial state ρ_A that Alice sent through the channel by doing measurements on his part of the system. But since Eve *also* has access to part of the system, she can also gain some knowledge. A natural question is to ask how much information Alice can send to Bob in such a way that Eve cannot learn anything about the message Alice wants to send to Bob by measurements on her part of the system. From the discussion in Section 3 this can be bounded by

$$\chi(\{p_x\}, \{\Phi_B(\rho_A^x)\}) - \chi(\{p_x\}, \{\Phi_E(\rho_A^x)\}),$$

where $\rho_A = \sum_x p_x \rho_A^x$. This quantity is sometimes called the *quantum privacy.* If each ρ_A^x is a *pure* state, it can be shown to be equal to a quantity called the *coherent information* [**42**].

We now come back to our subfactor setting, with $\mathfrak{N} \subset \mathfrak{M}$ a subfactor, and want to relate it to the wiretapping scenario above. In the Heisenberg picture we can model Bob and Eve with two observable algebras that mutually commute. However, in the general von Neumann algebra setting, there need not be a corresponding tensor product decomposition as above, so it is not clear how to define the channels Φ_B and Φ_E as before. Nevertheless, we can consider a scenario that is similar: we suppose that Eve has control over all observables in $\mathfrak{N}$. Bob, on the other hand, is more powerful, and can access $\mathfrak{M}$. Again Alice wants to encode information for Bob that Eve cannot recover. We consider the simplest scenario possible, where Alice encodes the information directly in normal states ω on $\mathfrak{M}$ and does not send them through a channel first. Then a bound on the amount that Bob can hide from Eve is given by equation (3.4). Note that the restriction to $\mathfrak{N}$ is a quantum channel: it is the adjoint of the channel (in the Heisenberg picture) $\iota : \mathfrak{N} \to \mathfrak{M}$, the inclusion homomorphism. This channel plays the role of Φ_E. But using subfactor theory we can say more: if Alice is restricted to *ensembles* $\omega = \omega \circ \mathcal{E}$, the optimum value is given by the (log of the) index $[\mathfrak{M} : \mathfrak{N}]$ by Theorem 4.5! Again, it is *not* necessary that the individual states in the ensemble are also invariant with respect to $\mathcal{E}$.

Let us also briefly comment on the condition that the ensembles are invariant with respect to $\mathcal{E}$. Let $\omega \circ \mathcal{E} = \omega$ be a normal state. If $\varphi_{\mathfrak{N}}$ is a normal state on $\mathfrak{N}$, we can extend it to a normal state on $\mathfrak{M}$ by $\varphi_{\mathfrak{N}} \circ \mathcal{E}$. But there can be (and in general, are) many extensions of $\varphi_{\mathfrak{N}}$ to $\mathfrak{M}$. Then $\varphi_{\mathfrak{N}} \circ \mathcal{E}$ is the *unique* extension φ that minimizes $S(\omega, \varphi)$ (under the condition that $S(\omega, \varphi) < \infty$ for at least one extension), see the discussion after [**39**, Lemma 5.18]. So be restricting to such states, we isolate the contribution that is due to Bob being able to distinguish individual states in the ensemble, that cannot be distinguished from the ensemble itself.

What if we consider n copies of the system, can we do better? Alice can send states $\rho^{(n)} = \sum_x p_x^{(n)} \rho_x^{(n)}$, with $\rho_x^{(n)}$ states on $\mathcal{H}_A^{\otimes n}$ for some integer n. The states that Bob receive are described by the channels $\Phi_B \otimes \cdots \otimes \Phi_B$, and similarly for Eve. Note that the states $\rho^{(n)}$ need not be product states (or even separable states), and neither do the allowed measurements have to be of product form. The *private capacity* is then defined as the (average) amount of information that Alice can send to Bob using n channels, in the limit $n \to \infty$, such that the amount of information

Eve can recover is negligible. It turns out that this is in fact *equal* to

$$\lim_{n\to\infty} \frac{1}{n} \max_{\{p_x^{(n)},\rho_x^{(n)}\}} \left[\chi(\{p_x^{(n)}\},\{\Phi_B^{\otimes n}(\rho_x^{(n)})\})) - \chi(\{p_x^{(n)}\},\{\Phi_E^{\otimes n}(\rho_x^{(n)})\})\right],$$

a result proven by Devetak [**9**] and Cai, Winter and Young [**6**]. Note that this can also be interpreted as a coding theorem.

In the operator algebraic setting we can consider the von Neumann algebra $\mathfrak{M}_1\overline{\otimes}\mathfrak{M}_2$ generated by the algebraic tensor product $\mathfrak{M}_1 \odot \mathfrak{M}_2$. If $\Phi_i : \mathfrak{M}_i \to \mathfrak{N}_i$, $i = 1, 2$ are normal maps, there is a normal map $\Phi_1 \otimes \Phi_2 : \mathfrak{M}_1\overline{\otimes}\mathfrak{M}_2 \to \mathfrak{N}_1\overline{\otimes}\mathfrak{N}_2$ defined in the natural way. Hence we can ask the same question as before, but not with $\mathfrak{M}^{\otimes n}$ and $\mathfrak{N}^{\otimes n}$, respectively:

THEOREM 5.1. *Let $\mathfrak{N} \subset \mathfrak{M}$ be an irreducible subfactor. Then for each $n \in \mathbb{N}$ we have*

$$\sup_{\varphi:\varphi\circ\mathcal{E}^{\otimes n}=\varphi} \sup_{\{p_x^{(n)},\varphi_x^{(n)}\}} \left[\chi(\{p_x^{(n)}\},\{\varphi_x^{(n)}\}) - \chi(\{p_x^{(n)}\},\{\varphi_x^{(n)}|\mathfrak{N}^{\otimes n}\})\right] = n \log[\mathfrak{M} : \mathfrak{N}].$$

A similar statement is true for reducible subfactors if one takes $\mathcal{E}$ to be the conditional expectation minimizing the index.

PROOF. First note that $\mathfrak{N}^{\otimes n} \subset \mathfrak{M}^{\otimes n}$ is again a subfactor. Hence we can apply Theorem 4.5 with $\mathcal{E}^{(n)} : \mathfrak{M}^{\otimes n} \to \mathfrak{N}^{\otimes n}$ the minimal conditional expectation for this subfactor. But it turns out that $\mathcal{E}^{(n)} = \mathcal{E}^{\otimes n}$, see the proof of Corollary 5.6 of [**32**]. The claim then follows since $[\mathfrak{M}^{\otimes n} : \mathfrak{N}^{\otimes n}] = [\mathfrak{M} : \mathfrak{N}]^n$, by the same corollary. □

This shows that we do not gain any advantage from allowing the use of multiple copies of the system simultaneously, not even when we allow states which are entangled along these copies.

In general the private capacity is not additive, in the sense that the capacity of $\Phi \otimes \Phi$ is not twice the capacity of Φ [**31**]. Here on the other hand it appears that it *is* additive, and it is interesting to see that we can use the special structure of subfactors to explicitly calculate the quantum privacy. It should be noted however that this does not quite prove the private classical capacity of the channel, since it would require an analog of the results in [**6, 9**] (in particular on how to encode information) that is valid in this operator algebraic setting. It nevertheless suggests that using subfactor theory can be very powerful to study certain quantum channels, which warrants further study.

6. Examples: abelian quantum double models and CFT

Examples of subfactors arise naturally in physical systems. Here we recall how they appear in the study of superselection sectors in local quantum physics. A more detailed introduction to sector theory and local quantum physics can be found in [**14**]. The goal of the theory is to describe all properties of the charges (or superselection sectors). We will consider two examples here: rational conformal field theory on the circle [**23**], and a class of topologically ordered quantum spin systems, known as *Kitaev's quantum double* [**26**]. The connection between the latter and the Jones index, together with some applications, was discussed before in [**13**].

The interesting thing is that there is a direct physical interpretation of the index in these models. In particular, in both cases the different superselection sectors can be described by a *braided tensor C^*-category*. This category encodes all physical properties of the superselection sectors (or charges, anyons), such as fusion and braiding rules. To each such anyon i one can assign a *quantum dimension* (or just dimension) d_i. The total quantum dimension, an invariant of the tensor category, is then defined via $\mathcal{D}^2 = \sum_i d_i^2$, where the sum is over a set of representatives of the anyons (or in the language of category theory, of irreducible objects). It plays an important role in topologically ordered systems, where it is believed to be related to an area law for the ground states of such systems [**27, 30**].

The starting point is a net of local algebras, representing the observables that can be measured in finite parts of the system of interest. More precisely, let Γ be the space on which the model is defined (for example Minkowksi space for relativistic theories or a discrete lattice for spin systems). Then to suitable subsets $\Lambda \subset \Gamma$ we assign a C^*-algebra $\mathfrak{A}(\Lambda)$, representing all observables in the region Λ. There are two natural properties that we require, the first being isotony: if $\Lambda_1 \subset \Lambda_2$, there should be a unital inclusion $\mathfrak{A}(\Lambda_1) \subset \mathfrak{A}(\Lambda_2)$. The second is locality: if $\Lambda_1 \perp \Lambda_2$ then $\mathfrak{A}(\Lambda_1)$ and $\mathfrak{A}(\Lambda_2)$ commute. Here the meaning of $\Lambda_1 \perp \Lambda_2$ depends on the context: for our purposes it is enough for this to mean that Λ_1 and Λ_2 are disjoint. In relativistic theories we require that they are spacelike separated. We also introduce the notation $\Lambda^c := \Gamma \setminus \Lambda$ (again, in relativistic theories one would take the spacelike complement). The assignment $\Lambda \mapsto \mathfrak{A}(\Lambda)$ defines a local net of C^*-algebras. The *quasi-local algebra* $\mathfrak{A}$ is the inductive limit of this net (in the category of C^*-algebras). It can be interpreted as the set of all observables that can be approximated arbitrarily well by strictly local observables. We also require that $\mathfrak{A}$ is represented on some Hilbert space $\mathcal{H}$ using a representation π_0 (typically the vacuum representation).

The type of charges that we want to describe are localized, in the sense that outside of the localization region the system looks like the vacuum or ground state. Hence we need a set $\mathcal{C}$ of localization regions. The elements of $\mathcal{C}$ are subsets of the space Γ. The choice of admissible regions depends on the type of charges that we want to describe. For example, using Gauss' law we can always detect an electrical charge from arbitrarily far away by measuring the flux through a sphere. Hence such charges cannot be localized in a compact region. Once the set $\mathcal{C}$ is fixed, we consider irreducible representations π that look like π_0 outside the localization region. More precisely, we demand that

$$\pi|\mathfrak{A}(C^c) \cong \pi_0|\mathfrak{A}(C^c) \tag{6.1}$$

for *all* $C \in \mathcal{C}$. Here $\cong$ means unitary equivalence of representations. Note that they only have to be unitary equivalent when restricted to observables *outside* of C. Nevertheless, this is quite a strong condition, since it has to hold for all regions C. Physically this means that we are able to move charges from one region $C_1 \in \mathcal{C}$ to another region C_2. A *sector* is then an equivalence class of irreducible representations satisfying equation (6.1).

The set of such representations has a surprisingly rich structure, which can be analyzed with an additional technical assumption: Haag duality. This says that for each $C \in \mathcal{C}$ we have $\pi_0(\mathfrak{A}(C))'' = \pi_0(\mathfrak{A}(C^c))'$. One inclusion follows from locality, but the other is non-trivial. The main utility of Haag duality is that it allows one to pass from representations to endomorphisms of $\mathfrak{A}$. That is, each representation π as above is equivalent to $\pi_0 \circ \rho$, where ρ is an endomorphism of $\mathfrak{A}$. Moreover, it follows that $\rho(A) = A$ for all $A \in \mathfrak{A}(C^c)$ for some $C \in \mathcal{C}$. That is, it only acts non-trivially in the localization region C. These endomorphisms can be endowed with the structure of a *braided tensor category*, which encodes all physical properties of the charges. For example, it is possible to define a braiding, which tells us what happens if we interchange to charges. For bosons and fermions, interchanging two particles *twice* is always trivial, but in lower dimensional systems there are other possibilities. This happens in fact in both of the examples discussed below.

The final piece of structure that we need is a set of *double localization regions* $\mathcal{C}_2$, which consists of pairs C_A, C_B of regions in $\mathcal{C}$, with the condition that C_A and C_B are sufficiently far apart (the precise notion of which is model dependent). We will also write $AB \in \mathcal{C}_2$. With such a choice we can define the von Neumann algebra $\mathcal{R}_{AB} := \pi_0(\mathfrak{A}(C_A \cup C_B))''$, that is, the von Neumann algebra generated by $\pi_0(\mathfrak{A}(C_A))''$ and $\pi_0(\mathfrak{A}(C_B))''$. Our final assumption is that the *split property* holds. This says that $\mathcal{R}_{AB} \simeq \pi_0(\mathfrak{A}(C_A))''\overline{\otimes}\pi_0(\mathfrak{A}(C_B))''$. Here $\simeq$ means that the two algebras are naturally isomorphic, in the sense that the map $A \otimes B \mapsto AB$ extends to an isomorphism of von Neumann algebras.[5] Finally, we can define an algebra $\widehat{\mathcal{R}}_{AB} := \pi_0(\mathfrak{A}((C_A \cup C_B)^c))'$. Note that by locality we have $\mathcal{R}_{AB} \subset \widehat{\mathcal{R}}_{AB}$.

It will become clear in a moment why this is an interesting inclusion of algebras to study, but first we give two examples of physical theories that satisfy all the conditions.

EXAMPLE 6.1 (Rational conformal field theory [**23**]). Consider $\Gamma = S^1$, the circle. To each interval we associate a local algebra $\mathfrak{A}(I)$ in such a way that if $I \subset J$, then $\mathfrak{A}(I) \subset \mathfrak{A}(J)$ and all algebras are represented on the same Hilbert space $\mathcal{H}$. In addition we assume that the net is local, in the sense that $\mathfrak{A}(I)$ and $\mathfrak{A}(J)$ commute if $\overline{I} \cap \overline{J} = \emptyset$ and that the net is irreducible, in the sense that the algebra generated by all $\mathfrak{A}(I)$ is equal to $\mathfrak{B}(\mathcal{H})$. The representation π_0 is the identity representation and will be omitted. We also assume that the net is conformal, in the sense that it is covariant with respect to a positive energy representation of the Möbius group and there is a cyclic vacuum vector. In that case, Haag duality holds for intervals, and so does the split property under a mild additional assumption. Hence all prerequisites to analyze the sectors of the theory are fulfilled.

The localization regions are the open intervals on the circle. Let I_A and I_B be two intervals with disjoint closure. Then we get a subfactor $\mathcal{R}_{AB} = \mathfrak{A}(I_A \cup I_B)'' \subset \widehat{\mathcal{R}}_{AB} = \mathfrak{A}((I_A \cup I_B)^c)'$. If this subfactor has finite index, we call the conformal field theory *rational*. One can show that the index does not depend on the choice of intervals.

EXAMPLE 6.2 (Kitaev's quantum double model for abelian groups in the thermodynamic limit [**34**]). Consider the space Γ to be the edges of a square $\mathbb{Z}^2$ lattice and let G be a finite abelian group. At each edge we have a $|G|$-dimensional system. The local algebras are given by $\mathfrak{A}(\Lambda) := \bigotimes_{x \in \Lambda} M_{|G|}(\mathbb{C})$ with $\Lambda \subset \Gamma$ a finite subset. This gives a local net as before.

Dynamics on the system can be introduced by defining local Hamiltonians. For the quantum double model these were first introduced in [**26**]. Once these are defined it is possible to talk about ground states. The model has many ground states, but only one of them is translation invariant [**7**], which we will denote by ω_0. The corresponding GNS representation will be denoted by π_0. This will be

[5]This is not automatic if the algebras are not Type I.

taken as the reference representation. As localization regions $\mathcal{C}$ we consider *cones*. They can be obtained by taking a point in the plane and draw two straight lines from this point going to infinity. Then we identify $C \subset \Gamma$ with all edges that are contained in the region bounded by the two lines (corresponding to the smallest angle between them), or intersect the two lines. See Figure 1 for an example of two cones. For the double localization regions $\mathcal{C}_2$ we consider all pairs of disjoint cones such that their distance is at least two. With this choice of localization regions Haag duality and the split property hold [**11**, **12**, **35**].

We come back to the inclusions $\mathcal{R}_{AB} \subset \widehat{\mathcal{R}}_{AB}$. As mentioned before, a representation π satisfying the criterion (6.1) can be represented by a localized endomorphism. That is, we can find an endomorphism ρ_1 of $\mathfrak{A}$ such that $\rho_1(A) = A$ for all $A \in \mathfrak{A}(C_A^c)$ and $\pi_0 \circ \rho_1 \cong \pi$. But using again (6.1), the same is true for some ρ_2 localized in C_B, and there is a unitary V such that $V\pi_0 \circ \rho_1(A) = \pi_0 \circ \rho_2(A)V$. Such a V is called a *charge transporter*, since it moves a charge from C_A to C_B. Moreover, using locality it follows that $V \in \widehat{\mathcal{R}}_{AB}$. However, unless ρ_1 is trivial, V is *not* in $\mathcal{R}_{AB}$. Hence it is reasonable to conjecture that $\widehat{\mathcal{R}}_{AB}$ is bigger than $\mathcal{R}_{AB}$ precisely because of the charge transporters, and that we can learn something about the charges if we can see how much "bigger" it is.

This is the motivation behind the following definition.

DEFINITION 6.3. Consider a local theory as described above, together with a suitable choice of double localization regions $\mathcal{C}_2$ such that $\mathcal{R}_{AB} \subset \widehat{\mathcal{R}}_{AB}$ is a subfactor for every choice of two localization regions $AB \in \mathcal{C}_2$. The we define $\mu_{\pi_0} := \inf_{AB \in \mathcal{C}_2} [\widehat{\mathcal{R}}_{AB} : \mathcal{R}_{AB}]$.

Typically the index does not depend on the choice of the two localization regions, and one can forget about the infimum.

With this notation we can make the connection between the inclusion $\mathcal{R}_{AB} \subset \widehat{\mathcal{R}}_{AB}$ and the sectors of the theory.

THEOREM 6.4 ([**23**, **36**]). *If the category of superselection sectors comes from a rational conformal net as in Example* 6.1 *we have* $\mathcal{D}^2 = \mu_{\pi_0}$. *In the case of Example* 6.2, *we have* $\mathcal{D}^2 \leq \mu_{\pi_0}$. *Here* $\mathcal{D}^2$ *is the total quantum dimension of the tensor category of the sectors of the theory, as before.*

Even though the statement for the cone-localized charges in quantum spin systems is somewhat weaker, in practice it turns out that the bound is actually saturated and we have an equality. For example one can consider the class of models defined by Kitaev [**26**] for a finite abelian group G. This can be considered in the setting of Example 6.2, where π_0 is the GNS representation of the translation invariant ground state (which is unique). The index can be calculated explicitly in these models, and is equal to $|G|^2$, which is equal to $\mathcal{D}^2$ for these models [**11**, **36**].

Let us consider the case $G = \mathbb{Z}_2$ in more detail [**34**]. Recall that it is defined on the edges of $\mathbb{Z}^2$. An important role in the theory is played by *path* or *string* operators. To a finite path ξ of edges we can associate an operator F_ξ, by acting with σ_z on each edge. Similarly, we can take a path on the dual lattice $\widehat{\xi}$ and identify it with all the edges it crosses. To such a dual path we assign the path operator $F_{\widehat{\xi}}$ by acting with σ_x on the edges of the dual path. Note that both operators are self-adjoint and square to the identity (since the Pauli matrices do). These operators play a fundamental role because they create excitations: if Ω is

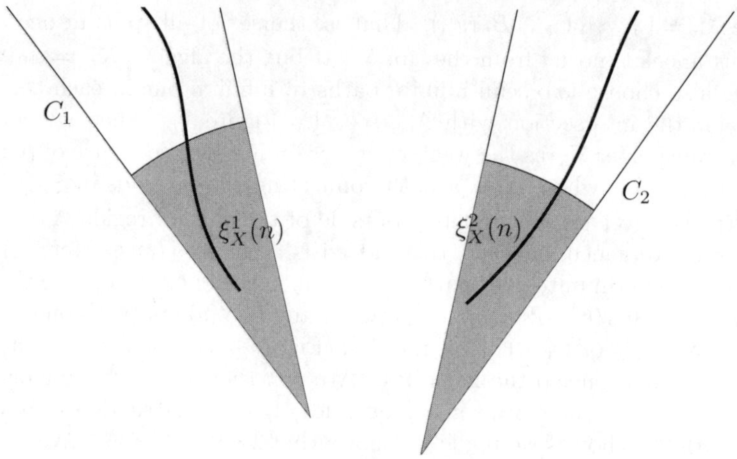

FIGURE 1. Two cones C_1 and C_2. The shaded part is an example of all sites in $\mathcal{R}_n$ for some n. Also shown are semi-infinite strings used in the definition of V_X. The paths $\xi_X^1(n)$ and $\xi_X^2(n)$ are the finite parts of these paths obtained by taking the intersection with the shaded parts.

the ground state vector, $F_\xi\Omega$ is a state with two excitations, which can be thought of as being located at the end of the string. A similar statement is true for $F_{\widehat{\xi}}$. To get examples of representations satisfying (6.1), we can choose a semi-infinite string ξ. If $\xi(n)$ is the finite path consisting of the first n edges, one can show that $\rho(A) = \lim_{n\to\infty} F_{\xi(n)} A F_{\xi(n)}$ defined an automorphism of $\mathfrak{A}$, and $\pi_0 \circ \rho$ satisfies equation (6.1). The charge transporters can be constructed explicitly as well.

With notation as before, it can be shown that $\widehat{\mathcal{R}}_{AB}$ is generated by $\mathcal{R}_{AB}$ and two charge transporters V_X and V_Z, which both square to the identity and can be chosen to commute. We set $V_0 := I$ and $V_Y := V_X V_Z$, so that we get a unitary representation $g \mapsto V_g$ of $\mathbb{Z}_2 \times \mathbb{Z}_2$ in a natural way. With this notation, it can be shown that every element $X \in \widehat{\mathcal{R}}_{AB}$ can be uniquely written as $X = \sum_i A_i V_i$ with $A_i \in \mathcal{R}_{AB}$. The conditional expectation $\mathcal{E} : \widehat{\mathcal{R}}_{AB} \to \mathcal{R}_{AB}$ is then given by $\mathcal{E}\left(\sum_i A_i V_i\right) = A_0$. As mentioned, the index is equal to $|G|^2 = 4$ [**36**].

7. Finite dimensional approximation

Consider again the setting of the toric code in the thermodynamic limit. We will continue to use the notation of Section 6. Recall that we are interested in relative entropies $S(\omega, \varphi)$, where ω and φ are both normal states of $\mathcal{R}_{AB}$ or $\widehat{\mathcal{R}}_{AB}$. Unfortunately, in this setting the definition of the relative entropy is rather technical, making it difficult to do concrete calculations. In addition, for the toric code and similar systems on finite lattices we have a much better understanding of, for example, the entanglement entropy in ground states. Hence it is desirable to be able to at least find an approximation of $S(\omega, \varphi)$ in terms of finite dimensional systems.

We show how this can be done in our standing example of the toric code. Consider two cones C_k as before, and write x_k for the tips of the cones.

Define $\Lambda_n = \bigcup_{k=1,2} C_k \cap B_n(x_k)$. That is, the set of all spins in one of the cones with distance at most i from the tip. We define the algebra $\mathcal{R}_n := \pi_0(\mathfrak{A}(\Lambda_n))$.

We also choose two semi-infinite paths to infinity, one in each cone. We write $\xi_X^i(n)$ for the intersection with Λ_n (see also Figure 1). Then we can define the charge transporter V_X as the weak-operator limit of the sequence of path operators $F_{\xi_X^1(n)} F_{\xi_X^2(n)} F_{\xi_n'}$, where ξ_n' is a path connecting the far ends of $\xi_X^1(n)$ and $\xi_X^2(n)$ using the shortest path completely outside of the shaded region Λ_n (see [**34**] for a proof of convergence and that this indeed is a charge transporter). Similarly, we choose two semi-infinite dual paths ξ_Z^i, one in each cone. We choose them in such a way that the shortest (dual) path between any two points (with one on each path) never crosses ξ_X^1 or ξ_X^2. For example, in Figure 1 we could choose a path to the right of ξ_X^1, and one to the left of ξ_X^2. We can then define V_Z analogously to the definition of V_X. There are some properties of this construction that we will need later: with the choices we made it follows that $V_X V_Z = V_Z V_X$. We also have that $V_X^2 = V_Z^2 = I$. Finally, note that $F_{\xi_X^1(n)} F_{\xi_X^2(n)} V_X$ commutes with *all* operators in $\mathcal{R}_n$, since the two string operators effectively cancel the action of V_X in the shaded region. A similar statement is true for V_Z.

With this notation we then define $\widehat{\mathcal{R}}_n := \pi_0(\mathfrak{A}(\Lambda_n)) \vee \{V_X, V_Z\}$. Together with $\mathcal{R}_n$ these two algebras will be used to approximate the index for $\mathcal{R}_{AB} \subset \widehat{\mathcal{R}}_{AB}$, using the following lemma.

LEMMA 7.1. *The sequence $\mathcal{R}_i$ is an increasing net of finite dimensional von Neumann algebras such that $(\bigcup_i \mathcal{R}_i)'' = \mathcal{R}_{AB}$, and similarly the weak* closure of the union of $\widehat{\mathcal{R}}_i$ is equal to $\widehat{\mathcal{R}}_{AB}$. Moreover, $\mathcal{E}$ restricts to a conditional expectation $\mathcal{E}_i : \widehat{\mathcal{R}}_i \to \mathcal{R}_i$.*

PROOF. Note that $\mathcal{R}_i$ is by definition the tensor product of finitely many spin-$\frac{1}{2}$ algebras, and hence finite dimensional. By definition $\bigcup_i \mathcal{R}_i$ is dense in $\pi(\mathfrak{A}(C_1 \cup C_2))$ in the norm topology. This algebra in turn is by definition weak*-dense in $\mathcal{R}_{AB}$.

Recall that $\widehat{\mathcal{R}}_{AB}$ is equal to $\mathcal{R}_{AB} \vee \{V_X, V_Z\}$ (by [**36**, Lemma 4.1]), from which the claim that $\bigcup \widehat{\mathcal{R}}_i$ is weak* dense in $\widehat{\mathcal{R}}_{AB}$ follows. In addition, since V_X and V_Z square to the identity and mutually commute, the algebra that they generate is finite dimensional. Also note that if X is a product of path operators, we have $XV_X = \pm V_X X$, and similarly for V_Z (see the proof of [**36**, Lemma 4.4]). It follows that $\widehat{\mathcal{R}}_i$ is finite dimensional.

The claim on the conditional expectations is clear from the definition of $\mathcal{E}$. □

The key point is that Lemma 7.1 gives us a way, through the relation of the index to an entropic quantity, to obtain an approximation in terms of finite dimensional systems. More precisely, under these conditions we have that the sequence $S(\omega|\mathcal{R}_i, \varphi|\mathcal{R}_i)$ converges to $S(\omega, \varphi)$ by Corollary 5.12(iv) of [**39**].

The advantage of using a finite dimensional approximation is that it is possible to use density matrix techniques instead of spatial derivatives. In general this makes the problem much more amenable to direct calculation. To this end it will be useful to have a more explicit description of $\widehat{\mathcal{R}}_i$. Since it is a finite dimensional C^*-algebra, it is isomorphic to an algebra of the form $\bigoplus_k M_{n_k}(\mathbb{C})$ for some integers n_k. The next lemma gives this decomposition. As we will see later, this decomposition is related to the superselection sectors of the model.

LEMMA 7.2. *For each i there is an integer n_i such that $\mathcal{R}_i \cong M_{n_i}(\mathbb{C})$ and $\widehat{\mathcal{R}}_i \cong \bigoplus_{j=1}^4 M_{n_i}(\mathbb{C})$. In this representation the algebra $\mathcal{R}_i$ is embedded into $\widehat{\mathcal{R}}_i$ via $A \mapsto \operatorname{diag}(A,A,A,A)$. Together with $\widehat{V}_X = \operatorname{diag}(I,-I,I,-I)$ and $\widehat{V}_Z = \operatorname{diag}(I,I,-I,-I)$ they generate the whole algebra.*

PROOF. The algebra $\mathcal{R}_i$ is a tensor product of finitely many matrix algebras $M_2(\mathbb{C})$, and hence is isomorphic to $M_{n_i}(\mathbb{C})$ for some n_i. As for $\widehat{\mathcal{R}}_i$, first write $F_X := F_{\xi_X^1(i)} F_{\xi_X^2(i)}$, and similarly for F_Z. Note that $F_X V_X$ and $F_Z V_Z$, together with $\mathcal{R}_i$, still generate $\widehat{\mathcal{R}}_i$. But using the discussion before Lemma 7.1, $\frac{1}{2}(I \pm F_X V_X)$ and $\frac{1}{2}(I \pm F_Z V_Z)$ are easily checked to be projections with mutually orthogonal ranges in the center of $\widehat{\mathcal{R}}_i$. Hence we have four central projections that we can use to decompose $\widehat{\mathcal{R}}_i$ into four blocks.

This can be done explicitly: setting $\pi(A) = \operatorname{diag}(A,A,A,A)$ and $\pi(F_X V_X) = \widehat{V}_X$ for $k = X, Z$ we obtain an injective map from $\widehat{\mathcal{R}}_i$ into $\bigoplus_{i=1}^4 M_n(\mathbb{C})$. It is straightforward to check that it preserves the algebraic relations of $\mathcal{R}_i$. We also note that $\frac{1}{4}\pi((I+F_X V_X)(I+F_Z V_Z)) = \operatorname{diag}(I,0,0,0)$, that is, the projection onto the first block. With similar expressions we obtain the projections onto the other blocks, and we see that π is also surjective. This completes the proof. □

REMARK 7.3. Note that the lemma in particular shows that $\widehat{\mathcal{R}}_i$ is *not* a factor. The reason is that we can conjugate the charge transporters V_k with a unitary in $\mathcal{R}_i$, such that this conjugated operator commutes with all operators in $\widehat{\mathcal{R}}_i$, hence the center is non-trivial. This is no longer true for the infinite algebra $\widehat{\mathcal{R}}_{AB}$. Hence $\mathcal{R}_i \subset \widehat{\mathcal{R}}_i$ is not a subfactor, and even though we have a conditional expectation between them, we therefore cannot apply the theory of subfactors directly. In addition, for finite Type I factors the theory behaves a bit differently (see for example [**29**, Thm 3.8]).

We now come to the definition of a state $\omega = \omega \circ \mathcal{E}$ that will be used to approximate the index. First define projections $P_\pm := \frac{1}{2}(I \pm V_X)$ and $Q_\pm := \frac{1}{2}(I \pm V_Z)$. Note that they all mutually commute. Moreover, $P_+P_- = Q_+Q_- = 0$.

It will be useful to calculate expectation values $\omega_0(P_j Q_k A P_j Q_k)$ with $A \in \mathcal{R}_i$. To this end, first note that $\omega_0(V_j A V_k) = 0$ if $j \neq k$. Heuristically this can be understood in terms of superselection sectors: the operators V_j create a pair of excitations, one in each of the cones C_i. Hence they can interpolate between the different sectors in each cone. The local operators A, on the other hand, can only create pairs of excitations in each cone individually, i.e. after acting with A you stay in the same sector. It follows that $\langle V_j^* \Omega, A V_k \Omega \rangle = 0$, where Ω is the GNS vector of ω_0. A rigorous argument can be distilled from the results in [**7**]. With this observation, we find

$$\begin{aligned}(7.1)\qquad &\omega_0(P_i Q_j A V_k P_i Q_j) = \\ &\qquad \frac{c(i,j,k)}{16}\left(\omega_0(A) + \omega_0(V_X A V_X) + \omega_0(V_Y A V_Y) + \omega_0(V_Z A V_Z)\right),\end{aligned}$$

for $A \in \mathfrak{A}(C_1 \cup C_2)$. Here $c(i,j,k)$ is given by

$$c(i,j,0) = 1, \quad c(i,j,X) = i, \quad c(i,j,Y) = i \cdot j, \quad c(i,j,Z) = j,$$

where $i, j \in \{+,-\}$ and we identify $\pm$ with ± 1.

We can now define a state ω on $\widehat{\mathcal{R}}_{AB}$ that will be used to estimate the relative entropy of $\mathcal{R}_{AB}$ and $\widehat{\mathcal{R}}_{AB}$:

$$\omega(X) := \sum_{j,k\in\{+,-\}} \omega_0(P_j Q_k X P_j Q_k). \tag{7.2}$$

From the previous calculation it follows that $\omega(I) = 1$, and hence it is a state. Moreover, by collecting terms one sees that $\omega(AV_k) = 0$ if $k \neq 0$, and hence $\omega \circ \mathcal{E} = \omega$. Importantly, this is *not* true for the individual states in the decomposition (7.2).

LEMMA 7.4. *The state ω defined in equation is a normal state on $\widehat{\mathcal{R}}_{AB}$ such that $\omega \circ \mathcal{E} = \omega$. It can be written as an equal weight superposition of four* distinct *states, but whose restrictions to $\mathcal{R}_{AB}$ are equal. The same is true when we restrict ω to $\widehat{\mathcal{R}}_i$.*

PROOF. Since ω_0 is implemented as a vector state for $\widehat{\mathcal{R}}_{AB}$ (the GNS vector for π_0), normality is clear. Using equation (7.1) we see that the states are distinct, and that the observables V_k can be used to distinguish them. However, the same equation shows that when restricted to $\mathcal{R}_{AB}$, all states have the same expectation values. The last claim follows directly from the construction. □

By Lemma 7.2, the algebra $\widehat{\mathcal{R}}_i$ is a direct sum of four copies of $M_n(\mathbb{C})$. To calculate the relative entropies that we need, it will be useful to explicitly find the density operators in this representation. First define $\omega_{jk}(X) := 4\omega_0(P_j Q_k X Q_k P_j)$ with $j, k = \pm$. The factor of four ensures that $\omega_{jk}(I) = 1$, so that it is in fact a state. The corresponding density matrices can then be written as:

LEMMA 7.5. *The density matrix for ω_{++} is given by*

$$\rho_{++} = \frac{1}{16}\begin{bmatrix} F_{++}\rho_0 F_{++} & & & \\ & F_{-+}\rho_0 F_{-+} & & \\ & & F_{+-}\rho_0 F_{+-} & \\ & & & F_{--}\rho_0 F_{--} \end{bmatrix}.$$

Here ρ_0 is the density matrix of ω_0 restricted to $\mathcal{R}_i$ and $F_{\pm\pm} = (I \pm F_X)(I \pm F_Z)$ (the first $\pm$ in the subscript corresponds to the first $\pm$ in the product, and similarly for the second). The same is true for the other combinations of $P_\pm$ and $Q_\pm$ by flipping the appropriate signs.

PROOF. Let ρ_0 be as in the statement. Then $\rho := \frac{1}{4}(\rho_0 \oplus \rho_0 \oplus \rho_0 \oplus \rho_0)$ is the density matrix of ω_0 restricted to $\widehat{\mathcal{R}}_i$ in the representation of Lemma 7.2: it is easy to check that $\operatorname{Tr}(\rho \operatorname{diag}(A, A, A, A)) = \omega_0(A)$ for all $A \in \mathcal{R}_i$, while $\operatorname{Tr}(\rho \operatorname{diag}(A, A, A, A)\widehat{V}_k) = 0$ for $k \neq 0$. Hence $4P_+Q_+\rho P_+Q_+$ is the density operator for ω_{++}. By writing again $V_X = F_X \widehat{V}_X$ as in the proof of Lemma 7.2 we can write P_+ and Q_+ in the representation of Lemma 7.2, from which we obtain ρ_{++}.

Alternatively the correctness can be be verified directly using equation (7.1). The other choices of $\pm$ are shown in the same way. □

We are now in a position to approximate $[\widehat{\mathcal{R}}_{AB} : \mathcal{R}_{AB}]$ via a limiting procedure. The result below coincides with the value of the index $[\widehat{\mathcal{R}}_{AB} : \mathcal{R}_{AB}]$ which has been obtained before using different methods [**36**]. The main interest is in providing a specific state that realizes the equality, as well as in the proof method of taking the limit of finite dimensional systems. This concrete procedure provides an insight into the physical mechanisms behind the equality.

THEOREM 7.6. *Let $\mathcal{E} : \widehat{\mathcal{R}}_{AB} \to \mathcal{R}_{AB}$ be the conditional expectation given above. Then*

$$\sup_{\varphi:\varphi\circ\mathcal{E}=\varphi}\ \sup_{\{p_x\},\{\varphi_x\}} S_{\widehat{\mathcal{R}}_{AB}|\mathcal{R}_{AB}}(\{p_x\},\{\varphi_x\}) = 2\log 2 = \log \mathcal{D}^2.$$

The supremum is over all normal states leaving $\mathcal{E}$ invariant. It is attained on the state ω defined in equation (7.2), decomposed as an equal-weight convex combination of all four possibilities for $\omega_{\pm\pm}$.

PROOF. For the toric code there are precisely four abelian sectors, so $\mathcal{D}^2 = \sum_{i=1}^4 d_i^2 = 4$ [**34, 36**], from which the second equality follows. To show the first equality, we will calculate $S_{\widehat{\mathcal{R}}_{AB}|\mathcal{R}_{AB}}(\{p_x\},\{\varphi_x\})$ for suitable decompositions. We do this by first restricting all states to $\widehat{\mathcal{R}}_i$ (and $\mathcal{R}_i$, respectively) and using the limiting procedure of [**39**, Corollary 5.12(iv)].

We first do this for $\varphi = \omega$ defined earlier, to show that the value $2\log 2$ can be attained. Set $p_x = 1/4$ for $x = 1,\ldots,4$ and choose each corresponding ω_i to be a distinct ω_{jk}, $j,k = \pm$, as above. We have already observed that $\omega(A) = \omega_{jk}(A)$ for all $A \in \mathcal{R}_i$ and $j,k \in \{+,-\}$, hence $S(\omega_x|\mathcal{R}_i, \omega|\mathcal{R}_i) = 0$. To calculate $S(\omega_x,\omega)$, write ρ_x and ρ for the corresponding density matrices in the representation of Lemma 7.2. Since $(I+F_X)(I-F_X) = (I+F_Z)(I-F_Z) = 0$, Lemma 7.5 implies that the density matrices ρ_x have mutually disjoint support projections. Because $\rho = \frac{1}{4}\sum \rho_x$ by definition, it follows that

$$S(\rho_x,\rho) = S\left(\rho_x, \frac{1}{4}\rho_x\right) = -\log\left(\frac{1}{4}\right) + S(\rho_x,\rho_x) = 2\log 2,$$

independent of x. This gives a lower bound on the supremum.

To complete the proof we will show that $2\log 2$ is the *maximum* value that equation (3.4) can attain, with the extra condition that $\varphi := \sum_x p_x\varphi_x$ satisfies $\varphi\circ\mathcal{E} = \varphi$. We first characterize such states. By Lemma 7.2 the density operator ρ_φ for φ can be written in the form $\rho_\varphi^1 \oplus\cdots\oplus \rho_\varphi^4$, with each component a positive operator in $\mathcal{R}_n$. Since $\varphi\circ\mathcal{E} = \varphi$, we must have that $\varphi(AV_i) = 0$ for $i \neq 0$ and $A \in \mathcal{R}_i$. But this implies that $\mathrm{Tr}(\rho_\varphi \operatorname{diag}(A,A,A,A)\widehat{V}_X) = 0$ for all $A \in \mathcal{R}_i$ and $\widehat{V}_X$ as in Lemma 7.2. This can only be true if

$$\rho_\varphi^1 - \rho_\varphi^2 + \rho_\varphi^3 - \rho_\varphi^4 = 0.$$

A similar argument for $\widehat{V}_Z$ and $\widehat{V}_Y := \widehat{V}_X\widehat{V}_Z$ gives a system of linear equations, which has as only solution $\rho_\varphi^i = \rho_\varphi^1$ for $i = 1,\ldots,4$.

Similarly as for the state φ, the density operators for the states φ_x can be decomposed as $\rho_{\varphi_x} =: \rho_x^1 \oplus\cdots\oplus\rho_x^4$. With this notation we find

$$S(\varphi_x|\mathcal{R}_i, \varphi|\mathcal{R}_i) = \sum_{i=1}^4 \mathrm{Tr}_0\left[\rho_x^i\left(\log\left(\sum_i \rho_x^i\right) - \log\left(4\rho_\varphi^1\right)\right)\right],$$

where Tr_0 is the canonical trace on $\mathcal{R}_i$. Note that $\log(4\rho_\varphi^1) = 2\log 2 + \log(4\rho_\varphi^1)$. Again using that the density operators for φ_x and φ are block-diagonal, this leads to

$$\begin{aligned} S(\varphi_x, \varphi) &- S(\varphi_x|\mathcal{R}_i, \varphi|\mathcal{R}_i) \\ &= \sum_{i=1}^{4} \mathrm{Tr}_0 \left[\rho_x^i \left(\log\left(\rho_x^i\right) - \log\left(\sum_i \rho_x^i \right) + 2\log 2 \right) \right] \\ &= 2\log 2 + \sum_i S_0 \left(\rho_x^i, \sum_i \rho_x^i \right). \end{aligned}$$

To conclude the argument it is enough to show that the summands are always less than or equal to zero (note that ρ_x^i is not a state since it is not normalized, so this is not a contradiction with the relative entropy of two states always being positive). Clearly $\rho_x^i \leq \sum_k \rho_x^k$, since each term is a positive operator. We may assume that $\rho_x^i > 0$ since otherwise $S_0(\rho_x^i, \sum_k \rho_x^k) = 0$. Because log is operator monotone it follows that $\log(\rho_x^i) \leq \log\left(\sum_k \rho_x^k\right)$, so that

$$S_0 \left(\rho_x^i, \sum_k \rho_x^k \right) = \mathrm{Tr}_0 \left[\rho_x^i \left(\log\left(\rho_x^i\right) - \log\left(\sum_k \rho_x^k \right) \right) \right] \leq 0$$

Taking the sum over p_x gives the desired result. The proof is complete by noting that Theorem 6.4 gives an upper bound to $\log \mathcal{D}^2$, but from the remark at the beginning of the proof and the calculation here, the bound is an equality. □

REMARK 7.7. In the proof we directly calculated the relative entropies, since they have a clear physical meaning in quantum information theory. Since $\mathcal{R}_i \subset \widehat{\mathcal{R}}_i$ is an inclusion of C^*-algebras, and we have a faithful conditional expectation $\mathcal{E}_i$, it is interesting to compare this with Watatani's index theory for C^*-algebras [**50**]. Note that the inclusion maps $M_n(\mathbb{C})$ onto four copies of itself. Moreover, the standard trace on $\widehat{\mathcal{R}}_i$ satisfies $\mathrm{Tr}(\mathcal{E}(A)B) = \mathrm{Tr}(AB)$ for all $A \in \widehat{\mathcal{R}}_i$ and $B \in \mathcal{R}_i$. It follows from the analysis in Section 2.4 of [**50**] that $\mathrm{Index}\, \mathcal{E}_i = 4(I \oplus I \oplus I \oplus I) \in \widehat{\mathcal{R}}_i$.

For quantum double models for general finite abelian groups G (the toric code corresponds to $G = \mathbb{Z}_2$) the structure is very similar [**11, 12**], and with the appropriate modifications the same proof goes through with $\mathcal{D}^2 = |G|^2$, although carrying it out explicitly is much more involved. For non-abelian G we expect the structure to be a bit different: there the irreducible representations of the quantum double $\mathcal{D}(G)$, the symmetry algebra behind the model, are no longer one dimensional. These irreducible representations are in correspondence with the different charges of the model. As a consequence, we expect that $\widehat{\mathcal{R}}_i$ no longer decomposes into blocks of equal size, but rather dependent on the dimension of the irreducible representations of $\mathcal{D}(G)$.

We conclude with a discussion of the properties that made the finite dimensional approximation work. First of all, Lemma 7.2 is a direct manifestation of superselection sectors, in the sense that local operators cannot interpolate between the different sectors. Such a structure is expected for all topologically ordered models. To do the explicit computation here, it was helpful that the sequence of approximating algebras is essentially the same for each n. This is due to the convenient choice of V_X and V_Z, in such a way that we can easily cancel their action inside the shaded regions in Figure 1. It seems reasonable to conjecture that as long as excitations are created by string-like operators, a similar structure is true.

Finally, one needs to find a suitable state $\omega = \omega \circ \mathcal{E}$. In particular, it is useful if $\mathcal{E}$ easily restricts to the finite dimensional algebras. Indeed, this was necessary to characterize all invariant states in the theorem.

We also note that the structure compares closely to a construction due to Haah [**16**]. He characterizes operators that do not change the charge within an annulus, while here we consider the dual viewpoint: we take those operators that create a pair of excitations, one in each annulus. Indeed, the operators V_X, V_Y and V_Z play that role. In the finite setting we can replace these by suitable string operators. These are precisely characterized by the condition that they do not create any excitations outside of the two parts of the cone (the path outside the cone is not important, and any choice will generate the same algebra $\widehat{\mathcal{R}}_i$). This is reminiscent of Haah's condition that certain operators should commute with the projections in the Hamiltonians he considers. As a matter of fact, in the end Haah obtains a similar algebraic structure as we do (in particular, equation (3) of [**16**]).

References

[1] Huzihiro Araki, *Relative entropy of states of von Neumann algebras*, Publ. Res. Inst. Math. Sci. **11** (1975/76), no. 3, 809–833, DOI 10.2977/prims/1195191148. MR0425631

[2] Huzihiro Araki, *Relative entropy for states of von Neumann algebras. II*, Publ. Res. Inst. Math. Sci. **13** (1977/78), no. 1, 173–192, DOI 10.2977/prims/1195190105. MR0454656

[3] Cédric Bény, Achim Kempf, and David W. Kribs, *Quantum error correction on infinite-dimensional Hilbert spaces*, J. Math. Phys. **50** (2009), no. 6, 062108, 24, DOI 10.1063/1.3155783. MR2541204

[4] B. Blackadar, *Operator algebras*, Encyclopaedia of Mathematical Sciences, vol. 122, Springer-Verlag, Berlin, 2006. Theory of C^*-algebras and von Neumann algebras; Operator Algebras and Non-commutative Geometry, III. MR2188261

[5] Ola Bratteli and Derek W. Robinson, *Operator algebras and quantum statistical mechanics. 1*, 2nd ed., Texts and Monographs in Physics, Springer-Verlag, New York, 1987. C^*- and W^*-algebras, symmetry groups, decomposition of states. MR887100

[6] N. Kaĭ, A. Vinter, and R. V. Ĭong, *Quantum privacy and quantum wiretap channels* (Russian, with Russian summary), Problemy Peredachi Informatsii **40** (2004), no. 4, 26–47, DOI 10.1007/s11122-004-0002-2; English transl., Probl. Inf. Transm. **40** (2004), no. 4, 318–336. MR2105852

[7] Matthew Cha, Pieter Naaijkens, and Bruno Nachtergaele, *The complete set of infinite volume ground states for Kitaev's abelian quantum double models*, Comm. Math. Phys. **357** (2018), no. 1, 125–157, DOI 10.1007/s00220-017-2989-4. MR3764565

[8] A. Connes, *On the spatial theory of von Neumann algebras*, J. Funct. Anal. **35** (1980), no. 2, 153–164, DOI 10.1016/0022-1236(80)90002-6. MR561983

[9] Igor Devetak, *The private classical capacity and quantum capacity of a quantum channel*, IEEE Trans. Inform. Theory **51** (2005), no. 1, 44–55, DOI 10.1109/TIT.2004.839515. MR2234571

[10] J. Dixmier, *Formes linéaires sur un anneau d'opérateurs* (French), Bull. Soc. Math. France **81** (1953), 9–39. MR0059485

[11] Leander Fiedler, *Haag duality and Jones-Kosaki-Longo index in Kitaev's quantum double models for finite abelian groups*, Ph.D. thesis, Leibniz Universität Hannover, 2017.

[12] Leander Fiedler and Pieter Naaijkens, *Haag duality for Kitaev's quantum double model for abelian groups*, Rev. Math. Phys. **27** (2015), no. 9, 1550021, 43, DOI 10.1142/S0129055X1550021X. MR3426207

[13] Leander Fiedler, Pieter Naaijkens, and Tobias J. Osborne, *Jones index, secret sharing and total quantum dimension*, New J. Phys. **19** (2017), 023039.

[14] Rudolf Haag, *Local quantum physics*, 2nd ed., Texts and Monographs in Physics, Springer-Verlag, Berlin, 1996. Fields, particles, algebras. MR1405610

[15] Uffe Haagerup, *The standard form of von Neumann algebras*, Math. Scand. **37** (1975), no. 2, 271–283, DOI 10.7146/math.scand.a-11606. MR0407615

[16] Jeongwan Haah, *An invariant of topologically ordered states under local unitary transformations*, Comm. Math. Phys. **342** (2016), no. 3, 771–801, DOI 10.1007/s00220-016-2594-y. MR3465431

[17] Fumio Hiai, *Minimum index for subfactors and entropy*, J. Operator Theory **24** (1990), no. 2, 301–336. MR1150623

[18] Fumio Hiai, *Minimum index for subfactors and entropy. II*, J. Math. Soc. Japan **43** (1991), no. 2, 347–379, DOI 10.2969/jmsj/04320347. MR1096438

[19] A. S. Holevo, *The capacity of the quantum channel with general signal states*, IEEE Trans. Inform. Theory **44** (1998), no. 1, 269–273, DOI 10.1109/18.651037. MR1486663

[20] Alexander S. Holevo, *Quantum systems, channels, information*, De Gruyter Studies in Mathematical Physics, vol. 16, De Gruyter, Berlin, 2012. A mathematical introduction. MR2986302

[21] Jinchuan Hou, *A characterization of positive linear maps and criteria of entanglement for quantum states*, J. Phys. A **43** (2010), no. 38, 385201, 20, DOI 10.1088/1751-8113/43/38/385201. MR2718321

[22] V. F. R. Jones, *Index for subfactors*, Invent. Math. **72** (1983), no. 1, 1–25, DOI 10.1007/BF01389127. MR696688

[23] Yasuyuki Kawahigashi, Roberto Longo, and Michael Müger, *Multi-interval subfactors and modularity of representations in conformal field theory*, Comm. Math. Phys. **219** (2001), no. 3, 631–669, DOI 10.1007/PL00005565. MR1838752

[24] M. Keyl, D. Schlingemann, and R. F. Werner, *Infinitely entangled states*, Quantum Inf. Comput. **3** (2003), no. 4, 281–306. MR2003567

[25] A. S. Kholevo, *Information-theoretical aspects of quantum measurement* (Russian), Problemy Peredači Informacii **9** (1973), no. 2, 31–42; English transl., Problems of Information Transmission **9** (1973), no. 2, 110–118 (1975). MR0398304

[26] A. Yu. Kitaev, *Fault-tolerant quantum computation by anyons*, Ann. Physics **303** (2003), no. 1, 2–30, DOI 10.1016/S0003-4916(02)00018-0. MR1951039

[27] Alexei Kitaev and John Preskill, *Topological entanglement entropy*, Phys. Rev. Lett. **96** (2006), no. 11, 110404, 4, DOI 10.1103/PhysRevLett.96.110404. MR2214523

[28] Hideki Kosaki, *Extension of Jones' theory on index to arbitrary factors*, J. Funct. Anal. **66** (1986), no. 1, 123–140, DOI 10.1016/0022-1236(86)90085-6. MR829381

[29] Hideki Kosaki, *Type III factors and index theory*, Lecture Notes Series, vol. 43, Seoul National University, Research Institute of Mathematics, Global Analysis Research Center, Seoul, 1998. MR1662525

[30] Michael Levin and Xiao-Gang Wen, *Detecting topological order in a ground state wave function*, Phys. Rev. Lett. **96** (2006), 110405.

[31] Ke Li, Andreas Winter, XuBo Zou, and GuangCan Guo, *Private capacity of quantum channels is not additive*, Phys. Rev. Lett. **103** (2009), 120501.

[32] Roberto Longo, *Index of subfactors and statistics of quantum fields. I*, Comm. Math. Phys. **126** (1989), no. 2, 217–247. MR1027496

[33] Roberto Longo, *Index of subfactors and statistics of quantum fields. II. Correspondences, braid group statistics and Jones polynomial*, Comm. Math. Phys. **130** (1990), no. 2, 285–309. MR1059320

[34] Pieter Naaijkens, *Localized endomorphisms in Kitaev's toric code on the plane*, Rev. Math. Phys. **23** (2011), no. 4, 347–373, DOI 10.1142/S0129055X1100431X. MR2804555

[35] Pieter Naaijkens, *Haag duality and the distal split property for cones in the toric code*, Lett. Math. Phys. **101** (2012), no. 3, 341–354, DOI 10.1007/s11005-012-0572-7. MR2956822

[36] Pieter Naaijkens, *Kosaki-Longo index and classification of charges in 2D quantum spin models*, J. Math. Phys. **54** (2013), no. 8, 081901, 17, DOI 10.1063/1.4818272. MR3135456

[37] Sergey Neshveyev and Erling Størmer, *Dynamical entropy in operator algebras*, Ergebnisse der Mathematik und ihrer Grenzgebiete. 3. Folge. A Series of Modern Surveys in Mathematics [Results in Mathematics and Related Areas. 3rd Series. A Series of Modern Surveys in Mathematics], vol. 50, Springer-Verlag, Berlin, 2006. MR2251116

[38] Michael A. Nielsen and Isaac L. Chuang, *Quantum computation and quantum information*, Cambridge University Press, Cambridge, 2000. MR1796805

[39] Masanori Ohya and Dénes Petz, *Quantum entropy and its use*, Texts and Monographs in Physics, Springer-Verlag, Berlin, 1993. MR1230389

[40] Mihai Pimsner and Sorin Popa, *Entropy and index for subfactors*, Ann. Sci. École Norm. Sup. (4) **19** (1986), no. 1, 57–106. MR860811

[41] Benjamin Schumacher and Michael D. Westmoreland, *Sending classical information via noisy quantum channels*, Phys. Rev. A **56** (1997), 131–138.

[42] ———, *Quantum privacy and quantum coherence*, Phys. Rev. Lett. **80** (1998), 5695–5697.

[43] M. E. Shirokov, *The Holevo capacity of infinite dimensional channels and the additivity problem*, Comm. Math. Phys. **262** (2006), no. 1, 137–159, DOI 10.1007/s00220-005-1457-8. MR2200885

[44] M. E. Shirokov and A. S. Kholevo, *On lower semicontinuity of the Δ_χ-value and its consequences in quantum information theory* (Russian, with Russian summary), Izv. Ross. Akad. Nauk Ser. Mat. **81** (2017), no. 5, 165–182, DOI 10.4213/im8609; English transl., Izv. Math. **81** (2017), no. 5, 1044–1060. MR3706864

[45] W. Forrest Stinespring, *Positive functions on C^*-algebras*, Proc. Amer. Math. Soc. **6** (1955), 211–216, DOI 10.2307/2032342. MR0069403

[46] Şerban Strătilă, *Modular theory in operator algebras*, Editura Academiei Republicii Socialiste România, Bucharest; Abacus Press, Tunbridge Wells, 1981, Translated from the Romanian by the author. MR696172

[47] Jun Tomiyama, *On the projection of norm one in W^*-algebras*, Proc. Japan Acad. **33** (1957), 608–612. MR0096140

[48] Hisaharu Umegaki, *Conditional expectation in an operator algebra*, Tôhoku Math. J. (2) **6** (1954), 177–181, DOI 10.2748/tmj/1178245177. MR0068751

[49] Hisaharu Umegaki, *Conditional expectation in an operator algebra. IV. Entropy and information*, Kōdai Math. Sem. Rep. **14** (1962), 59–85. MR0142006

[50] Yasuo Watatani, *Index for C^*-subalgebras*, Mem. Amer. Math. Soc. **83** (1990), no. 424, vi+117, DOI 10.1090/memo/0424. MR996807

[51] Mark M. Wilde, *Quantum information theory*, Cambridge University Press, Cambridge, 2013. MR3088659

[52] A. D. Wyner, *The wire-tap channel*, Bell System Tech. J. **54** (1975), no. 8, 1355–1387, DOI 10.1002/j.1538-7305.1975.tb02040.x. MR0408979

[53] Jakob Yngvason, *Localization and entanglement in relativistic quantum physics*, The Message of Quantum Science: Attempts Towards a Synthesis (Philippe Blanchard and Jürg Fröhlich, eds.), Springer Berlin Heidelberg, Berlin, Heidelberg, 2015, pp. 325–348.

Department of Mathematics, University of California, Davis, One Shields Avenue, Davis, California 95616—and—JARA Institute for Quantum information, RWTH Aachen University, Otto-Blumenthal-Strasse 20, 52074 Aachen, Germany

Email address: `pieter@naaijkens.nl`

Contemporary Mathematics
Volume **717**, 2018
https://doi.org/10.1090/conm/717/14454

Dressing up for length gauge: Mathematical aspects of a debate in quantum optics

Rainer Dick

ABSTRACT. We address the surprising issue that different gauges in quantum optics can yield very different results in the description of physical systems. This has been observed for many years for the description of the electromagnetic couplings of atoms and molecules in velocity gauge and length gauge. Indeed, we find that it is mathematically consistent to mix velocity gauge for some components of a material with length gauge for other components, although this should not be possible for a *bona fide* gauge transformation. For many-particle systems in a Hartree approximation, it is even possible to mix velocity gauge and length gauge for different particles of the same kind.

We identify four main sources of analytic differences between velocity gauge and length gauge. These sources imply differences between velocity gauge and length gauge in particular for subfemtosecond spectroscopy and for experiments involving strong fields. Finally, we also emphasize that the transformation from velocity gauge to length gauge is just a particular example of a picture changing unitary transformation. However, all these transformations lead to non-unitary shifts of the Hamiltonian, irrespective of whether the transformation can be described as a gauge transformation or not. Therefore, all these descriptions of quantum optics in dipole approximation are formally equivalent if agreement is achieved which particular formulation of the time-dependent interaction terms perturbs the "true" energy eigenstates of a system. This is where the discrepancies between velocity gauge, length gauge, and also other formulations like acceleration gauge ultimately originate. On the basis of a reductionist approach, we argue that velocity gauge should be considered as the basic description of a quantum system in dipole approximation, and use of other gauges requires careful evaluations of the accompanying transformations of the basis of states. In particular, we find that atomic states in length gauge are dressed by a coherent photon state.

1. Introduction

1.1. Preliminaries on quantum optics. We define the free quantum field theory of quantum optics as a set of free field operators $c_i(\boldsymbol{k})$, $c_i^+(\boldsymbol{k})$, $a_\alpha(\boldsymbol{k})$, and $a_\alpha^+(\boldsymbol{k})$, together with a specific representation, *viz.* the free Fock space representation. The operators are postulated to satisfy the usual (anti-)commutation relations

---

2010 *Mathematics Subject Classification.* Primary 81V80, 81Txx, 81V10.

This work was supported in part by NSERC Canada. The hospitality of the Kavli Institute for Cosmological Physics during completion of this work is gratefully acknowledged.

of a free field theory,

$$[a_\alpha(\boldsymbol{k}), a_\gamma^+(\boldsymbol{k}')] \equiv a_\alpha(\boldsymbol{k})a_\gamma^+(\boldsymbol{k}') - a_\gamma^+(\boldsymbol{k}')a_\alpha(\boldsymbol{k}) = \delta_{\alpha\beta}\delta(\boldsymbol{k}-\boldsymbol{k}'), \tag{1.1}$$

$$[c_i(\boldsymbol{k}), c_j^+(\boldsymbol{k}')]_\pm \equiv c_i(\boldsymbol{k})c_j^+(\boldsymbol{k}') - (-)^{f(i)f(j)}c_j^+(\boldsymbol{k}')c_i(\boldsymbol{k}) = \delta_{ij}\delta(\boldsymbol{k}-\boldsymbol{k}'), \tag{1.2}$$

$$[a_\alpha(\boldsymbol{k}), c_i(\boldsymbol{k}')] = 0, \quad [a_\alpha(\boldsymbol{k}), c_i^+(\boldsymbol{k}')] = 0,$$

$$[c_i(\boldsymbol{k}), c_j(\boldsymbol{k}')]_\pm \equiv c_i(\boldsymbol{k})c_j(\boldsymbol{k}') - (-)^{f(i)f(j)}c_j(\boldsymbol{k}')c_i(\boldsymbol{k}) = 0,$$

with the corresponding adjoint relations. The operators $a_\alpha(\boldsymbol{k})$, $a_\alpha^+(\boldsymbol{k})$ are denoted as photon operators, and the greek label $\alpha \in \{1,2\}$ refers to photon polarizations. The operators $c_i(\boldsymbol{k})$, $c_i^+(\boldsymbol{k})$, are colloquially often denoted as "matter fields", and the label i refers to a particle species with mass m_i and electric charge q_i. The map $f(i) \in \{0,1\}$ takes values $f(i) = 0$ if the particle species i consists of bosons (e.g. mesons or atomic nuclei with even total number of nucleons), or $f(i) = 1$ if the particle species i consists of fermions (e.g. electrons or atomic nuclei with an odd number of nucleons). If the label i takes the value e.g. for an electron, the corresponding operators $c_i(\boldsymbol{k})$, $c_i^+(\boldsymbol{k})$ are denoted as electron operators. Spin labels are suppressed here, although the fermion operators of course carry them.

The variable $\boldsymbol{k} \in \mathbb{R}^3$ refers to wavenumber or momentum $\boldsymbol{p} = \hbar\boldsymbol{k}$. The field operators are apparently distributions, but for the purposes of this paper we could avoid that notion by using a lattice discretization of $\boldsymbol{k}$. Indeed, if the corresponding construction is employed in condensed matter physics, statistical physics, or mathematical physics, the operators are often defined in discretized $\boldsymbol{x}$-space rather than $\boldsymbol{k}$-space.

The free Fock space is defined as the linear span of states of the form

$$\prod_{a=1}^{N_a} c_{i_a}^+(\boldsymbol{k}_a) \prod_{b=1}^{P_b} a_{\alpha_b}^+(\boldsymbol{k}_b)|0\rangle,$$

with $c_i(\boldsymbol{k})|0\rangle = 0$, $a_\alpha(\boldsymbol{k})|0\rangle = 0$. We also define free particle energies

$$E_i(\boldsymbol{k}) = \hbar\omega_i(\boldsymbol{k}) = \frac{\hbar^2\boldsymbol{k}^2}{2m_i}, \quad E_\gamma(\boldsymbol{k}) = \hbar c|\boldsymbol{k}|, \tag{1.3}$$

and the free field operators in $\boldsymbol{x}$-space,

$$\psi_i(\boldsymbol{x},t) = \frac{1}{\sqrt{2\pi}^3}\int d^3\boldsymbol{k}\, c_i(\boldsymbol{k})\exp[\mathrm{i}\boldsymbol{k}\cdot\boldsymbol{x} - \mathrm{i}\omega_i(\boldsymbol{k})t], \tag{1.4}$$

$$\psi_i^+(\boldsymbol{x},t) = \frac{1}{\sqrt{2\pi}^3}\int d^3\boldsymbol{k}\, c_i^+(\boldsymbol{k})\exp[-\mathrm{i}\boldsymbol{k}\cdot\boldsymbol{x} + \mathrm{i}\omega_i(\boldsymbol{k})t], \tag{1.5}$$

$$\begin{aligned}\boldsymbol{A}(\boldsymbol{x},t) &= \sqrt{\frac{\hbar\mu_0 c}{(2\pi)^3}}\int\frac{d^3\boldsymbol{k}}{\sqrt{2k}}\sum_{\alpha=1}^{2}\boldsymbol{\epsilon}_\alpha(\boldsymbol{k})\Big(a_\alpha(\boldsymbol{k})\exp[\mathrm{i}(\boldsymbol{k}\cdot\boldsymbol{x} - ckt)] \\ &\quad + a_\alpha^+(\boldsymbol{k})\exp[-\mathrm{i}(\boldsymbol{k}\cdot\boldsymbol{x} - ckt)]\Big),\end{aligned} \tag{1.6}$$

with normalized transverse polarization vectors, $\boldsymbol{k}\cdot\boldsymbol{\epsilon}_\alpha(\boldsymbol{k}) = 0$ and $\boldsymbol{\epsilon}_\alpha(\boldsymbol{k})\cdot\boldsymbol{\epsilon}_\beta(\boldsymbol{k}) = \delta_{\alpha\beta}$. The derived photon operators in $\boldsymbol{x}$-space are the transverse electric and magnetic field operators, $\boldsymbol{E}_\perp(\boldsymbol{x},t) = -\partial\boldsymbol{A}(\boldsymbol{x},t)/\partial t$, $\boldsymbol{B}(\boldsymbol{x},t) = \boldsymbol{\nabla}\times\boldsymbol{A}(\boldsymbol{x},t)$.

Products of three or more field operators, as well as linear combinations of such products, are denoted as *interaction operators*. Two important examples of interaction operators include the Coulomb operator

$$H_C(t) = \sum_{ij}\int d^3\boldsymbol{x}\int d^3\boldsymbol{x}'\,\psi_i^+(\boldsymbol{x},t)\psi_j^+(\boldsymbol{x}',t)\frac{q_iq_j}{8\pi\epsilon_0|\boldsymbol{x}-\boldsymbol{x}'|}\psi_j(\boldsymbol{x}',t)\psi_i(\boldsymbol{x},t), \tag{1.7}$$

and the photon-matter interaction operator,

$$\begin{aligned} H_I(t) &= \int d^3\boldsymbol{x}\sum_i\frac{1}{2m_i}\Big[\mathrm{i}q_i\hbar\boldsymbol{A}(\boldsymbol{x},t)\cdot\Big(\psi_i^+(\boldsymbol{x},t)\overset{\leftrightarrow}{\boldsymbol{\nabla}}\psi_i(\boldsymbol{x},t)\Big) \\ &\quad + q_i^2\psi_i^+(\boldsymbol{x},t)\boldsymbol{A}^2(\boldsymbol{x},t)\psi_i(\boldsymbol{x},t)\Big], \end{aligned} \tag{1.8}$$

where

$$\psi_i^+(\boldsymbol{x},t)\overset{\leftrightarrow}{\boldsymbol{\nabla}}\psi_i(\boldsymbol{x},t) \equiv \psi_i^+(\boldsymbol{x},t)\boldsymbol{\nabla}\psi_i(\boldsymbol{x},t) - (\boldsymbol{\nabla}\psi_i^+(\boldsymbol{x},t))\psi_i(\boldsymbol{x},t).$$

In the common language of applied quantum field theory, the operators (1.7,1.8) are interaction operators in the "interaction picture", since we expressed these operators in terms of the free field operators. However, applied quantum field theory never needs to switch into any other picture for the actual calculation e.g. of atomic energy levels, eigenstates, scattering cross sections, or photon emission or absorption rates. At an operational level, it is completely sufficient to define quantum field theory within the interaction picture.

It is often helpful at a mnemonic level, and therefore convenient, to formally switch into the Schrödinger picture, where the field operators are time-independent and the full time evolution of a physical system is carried by the states, or to switch into the Heisenberg picture, where states do not evolve and the full time evolution of a physical system is carried by the field operators.

Before specifying to dipole approximation, we also note that the time evolution of the field operators (1.4,1.5,1.6) can be gotten in the form,

$$\psi_i(\boldsymbol{x},t) = \exp(\mathrm{i}H_0t/\hbar)\psi_i(\boldsymbol{x})\exp(-\mathrm{i}H_0t/\hbar), \tag{1.9}$$

$$\boldsymbol{A}(\boldsymbol{x},t) = \exp(\mathrm{i}H_0t/\hbar)\boldsymbol{A}(\boldsymbol{x})\exp(-\mathrm{i}H_0t/\hbar), \tag{1.10}$$

with the "free" Hamiltonian (in the interaction picture)

$$H_0 = \int d^3\boldsymbol{x}\left(\sum_i\frac{\hbar^2}{2m_i}\boldsymbol{\nabla}\psi_i^+(\boldsymbol{x},t)\cdot\boldsymbol{\nabla}\psi_i(\boldsymbol{x},t) + \frac{\epsilon_0}{2}\boldsymbol{E}_\perp^2(\boldsymbol{x},t) + \frac{1}{2\mu_0}\boldsymbol{B}^2(\boldsymbol{x},t)\right).$$

The free Hamiltonian is time-independent in the interaction picture (and in the Schrödinger picture) and can therefore also be expressed in terms of the $t=0$ operators $\psi_i(\boldsymbol{x}), \psi_i^+(\boldsymbol{x}), \boldsymbol{A}(\boldsymbol{x})$.

This provides us with the basic tools to discuss quantum optics in length gauge and velocity gauge.

1.2. Dipole approximation, length gauge, and velocity gauge. The description of interactions of atoms, molecules, and excitations in condensed materials, with electromagnetic waves in the photon energy range below 10 eV is the realm of quantum optics in *dipole approximation*, where the electric and magnetic fields can be well approximated in terms of time-dependent fields,

$$\boldsymbol{E}(\boldsymbol{x},t) \simeq \boldsymbol{E}(t), \quad \boldsymbol{B}(\boldsymbol{x},t) \simeq \boldsymbol{B}(t).$$

The reason is that for photon energies in the ultraviolet or lower energy ranges, $E_\gamma \lesssim 10$ eV, the corresponding photon wavelengths $\lambda = hc/E_\gamma \gtrsim 100$ nm exceed the extensions of atomic or molecular wavefunctions by two to three orders of magnitude, and therefore the spatial variation of the electromagnetic fields has negligible impact on the calculation of quantum mechanical matrix elements, e.g.

$$\int d^3\boldsymbol{x}\,\phi_1^+(\boldsymbol{x},t)\boldsymbol{x}\cdot\boldsymbol{E}(\boldsymbol{x},t)\phi_2(\boldsymbol{x},t) \simeq \int d^3\boldsymbol{x}\,\phi_1^+(\boldsymbol{x},t)\boldsymbol{x}\cdot\boldsymbol{E}(t)\phi_2(\boldsymbol{x},t),$$

where $\phi_1(\boldsymbol{x},t)$ and $\phi_2(\boldsymbol{x},t)$ are wavefunctions. Indeed, the use of dipole approximation is even commonplace for photon energies in the sub-keV energy range, where the approximation is not as good.

There exist two common Hamilton operators for the description of photon-matter interactions in dipole approximation. For brevity, we will focus on the interaction pieces of those Hamiltonians, i.e. the parts of the total Hamilton operators which describe the couplings of the photons to charged particles. Those interaction Hamiltonians are the pieces which differ between the full Hamiltonians of quantum optics in dipole approximation.

Historically, the *length gauge* interaction Hamiltonian,

$$H_{Il} = -\int d^3\boldsymbol{x}\sum_i q_i\psi_i^+(\boldsymbol{x},t)\boldsymbol{x}\cdot\boldsymbol{E}(t)\psi_i(\boldsymbol{x},t), \tag{1.11}$$

was the first Hamiltonian which was extensively used alreay in the 1920s to calculate photon scattering, emission and absorption. The phrase *length gauge* refers to the appearance of the particle location $\boldsymbol{x}$ in the product $\boldsymbol{x}\cdot\boldsymbol{E}(t)$. The length gauge Hamiltonian also motivated the phrase *dipole approximation*, because the term $V = -q_i\boldsymbol{x}\cdot\boldsymbol{E}(t) = -\boldsymbol{d}_i\cdot\boldsymbol{E}(t)$ is the energy of the electric dipole moment $\boldsymbol{d}_i = q_i\boldsymbol{x}$ in the field $\boldsymbol{E}(t)$.

On the other hand, the *velocity gauge* interaction Hamiltonian

$$\begin{aligned} H_{Iv} &= \int d^3\boldsymbol{x}\sum_i\left(\frac{\mathrm{i}q_i\hbar}{2m_i}\boldsymbol{A}(t)\cdot\left(\psi_i^+(\boldsymbol{x},t)\overset{\leftrightarrow}{\boldsymbol{\nabla}}\psi_i(\boldsymbol{x},t)\right)\right.\\ &\quad\left.+\frac{q_i^2}{2m_i}\psi_i^+(\boldsymbol{x},t)\boldsymbol{A}^2(t)\psi_i(\boldsymbol{x},t)\right) \end{aligned} \tag{1.12}$$

owes its name to the appearance of the velocity in the form $\boldsymbol{p}/m_i \to -\mathrm{i}\hbar\boldsymbol{\nabla}/m_i$, where m_i is the mass of a particle with charge q_i.

It has been known for many years [**1**] that these two interaction Hamiltonians yield very different results for basic quantum optical observables. A textbook example is provided by the differential electron-photon scattering cross section for scattering of photons with initial momentum $\hbar\boldsymbol{k}$ and polarization $\boldsymbol{\epsilon}(\boldsymbol{k})$ off electronic states $|n\rangle$. the length gauge Hamiltonian (1.11) yields the original Kramers-Heisenberg formula [**2**],

$$\begin{aligned} \frac{d\sigma}{d\Omega} &= \alpha_S^2 c^2 k k'^3\left|\sum_{n''}\left(\frac{\langle n'|\boldsymbol{\epsilon}'(\boldsymbol{k}')\cdot\mathbf{x}|n''\rangle\langle n''|\boldsymbol{\epsilon}(\boldsymbol{k})\cdot\mathbf{x}|n\rangle}{\omega_{n'',n}-ck-\mathrm{i}\epsilon}\right.\right.\\ &\quad\left.\left.+\frac{\langle n'|\boldsymbol{\epsilon}(\boldsymbol{k})\cdot\mathbf{x}|n''\rangle\langle n''|\boldsymbol{\epsilon}'(\boldsymbol{k}')\cdot\mathbf{x}|n\rangle}{\omega_{n'',n}+ck'-\mathrm{i}\epsilon}\right)\right|^2, \end{aligned} \tag{1.13}$$

where $k' = k - (\omega_{n',n}/c)$ and the sum over virtual states $|n''\rangle$ includes integration over continuous quantum numbers. The constant $\alpha_S = e^2/4\pi\epsilon_0\hbar c$ is Sommerfeld's fine structure constant.

On the other hand, the velocity gauge Hamiltonian (1.12) yields [**3**, **4**]

$$\begin{aligned}\frac{d\sigma}{d\Omega} &= \frac{\alpha_S^2}{c^2}\frac{k'}{k}\left|\frac{\hbar}{m_e}\delta_{n'n}\boldsymbol{\epsilon}'(\boldsymbol{k}')\cdot\boldsymbol{\epsilon}(\boldsymbol{k}) + \sum_{n''}\omega_{n',n''}\omega_{n'',n}\right.\\ &\quad\times\left(\frac{\langle n'|\boldsymbol{\epsilon}'(\boldsymbol{k}')\cdot\mathbf{x}|n''\rangle\langle n''|\boldsymbol{\epsilon}(\boldsymbol{k})\cdot\mathbf{x}|n\rangle}{\omega_{n'',n} - ck - \mathrm{i}\epsilon}\right.\\ &\quad\left.\left.+\frac{\langle n'|\boldsymbol{\epsilon}(\boldsymbol{k})\cdot\mathbf{x}|n''\rangle\langle n''|\boldsymbol{\epsilon}'(\boldsymbol{k}')\cdot\mathbf{x}|n\rangle}{\omega_{n'',n} + ck' - \mathrm{i}\epsilon}\right)\right|^2.\end{aligned} \tag{1.14}$$

These two expressions are practically equivalent if near resonance conditions can be satisfied in the sense that there are nonvanishing transition matrix elements $\langle n'|\boldsymbol{\epsilon}'(\boldsymbol{k}')\cdot\mathbf{x}|n''\rangle$ and $\langle n''|\boldsymbol{\epsilon}(\boldsymbol{k})\cdot\mathbf{x}|n\rangle$ with the properties $\omega_{n'',n} \simeq ck$ and $\omega_{n',n''} \simeq -ck'$, or if there are nonvanishing matrix elements $\langle n'|\boldsymbol{\epsilon}(\boldsymbol{k})\cdot\mathbf{x}|n''\rangle$ and $\langle n''|\boldsymbol{\epsilon}'(\boldsymbol{k}')\cdot\mathbf{x}|n\rangle$ with the properties $\omega_{n'',n} \simeq -ck'$ and $\omega_{n',n''} \simeq ck$. There will be observational bias towards observation of scattered photons where these near resonance conditions are met, while at the same time electronic energy level systems are generically dense enough to meet these requirements. In spite of the glaring differences between the two expressions, the domination by nearly resonant scattering implies equivalence of (1.13) and (1.14) for practical purposes in the sense of yielding comparable numerical results for scattering experiments. However, discrepancies between the predictions from velocity and length gauge Hamiltonians have become observationally relevant in strong field systems. Indeed, it is known that the two Hamiltonians (1.11) and (1.12) can yield very different and occasionally contradictory results in strong electromagnetic fields [**5**–**13**]. In particular, for electron detachment in strong fields, authors have argued for velocity gauge on the basis of analytic advantages [**14**, **15**] and momentum conservation [**16**], whereas Schlicher *et al.* [**17**] and also Cohen-Tannoudji *et al.* [**18**] had emphasized the advantage of the length gauge formulation in terms of kinetic momentum $m\boldsymbol{v}$ and electric field $\boldsymbol{E}(t)$. Furthermore, many studies in recent years suggest that length gauge yields results in better agreement with observations [**5**, **6**, **13**, **19**–**24**]. Yet again, for high-order harmonic generation in diatomic molecules at large internuclear separation, Chirilă and Lein found good agreement with the semiclassical three-step mechanism only in velocity gauge [**25**].

For another example, Zhang and Nakajima report discrepancies between velocity and length gauge predictions for the photoelectron angular distribution in hydrogen, in particular for elliptical polarization [**7**], while Majety *et al.* [**26**] report advantages in using length gauge at short distances and velocity gauge at longer distances for the calculation of photoelectron spectra from single-electron systems, helium, and hydrogen molecules.

Furthermore, Dong *et al.* find unphysical oscillations in the photoinduced carrier densities in graphene if they use length gauge, whereas velocity gauge does not exhibit that problem [**10**]. On the other hand, theoretical investigations both in velocity gauge [**27**] and in length gauge [**28**] show the possibility of strong third harmonic generation in graphene, but Ishikawa [**27**] reports a reduction of the nonlinear response due to the interplay of interband and intraband dynamics in velocity gauge, whereas Al-Naib *et al.* find an increase in length gauge [**28**]. These observations shed light on another important difference between length gauge and velocity gauge in systems with energy bands: velocity gauge is diagonal in quasi-momentum and therefore *prima facie* less sensitive to intraband effects, whereas length gauge is less sensitive to interband interactions. The choice of length gauge [**29–31**] versus velocity gauge [**32, 33**] for the study of optical transitions and higher harmonic generation in solids therefore also depends on a judicious estimate of relative importance of intraband and interband interactions.

The choice of the interaction term naturally affects numerical implementations and calculational efficiencies for integrations of the time-dependent Schrödinger equation [**8, 9, 34**], but it is also important to recognize physical differences between the interaction Hamiltonians (1.11) and (1.12). Close inspection of the transformation which maps (1.12) into (1.11) demonstrates that the transformation can also be understood as a redefinition of the Schrödinger field without any transformation of the gauge fields. As a consequence, we will find that the transformation can be selectively applied to different particle species, i.e. we could perform the transformation e.g. only for electrons, but not for protons in a many-particle system, without violating any mathematical consistency conditions. Indeed, the transition from velocity gauge to length gauge is only a subset of picture changing operations which also include e.g. the transformation into "acceleration gauge" [**18, 35, 36**], and understanding the transition from velocity gauge to length gauge in this wider framework also helps to understand sources of differences between velocity gauge and length gauge. The advantage of an improved understanding of analytic sources of differences between those different pictures of time evolution illuminates the question why one picture can be better or worse than another picture in describing specific physical systems or effects.

The transformation between velocity and length gauge for a nonrelativistic many-particle system will be re-examined in Sec. 2 from a first principles perspective, which emphasizes the separation of internal short wavelength components and external long wavelength components in the dipole approximation. Selective application to different particle species will be discussed in Sec. 3. The appearance of analytic differences between velocity gauge and length gauge in physical systems is first discussed in the single electron approximation in Sec. 4. Sec. 5 revisits the question of analytic differences between velocity gauge and length gauge in the wider scope of picture changing operations, which also includes the transition to acceleration gauge. Sec. 6 summarizes our conclusions.

2. Dipole approximation and the transition to length gauge

The transition from velocity gauge to length gauge actually employs a residual gauge transformation within Coulomb gauge. Therefore Coulomb gauge will be reviewed first before we discuss the transformation from velocity gauge to length gauge for the interaction Hamiltonian.

The preferred gauge for the electromagnetic potentials in quantum optics is Coulomb gauge,

$$\nabla \cdot \boldsymbol{A}(\boldsymbol{x}, t) = 0, \quad \Phi(\boldsymbol{x}, t) = \frac{1}{4\pi\epsilon_0} \int d^3x' \, \frac{\varrho(\boldsymbol{x}', t)}{|\boldsymbol{x} - \boldsymbol{x}'|}, \tag{2.1}$$

$$\left(\frac{1}{c^2}\frac{\partial^2}{\partial t^2}-\Delta\right)\boldsymbol{A}(\boldsymbol{x},t)=\mu_0\boldsymbol{J}(\boldsymbol{x},t), \tag{2.2}$$

where Φ is the electric potential. The current $\boldsymbol{J}$ includes the contributions from the longitudinal components $\boldsymbol{E}_\parallel = -\boldsymbol{\nabla}\Phi$ of the electric field, which cancel the longitudinal components from the electric current $\boldsymbol{j}$,

$$\boldsymbol{J}(\boldsymbol{x},t)=\boldsymbol{j}(\boldsymbol{x},t)-\epsilon_0\frac{\partial}{\partial t}\boldsymbol{\nabla}\Phi(\boldsymbol{x},t),\quad \boldsymbol{\nabla}\cdot\boldsymbol{J}(\boldsymbol{x},t)=0, \tag{2.3}$$

see e.g. Sec. 18.1 in Ref. [**4**]. The vector potential is then

$$\boldsymbol{A}(\boldsymbol{x},t)=\boldsymbol{A}_J(\boldsymbol{x},t)+\boldsymbol{A}_\gamma(t), \tag{2.4}$$

where the contribution from the local charges is

$$\boldsymbol{A}_J(\boldsymbol{x},t)=\int d^3\boldsymbol{x}'\,\frac{\mu_0}{4\pi|\boldsymbol{x}-\boldsymbol{x}'|}\boldsymbol{J}\left(\boldsymbol{x}',t-\frac{|\boldsymbol{x}-\boldsymbol{x}'|}{c}\right), \tag{2.5}$$

and the freely evolving part $\boldsymbol{A}_\gamma(t)$ accounts for the external radiation fields

$$\boldsymbol{E}_\gamma(t)=-\,d\boldsymbol{A}_\gamma(t)/dt \tag{2.6}$$

and

$$\boldsymbol{B}_\gamma(t)=\boldsymbol{\nabla}\times\boldsymbol{A}_\gamma(\boldsymbol{x},t)|_{\boldsymbol{k}\cdot\boldsymbol{x}\to 0}. \tag{2.7}$$

The advantage of Coulomb gauge is therefore twofold: First, Coulomb gauge explicitly accounts for atomic or molecular potentials and currents without the potential pitfall of double-counting of electromagnetic interactions through longitudinal photon operators. Furthermore, Coulomg gauge separates the long wavelength radiation contribution $\boldsymbol{A}_\gamma(t)$ on the one hand from the local contributions $\Phi(\boldsymbol{x},t)$ and $\boldsymbol{A}_J(\boldsymbol{x},t)$ due to the charges in the atoms or molecules on the other hand.

We also note that the contributions from $\boldsymbol{A}_J(\boldsymbol{x},t)$ to Hamilton operators can usually be neglected: Atomic orbitals scale with electron charge at least like e^3, which implies that terms $e\boldsymbol{A}_J(\boldsymbol{x},t)$ from local vector potentials scale at least like e^8. Therefore we omit the index γ for the radiation contributions from now on, $\boldsymbol{A}_\gamma(t)\equiv\boldsymbol{A}(t)$.

The semiclassical Hamiltonian in Coulomb gauge and in dipole approximation then takes the form

$$\begin{aligned}
H_v \;=\; & \int d^3\boldsymbol{x}\sum_i\Bigg(\frac{1}{2m_i}\Big[\hbar^2\boldsymbol{\nabla}\psi_i^+(\boldsymbol{x},t)\cdot\boldsymbol{\nabla}\psi_i(\boldsymbol{x},t)+\mathrm{i}q_i\hbar\boldsymbol{A}(t)\cdot\Big(\psi_i^+(\boldsymbol{x},t)\overleftrightarrow{\boldsymbol{\nabla}}\psi_i(\boldsymbol{x},t)\Big)\\
& +q_i^2\psi_i^+(\boldsymbol{x},t)\boldsymbol{A}^2(t)\psi_i(\boldsymbol{x},t)\Big]+\psi_i^+(\boldsymbol{x},t)V_i(\boldsymbol{x},t)\psi_i(\boldsymbol{x},t)\Bigg)\\
& +\frac{1}{2}\sum_{ij}\int d^3\boldsymbol{x}\int d^3\boldsymbol{x}'\,\psi_i^+(\boldsymbol{x},t)\psi_j^+(\boldsymbol{x}',t)V_{ij}(\boldsymbol{x}-\boldsymbol{x}',t)\psi_j(\boldsymbol{x}',t)\psi_i(\boldsymbol{x},t).
\end{aligned} \tag{2.8}$$

This is the second quantized Hamiltonian with respect to the matter fields $\psi_i(\boldsymbol{x},t)$ in the Heisenberg picture,

$$[\psi_i(\boldsymbol{x},t),\psi_j^+(\boldsymbol{x}',t)]_\pm=\delta_{ij}\delta(\boldsymbol{x}-\boldsymbol{x}'), \tag{2.9}$$

while the radiation fields are treated classically, e.g. as expectation values of photon operators with long-wavelength coherent photon states. Spin labels are suppressed, since the subleading Pauli terms are (as usual) not included in (2.8).

The single particle potentials $V_i(\boldsymbol{x},t)$ include the contributions to the scalar potential $\Phi(\boldsymbol{x},t)$ (2.1) due to fixed charges (e.g. massive ion cores),

$$V_i(\boldsymbol{x},t) \supseteq \sum_I \frac{q_i Q_I}{4\pi\epsilon_0|\boldsymbol{x}-\boldsymbol{X}_I|}, \tag{2.10}$$

while the two-particle interaction potentials contain the contributions to $\Phi(\boldsymbol{x},t)$ from the dynamical charged particles,

$$V_{ij}(\boldsymbol{x}-\boldsymbol{x}',t) \supseteq \frac{q_i q_j}{4\pi\epsilon_0|\boldsymbol{x}-\boldsymbol{x}'|}. \tag{2.11}$$

The quantum optics Hamiltonian (2.8) with vector potentials $\boldsymbol{A}(\boldsymbol{x},t)$ would yield equations of motion

$$\mathrm{i}\hbar\frac{\partial}{\partial t}\psi_i(\boldsymbol{x},t) = [\psi_i(\boldsymbol{x},t), H],$$

which are form invariant under gauge transformations

$$\psi_i'(\boldsymbol{x},t) = \exp[\mathrm{i}q_i\varphi(\boldsymbol{x},t)/\hbar]\psi_i(\boldsymbol{x},t), \quad \boldsymbol{A}'(\boldsymbol{x},t) = \boldsymbol{A}(\boldsymbol{x},t) + \boldsymbol{\nabla}\varphi(\boldsymbol{x},t),$$

$$V_i'(\boldsymbol{x},t) = V_i(\boldsymbol{x},t) - q_i\frac{\partial}{\partial t}\varphi(\boldsymbol{x},t), \quad V_{ij}'(\boldsymbol{x}-\boldsymbol{x}',t) = V_{ij}(\boldsymbol{x}-\boldsymbol{x}',t).$$

However, as a consequence of the dipole approximation, the Schrödinger equations

$$\begin{aligned}\mathrm{i}\hbar\frac{\partial}{\partial t}\psi_i(\boldsymbol{x},t) &= [\psi_i(\boldsymbol{x},t), H_v] \\ &= -\frac{[\hbar\boldsymbol{\nabla} - \mathrm{i}q_i\boldsymbol{A}(t)]^2}{2m_i}\psi_i(\boldsymbol{x},t) + V_i(\boldsymbol{x},t)\psi_i(\boldsymbol{x},t) \\ &\quad + \sum_j \int d^3\boldsymbol{x}'\,\psi_j^+(\boldsymbol{x}',t)V_{ij}(\boldsymbol{x}-\boldsymbol{x}',t)\psi_j(\boldsymbol{x}',t)\psi_i(\boldsymbol{x},t)\end{aligned} \tag{2.12}$$

following from the Hamiltonian (2.8) are manifestly form invariant only under restricted gauge transformations where the gauge function $\varphi(\boldsymbol{x},t)$ is constrained to be at most linear in spatial coordinates,

$$\varphi(\boldsymbol{x},t) = \boldsymbol{a}(t)\cdot\boldsymbol{x} + b(t) \Rightarrow \boldsymbol{A}'(t) = \boldsymbol{A}(t) + \boldsymbol{a}(t). \tag{2.13}$$

These residual gauge transformations trivially preserve the Coulomb gauge.

The minimal coupling terms in the effective dipole Hamiltonian can therefore be absorbed into the Schrödinger fields through the particular transformation with gauge function $\varphi(\boldsymbol{x},t) = -\boldsymbol{x}\cdot\boldsymbol{A}(t)$,

$$\psi_i(\boldsymbol{x},t) \Rightarrow \psi_i^{(l)}(\boldsymbol{x},t) = \exp\left(-\frac{\mathrm{i}}{\hbar}q_i\boldsymbol{x}\cdot\boldsymbol{A}(t)\right)\psi_i(\boldsymbol{x},t), \tag{2.14}$$

$$\boldsymbol{A}(t) \Rightarrow \boldsymbol{A}'(t) = \boldsymbol{A}(t) - \boldsymbol{\nabla}(\boldsymbol{x}\cdot\boldsymbol{A}(t)) = 0, \tag{2.15}$$

$$V_i(\boldsymbol{x},t) \Rightarrow V_i'(\boldsymbol{x},t) = V_i(\boldsymbol{x},t) + q_i\frac{\partial}{\partial t}\boldsymbol{x}\cdot\boldsymbol{A}(t) = V_i(\boldsymbol{x},t) - q_i\boldsymbol{x}\cdot\boldsymbol{E}(t). \tag{2.16}$$

The transformation (2.14) preserves the canonical (anti-)commutation relations (2.9).

The resulting Hamiltonian in terms of the *new* Schrödinger fields and the *old* gauge fields and potentials is (after dropping the label (l) for the Schrödinger fields in length gauge)

$$\begin{aligned} H_l &= \int d^3\boldsymbol{x} \sum_i \left(\frac{\hbar^2}{2m_i} \nabla\psi_i^+(\boldsymbol{x},t) \cdot \nabla\psi_i(\boldsymbol{x},t) \right. \\ &\quad \left. + \psi_i^+(\boldsymbol{x},t) V_i(\boldsymbol{x},t)\psi_i(\boldsymbol{x},t) - q_i\psi_i^+(\boldsymbol{x},t)\boldsymbol{x}\cdot\boldsymbol{E}(t)\psi_i(\boldsymbol{x},t) \right) \\ &\quad + \frac{1}{2}\sum_{ij}\int d^3\boldsymbol{x}\int d^3\boldsymbol{x}'\, \psi_i^+(\boldsymbol{x},t)\psi_j^+(\boldsymbol{x}',t)V_{ij}(\boldsymbol{x}-\boldsymbol{x}',t)\psi_j(\boldsymbol{x}',t)\psi_i(\boldsymbol{x},t). \end{aligned} \tag{2.17}$$

Equation (2.17) is the second quantized Hamiltonian in length gauge, whereas (2.8) is the Hamiltonian in velocity gauge.

In spite of our previous observation that the velocity gauge Hamiltonian (2.8) preserves form invariance of the equations of motion (2.12) under the restricted gauge transformations (2.13), the equations of motion resulting from (2.17),

$$\begin{aligned} \mathrm{i}\hbar\frac{\partial}{\partial t}\psi_i(\boldsymbol{x},t) &= [\psi_i(\boldsymbol{x},t), H_l] \\ &= -\frac{\hbar^2}{2m_i}\Delta\psi_i(\boldsymbol{x},t) + V_i(\boldsymbol{x},t)\psi_i(\boldsymbol{x},t) - q_i\boldsymbol{x}\cdot\boldsymbol{E}(t)\psi_i(\boldsymbol{x},t) \\ &\quad + \sum_j \int d^3\boldsymbol{x}'\, \psi_j^+(\boldsymbol{x}',t)V_{ij}(\boldsymbol{x}-\boldsymbol{x}',t)\psi_j(\boldsymbol{x}',t)\psi_i(\boldsymbol{x},t), \end{aligned} \tag{2.18}$$

do *not* have the same form as the equations resulting from (2.8) because we explicitly substituted $\boldsymbol{A}'(t) = 0$ and $V_i'(\boldsymbol{x},t) = V_i(\boldsymbol{x},t) - q_i\boldsymbol{x}\cdot\boldsymbol{E}(t)$.

There are different ways to understand the discrepancy between (2.12) and (2.18). In a geometric analogy, we can think of the general formulation of equations of motion of a gauge theory as the form invariant formulation of the theory, akin to the tensor formulation of the Einstein equation in General Relativity. However, formulating the equations of motion in a particular gauge is akin to choosing a particular coordinate system, and manifest form invariance is lost once the equations are formulated in that system.

Alternatively, we can think of the transformation that led to (2.18) as an incomplete gauge transformation which mixes the *transformed* Schrödinger fields with the *original* potentials through $V_i(\boldsymbol{x},t)$ and

$$-q_i\boldsymbol{x}\cdot\boldsymbol{E}(t) = q_i\boldsymbol{x}\cdot d\boldsymbol{A}(t)/dt.$$

In a geometric analog, this is like using mixed coordinates $x' = x\cos\phi + y\sin\phi$ and y, but not the transformed coordinate $y' = y\cos\phi - x\sin\phi$ after a rotation in two dimensions. In that picture, the transformation that led from (2.12) to (2.18) is rather the analog of an affine transformation in field space instead of a full gauge transformation: the transition from (2.8) to (2.17) only involved the field redefinition (2.14), wich we can think of as a basic dressing operation, but not a complete gauge transformation. We also note that contrary to the original Hamiltonian (2.8), the length gauge Hamiltonian (2.17) yields equations of motion (2.18) which are manifestly form invariant only under even more restricted gauge transformations $\varphi(\boldsymbol{x},t) = b(t)$. This point of view sheds new light on the gauge problem from a different angle. In particular, we will see explicitly in Sec. 3 that contrary to an ordinary full gauge transformation, the transition to length gauge can be selectively applied only to particular Schrödinger field operators, thus implying the possibility of selective transformations only for particular particle species.

The loss of manifest form invariance cannot affect the underlying physics, and ultimately the Hamiltonians (2.8) and (2.17) should yield the same physical results if evaluated analytically or with sufficiently high precision, and yet careful evaluations have led to very different results in many instances.

3. Switching to length gauge only for particular species

The fact that the transition between velocity gauge and length gauge can be viewed as an incomplete gauge transformation, *viz.* acting only on the matter fields, has the interesting implication that we can *separately choose for each matter field* $\psi_i(\boldsymbol{x},t)$ whether we wish to repesent its couplings in velocity gauge or in length gauge. Recall that the operators $\psi_i(\boldsymbol{x},t)$ for different values of the index i, $1 \le i \le n$, represent different particle species, i.e. the assertion above implies that in a system of electrons, protons and α particles, we may e.g. choose to represent the electrons and α particles in velocity gauge and the protons in length gauge. Suppose we have n different species of charged particles, and we retain velocity gauge for n_v of them, labeled by indices $1 \le i \le n_v$, while we switch to length gauge for $n_l = n - n_v$. We will label the latter field operators with indices $1 \le \bar{\imath} \equiv i - n_v \le n_l$. The Hamiltonian in the mixed representation of electromagnetic interactions then takes the form

$$\begin{aligned} H_{vl} &= \int\! d^3\boldsymbol{x} \sum_{i=1}^{n_v} \frac{1}{2m_i}\Big(\hbar^2 \boldsymbol{\nabla}\psi_i^+(\boldsymbol{x},t)\cdot\boldsymbol{\nabla}\psi_i(\boldsymbol{x},t) + \mathrm{i}q_i\hbar\boldsymbol{A}(t)\cdot\Big(\psi_i^+(\boldsymbol{x},t)\overset{\leftrightarrow}{\boldsymbol{\nabla}}\psi_i(\boldsymbol{x},t)\Big) \\ &\quad + q_i^2\psi_i^+(\boldsymbol{x},t)\boldsymbol{A}^2(t)\psi_i(\boldsymbol{x},t)\Big) + \int\! d^3\boldsymbol{x}\sum_{\bar{\imath}=1}^{n_l}\left(\frac{\hbar^2}{2m_{\bar{\imath}}}\boldsymbol{\nabla}\psi_{\bar{\imath}}^+(\boldsymbol{x},t)\cdot\boldsymbol{\nabla}\psi_{\bar{\imath}}(\boldsymbol{x},t)\right. \\ &\quad \left. - q_{\bar{\imath}}\psi_{\bar{\imath}}^+(\boldsymbol{x},t)\boldsymbol{x}\cdot\boldsymbol{E}(t)\psi_{\bar{\imath}}(\boldsymbol{x},t)\right) + \sum_{i=1}^{n}\psi_i^+(\boldsymbol{x},t)V_i(\boldsymbol{x},t)\psi_i(\boldsymbol{x},t) \\ &\quad + \frac{1}{2}\sum_{i,j=1}^{n}\int\! d^3\boldsymbol{x}\int\! d^3\boldsymbol{x}'\,\psi_i^+(\boldsymbol{x},t)\psi_j^+(\boldsymbol{x}',t)V_{ij}(\boldsymbol{x}-\boldsymbol{x}',t)\psi_j(\boldsymbol{x}',t)\psi_i(\boldsymbol{x},t). \end{aligned} \tag{3.1}$$

If we are using a semiclassical approximation by replacing the photon operators $\boldsymbol{A}(t)$, $\boldsymbol{E}(t) = -d\boldsymbol{A}(t)/dt$, with classical fields, the operator H_{vl} preserves particle numbers. In this case we can map the second quantized Schrödinger equation for states in Fock space,

$$\mathrm{i}\hbar\frac{d}{dt}|\Psi(t)\rangle = H_{vl}|\Psi(t)\rangle \tag{3.2}$$

into decoupled wave equations for many-particle wavefunctions within each sector of Fock space by using the *ansatz*

$$|\Psi(t)\rangle = \int\! d^3\boldsymbol{x}_1\,\psi_{i_1}^+(\boldsymbol{x}_1)\ldots\int\! d^3\boldsymbol{x}_N\,\psi_{i_N}^+(\boldsymbol{x}_N)|0\rangle\Psi_{i_1\ldots i_N}(\boldsymbol{x}_1,\ldots,\boldsymbol{x}_N,t), \tag{3.3}$$

where $\psi_i^+(\boldsymbol{x})$ are the Schrödinger picture creation operators. We assume ordering of the particle species labels i_k in such a way that all labels $i_k \le n_v$ correspond to field operators in velocity gauge, while all lables $i_k > n_v$ correspond to field operators in length gauge. Use of the canonical (anti-)commutation relations of the field operators and linear independence of the N-particle states then reduces the second quantized Schrödinger equation (3.2) to the wave equation

$$\mathrm{i}\hbar\frac{\partial}{\partial t}\Psi_{i_1\ldots i_N}(\boldsymbol{x}_1,\ldots,\boldsymbol{x}_N,t) = H_{vl}^{(1)}\Psi_{i_1\ldots i_N}(\boldsymbol{x}_1,\ldots,\boldsymbol{x}_N,t) \tag{3.4}$$

with the first quantized Hamiltonian

$$\begin{aligned} H_{vl}^{(1)} &= -\sum_{k|i_k\leq n_v}\frac{\hbar^2}{2m_{i_k}}\left(\frac{\partial}{\partial \boldsymbol{x}_k} - \frac{\mathrm{i}}{\hbar}q_{i_k}\boldsymbol{A}(t)\right)^2 - \sum_{k|i_k>n_v}\left(\frac{\hbar^2}{2m_{i_k}}\frac{\partial^2}{\partial \boldsymbol{x}_k^2} + q_{i_k}\boldsymbol{x}_k\cdot\boldsymbol{E}(t)\right) \\ &\quad + \frac{1}{2}\sum_{j,k=1}^{N} V_{i_j,i_k}(\boldsymbol{x}_j - \boldsymbol{x}_k,t) + \sum_{k=1}^{N} V_{i_k}(\boldsymbol{x}_k,t). \end{aligned} \tag{3.5}$$

Furthermore, note that within the framework of a Hartree approximation, an *ansatz*

$$\Psi_{i_1\ldots i_N}(\boldsymbol{x}_1,\ldots,\boldsymbol{x}_N,t) = \prod_{k=1}^{N}\Psi_{i_k}(\boldsymbol{x}_k,t)$$

is used for the many-particle wavefunctions without imposing symmetry or anti-symmetry for bosons or fermions of the same species. In this case the transformation between velocity gauge and length gauge wavefunctions,

$$\Psi_{i_k}^{(v)}(\boldsymbol{x}_k,t) \Leftrightarrow \Psi_{i_k}^{(l)}(\boldsymbol{x}_k,t) = \exp\!\left(-\frac{\mathrm{i}}{\hbar}q_{i_k}\boldsymbol{x}_k\cdot\boldsymbol{A}(t)\right)\Psi_{i_k}^{(v)}(\boldsymbol{x}_k,t), \tag{3.6}$$

can even be *separately imposed for each particle in the system*, i.e. within the limitations of Hartree approximations to many-particle systems, we could even treat e.g. electrons in outer orbitals of an atom in velocity gauge and electrons in inner orbitals in length gauge!

4. Analytic differences between velocity gauge and length gauge

The different form of interaction terms in the equations of motion (2.12,2.18) and the corresponding Hamiltonians (2.8,2.17) affects numerical procedures which can contribute to different results for the two descriptions of quantum optical interactions [**8**,**9**,**34**]. However, in spite of the formal analytic equivalence of the theories through the transformation (2.14), there are also important sources of analytical *inequivalence* between the two formalisms.

We will further pursue this question for an analytic origin of differences between velocity gauge and length gauge in a single electron approximation. Examinations of atoms and molecules in strong electromagnetic fields also use this approximation, and it will simplify the equations considerably without limiting the conclusions on the origin of differences between velocity gauge or length gauge. The corresponding Hamiltonians are

$$\begin{aligned} H_v &= \int d^3\boldsymbol{x}\left(\frac{1}{2m_e}\left[\hbar^2\boldsymbol{\nabla}\psi^+(\boldsymbol{x},t)\cdot\boldsymbol{\nabla}\psi(\boldsymbol{x},t) - \mathrm{i}e\hbar\boldsymbol{A}(t)\cdot\left(\psi^+(\boldsymbol{x},t)\overset{\leftrightarrow}{\boldsymbol{\nabla}}\psi(\boldsymbol{x},t)\right)\right.\right. \\ &\quad \left.\left. + e^2\psi^+(\boldsymbol{x},t)\boldsymbol{A}^2(t)\psi(\boldsymbol{x},t)\right] + \psi^+(\boldsymbol{x},t)V(\boldsymbol{x})\psi(\boldsymbol{x},t)\right), \end{aligned} \tag{4.1}$$

and

$$H_l = \int d^3\boldsymbol{x} \left(\frac{\hbar^2}{2m_e} \boldsymbol{\nabla}\Psi^+(\boldsymbol{x},t) \cdot \boldsymbol{\nabla}\Psi(\boldsymbol{x},t) + \Psi^+(\boldsymbol{x},t)V(\boldsymbol{x})\Psi(\boldsymbol{x},t) \right.$$
$$\left. + e\Psi^+(\boldsymbol{x},t)\boldsymbol{x} \cdot \boldsymbol{E}(t)\Psi(\boldsymbol{x},t) \right), \tag{4.2}$$

where $V(\boldsymbol{x})$ is the effective potential experienced by the electron. The transformation of the electron field operators to length gauge is

$$\Psi(\boldsymbol{x},t) = \exp[\mathrm{i}e\boldsymbol{x} \cdot \boldsymbol{A}(t)/\hbar]\psi(\boldsymbol{x},t), \tag{4.3}$$

or equivalently in terms of the $\boldsymbol{k}$-space operators

$$\Psi(\boldsymbol{k},t) = \psi(\boldsymbol{k} - e\boldsymbol{A}(t)/\hbar, t). \tag{4.4}$$

Strong electric field situations are usually analyzed in terms of numerical solutions of the corresponding Schrödinger wave equations. The equations (2.12,2.18) for the operators become the corresponding Schrödinger wave equations for wavefunctions $\tilde{\psi}(\boldsymbol{x},t)$ and $\tilde{\Psi}(\boldsymbol{x},t)$, respectively, after switching into the Schrödinger picture and expanding the states in the one-particle sector of Fock space. The velocity gauge Schrödinger equation in Fock space,

$$\mathrm{i}\hbar\frac{d}{dt}|\Phi(t)\rangle = H_v|\Phi(t)\rangle,$$

applied to single-particle states

$$|\Phi(t)\rangle = \int d^3\boldsymbol{x}\, \psi^+(\boldsymbol{x})|0\rangle\tilde{\psi}(\boldsymbol{x},t) \tag{4.5}$$

yields (after dropping the tilde from the wavefunction)

$$\mathrm{i}\hbar\frac{\partial}{\partial t}\psi(\boldsymbol{x},t) = -\frac{[\hbar\boldsymbol{\nabla} + \mathrm{i}e\boldsymbol{A}(t)]^2}{2m_e}\psi(\boldsymbol{x},t) + V(\boldsymbol{x})\psi(\boldsymbol{x},t), \tag{4.6}$$

and similarly in length gauge

$$\mathrm{i}\hbar\frac{\partial}{\partial t}\Psi(\boldsymbol{x},t) = -\frac{\hbar^2}{2m_e}\Delta\Psi(\boldsymbol{x},t) + V(\boldsymbol{x})\Psi(\boldsymbol{x},t) + e\boldsymbol{x}\cdot\boldsymbol{E}(t)\Psi(\boldsymbol{x},t). \tag{4.7}$$

Nevertheless, equations (4.3,4.4) now apply to the wavefunctions and map the solutions of (4.6) bijectively and unitarily into the solutions of (4.7). So how can we get different physical results from equations (4.6) and (4.7)?

There are several effects here that need to be taken into account. On the one hand, since the exponent in (4.3) is only first order in e, any perturbative results from (4.6) and (4.7) will differ beyond first order, or stated differently, what is formally second order perturbation theory in (4.7) is not actually second order perturbation theory in (4.6). For first order matrix elements, the familiar identity

$$\hbar\mathbf{p} = \mathrm{i}m_e[H, \mathbf{x}] \tag{4.8}$$

ensures

$$\langle f|\mathbf{p}|i\rangle = \mathrm{i}m_e\omega_{fi}\langle f|\mathbf{x}|i\rangle \tag{4.9}$$

between energy eigenstates. This implies equivalence of velocity and length forms of transition matrix elements in the limit of large transition times, which in terms of scattering matrix elements reads

$$S_{fi} = \lim_{t\to\infty} \langle f|U_D(t,-t)|i\rangle.$$

Here $U_D(t,t')$ is the time evolution operator in the interaction picture. Equation (4.9) then takes the form which ensures strict equivalence between first order matrix elements in velocity and length form[1],

$$\langle f|\mathbf{p}|i\rangle \to \pm i m_e ck\langle f|\mathbf{x}|i\rangle. \tag{4.10}$$

At the level of second order matrix elements, Jentschura and Pachucki showed that (4.8) can also be used to demonstrate equivalence between *velocity gauge* and *length gauge* results for polarizabilities of isotropic states [**39**]. This is remarkable, because we have to distinguish between *length form* and *velocity form* of matrix elements on the one hand, and *length gauge* and *velocity gauge* results on the other hand. The identity (4.9) can always be used to transform matrix elements in either gauge, without ever invoking the gauge transformation (2.14), and at the level of first order perturbation theory Eq. (4.10) can then be used to demonstrate equivalence of length gauge and velocity gauge results if the amplitudes come with an energy conserving factor $\delta(\omega_{fi} \mp ck)$. However, this does not generically work at second or higher oder. E.g. the differential scattering cross section (1.14) is the *velocity gauge* electron-photon scattering cross section in *length form*, and the comparison with (1.13) shows the differences with the *length gauge* electron-photon scattering cross section in *length form*.

The transition from (4.9) to (4.10) is justified by the fact that these matrix elements appear with factors $\delta(\omega_{fi} \mp ck)$ for large observation times $\Delta t = t - t'$. However, for monochromatic perturbation operators with frequency $\omega = \pm ck$ and small pulse duration Δt, these factors are replaced already in first order by a Dirichlet kernel $\sin[(\omega_{fi} \mp ck)\Delta t/2]/\pi(\omega_{fi} \mp ck)$. This is a matter of concern for the description of subfemtosecond spectroscopy in velocity or length gauge, since the resulting energy uncertainty for time resolution $\Delta t \lesssim 10^{-15}$ s is already of order $\hbar/\Delta t \gtrsim 0.7$ eV. This implies discrepancies between first order matrix elements in length gauge and velocity gauge at least at the several percent level or higher, if the subfemtosecond experiments are performed with UV laser pulses, and even higher uncertainties for subfemtosecond experiments at lower wavelenghts [**40, 41**].

Furthermore, we should expect even larger differences between velocity gauge and length gauge for electron detachment in strong fields. Before the detachment, energy eigenstates would be defined in terms of atomic or molecular wavefunctions. However, after the detachment, wavefunctions are usually described in terms of dominant electric field and subdominant potential terms, i.e. as Volkov solutions. This implies that equation (4.8) and therefore also equations (4.9,4.10) do not apply and cannot be used to infer equivalence of velocity gauge and length gauge matrix elements.

A third analytic caveat regarding the equivalence of velocity gauge and length gauge concerns the fact that the mapping (4.3), $|\Psi(t)\rangle = \exp[ie\mathbf{x}\cdot\boldsymbol{A}(t)/\hbar]|\psi(t)\rangle$, is *not a unitary mapping of the Hamiltonian* [**42**]. To elucidate this point, note that

[1] See [**37, 38**] for the inclusion of relativistic corrections in Eq. (4.8).

the mapping of the Hamiltonians in

$$\mathrm{i}\hbar\frac{d}{dt}|\psi(t)\rangle = H_v(t)|\psi(t)\rangle \;\Rightarrow\; \mathrm{i}\hbar\frac{d}{dt}|\Psi(t)\rangle = H_l(t)|\Psi(t)\rangle$$

transforms

$$H_v(t) = \frac{[\mathbf{p}+e\boldsymbol{A}(t)]^2}{2m_e} + V(\mathbf{x})$$

into

$$\begin{aligned} H_l(t) &= \exp[\mathrm{i}e\mathbf{x}\cdot\boldsymbol{A}(t)/\hbar]\left(H_v(t) - e\mathbf{x}\cdot\frac{d\boldsymbol{A}(t)}{dt}\right)\exp[-\mathrm{i}e\mathbf{x}\cdot\boldsymbol{A}(t)/\hbar] \\ &= \frac{\mathbf{p}^2}{2m_e} + V(\mathbf{x}) - e\mathbf{x}\cdot\frac{d\boldsymbol{A}(t)}{dt}, \end{aligned} \tag{4.11}$$

and therefore the matrix elements of the Hamiltonians in velocity gauge and length gauge differ,

$$\langle\Phi(t)|H_l(t)|\Psi(t)\rangle = \langle\phi(t)|H_v(t)|\psi(t)\rangle - e\langle\phi(t)|\mathbf{x}|\psi(t)\rangle\cdot\frac{d\boldsymbol{A}(t)}{dt}. \tag{4.12}$$

This implies that $H_l(t)$ and $H_v(t)$ evolve the system differently through the eigenstates of the unperturbed system during an electromagnetic pulse. We also note that (4.12) can yield different energy expectation values for systems which are not invariant under parity, or which have degeneracy between even and odd states. Note that the question for equivalence of energy expectation values for systems with e.g. a single mirror symmetry but no other parity invariance hinges on the polarization of the incident radiation.

These observations are of no concern if we only care about time evolution of a system from a state before a classical electromagnetic pulse was applied to a state after the pulse has ceased. However, if we wish to derive information about optical responses of the system from shifts of energy levels and eigenstates during application of a strong external field, then we have to anticipate principally different results from velocity and length gauge descriptions. These results can ultimately be mapped through application of (4.3). However, this will yield unique, generally agreed upon results only if we agree whether the velocity gauge states $|\psi(t)\rangle$ or the length gauge states $|\Psi(t)\rangle$ correspond to unperturbed system eigenstates in the evaluation of any matrix elements of $H_v(t)$ or $H_l(t)$ or their corresponding interaction terms. Stated differently, the question is which state we consider as a basic atomic or molecular state, and which state we consider as a kind of dressed state through application of a basic dressing operation $\exp[\pm\mathrm{i}e\boldsymbol{x}\cdot\boldsymbol{A}(t)/\hbar]$. After all, the transformation (4.3) impresses the Fourier components $\boldsymbol{A}(\omega)$ along with all higher harmonics onto the dressed wavefunction. Integrations of equations (4.6) and (4.7) starting with the same atomic or molecular state in a laser field of frequency ω_0 would therefore describe different physical situations. This confirms the observation of Galstyan *et al.* of two principally different families of solutions in quantum optical systems [**34**].

We can also generate the transformation (4.3) of the Schrödinger field operators from velocity gauge to length gauge as a unitary transformation in the form

$$\Psi(\boldsymbol{x},t) = \Sigma(t)\cdot\psi(\boldsymbol{x},t)\cdot\Sigma^{+}(t), \tag{4.13}$$

with

$$\Sigma(t) = \exp\Big(\frac{e}{\mathrm{i}\hbar}\int d^3\boldsymbol{x}\,\psi^+(\boldsymbol{x},t)\boldsymbol{x}\cdot\boldsymbol{A}(t)\psi(\boldsymbol{x},t)\Big). \tag{4.14}$$

To fully appreciate the implications of this observation, we have to go beyond the semiclassical approximation and restore the photon operators in the dipole approximation,

$$\boldsymbol{A}(t) = \sqrt{\frac{\hbar\mu_0 c}{(2\pi)^3}}\int\frac{d^3\boldsymbol{k}}{\sqrt{2k}}\sum_{\alpha=1}^{2}\boldsymbol{\epsilon}_\alpha(\boldsymbol{k})\Big(a_\alpha(\boldsymbol{k})\exp(-\mathrm{i}ckt) + a_\alpha^+(\boldsymbol{k})\exp(\mathrm{i}ckt)\Big). \tag{4.15}$$

Even after restoring $\boldsymbol{A}(t)$ as a photon operator, the unitary transformation (4.14) leaves the vacuum invariant, but charged particle states pick up a photon cloud either in length gauge or in velocity gauge. Suppose we start with the single charged particle state in velocity gauge

$$|\phi(t)\rangle = \int d^3\boldsymbol{x}\,\psi^+(\boldsymbol{x})|0\rangle\phi(\boldsymbol{x},t). \tag{4.16}$$

The corresponding length gauge state is

$$\begin{aligned} |\Phi(t)\rangle &= \int d^3\boldsymbol{x}\,\Psi^+(\boldsymbol{x})|0\rangle\phi(\boldsymbol{x},t) = \Sigma(0)|\phi(t)\rangle \\ &= \int d^3\boldsymbol{x}\,\exp\Big(\frac{e}{\mathrm{i}\hbar}\boldsymbol{x}\cdot\boldsymbol{A}\Big)\psi^+(\boldsymbol{x})|0\rangle\phi(\boldsymbol{x},t), \end{aligned} \tag{4.17}$$

where $\boldsymbol{A} \equiv \boldsymbol{A}(t=0)$ is the photon operator in the Schrödinger picture. The length gauge state would be dressed by a coherent photon state.

On the other hand, if we suppose that the length gauge state is a charged single particle state, we would infer a dressed velocity gauge state

$$|\phi(t)\rangle = \int d^3\boldsymbol{x}\,\exp(\mathrm{i}e\boldsymbol{x}\cdot\boldsymbol{A}/\hbar)\Psi^+(\boldsymbol{x})|0\rangle\phi(\boldsymbol{x},t). \tag{4.18}$$

Note that the photon components in the dressed length gauge states (4.17) or dressed velocity gauge states (4.18) correspond to coherent states (see e.g. Sec. 18.3 in Ref. [**4**] for coherent states in quantum optics) $|\boldsymbol{\zeta}\rangle = \exp(\pm\mathrm{i}e\boldsymbol{x}\cdot\boldsymbol{A}/\hbar)|0\rangle$, with amplitudes

$$\zeta_\alpha(\boldsymbol{k},\boldsymbol{x}) = \pm\mathrm{i}\frac{e}{4\pi}\sqrt{\frac{\mu_0 c}{\pi\hbar k}}\boldsymbol{x}\cdot\boldsymbol{\epsilon}_\alpha(\boldsymbol{k}).$$

5. Acceleration gauge and shifts in phase space

The apparent differences in velocity gauge and length gauge results have always been puzzling, since the transformation (2.14) can be implemented as a gauge transformation (2.15,2.16). On the other hand, every unitary time-dependendent transformation of wavefunctions or quantum states constitutes a change of the picture of quantum dynamics. $U(1)$ gauge transformations can be expressed in terms of a hermitian phase factor $\varphi(\mathbf{x},t)$, like the transformation (2.14) expressed on the states,

$$\begin{aligned} |\psi^{(l)}(t)\rangle &= \Upsilon_{lv}(t)|\psi^{(v)}(t)\rangle, \\ \Upsilon_{lv}(t) &= \exp[-\mathrm{i}q\mathbf{x}\cdot\boldsymbol{A}(t)/\hbar]\,. \end{aligned} \tag{5.1}$$

However, more general $U(1)$ transformations with hermitian phase factors $\varphi(\mathbf{p},\mathbf{x},t)$ can just as well be used to change the picture of quantum dynamics, and we must just as well be able to recover the correct physical predictions from any unitarily related picture of quantum dynamics. The only special property of the particular subgroup of picture changing operations which are gauge transformations is the fact that *the equations of the corresponding gauge theory remain form invariant as long as we do not commit to any particular gauge* (just like the equations of Geometry or General Relativity remain form invariant under diffeomorphisms as long as we do not commit to any particular coordinate system). Therefore it helps to illuminate the problem of apparent inequivalence of velocity gauge and length gauge from the more general perspective of picture changing transformations, which we will limit to the case of time-dependent $U(1)$ transformations $\Upsilon(t)=\exp[\mathrm{i}q\varphi(\mathbf{p},\mathbf{x},t)/\hbar]$.

One example is provided by the Galilei transformation with boost parameter $\boldsymbol{v}_G$,

$$\Upsilon_G(t)=\exp[-\mathrm{i}\mathbf{p}\cdot\boldsymbol{v}_G t/\hbar]\cdot\exp[\mathrm{i}m\mathbf{x}\cdot\boldsymbol{v}_G/\hbar]\,, \tag{5.2}$$

which transforms the velocity gauge Hamiltonian

$$H_v(t)=\frac{[\mathbf{p}-q\boldsymbol{A}(t)]^2}{2m}+V(\mathbf{x}) \tag{5.3}$$

into

$$H_G(t)=\frac{[\mathbf{p}-m\boldsymbol{v}_G-q\boldsymbol{A}(t)]^2}{2m}+V(\mathbf{x}-\boldsymbol{v}_G t)+\boldsymbol{v}_G\cdot\mathbf{p}.$$

Another well known example is provided by the acceleration frame transformation [**18**, **35**, **36**] (see also [**43**] for references to earlier occurences of this transformation)

$$\Upsilon_{av}(t)=\exp\!\left(-\mathrm{i}\frac{q}{\hbar}\frac{\mathbf{p}}{m}\cdot\int_{t_0}^{t}d\tau\,\boldsymbol{A}(\tau)\right)\exp\!\left(\mathrm{i}\frac{q^2}{2m\hbar}\int_{t_0}^{t}d\tau\,\boldsymbol{A}^2(\tau)\right), \tag{5.4}$$

which shifts the operator for particle location according to

$$\mathbf{x}\to\mathbf{X}(t)=\Upsilon_{av}(t)\cdot\mathbf{x}\cdot\Upsilon_{av}^{+}(t)=\mathbf{x}-\frac{q}{m}\int_{t_0}^{t}d\tau\,\boldsymbol{A}(\tau),$$

and transforms the velocity gauge Hamiltonian (5.3) into the "acceleration gauge" (or "Henneberger frame" [**36**]) Hamiltonian

$$H_a(t)=\frac{\mathbf{p}^2}{2m}+V(\mathbf{X}(t)).$$

The transformation which maps the length gauge Hamiltonian

$$H_l(t)=\frac{\mathbf{p}^2}{2m}+V(\mathbf{x})+q\mathbf{x}\cdot\frac{d\boldsymbol{A}(t)}{dt}$$

into $H_a(t)$ is then given by

$$\begin{aligned}\Upsilon_{al}(t)=\Upsilon_{av}(t)\cdot\Upsilon_{lv}^{+}(t)&=\exp\!\left[\mathrm{i}\frac{q^2}{2m\hbar}\int_{t_0}^{t}d\tau\,\left(\boldsymbol{A}^2(\tau)-\boldsymbol{A}(t)\cdot\boldsymbol{A}(\tau)\right)\right]\\&\quad\times\exp\!\left[\mathrm{i}\frac{q}{\hbar}\left(\mathbf{x}\cdot\boldsymbol{A}(t)-\frac{\mathbf{p}}{m}\cdot\int_{t_0}^{t}d\tau\,\boldsymbol{A}(\tau)\right)\right].\end{aligned} \tag{5.5}$$

All these cases are captured in the group of shift operators in phase space,

$$\Upsilon(t)=\exp[\mathrm{i}q\mathbf{p}\cdot\boldsymbol{c}(t)/\hbar]\cdot\exp[\mathrm{i}q\mathbf{x}\cdot\boldsymbol{a}(t)/\hbar]=\exp[\mathrm{i}q\varphi(\mathbf{p},\mathbf{x},t)/\hbar] \tag{5.6}$$

with

$$\varphi(\mathbf{p},\mathbf{x},t)=\mathbf{x}\cdot\boldsymbol{a}(t)+\mathbf{p}\cdot\boldsymbol{c}(t)+\frac{q}{2}\boldsymbol{a}(t)\cdot\boldsymbol{c}(t), \tag{5.7}$$

and these reduce to $U(1)$ gauge transformations if $\boldsymbol{c}(t) = 0$.

Besides the advantage of (5.6) of also including Galilei transformations (5.2) and the Henneberger transformation (5.4), it is also instructive to think about the question of equivalence of time-dependent gauge transformations in the larger framework of time-dependent unitary transformations. A time-dependent unitary transformation

$$|\psi'(t)\rangle = \Upsilon(t)|\psi(t)\rangle \tag{5.8}$$

shifts the Hamiltonian according to

$$\begin{aligned} H'(t) &= \Upsilon(t)\cdot\left[H(t) - \mathrm{i}\hbar\frac{d}{dt}\right]\cdot\Upsilon^{+}(t) \\ &= \Upsilon(t)\cdot H(t)\cdot\Upsilon^{+}(t) - \mathrm{i}\hbar\sum_{n=1}^{\infty}\frac{1}{n!}\left(\frac{\mathrm{i}q}{\hbar}\right)^{n}\overset{n}{[}\varphi(t), d/dt] \\ &= \Upsilon(t)\cdot H(t)\cdot\Upsilon^{+}(t) - q\sum_{n=1}^{\infty}\frac{1}{n!}\left(\frac{\mathrm{i}q}{\hbar}\right)^{n-1}\overset{n-1}{[}\varphi(t), \dot{\varphi}(t)]. \end{aligned} \tag{5.9}$$

This is the necessary and sufficient condition for the correct transformation law

$$U'(t,t') = \Upsilon(t)U(t,t')\Upsilon^{+}(t') \tag{5.10}$$

of the time evolution operators in the two pictures.

Eq. (5.9) implies for the phase space shift operators (5.6,5.7) the following transformation of Hamiltonians,

$$\begin{aligned} H'(t) &= \Upsilon(t)\cdot H(t)\cdot\Upsilon^{+}(t) - q\dot{\varphi}(t) - \frac{\mathrm{i}q^2}{2\hbar}[\varphi(t), \dot{\varphi}(t)] \\ &= \Upsilon(t)\cdot H(t)\cdot\Upsilon^{+}(t) - q\mathbf{x}\cdot\dot{\boldsymbol{a}}(t) - q\mathbf{p}\cdot\dot{\boldsymbol{c}}(t) - q^2\dot{\boldsymbol{a}}(t)\cdot\boldsymbol{c}(t). \end{aligned} \tag{5.11}$$

Therefore, while Eq. (5.10) of course implies

$$\langle\phi'(t)|U'(t,t')|\psi'(t)\rangle = \langle\phi(t)|U(t,t')|\psi(t)\rangle, \tag{5.12}$$

the matrix elements of the Hamiltonians in the two representations are related by

$$\begin{aligned} \langle\phi'(t)|H'(t)|\psi'(t)\rangle &= \langle\phi(t)|H(t)|\psi(t)\rangle - q\dot{\boldsymbol{a}}(t)\cdot\langle\phi(t)|\mathbf{x}|\psi(t)\rangle \\ &\quad - q\dot{\boldsymbol{c}}(t)\cdot\langle\phi(t)|\mathbf{p}|\psi(t)\rangle - q^2\boldsymbol{a}(t)\cdot\dot{\boldsymbol{c}}(t)\langle\phi(t)|\psi(t)\rangle, \end{aligned} \tag{5.13}$$

which generalizes (4.12). Equivalence e.g. of velocity and length gauge has been argued on the basis of the identity (5.12), see e.g. [**18**]. However, actual calculations of transition probabilities, decay rates, or scattering cross sections are based on the matrix elements of the interaction picture (or Dirac picture) time evolution operator between energy eigenstates,

$$S_{nm}(t,t') = \langle\psi_n|U_D(t,t')|\psi_m\rangle, \tag{5.14}$$

with the interaction picture time evolution operator

$$\begin{aligned} U_D(t,t') &= \exp(\mathrm{i}H_0 t/\hbar)U(t,t')\exp(-\mathrm{i}H_0 t'/\hbar) \\ &= \mathrm{T}\exp\left(-\frac{\mathrm{i}}{\hbar}\int_{t'}^{t} d\tau\, H_D(\tau)\right), \end{aligned} \tag{5.15}$$

the Hamiltonian

$$H_D(t) = \exp(\mathrm{i}H_0 t/\hbar)W(t)\exp(-\mathrm{i}H_0 t/\hbar),$$

and the identification of $W(t)$ e.g. from

$$H_0(\mathbf{p}, \mathbf{x}) = H(\mathbf{p}, \mathbf{x}, t) - W(\mathbf{p}, \mathbf{x}, t) = \frac{\mathbf{p}^2}{2m} + V(\mathbf{x}) \tag{5.16}$$

in a single (quasi-)particle approximation.

The initial and final times in (5.14) are usually taken as $t \to \infty$, $t' \to -\infty$, but the scattering matrix elements both for finite and infinite times are principally just a matrix representation of the integration of the Schrödinger equation in the interaction picture. However, applications of the different pictures (5.8,5.9) to the same system are based on the same identification (5.16) of the unperturbed Hamiltonian, and therefore use the same eigenstates in (5.14), although $U'_D(t, t') \neq U_D(t, t')$. This is in short the origin of the principal difference of the scattering cross sections (1.14) and (1.13). More specifically, scattering matrix elements in the two pictures (5.8,5.9) are related by

$$\begin{aligned} S'_{nm}(t, t') &= \langle \psi_n | U'_D(t, t') | \psi_m \rangle \\ &= \sum_{kl} \exp[\mathrm{i}(\omega_{nk} t + \omega_{lm} t')] \langle \psi_n | \Upsilon(t) | \psi_k \rangle S_{kl}(t, t') \langle \psi_l | \Upsilon^+(t') | \psi_m \rangle. \end{aligned} \tag{5.17}$$

This is a more precise formulation of the observation from Eq. (4.12) that the different formulations evolve a system differently through the eigenstates of the unperturbed system. Eq. (5.17) implies that transition probabilites, decay rates, and scattering cross sections will generically be different in the different pictures (5.8,5.9), as explicitly demonstrated for the case of velocity gauge versus length gauge in equations (1.14,1.13). On the other hand, if the picture changing operation $\Upsilon(t)$ involves semiclassical electromagnetic potentials which correspond to pulses of finite duration $\Delta t < t - t'$, then the scattering matrix elements $S_{nm}(t, t')$, with initial time t' before the time t_0 of onset of the pulse and final time t after cessation of the pulse, are *invariant* with respect to (5.8), and therefore *any observable computed from those scattering matrix elements would also be invariant under the picture changing operation* (5.8),

$$S'_{nm}(t, t') = S_{nm}(t, t'), \quad t' < t_0 < t_0 + \Delta t < t. \tag{5.18}$$

This does not apply to the scattering matrices leading to the scattering cross sections (1.14,1.13), since their calculation involves photon operators which describe spontaneous photon emission or absorption at any time, as opposed to time-constrained semiclassical pulses.

However, if Eq. (5.18) holds for semiclassical pulses, why then do many researchers find strikingly different results in studies of strong laser pulses, which are used e.g. for electron detachment and higher harmonic generation? From the previous observations, we can easily identify four sources for different results in different pictures:

1. The scattering matrix formalism will generically yield different results in the different pictures (5.8,5.9) due to the transformation law Eq. (5.17), unless it is applied to semiclassical pulses of finite duration (5.18).

2. Due to the nonperturbative nature of the strong fields involved, theoretical investigations of strong field experiments do not use the scattering matrix formalism, but employ numerical integrations of the time-dependent Schrödinger equation. Given the inhomogeneous transformation law (5.9) of the Hamiltonian during picture changing, it is not surprising that numerical integration of the Schrödinger equations for $H(t)$ and $H'(t)$ in the different pictures can lead to very different conclusions.

3. The modeling of electron detachment in strong fields involves a change of H_0 during the experiment, since the bound electron state is dominated by the attractive atomic or molecular potentials, whereas the ionized state is presumed to be dominated by the strong external radiation field. This renders the relations (4.9,4.10) useless for trying to establish equivalence even only at first order.

4. As pointed out already before, the replacement of the energy preserving δ function through the Dirichlet kernel

$$\delta(\omega_{nm} \mp ck) \to \frac{\sin[(\omega_{nm} \mp ck)\Delta t/2]}{\pi(\omega_{nm} \mp ck)}$$

for subfemtosecond pulses of frequency $\pm ck$ invalidates the relation (4.10) for equivalence between velocity and length gauge matrix elements,

$$\langle\psi_f|\mathbf{p}|\psi_i\rangle = im\omega_{fi}\langle\psi_f|\mathbf{x}|\psi_i\rangle \neq \pm imck\langle\psi_f|\mathbf{x}|\psi_i\rangle.$$

Note that this effect alone cannot change the *numerical equivalence* (5.18) of scattering matrix elements for times t' and t before and after semiclassical pulses, but we cannot expect to be able to transform the scattering matrix elements any more into the same analytic form.

6. Conclusions

Four sources of analytic differences between velocity gauge and length gauge have been identified. The difference between $|\omega_{fi}|\langle f|\mathbf{x}|i\rangle$ and $ck\langle f|\mathbf{x}|i\rangle$ for short observation times or short pulses can be interpreted as a manifestation of energy-time uncertainty, but it implies a discrepancy between theoretical calculations in length gauge versus velocity gauge at the several percent level or higher for subfemtosecond spectroscopy.

Furthermore, equation (4.8) does not hold if the initial and final states are eigenstates of different Hamiltonians, as often appears in strong field electron detachment. Therefore no corresponding equivalence of velocity gauge matrix elements and length gauge matrix elements can be inferred from (4.8) in these cases.

In addition the scattering matrix formalism in the different pictures (5.8,5.9) will generically yield different results due to the transformation law Eq. (5.17), unless it is applied to semiclassical pulses of finite duration (5.18).

Finally, the inequivalence (4.12) between matrix elements of the Hamilton operators implies potentially large analytic uncertainties for systems without parity invariance or with accidental degeneracies between even and odd states. This is in agreement with the observations of a high level of agreement between velocity and length gauge calculations for helium atoms [**38**] and the observation of equivalence of second order matrix elements for polarizabilities [**39**] and two-photon transitions [**42**] in rotationally symmetric systems, while at the same time numerically evaluated matrix elements in velocity gauge and length gauge can be very different for chiral compounds [**11, 12**], and calculated ionization properties can be very different for hydrogen [**7, 13, 15, 26**].

From an analytic perspective, we would expect to find strong differences between velocity gauge and length gauge both for systems without parity invariance, and for strong field electron detachment. This seems to be corroborated by numerical calculations in both cases, and also by analytic work on strong field electron detachment.

So far everything has been formulated in a neutral way without stating a preference for velocity gauge or length gauge or any of the other pictures implied in (5.6)-(5.9). The transformation between velocity gauge and length gauge has always been formulated as a transition from velocity gauge to length gauge, see e.g. (4.3,4.4), but everything can just as well be formulated in the opposite direction. However, since the different pictures can produce different results even at analytic level, for the reasons outlined above, is there a picture that should be preferred for principal reasons? This question has been asked time and again especially for velocity gauge and length gauge. The results (4.13)-(4.18) indicate that if the wavefunction in one of the pictures is considered as fundamental, the wavefunction in the other picture corresponds to a state which is dressed with a coherent photon component. However, this does not break the tie with respect to the question which picture should be considered as "fundamental". In light of the fact that both velocity gauge and length gauge have proven to be particularly successful in their own right in different physical situations, the practical aspect of this question is rather: which of these different situations are better described by dressed states, and why? An argument which has often been made in favor of length gauge (besides the simpler form of the interaction term) is the apparent formulation in terms of gauge invariant quantities $\boldsymbol{p} = m\boldsymbol{v}$ and $\boldsymbol{E}(t)$, see e.g. [**17**, **18**]. Indeed, the momentum in length gauge wavefunctions $\Psi(\boldsymbol{x}, t)$ and velocity gauge wavefunctions $\psi(\boldsymbol{x}, t)$ is

$$\boldsymbol{P} = \int d^3\boldsymbol{x}\, \frac{\hbar}{\mathrm{i}} \sum_i \Psi_i^+(\boldsymbol{x}, t) \boldsymbol{\nabla} \Psi_i(\boldsymbol{x}, t) \tag{6.1}$$

$$= \int d^3\boldsymbol{x}\, \frac{\hbar}{\mathrm{i}} \sum_i \psi_i^+(\boldsymbol{x}, t) \boldsymbol{\nabla} \psi_i(\boldsymbol{x}, t) - \int d^3\boldsymbol{x} \sum_i q_i \psi_i^+(\boldsymbol{x}, t) \boldsymbol{A}(t) \psi_i(\boldsymbol{x}, t), \tag{6.2}$$

see also the Appendix, where these relations are derived on the level of Schrödinger field operators and taking into account the contributions from the electromagnetic fields to energy and momentum conservation. These relations and the equations (2.8,2.17) for Hamiltonians in velocity gauge and length gauge show that in length gauge we can identify momentum with particle velocity $\boldsymbol{p} = m\boldsymbol{v}$, whereas velocity gauge is of course based on the canonical relation $\boldsymbol{p} = m\boldsymbol{v} + q\boldsymbol{A}$ for particle momentum in external fields. As noted above, this apparent formulation in terms of gauge invariant quantities has been alluded to as one advantage of the length gauge formulation. Furthermore, the direct identification of momentum and velocity also makes length gauge advantageous for investigations of electron detachment [**5**, **6**, **19**–**24**] and applications of the Keldysh formalism [**9**, **13**], since intuitive methods using electron motion are better captured in that formalism. This advantage disappears for large internuclear separation in molecules, when the dipole operator during electron transfer can become too large an inherent perturbation of the system [**25**, **26**].

On the other hand, reductionism breaks the tie in favor of the velosity gauge. Recall that the Dirac Hamiltonian for quantum electrodynamics in Coulomb gauge has the form (see e.g. Sec. 21.4 in Ref. [**4**])

$$H = \int d^3\boldsymbol{x} \left(\frac{\epsilon_0}{2} \boldsymbol{E}_\perp^2(\boldsymbol{x},t) + \frac{1}{2\mu_0} \boldsymbol{B}^2(\boldsymbol{x},t) \right) + \sum_i H_i + \sum_{ij} V_{ij}, \tag{6.3}$$

with $\boldsymbol{E}_\perp(\boldsymbol{x},t) = -\partial\boldsymbol{A}(\boldsymbol{x},t)/\partial t$, $\boldsymbol{B}(\boldsymbol{x},t) = \boldsymbol{\nabla}\times\boldsymbol{A}(\boldsymbol{x},t)$, and the kinetic and Coulomb terms for fermion species with labels i,j (summations over the Dirac indices s are implicitly included in H_i)

$$\begin{aligned} H_i &= -c\int d^3\boldsymbol{x}\,\overline{\psi}_i(\boldsymbol{x},t)\gamma\cdot[\mathrm{i}\hbar\boldsymbol{\nabla} + q_i\boldsymbol{A}(\boldsymbol{x},t)]\,\psi_i(\boldsymbol{x},t) \\ &\quad + m_i c^2\int d^3\boldsymbol{x}\,\overline{\psi}_i(\boldsymbol{x},t)\psi_i(\boldsymbol{x},t), \end{aligned} \tag{6.4}$$

$$V_{ij} = \sum_{s,s'}\int d^3\boldsymbol{x}\int d^3\boldsymbol{x}'\,\frac{q_i q_j}{8\pi\epsilon_0|\boldsymbol{x}-\boldsymbol{x}'|}\psi_{is}^+(\boldsymbol{x},t)\psi_{js'}^+(\boldsymbol{x}',t)\psi_{js'}(\boldsymbol{x}',t)\psi_{is}(\boldsymbol{x},t). \tag{6.5}$$

This Hamiltonian reduces to the velocity gauge Hamiltonian (2.8) in the nonrelativistic semiclassical dipole approximation, and therefore from a *reductionist* point of view the velocity gauge Hamiltonian must be considered as more fundamental than the length gauge Hamiltonian or any other picture of quantum dynamics in the semiclassical dipole approximation. This is also directly manifested in the $\psi^+\boldsymbol{A}^2\psi$ term in (1.12) and the corresponding first term in the scattering cross section (1.14), which arise in the nonrelativistic limit from intermediate virtual positron states. Another case in point concerns the question which gauge naturally relates to standard atomic energy eigenstates. The $\boldsymbol{k}$-space representation of hydrogen eigenstates $|n,\ell,m_\ell\rangle$ is concentrated around $\boldsymbol{k}=0$ within a characteristic radius $k \lesssim (na)^{-1}$, where a is the Bohr radius, e.g. for the 1s state,

$$\langle \boldsymbol{k}|\Phi_{1,0,0}(t)\rangle = \frac{\sqrt{2a}^3}{\pi}\frac{\exp(-\mathrm{i}E_1 t/\hbar)}{[1+(ka)^2]^2}. \tag{6.6}$$

The transformation (4.3,4.4) therefore implies that the $\boldsymbol{k}$-space representation of the 1s state in velocity gauge, $\psi_{1,0,0}(\boldsymbol{k},t)$, would be concentrated around the wave number $-e\boldsymbol{A}(t)/\hbar$, if the length gauge wavefunction $\Psi_{1,0,0}(\boldsymbol{k},t)$ would agree with (6.6). On the other hand, the length gauge wavefunction $\Psi_{1,0,0}(\boldsymbol{k},t)$ would be concentrated in a radius $k \lesssim a^{-1}$ around the wave number $e\boldsymbol{A}(t)/\hbar$, if the velocity gauge wavefunction $\psi_{1,0,0}(\boldsymbol{k},t)$ agrees with (6.6). The latter result makes more sense due to the length gauge Schrödinger equation (4.7). Assume an extremely strong field such that (4.7) approximately reduces to

$$\frac{\partial}{\partial t}\Psi(\boldsymbol{x},t) \simeq \frac{\mathrm{i}e}{\hbar}\boldsymbol{x}\cdot\frac{d\boldsymbol{A}(t)}{dt}\Psi(\boldsymbol{x},t). \tag{6.7}$$

This has the solution

$$\Psi(\boldsymbol{x},t) \simeq \exp\!\Big(\mathrm{i}\frac{e}{\hbar}\boldsymbol{x}\cdot[\boldsymbol{A}(t)-\boldsymbol{A}(t_0)]\Big)\,\Psi(\boldsymbol{x},t_0), \tag{6.8}$$

which corresponds to propagation of the initial state with a plane wave factor with wave number $e\boldsymbol{A}(t)/\hbar$. *This leads to the conclusion that standard atomic eigenstates correspond to the velocity gauge picture.* This does not invalidate length gauge, but rephrases the question in the form: Why are those system which are better described in length gauge described by dressed states of the form (4.17)?

7. Appendix: Energy and momentum in Coulomb gauge

The Lagrange density for coupled electromagnetic and non-relativistic matter fields without any particular choice of gauge is given by

$$\begin{aligned}\mathcal{L} &= \sum_i \left[\frac{\mathrm{i}\hbar}{2}\left(\psi_i^+ \cdot \frac{\partial \psi_i}{\partial t} - \frac{\partial \psi_i^+}{\partial t} \cdot \psi_i\right) - q_i \psi_i^+ \Phi \psi_i - \frac{\hbar^2}{2m_i} \nabla \psi_i^+ \cdot \nabla \psi_i \right. \\ &\quad \left. - \mathrm{i}\frac{q_i \hbar}{2m_i} \boldsymbol{A} \cdot \left(\psi_i^+ \overset{\leftrightarrow}{\nabla} \psi_i\right) - \frac{q_i^2}{2m_i} \psi_i^+ \boldsymbol{A}^2 \psi_i \right] - \frac{1}{4\mu_0} F_{\mu\nu} F^{\mu\nu}. \end{aligned} \tag{7.1}$$

Here $\Phi = -cA_0$ is the electric potential, and summation over spin labels is implicitly understood.

The canonical energy-momentum tensor following from the Lagrange density (7.1),

$$\Theta_\mu{}^\nu = \eta_\mu{}^\nu \mathcal{L} + \frac{1}{\mu_0} \partial_\mu A_\lambda \cdot F^{\nu\lambda} - \sum_i \left(\partial_\mu \psi_i \frac{\partial \mathcal{L}}{\partial(\partial_\nu \psi_i)} + \partial_\mu \psi_i^+ \frac{\partial \mathcal{L}}{\partial(\partial_\nu \psi_i^+)}\right), \tag{7.2}$$

is rendered gauge invariant in the usual way by adding the trivially conserved tensor

$$\delta\Theta_\mu{}^\nu = -\frac{1}{\mu_0} \partial_\lambda (A_\mu F^{\nu\lambda}) = -A_\mu j^\nu - \frac{1}{\mu_0} \partial_\lambda A_\mu \cdot F^{\nu\lambda}, \quad \partial_\nu \delta\Theta_\mu{}^\nu \equiv 0.$$

The improved energy-momentum tensor $t_\mu{}^\nu = \Theta_\mu{}^\nu + \delta\Theta_\mu{}^\nu$ yields in particular the energy density $\mathcal{H} = -t_0{}^0$ for quantum optics,

$$\begin{aligned}\mathcal{H} &= \frac{\epsilon_0}{2} \boldsymbol{E}^2 + \frac{1}{2\mu_0} \boldsymbol{B}^2 + \sum_i \frac{1}{2m_i} \left[\hbar^2 \nabla \psi_i^+ \cdot \nabla \psi_i \right. \\ &\quad \left. + \mathrm{i} q_i \hbar \boldsymbol{A} \cdot \left(\psi_i^+ \overset{\leftrightarrow}{\nabla} \psi_i\right) + q_i^2 \psi_i^+ \boldsymbol{A}^2 \psi_i \right], \end{aligned} \tag{7.3}$$

and the gauge invariant momentum density $\mathcal{P}_a = t_a{}^0/c$,

$$\boldsymbol{\mathcal{P}} = \epsilon_0 \boldsymbol{E} \times \boldsymbol{B} + \frac{1}{2\mathrm{i}} \sum_i \left(\hbar \psi_i^+ \overset{\leftrightarrow}{\nabla} \psi_i - 2\mathrm{i} q_i \psi_i^+ \boldsymbol{A} \psi_i\right). \tag{7.4}$$

We split the electric field components in Coulomb gauge $\nabla \cdot \boldsymbol{A} = 0$ according to $\boldsymbol{E}_\| = -\nabla\Phi$ and $\boldsymbol{E}_\perp = -\partial \boldsymbol{A}/\partial t$. The equation for the electrostatic potential decouples from the vector potential,

$$\Delta\Phi = -\frac{1}{\epsilon_0} \sum_i q_i \psi_i^+ \psi_i,$$

and is solved by

$$\Phi(\boldsymbol{x}, t) = \frac{1}{4\pi\epsilon_0} \int d^3\boldsymbol{x}' \sum_i \frac{q_i}{|\boldsymbol{x} - \boldsymbol{x}'|} \psi_i^+(\boldsymbol{x}', t) \psi_i(\boldsymbol{x}', t).$$

Furthermore, the two components of the electric field are orthogonal in the Coulomb gauge,

$$\begin{aligned}\int d^3\boldsymbol{x}\, \boldsymbol{E}_\|(\boldsymbol{x}, t) \cdot \boldsymbol{E}_\perp(\boldsymbol{x}, t) &= \int d^3\boldsymbol{k}\, \boldsymbol{E}_\|(\boldsymbol{k}, t) \cdot \boldsymbol{E}_\perp(-\boldsymbol{k}, t) \\ &= -\int d^3\boldsymbol{x}\, \Phi(\boldsymbol{x}, t) \frac{\partial}{\partial t} \nabla \cdot \boldsymbol{A}(\boldsymbol{x}, t) = 0, \end{aligned} \tag{7.5}$$

and the contribution from $\boldsymbol{E}_{\|}$ to the Hamiltonian generates the Coulomb potentials in the Hamiltonian,

$$\begin{aligned} H_C &= \frac{\epsilon_0}{2}\int d^3\boldsymbol{x}\,\boldsymbol{E}_{\|}^2(\boldsymbol{x},t) = -\frac{\epsilon_0}{2}\int d^3\boldsymbol{x}\,\Phi(\boldsymbol{x},t)\Delta\Phi(\boldsymbol{x},t) = \frac{1}{2}\int d^3\boldsymbol{x}\,\Phi(\boldsymbol{x},t)\varrho(\boldsymbol{x},t) \\ &= \sum_{ij}\int d^3\boldsymbol{x}\int d^3\boldsymbol{x}'\,\psi_i^+(\boldsymbol{x},t)\psi_j^+(\boldsymbol{x}',t)\frac{q_iq_j}{8\pi\epsilon_0|\boldsymbol{x}-\boldsymbol{x}'|}\psi_j(\boldsymbol{x}',t)\psi_i(\boldsymbol{x},t). \end{aligned} \tag{7.6}$$

The resulting Hamiltonian in Coulomb gauge therefore has the form

$$\begin{aligned} H &= \int d^3\boldsymbol{x}\left(\sum_i \frac{1}{2m_i}\Big[\hbar^2\boldsymbol{\nabla}\psi_i^+(\boldsymbol{x},t)\cdot\boldsymbol{\nabla}\psi_i(\boldsymbol{x},t)\right. \\ &\quad + \mathrm{i}q_i\hbar\boldsymbol{A}(\boldsymbol{x},t)\cdot\Big(\psi_i^+(\boldsymbol{x},t)\overleftrightarrow{\boldsymbol{\nabla}}\psi_i(\boldsymbol{x},t)\Big) \\ &\quad \left. + q_i^2\psi_i^+(\boldsymbol{x},t)\boldsymbol{A}^2(\boldsymbol{x},t)\psi_i(\boldsymbol{x},t)\Big] + \frac{\epsilon_0}{2}\boldsymbol{E}_\perp^2(\boldsymbol{x},t) + \frac{1}{2\mu_0}\boldsymbol{B}^2(\boldsymbol{x},t)\right) \\ &\quad + \sum_{ij}\int d^3\boldsymbol{x}\int d^3\boldsymbol{x}'\,\psi_i^+(\boldsymbol{x},t)\psi_j^+(\boldsymbol{x}',t)\frac{q_iq_j}{8\pi\epsilon_0|\boldsymbol{x}-\boldsymbol{x}'|}\psi_j(\boldsymbol{x}',t)\psi_i(\boldsymbol{x},t). \end{aligned} \tag{7.7}$$

The momentum operator in Coulomb gauge follows from Eq. (7.4) and

$$\int d^3\boldsymbol{x}\,\epsilon_0\boldsymbol{E}_{\|}\times\boldsymbol{B} = -\int d^3\boldsymbol{x}\,\epsilon_0\Phi\Delta\boldsymbol{A} = \int d^3\boldsymbol{x}\,\varrho\boldsymbol{A} = \int d^3\boldsymbol{x}\sum_i q_i\psi_i^+\boldsymbol{A}\psi_i$$

as

$$\begin{aligned} \boldsymbol{P} &= \int d^3\boldsymbol{x}\left(\frac{\hbar}{\mathrm{i}}\sum_i\psi_i^+\boldsymbol{\nabla}\psi_i - \sum_i q_i\psi_i^+\boldsymbol{A}\psi_i + \epsilon_0\boldsymbol{E}\times\boldsymbol{B}\right) \\ &= \int d^3\boldsymbol{x}\left(\frac{\hbar}{\mathrm{i}}\sum_i\psi_i^+\boldsymbol{\nabla}\psi_i + \epsilon_0\boldsymbol{E}_\perp\times\boldsymbol{B}\right). \end{aligned} \tag{7.8}$$

All these results hold at the operator level within the full second quantized theory [**4**], with the only approximation of nonrelativistic charged fields. If we now specify to the semiclassical theory in dipole approximation, the transformation to Schrödinger field operators in length gauge

$$\Psi_i(\boldsymbol{x},t) = \exp\left(-\frac{\mathrm{i}}{\hbar}q_i\boldsymbol{x}\cdot\boldsymbol{A}(t)\right)\psi_i(\boldsymbol{x},t)$$

yields expressions (after leaving out the formally divergent kinetic contributions from the semiclassical fields $\boldsymbol{E}(t) = -d\boldsymbol{A}(t)/dt$ and $\boldsymbol{B}(t) = [\boldsymbol{\nabla}\times\boldsymbol{A}(\boldsymbol{x},t)]_{\boldsymbol{k}\cdot\boldsymbol{x}\to 0}$ and neglecting the contributions from $\boldsymbol{A}_{\boldsymbol{J}}(\boldsymbol{x},t)$ (2.5) for the reasons pointed out after Eq. (2.7))

$$\begin{aligned} H &= \int d^3\boldsymbol{x}\left(\sum_i\frac{\hbar^2}{2m_i}\boldsymbol{\nabla}\Psi_i^+(\boldsymbol{x},t)\cdot\boldsymbol{\nabla}\Psi_i(\boldsymbol{x},t) - q_i\Psi_i^+(\boldsymbol{x},t)\boldsymbol{x}\cdot\boldsymbol{E}(t)\Psi_i(\boldsymbol{x},t)\right) \\ &\quad + \sum_{ij}\int d^3\boldsymbol{x}\int d^3\boldsymbol{x}'\,\Psi_i^+(\boldsymbol{x},t)\Psi_j^+(\boldsymbol{x}',t)\frac{q_iq_j}{8\pi\epsilon_0|\boldsymbol{x}-\boldsymbol{x}'|}\Psi_j(\boldsymbol{x}',t)\Psi_i(\boldsymbol{x},t) \end{aligned} \tag{7.9}$$

and

$$\begin{aligned}
\boldsymbol{P} &= \int d^3\boldsymbol{x}\Big(- \mathrm{i}\hbar \sum_i \Psi_i^+(\boldsymbol{x},t)\boldsymbol{\nabla}\Psi_i(\boldsymbol{x},t) + \epsilon_0 \boldsymbol{E}_{\|}(\boldsymbol{x},t) \times \boldsymbol{B}(t)\Big) \\
&= \int d^3\boldsymbol{x}\, \frac{\hbar}{\mathrm{i}} \sum_i \Psi_i^+(\boldsymbol{x},t)\boldsymbol{\nabla}\Psi_i(\boldsymbol{x},t) \qquad (7.10) \\
&= \int d^3\boldsymbol{x}\, \frac{\hbar}{\mathrm{i}} \sum_i \psi_i^+(\boldsymbol{x},t)\boldsymbol{\nabla}\psi_i(\boldsymbol{x},t) \int d^3\boldsymbol{x} \sum_i q_i \psi_i^+(\boldsymbol{x},t)\boldsymbol{A}(t)\psi_i(\boldsymbol{x},t), \qquad (7.11)
\end{aligned}$$

since $\int d^3\boldsymbol{x}\boldsymbol{E}_{\|}(\boldsymbol{x},t) \times \boldsymbol{B}(t)$ is a surface term.

Both the momentum (7.10) in terms of the length gauge Schrödinger operators and the kinetic terms in (7.9) support the identification $\boldsymbol{p} = m\boldsymbol{v}$, whereas the kinetic terms in (7.7) and the expression (7.11) agree with the general classical relation $\boldsymbol{p} = m\boldsymbol{v} + q\boldsymbol{A}$ for the canonical particle momentum in external fields.

References

[1] W. E. Lamb, Phys. Rev. **85**, 259 (1952).
[2] L. D. Landau and E. M. Lifshitz, *Mechanics*, Course of Theoretical Physics, Vol. 1. Translated from the Russian by J. B. Bell, Pergamon Press, Oxford-London-New York-Paris; Addison-Wesley Publishing Co., Inc., Reading, Mass., 1960. MR0120782
[3] J. J. Sakurai, *Advanced Quantum Mechanics* (Addison-Wesley, Reading, 1967).
[4] Rainer Dick, *Advanced quantum mechanics*, 2nd ed., Graduate Texts in Physics, Springer, [Cham], 2016. Materials and photons. MR3497551
[5] D. Bauer, D. B. Milošević, and W. Becker, Phys. Rev. A **72**, 023415 (2005).
[6] B. Bergues, Z. Ansari, D. Hanstorp, and I. Yu. Kiyan, Phys. Rev. A **75**, 063415 (2007).
[7] J. Zhang and T. Nakajima, Phys. Rev. A **77**, 043417 (2008).
[8] Y.-C. Han and L. B. Madsen, Phys. Rev. A **81**, 063430 (2010).
[9] S. V. Popruzhenko, J. Phys. B **47**, 204001 (2014).
[10] H. M. Dong, K. Han, and W. Xu, J. Appl. Phys. **115**, 063503 (2014).
[11] M. Kamiński, J. Cukras, M. Pecul, A. Rizzo, and S. Coriani, Phys. Chem. Chem. Phys. **17**, 19079 (2015).
[12] D. H. Friese and K. Ruud, Phys. Chem. Chem. Phys. **18**, 4174 (2016).
[13] J. H. Bauer, J. Phys. B **49**, 145601 (2016).
[14] H. R. Reiss, Phys. Rev. A **22**, 1786 (1980).
[15] E. Cormier and P. Lambropoulos, J. Phys. B **29**, 1667 (1996).
[16] R. Shakeshaft, Z. Phys. D **8**, 47 (1988).
[17] R. R. Schlicher, W. Becker, J. Bergou, and M. O. Scully, in *Quantum Electrodynamics and Quantum Optics*, edited by A. O. Barut (Plenum Press, New York, 1984), p. 405.
[18] C. Cohen-Tannoudji, J. Dupont-Roc, and G. Grynberg, *Photons and Atoms: Introduction to Quantum Electrodynamics* (Wiley, New York, 1989).
[19] G. F. Gribakin and M. Yu. Kuchiev, Phys. Rev. A **55**, 3760 (1997).
[20] M. V. Frolov, N. L. Manakov, E. A. Pronin, and A. F. Starace, Phys. Rev. Lett. **91**, 053003 (2003).
[21] S. Beiser, M. Klaiber, and I. Yu. Kiyan, Phys. Rev. A **70**, 011402 (2004).
[22] T. K. Kjeldsen and L. B. Madsen, J. Phys. B **37**, 2033 (2004); Phys. Rev. A **71**, 023411 (2005).
[23] D. B. Milošević, Phys. Rev. A **74**, 063404 (2006).
[24] W. Becker, J. Chen, S. G. Chen, and D. B. Milošević, Phys. Rev. A **76**, 033403 (2007).
[25] C. C. Chirilă and M. Lein, Phys. Rev. A **73**, 023410 (2006).
[26] V. P. Majety, A. Zielinski, and A. Scrinzi, J. Phys. B **48**, 025601 (2015).
[27] K. L. Ishikawa, Phys. Rev. B **82**, 201402 (2010).
[28] I. Al-Naib, J. E. Sipe, and M. M. Dignam, Phys. Rev. B **90**, 245423 (2014); New J. Phys. **17**, 113018 (2015).
[29] S. Glutsch, Phys. Rev. B **69**, 235317 (2004).
[30] D. Golde, T. Meier, and S. W. Koch, Phys. Rev. B **77**, 075330 (2008).

[31] T. G. Pedersen, Phys. Rev. B **92**, 235432 (2015).
[32] M. Korbmann, S. Yu. Kruchinin, and V. S. Yakovlev, New J. Phys. **15**, 013006 (2013).
[33] M. Wu, S. Ghimire, D. A. Reis, K. J. Schafer, and M. B. Gaarde, Phys. Rev. A **91**, 043839 (2015).
[34] A. Galstyan, O. Chuluunbaatar, A. Hamido, Yu. V. Popov, F. Mota-Furtado, P. F. O'Mahony, N. Janssens, F. Catoire, and B. Piraux, Phys. Rev. A **93**, 023422 (2016).
[35] W. C. Henneberger, Phys. Rev. Lett. **21**, 838 (1968).
[36] L. B. Madsen, Phys. Rev. A **65**, 053417 (2002).
[37] G. W. F. Drake, Phys. Rev. A **5**, 1979 (1972); J. Phys. B **9**, L169 (1976).
[38] D. C. Morton, E. E. Schulhoff, and G. W. F. Drake, J. Phys. B **48**, 235001 (2015).
[39] U. D. Jentschura and K. Pachucki, Eur. Phys. J. D **69**, 118 (2015).
[40] P. B. Corkum and F. Krausz, Nature Physics **3**, 381 (2007).
[41] M. Th. Hassan, T. T. Luu, A. Moulet, O. Raskazovskaya, P. Zhokhov, M. Garg, N. Karpowicz, A. M. Zheltikov, V. Pervak, F. Krausz, and E. Goulielmakis, Nature **530**, 66 (2016).
[42] U. D. Jentschura, Phys. Rev. A **94**, 022117 (2016).
[43] A. D. Bandrauk, *Quantum and semiclassical electrodynamics*, in *Molecules in Laser Fields*, edited by A. D. Bandrauk (Dekker, New York, 1993), pp. 1-69.

Department of Physics and Engineering Physics, University of Saskatchewan, 116 Science Place, Saskatoon, Saskatchewan S7N 5E2, Canada

Current address: Kavli Institute for Cosmological Physics, The University of Chicago, 5640 South Ellis Avenue, Chicago, Illinois 60637

Email address: rainer.dick@usask.ca

Contemporary Mathematics
Volume **717**, 2018
https://doi.org/10.1090/conm/717/14455

Random phase infinite coherent states: Construction and dynamics

Alain Joye and Marco Merkli

Abstract. We consider an infinitely extended reservoir of Boson coherent states characterized by a given spatial density of modes and i.i.d. random phases. We construct its Hilbert space representation which has a random part and is expressed by means of Ito stochastic integrals. We study the open system dynamics of an N-level system coupled to the random infinite coherent state by an energy conserving interaction. We show that the coherent state reservoir induces faster system decoherence than a thermal reservoir.

1. Introduction

Coherent states have played an important role in quantum mechanics since the early days of the theory. Schrödinger discussed them as early as 1926 and noted that they behave in many respects like classical states [**20**]. Later on, thanks to the work of Glauber [**7**, **8**], it was realized that these states are particularly suited to describe optical coherence, which gave them their name. In particular, the electromagnetic radiation generated by a classical current is a multimode coherent state, and so is the light produced by a laser in certain regimes [**13**, **14**]. Therefore, coherent states are building blocks of modern quantum optics [**6**, **13**, **21**], they lie at the heart of semiclassical analysis, see *e.g.* [**4**, **12**], and more recently, they found applications in quantum information experiments [**9**].

We focus in the current text on the discussion of coherent states with random phases, which is just one aspect of the analysis of [**11**], where the infinite volume- (or thermodynamic) limit of the quantized radiation field in multimode coherent states is considered in more generality.

Consider a free bosonic quantum field confined to a finite box $\Lambda \subset \mathbb{R}^3$. As usual, we associate to each discrete mode k a creation and an annihilation operator a_k^* and a_k, respectively, satisfying the canonical commutation relations $[a_k, a_\ell^*] = \delta_{k,\ell}$ (Kronecker symbol). Also, to each k corresponds a family of normalized coherent states $|\alpha\rangle_k$, indexed by $\alpha \in \mathbb{C}$, defined as eigenvectors of the annihilation operator a_k: $a_k|\alpha\rangle_k = \alpha|\alpha\rangle_k$. The state $|\alpha\rangle_k$ is not an eigenvector of the number operator $\widehat{N} = \sum_k a_k^* a_k$ and the average of $\widehat{N}$ in $|\alpha\rangle_k$ is $|\alpha|^2$. The phase $\theta \in (-\pi, \pi]$ of the coherent state, defined by $\alpha = |\alpha|\,\mathrm{e}^{\mathrm{i}\,\theta}$, is considered to be a random variable. The action of the *displacement operator* $D_k(\alpha) = \mathrm{e}^{\alpha a_k^* - \bar{\alpha} a_k}$ on the vacuum vector Ω of the quantum field yields the coherent state: $|\alpha\rangle_k = D_k(\alpha)\Omega$. In order to

2010 *Mathematics Subject Classification.* Primary 81T27, 81V80, 82C10.

produce *multimode coherent states*, one selects N modes $k_1, \dots, k_N$, and defines $D_{k_1}(\alpha_1) \cdots D_{k_N}(\alpha_N)\Omega$, for any choice of $\alpha_j \in \mathbb{C}$, $j = 1, \dots, N$.

We consider the thermodynamic limit $\Lambda \to \mathbb{R}^3$ of the radiation field. Our main physical motivation for doing this is to thus construct a "large quantum system" which serves as a model for an environment, or reservoir. When we couple a "small system" of "ordinary size" (much smaller than Λ) to the reservoir, then we obtain an open quantum system. On the one hand, the infinite nature of the reservoir causes irreversible dynamical effects in the small system, such as thermalization, decoherence and disentanglement, see Section 1.1. On the other hand, taking the infinite volume limit of the quantum field alone is an interesting mathematical challenge, which uncovers new Hilbert space representations of the canonical commutation relations, see Section 1.2. Similar representations for quantum fields in *thermal states* (as opposed to coherent states) have been constructed before by Araki and Woods and Araki and Wyss [**1**, **2**]. The resulting thermal representations of the canonical commutation relations have found important applications in mathematical physics.

1.1. Irreversibility and limit of continuous modes. Quantum systems whose dynamics is generated by a Hamiltonian H having purely discrete spectrum

$$\mathrm{spec}(H) = \{E_j\}_{j\in\mathbb{N}}$$

show quasi-periodic dynamics. Namely, the expectation of an observable A in a state $\rho(t) = \mathrm{e}^{-\mathrm{i}tH}\rho(0)\,\mathrm{e}^{\mathrm{i}tH}$ is given by

$$\langle A\rangle_t = \mathrm{Tr}\rho(t)A = \sum_{j,k\in\mathbb{N}} \mathrm{e}^{-\mathrm{i}t(E_k-E_l)}\,c_{k,l},$$

which is a quasi-periodic function of time t. This setup cannot describe irreversible processes, such as the approach to a final state for large times $t \to \infty$. To capture irreversible effects, one has to consider a limiting situation. An illustrative example is a quantum particle moving freely within a one-dimensional interval $[-L, L]$. The Hamiltonian is $H_L = -\frac{d^2}{dx^2}$ with, say, periodic boundary conditions. The spectrum is purely discrete, $\mathrm{spec}(H_L) = \{\frac{\pi^2 n^2}{L^2}\}_{n\in\mathbb{N}}$ and has gaps of the order $1/L^2$. Averages of observables are oscillating in time and only in the limit $L \to \infty$ can one observe irreversibility. For instance, suppose that the particle is initially in the state $\rho(0) = |\psi\rangle\langle\psi|$, localized in a bounded region $[-l, l]$ of $\mathbb{R}$ (with $l < L$, so that the support of the wave function $\psi(x)$ is inside $[-l, l]$). Then the probability of finding the particle at time t inside a bounded interval $I \subset \mathbb{R}$ is given by

$$p_I(t) = \int_I \mathrm{d}\,x\,|\mathrm{e}^{-\mathrm{i}tH_L}\,\psi(x)|^2$$

and satisfies $\lim_{t\to\infty}\lim_{L\to\infty} p_I(t) \to 0$ (use the Fourier transform to diagonalize H_L). The irreversibility is physically due to the fact that the particle can escape to spatial $\pm\infty$ when $L \to \infty$, and it is accompanied by the fact that the spectrum of H_L becomes *continuous* in that limit (the gap size $1/L^2 \to 0$).

It becomes thus apparent that in order to describe irreversible effects in a quantum system, at least some component of that system should 'become infinitely extended' or show 'continuous energy values'. In the above example, taking the infinite volume limit $L \to \infty$ is a rather simple procedure. The Hilbert space for the confined system is $L^2([-L, L], dx)$ (with periodic boundary conditions) and as $L \to \infty$, the Hilbert space of the infinite system is $L^2(\mathbb{R}, \mathrm{d}\,x)$. The relation between the 'finite L' and the 'infinite L' Hilbert spaces can be very complicated though, for other systems of interest.

To see a simple example illustrating this phenomenon, consider a system composed of K independent harmonic quantum oscillators, each one described by its own Hilbert space of pure states $\mathcal{H}_{\rm osc}$. Such oscillator arrays are ubiquitous in the physics literature on open quantum systems. They model 'reservoirs', components of open systems having many degrees of freedom (K large). A pure state of the oscillator array is given by a normalized vector in the Hilbert space $\otimes_{j=1}^K \mathcal{H}_{\rm osc}$. Suppose now that the frequencies of the K oscillators are $\omega_1, \ldots, \omega_K$, and that we want to create a model which shows irreversibility. We must then take $K \to \infty$ and a frequency spectrum that becomes *continuous*. What is the Hilbert space of the oscillators in this limit of continuous modes? This question does not have an obvious answer at all. Indeed, in the wanted limit, the cardinality K of the number of modes (frequencies) will increase to $\mathfrak{c} = |\mathbb{R}|$, the cardinality of $\mathbb{R}$. Then we do not even have a candidate for a Hilbert space, since we do not know how to make sense of $\otimes_{j=1}^K \mathcal{H}_{\rm osc}$ in this limit! It turns out that one can construct a Hilbert space for the infinite system following a general procedure, but the Hilbert space will depend on a 'reference state' with respect to which the infinite mode limit is taken.

1.2. Continuous mode limit and reconstruction of a Hilbert space. As discussed in Section 1.1 it is not in general clear how to perform a continuous mode limit. The issue is that such a limit cannot be taken directly on the *states* (wave functions, density matrices) of a sequence of discrete mode systems. The idea is then to implement the following procedure, which we explain here heuristically and treat in more detail in Sections 2.3 and 2.4.

1. Consider observables A belonging to a *(quasi-)local algebra* $\mathcal{A}_0$ of observables and a sequence of states ρ_L, with L describing the system 'size' (spatial volume, 1/gap between energy values...).

2. Take the limit $L \to \infty$ of the average of A in ρ_L,

$$\lim_{L\to\infty} \langle A\rangle_L \equiv \lim_{L\to\infty} \mathrm{Tr}(\rho_L A) =: \langle A\rangle.$$

Assuming the existence of this limit, it defines a *state* $\langle\cdot\rangle$ on the algebra $\mathcal{A}_0$. (Here, a state is defined as a positive, linear, normalized functional on the C^*-algebra $\mathcal{A}_0$ as is usual in algebraic quantum theory.)

3. While we have constructed now the continuous mode limit state $\langle\cdot\rangle$ by the values it takes on the algebra $\mathcal{A}_0$ in step 2., *we have lost the notion of Hilbert space of the state.* However, the general *Gelfand-Naimark-Segal reconstruction theorem* [**3**] ensures that to the given pair $(\langle\cdot\rangle, \mathcal{A}_0)$ is assciated a unique triple (the 'GNS triple') $(\mathcal{H}, \pi, \Omega)$ of a Hilbert space $\mathcal{H}$, a $*$-representation map $\pi : \mathcal{A}_0 \to \mathcal{B}(\mathcal{H})$ (bounded operators on $\mathcal{H}$) and a unit vector $\Omega \in \mathcal{H}$, satisfying

$$\langle A\rangle = \langle \Omega, \pi(A)\Omega\rangle\,.$$

This restores the 'usual' picture of quantum theory, where the state is determined by a unit vector in a Hilbert space. Of course, it is not clear *a priori* what the concrete realization of the GNS representation Hilbert space is for a given system. For example, Araki and Woods [**1**] have found the following explicit representation for the infinite volume limit ($\mathbb{R}^3$) of a free bose gas in thermal equilibrium at temperature $T = 1/\beta$:

$$\begin{aligned} \mathcal{H}_{AW} &= \mathcal{F} \otimes \mathcal{F} \\ \pi_{AW}(a^*(f)) &= a^*(\sqrt{1+\mu}f) \otimes \mathbb{1} + \mathbb{1} \otimes a(\sqrt{\mu}\bar{f}) \\ \Omega_{AW} &= \Omega \otimes \Omega \end{aligned}$$

Here, $\mathcal{F}$ is the Fock space over the one-particle space $L^2(\mathbb{R}^3, d^3k)$ (Fourier space, see also Section 2.1), Ω is the vacuum vector in $\mathcal{F}$, $\mu = \mu(k) = (\mathrm{e}^{-\beta\varepsilon(k)} - 1)^{-1}$ with $k \mapsto \varepsilon(k)$ the dispersion relation, and $a(f)$, $a^*(f)$ are the annihilation and creation operators (smoothed out with a form factor $f \in L^2(\mathbb{R}^3)$). A detailed textbook-style derivation of the Araki-Woods representation is given in [**17**].

1.3. Why look for a Hilbert space representation? In the physics literature on open quantum systems, it is customary to calculate expressions of interest for discrete mode systems and then take the continuous mode limit of those expressions 'by hand' (e.g. by turning sums over frequencies or energies into corresponding integrals). In this way, the question of the Hilbert space for the continuous mode system is completely avoided. This procedure cannot be carried out rigorously as most often, considerations involve, on top of the continuous mode limit, a limit of weak coupling or long times, making it impossible to control remainder terms. Even more fundamentally, not all quantities of interest have a form for which the above-mentioned continuous mode limit 'by inspection' can be found (not all quantities of interest have the convenient form of a 'Riemann sum' over discrete energies which can be turned into an integral in the continuous mode limit, see e.g. [**15**]). Knowing the continuous mode *state* itself, by following the GNS procedure we outlined in Section 1.2 (rather than only the expectation of some observables in it), gives in principle full access to the analysis of all quantities. It enables a mathematically rigorous treatment of the dynamics of open quantum systems.

One of the main goals of [**11**] is to construct the Hilbert space representation of *coherent state reservoirs* (with continuous modes), similar to the work of Araki and Woods and Araki and Wyss [**1, 2**], who did that for thermal Bose and Fermi reservoirs.

1.4. Overview of some of the results in [11]. We provide in this section an informal overview of the results to be stated more precisely below. We consider the quantized radiation field in a finite box $\Lambda \subset \mathbb{R}^3$, choose modes $k_1, \dots, k_N$ and consider the multimode coherent state $D_{k_1}(\alpha_1) \cdots D_{k_N}(\alpha_N)\Omega$. Our goal is to analyze this state in the limit $\Lambda \to \mathbb{R}^3$. As Λ changes, the eigenmodes of the field change accordingly, and in the limit of infinite volume, the values of the modes becomes a continuum, $k \in \mathbb{R}^3$, similarly to the discussion in Section 1.1. It turns out that any state ω of the radiation field (in finite or infinite volume) is determined uniquely by its *expectation functional* $E(f) = \omega(W(f))$, where $W(f)$ is the *Weyl operator* smoothed out with a test function f. Thus, it is natural to consider convergence of a sequence of states ω_{Λ_n} (with $\Lambda_n \to \mathbb{R}^3$) to mean convergence of the associated $E_n(f)$, for all f.

(A) Thermodynamic limit.

- For N fixed modes $k_1, \dots, k_N \in \mathbb{R}^3$, we obtain an explicit formula for the limiting state when $\Lambda \to \mathbb{R}^3$ while keeping the particle densities $\rho_j = |\alpha_j|^2/|\Lambda|$, and the phases $\theta_j \in S^1$ fixed.

- A continuous mode limit is characterized by a mode density distribution $\rho(k)$, meaning that $\rho(k)\,\mathrm{d}^3 k$ is the spatial density of particles (number of particles per unit volume in direct space) having momenta in an infinitesimal volume $\mathrm{d}^3 k$ around k. We show that the infinite volume *and* continuous mode limit (taken jointly or not) of the multimode coherent state does *not* exist.
- Despite the lack of convergence of the expectation functional in the limits of infinite volume and continuous modes, we show that if the phases of the coherent states are random and iid, then the state *converges in distribution* for the above limits. Mathematically, the convergence is due to the central limit theorem. As explained in Subsection 1.2, we now have the infinite volume, continuous mode state as a functional on a suitable algebra of observables. The next step is to reconstruct a Hilbert space representation for this state.

(B) Hilbert space representation.
We construct explicit Gelfand-Naimark-Segal (GNS) Hilbert spaces of the finite-mode infinite volume states, and for the state of continuous modes and random phases. They are regular representations of the canonical commutation relations, defining (represented) field-, creation- and annihilation operators. As mentioned in Section 1.3, the explicit representations provide a rigorous way to analyze the dynamics of a small quantum system coupled to the coherent state reservoir (see point (D) below).

(C) Dynamics of the infinite coherent state.
The Heisenberg dynamics of the quantum field alone is a Bogoliubov transformation on Weyl operators, $W(f) \mapsto W(\mathrm{e}^{\mathrm{i}t\varepsilon} f)$, where t denotes time and $\varepsilon = \varepsilon(k)$ is the dispersion relation. Coherent states are not eigenstates of this dynamics and so the reservoir has a nontrivial dynamics on its own. We show that the random iid phase state is driven into a final state for large times. The latter corresponds to an iid random phase state in which the phases are *uniformly distributed.* In this sense, the uniform phase distribution is the stable one.

(D) Coupling to an open quantum system.
We consider an N-level quantum system in contact with the infinite volume coherent reservoir having uniformly randomly distributed phases, the dynamically stable reservoir state as explained in point (C) above. We consider an *energy conserving* coupling between the N-level system and the reservoir so that the interaction term in the Hamiltonian describing the coupled evolution commutes with the system Hamiltonian. This model is *explicitly solvable* in the sense that we can calculate the exact density matrix of the N-level system at all times. We find that the expectation of the off-diagonal density matrix elements (in the energy basis) have *Gaussian* time decay due to the coupling with the coherent reservoir. This is a striking difference relative to the thermal reservoir case, where this decay is only exponential. We conclude that a small system placed in a random phase coherent reservoir undergoes *much faster decoherence* than in a thermal environment.[1]

---

[1]The name "coherent states" refers to the quantum field and is motivated by the fact that correlation functions of the field factorize in those states, which is the same as for classical coherent fields [**13**]. On the other hand "decoherence" of a quantum system is an entirely different notion, which refers to a system losing quantum correlations and becoming close to a classical one [**10**].

The very rapid loss of "quantumness" encoded by coherence of the small system is yet another manifestation of the classical character of coherent states.

Besides the energy conserving coupling considered here, which serves as a benchmark to measure the decay of coherence in the small system, it is desirable to analyze the dynamics of an N-level system coupled in a generic, not necessarily energy conserving fashion to the infinite random coherent reservoir. This turns out to be a difficult task. However, if one is interested in specific properties of the coupled system in an average sense with respect to the randomness of the reservoir only, one might hope for certain simplifications to occur when taking expectation values over the randomness. This hope is substantiated by the fact that the randomness in the reservoir state is completely captured by a Gaussian complex valued random variable.

2. Setup and main results

We now turn to the description of the general setup and concrete statements. Consider non-interacting quantum particles confined to a box of sidelength L in d dimensions,

$$\Lambda = [-L/2, L/2]^d \subset \mathbb{R}^d.$$

The wave function of a single particle is an element of $L^2(\Lambda, \mathrm{d}\, x)$ with periodic conditions at the boundaries. The space of pure states of the system of particles is the symmetric Fock space [**3**]

$$\mathcal{F} \equiv \mathcal{F}\big(L^2(\Lambda, \mathrm{d}\, x)\big) = \bigoplus_{n\geq 0} L^2_{\mathrm{symm}}(\Lambda^n, \mathrm{d}^n\, x).$$

Elements of $\mathcal{F}$ have the form $\psi = \oplus_{n\geq 0}\psi^{(n)}$, where $\psi^{(n)}(x_1, \ldots, x_n)$ is a symmetric function of n variables $x_j \in \mathbb{R}^d$ which is periodic in Λ^n. The summand $n = 0$ of the Fock space is called the vacuum sector, spanned by the vacuum vector Ω characterized by $\Omega^{(0)} = 1$, $\Omega^{(n)} = 0$ for $n \geq 1$.

The generator of the Schrödinger equation

$$\psi_t = \mathrm{e}^{-\mathrm{i}\, Ht}\,\psi_0,$$

is the self-adjoint Hamiltonian given by the second quantization of a one-body Hamilton operator. For photons (massless relativistic particles), the Hamiltonian writes

$$(H\psi)^{(n)}(x_1, \ldots, x_n) = \sum_{j=1}^{n} \sqrt{-\Delta_{x_j}}\,\psi^{(n)}(x_1, \ldots, x_n),$$

understood as a self-adjoint operator with periodic boundary conditions.

The *creation operator* $a^*(f)$ is defined for $f \in L^2(\Lambda, \mathrm{d}\, x)$ by

$$(a^*(f)\psi)^{(n)}(x_1, \ldots, x_n) = \sqrt{n}\,\mathcal{S} f(x_1)\psi^{(n-1)}(x_2, \ldots, x_n),$$

where $\mathcal{S}$ is the operator of symmetrization over the variables $x_1, \ldots, x_n$. The *annihilation operator* $a(f)$, given by

$$(a(f)\psi)^{(n)}(x_1, \ldots, x_n) = \sqrt{n+1} \int_{\Lambda} \bar{f}(x)\psi^{(n+1)}(x, x_1, \ldots, x_n)\, \mathrm{d}\, x,$$

is the adjoint of the creation operator. We also introduce the the self-adjoint *field operators*

$$\Phi(f) = \frac{1}{\sqrt{2}}\big(a^*(f) + a(f)\big)$$

and the unitary *Weyl operators*

$$W(f) = \mathrm{e}^{\mathrm{i}\,\Phi(f)}.$$

The following equivalent *canonical commutation relations* hold:

$$\begin{aligned}
[a(f), a^*(g)] &= \langle f, g\rangle \\
[\Phi(f), \Phi(g)] &= \mathrm{i}\ \mathrm{Im}\,\langle f, g\rangle \\
W(f)W(g) &= \mathrm{e}^{-\frac{\mathrm{i}}{2}\mathrm{Im}\langle f,g\rangle}\, W(f+g).
\end{aligned} \tag{2.1}$$

A normalized vector ψ in $\mathcal{F}$ determines an expectation functional, defined by $E(f) := \langle \psi, W(f)\psi\rangle$, $f \in L^2(\Lambda, \mathrm{d}\,x)$. Conversely, any functional $E : L^2(\Lambda, \mathrm{d}\,x) \to \mathbb{C}$ satisfying the three conditions

(E1) $E(0) = 1$

(E2) $\overline{E(f)} = E(-f)$

(E3) $\sum_{k,k'=1}^{K} z_k \overline{z_{k'}}\, \mathrm{e}^{\frac{\mathrm{i}}{2}\mathrm{Im}\,\langle f_k, f_{k'}\rangle}\, E(f_k - f_{k'}) \geq 0$, for all $K \geq 1$, $z_k \in \mathbb{C}$, $f_k \in L^2(\Lambda, \mathrm{d}\,x)$

determines a state ρ on the C^*-algebra generated by the Weyl operators by the relation $\rho(W(f)) = E(f)$, see *e.g.* [**17**].

We pass to the momentum space representation, the Fourier transformation of the Fock space $\mathcal{F}$, in order to define *coherent* states of the particles in the volume Λ.

2.1. Momentum space representation. The single-particle Hilbert space $L^2(\Lambda, \mathrm{d}\,x)$ is unitarily equivalent to $l^2(\frac{2\pi}{L}\mathbb{Z}^d)$ via the Fourier transform $\mathfrak{F} : L^2(\Lambda, \mathrm{d}\,x) \to l^2(\frac{2\pi}{L}\mathbb{Z}^d)$,

$$(\mathfrak{F}f)(k) = \widehat{f}_k = L^{-d/2} \int_\Lambda \mathrm{e}^{-\mathrm{i}\,kx}\, f(x)\,\mathrm{d}\,x, \tag{2.2}$$

with inverse

$$(\mathfrak{F}^{-1}\widehat{f})(x) = L^{-d/2} \sum_{k\in\frac{2\pi}{L}\mathbb{Z}^d} \mathrm{e}^{\mathrm{i}\,kx}\, \widehat{f}_k.$$

The expression, kx denotes the dot product $k\cdot x$ and the factors $L^{-d/2}$ guarantee that $\mathfrak{F}$ is unitary. In turn, the Fock space $\mathcal{F}$ is unitarily equivalent to its momentum version

$$\widehat{\mathcal{F}} \equiv \mathcal{F}\big(l^2(\tfrac{2\pi}{L}\mathbb{Z}^d)\big) = \bigoplus_{n\geq 0} \Big(l^2(\tfrac{2\pi}{L}\mathbb{Z}^d)\Big)^{\otimes^n_{\mathrm{symm}}},$$

with Ω and $\widehat{\Omega}$, the vacua of the Fock spaces $\mathcal{F}$ and $\widehat{\mathcal{F}}$, respectively. The unitary map between the Fock spaces is characterized by

$$\mathfrak{F}a^*(f_1)\cdots a^*(f_\ell)\Omega = a^*(\widehat{f_1})\cdots a^*(\widehat{f_\ell})\widehat{\Omega}.$$

Accordingly, the creation operators transform as $\mathfrak{F}a^*(f)\mathfrak{F}^{-1} = a^*(\widehat{f})$ and we write

$$a^*(\widehat{f}) = \sum_{k\in\frac{2\pi}{L}\mathbb{Z}^d} \widehat{f}_k a_k^*.$$

The field- and Weyl operators are transported to the momentum space as

$$\mathfrak{F}\Phi(f)\mathfrak{F}^{-1} = \Phi(\widehat{f}) \qquad \text{and} \qquad \mathfrak{F}W(f)\mathfrak{F}^{-1} = W(\widehat{f}),$$

where $\Phi(\widehat{f}) = \frac{1}{\sqrt{2}}\sum_{k\in\frac{2\pi}{L}\mathbb{Z}^d}(\widehat{f}_k a_k^* + \overline{\widehat{f}_k} a_k)$ and $W(\widehat{f}) = \mathrm{e}^{\mathrm{i}\,\Phi(\widehat{f})}$.

2.2. N-mode coherent states in finite volume. In line with Section 1.4, the *coherent state* associated to the collection of N modes $k'_1, \dots, k'_N \in \frac{2\pi}{L}\mathbb{Z}^d$ and N complex numbers $\alpha_1, \dots, \alpha_N$ is the normalized vector

$$\widehat{\Psi} = \mathrm{e}^{\sum_{j=1}^N \alpha_j a^*_{k'_j} - \bar{\alpha}_j a_{k'_j}} \widehat{\Omega} \in \widehat{\mathcal{F}}. \tag{2.3}$$

The *expectation functional* corresponding to the coherent state (2.3) is thus

$$E_N^\Lambda(f) = \left\langle \widehat{\Psi}, W(\widehat{f})\widehat{\Psi} \right\rangle, \tag{2.4}$$

for all $\widehat{f} \in l^2(\frac{2\pi}{L}\mathbb{Z}^d)$. The number operator $a^*_{k'_j} a_{k'_j}$ of the mode $k'_j \in \frac{2\pi}{L}\mathbb{Z}^d$ has average in the coherent state given by

$$\left\langle \widehat{\Psi}, a^*_{k'_j} a_{k'_j} \widehat{\Psi} \right\rangle = |\alpha_j|^2,$$

which is interpreted as the intensity of the mode considered.

2.3. Infinite volume and continuous mode limits, random phases.

2.3.1. *Infinite volume.* The momenta in the finite-volume coherent state are of the form $k'_j = 2\pi n_j / L$, for $n_j \in \mathbb{Z}^d$, so that as L increases, their spacing becomes increasingly small. Consider now $k_1, \dots, k_N \in \mathbb{R}^d$, N arbitrary target momenta and take $n_j = n_j(L) \in \mathbb{Z}^d$ so that $k'_j(L) = 2\pi n_j(L)/L$ satisfies $\lim_{L\to\infty} k'_j(L) = k_j$, $j = 1, \dots, N$. We want to take the thermodynamic limit of (2.4),

$$\lim_{L\to\infty} E_N^\Lambda(f) \equiv E_N(f). \tag{2.5}$$

This means we take $\lim_{L\to\infty} k'_j = k'_j(L)$, while keeping fixed the particle densities $\rho_j \geq 0$ which count the (average) number of particles in mode k'_j per unit volume, for $j = 1, \dots, N$. This imposes $|\alpha_j|^2 = L^d \rho_j$, or

$$\alpha_j(L) = L^{d/2} \sqrt{\rho_j}\ \mathrm{e}^{\mathrm{i}\,\theta_j}, \tag{2.6}$$

where θ_j is the phase of the complex number α_j, which is arbitrary so far.

PROPOSITION 2.1 (Thermodynamic limit for N modes, [**11**]). *Let $k_1, \dots, k_N \in \mathbb{R}$ and $\rho_1, \dots, \rho_N \geq 0$ be arbitrary momenta and arbitrary particle densities and suppose that $f \in L^1(\mathbb{R}^d, \mathrm{d}\,x) \cap L^2(\mathbb{R}^d, \mathrm{d}\,x)$. Then the limit (2.5) exists and*

$$E_N(f) = E_{\mathrm{Fock}}(f)\ \mathrm{e}^{\mathrm{i}\,\mathrm{Re}\sum_{j=1}^N \mathrm{e}^{-\,\mathrm{i}\,\theta_j} \sqrt{2\rho_j}\ \widehat{f}(k_j)}, \tag{2.7}$$

where $E_{\mathrm{Fock}}(f) = \mathrm{e}^{-\frac{1}{4}\|f\|^2}$ and $\widehat{f}(k) = \int_{\mathbb{R}^d} \mathrm{e}^{-\,\mathrm{i}\,kx} f(x)\,\mathrm{d}\,x$.

Here, $E_{\mathrm{Fock}}(f)$ is the usual *Fock expectation functional*, determined by the vacuum state,

$$E_{\mathrm{Fock}}(f) = \langle \Omega, W(f)\Omega \rangle = \mathrm{e}^{-\frac{1}{4}\|f\|_2^2} = \mathrm{e}^{-\frac{1}{4}(2\pi)^{-d}\|\widehat{f}\|_2^2},$$

where $\|\cdot\|_2$ is the L^2-norm (of functions of $k \in \mathbb{R}^d$ or $x \in \mathbb{R}^d$),

$$\|f\|_2^2 = \int_{\mathbb{R}^d} |f(x)|^2\,\mathrm{d}\,x \quad \text{and} \quad \|\widehat{f}\|_2^2 = \int_{\mathbb{R}^d} |\widehat{f}(k)|^2\,\mathrm{d}\,k.$$

2.3.2. *Continuous modes.* One may perform the infinite-volume limit and the continuous mode limit simultaneously, or one can take the continuous mode limit of (2.7). The result is the same [**11**] and we do the latter. Let $\rho(k)$ be a prescribed mode density distribution. Therefore, given a cube $I \subset \mathbb{R}^d$, the integral $\int_I \rho(k)\,\mathrm{d}\,k$ is the spatial density of particles in the infinite volume state, having momenta in I. Assume ρ is be supported in a finite cube $[-R,R]^d$. Discretize the cube by setting $k_j = (-R + j_1\frac{2R}{N}, \dots, -R + j_d\frac{2R}{N}) \in \mathbb{R}^d$, where $j_1, \dots, j_d \in \{1, 2, \dots, N\}$, and N is large. One gets the following form for the sum in the phase of the infinite volume expectation functional (2.7),

(2.8)

$$\sum_{j\in\{1,\dots,N\}^d} \mathrm{e}^{-\mathrm{i}\,\theta_j}\sqrt{2\rho_j}\,\widehat{f}(k_j) = (2R/N)^{d/2} \sum_{j\in\{1,\dots,N\}^d} \mathrm{e}^{-\mathrm{i}\,\theta(k_j)}\sqrt{2\rho(k_j)}\,\widehat{f}(k_j).$$

Here, $\theta(k)$ is an arbitrary function which determines the phase of the mode k. For N large, a Riemann sum argument yields that (2.8) equals approximately

$$(N/2R)^{d/2}\int_{[-R,R]^d} \mathrm{e}^{-\mathrm{i}\,\theta(k)}\sqrt{2\rho(k)}\,\widehat{f}(k)\,\mathrm{d}\,k \;\sim\; N^{d/2}, \tag{2.9}$$

which diverges as $N \to \infty$, and so the infinite volume discrete mode expectation functional $E_N(f)$, (2.7), *does not have a continuous mode limit* in this simple sense.

2.3.3. *Random phases.* Considering the scaling involved, one sees that the central limit theorem might help in case the phases are independent random variables: Let $\theta_j = \theta_j(\omega)$, $j \in \{1, \dots, N\}^d$ be iid random variables, so that $\widehat{\Psi} = \widehat{\Psi}_\omega$ given in (2.3) is a random pure state defining the random expectation functional (in infinite volume, by Proposition 2.1)

$$E_{N,\omega}(f) = E_{\mathrm{Fock}}(f)\,e^{\mathrm{i}\,N^{-d/2}\sum_{j\in\{1,\dots,N\}^d}\xi_j(\omega)}, \tag{2.10}$$

where

$$\xi_j(\omega) = (2R)^{d/2}\sqrt{2\rho(k_j)}\,\mathrm{Re}\,\mathrm{e}^{-\mathrm{i}\,\theta_j(\omega)}\,\widehat{f}(k_j). \tag{2.11}$$

We know from (2.9) that $E_{N,\omega}(f)$ does not converge almost everywhere w.r.t. ω as $N \to \infty$. However, we have the following result.

PROPOSITION 2.2 ([**11**]). *Let μ be the distribution function of the phase and suppose that its Fourier transform satisfies $\widehat{\mu}(1) = 0$. Then, in the sense of convergence in distribution,*

$$N^{-d/2}\sum_{j\in\{1,\dots,N\}^d}\xi_j(\omega) \xrightarrow{\mathcal{D}} \mathcal{N}_\omega\big(0, \sigma_\mu(f)^2\big), \quad \text{as } N \to \infty. \tag{2.12}$$

The right hand side is a normal random variable with mean zero and variance

$$\sigma_\mu(f)^2 = \int_{\mathbb{R}^d}\rho(k)\Big(|\widehat{f}(k)|^2 + \mathrm{Re}\{\widehat{\mu}(2)\widehat{f}(k)^2\}\Big)\,\mathrm{d}\,k. \tag{2.13}$$

The function $x \mapsto \mathrm{e}^{\mathrm{i}\,x}$ is bounded and continuous and so (2.10) implies that

$$E_{N,\omega}(f) \xrightarrow{\mathcal{D}} E_\omega(f) \equiv E_{\mathrm{Fock}}(f)\,\mathrm{e}^{\mathrm{i}\,\mathcal{N}_\omega(0,\sigma_\mu(f)^2)}, \quad \text{as } N \to \infty. \tag{2.14}$$

Since the convergence in Proposition 2.2 is only in distribution, it is not guaranteed that E_ω defined in (2.14) satisfies properties (E1)-(E3) which are necessary for E_ω to define a state. Among all random functionals having the distribution given by the right hand side of (2.14), we must make sure that there is one which satisfies (E1)-(E3). We construct such an explicit representation by means of Itô stochastic integrals.

For $f \in L^2(\mathbb{R}^d, \mathrm{d}\,x)$, we define χ_ω as an Itô integral

$$\chi_\omega(f) = \int_{\mathbb{R}^d} \mathrm{d}\,B_1^\omega(k) S_1(k) \widehat{f}(k) + \mathrm{i} \int_{\mathbb{R}^d} \mathrm{d}\,B_2^\omega(k) S_2(k) \widehat{f}(k), \tag{2.15}$$

where

$$S_1(k) = \sqrt{\frac{\rho(k)}{1+\mathrm{Re}\widehat{\mu}(2)}}\,(1+\widehat{\mu}(2)), \quad S_2(k) = \sqrt{\frac{\rho(k)}{1+\mathrm{Re}\widehat{\mu}(2)}}\,\sqrt{1-|\widehat{\mu}(2)|^2}. \tag{2.16}$$

Here, B_1^ω and B_2^ω are two independent Brownian motions of dimension d and $\widehat{\mu}$ is the Fourier transform of the measure μ.

THEOREM 2.3 ([**11**]). *Suppose that* $\widehat{\mu}(1) = 0$. *Then*

(1) *For all* $f \in L^2(\mathbb{R}^d, \mathrm{d}\,k)$ *we have* $\mathrm{Re}\,\chi_\omega(f) \sim \mathcal{N}_\omega(0, \sigma_\mu(f)^2)$, *where* $\sigma_\mu(f)^2$ *is given in* (2.13).

(2) *Let* $E_{N,\omega}(f)$ *be the functional* (2.10). *Then we have, for all* $f \in L^2(\mathbb{R}^d, \mathrm{d}\,x)$,

$$E_{N,\omega}(f) \xrightarrow{\mathcal{D}} E_\omega(f) := E_{\mathrm{Fock}}(f)\,\mathrm{e}^{\mathrm{i}\,\mathrm{Re}\chi_\omega(f)}, \quad \text{as } N \to \infty.$$

(3) $E_\omega(\cdot)$ *satisfies (E1)-(E3) in the following sense:*

- $E_\omega(0) = 1$ *a.e.*(ω)
- *for all* $f \in L^2(\mathbb{R}^d, dx)$, $\overline{E_\omega(f)} = E_\omega(-f)$ *a.e.*(ω)
- *for all* $K \geq 1$, $z_k \in \mathbb{C}$, $f_k \in L^2(\mathbb{R}^d, dx)$, $k = 1, \ldots, K$, *we have* $\sum_{k,k'=1}^K z_k \overline{z_{k'}}\, \mathrm{e}^{\frac{\mathrm{i}}{2}\mathrm{Im}\langle f_k, f_{k'}\rangle} E_\omega(f_k - f_{k'}) \geq 0$ *a.e.*(ω)

Remark. For each f we have $E_\omega(f) \in L^2(\mathcal{O}, \mathrm{d}\,\mathbb{P})$, where we call the sample space $\mathcal{O}$. So there is an $\mathcal{O}_f \subseteq \mathcal{O}$ with $\mathbb{P}(\mathcal{O}_f) = 1$ s.t. $E_\omega(f) \in \mathbb{C}$ for all $\omega \in \mathcal{O}_f$. (That is, we can choose a representative of the L^2 function which is well defined and finite on a set of full measure.) Given $f_1, \ldots, f_K$ we thus find $\mathcal{O}_{f_1}, \ldots, \mathcal{O}_{f_K}$, all of full measure, so that $\sum_{j=1}^k E_\omega(f_j)$ is well defined and finite for all $\omega \in \cap_{j=1}^K \mathcal{O}_{f_j}$, again a set of full measure. The latter sum then defines again an element in $L^2(\mathcal{O}, \mathrm{d}\,\mathbb{P})$. In this sense, we can form finite (or countably infinite) linear combinations of $E_\omega(f_j)$. The set of ω of full measure on which (E3) above holds generally depends on the functions f_j.

2.4. Hilbert space representation. Given a state ρ on a C^*-algebra $\mathfrak{A}$, there exists a unique (up to unitary equivalence) GNS triple $(\mathcal{H}, \pi, \Psi)$ [**3**] consisting of a Hilbert space $\mathcal{H}$, a representation map $\pi : \mathfrak{A} \to \mathcal{B}(\mathcal{H})$ and a normalized and cyclic vector $\Psi \in \mathcal{H}$, such that for all $A \in \mathfrak{A}$,

$$\rho(A) = \langle \Psi, \pi(A)\Psi \rangle\,. \tag{2.17}$$

Consider now the family E_ω given in Theorem 2.3, (2). Being an element of $L^2(\mathcal{O}, \mathrm{d}\,\mathbb{P})$, $E_\omega(f)$ is only well defined (represented by a function with finite complex values) for $\omega \in \mathcal{O}_f \subseteq \mathcal{O}$ for some $\mathcal{O}_f$ with $\mathbb{P}(\mathcal{O}_f) = 1$. The range over which ω varies thus depends on f. Therefore, it is not clear that there is *any* $\omega \in \mathcal{O}$ for which one can define simultaneously $E_\omega(f)$ for all $f \in L^2(\mathbb{R}^d, \mathrm{d}\,x)$. However, we can restrict the range of f to a countable subset of "test functions" $\mathcal{D} \subset L^2(\mathbb{R}^d, \mathrm{d}\,x)$. For each $f \in \mathcal{D}$, there is an $\mathcal{O}_f \subseteq \mathcal{O}$, $\mathbb{P}(\mathcal{O}_f) = 1$, on which $E_\omega(f)$ is well defined, that is, for which one can choose a representative of the $L^2(\mathcal{O}, \mathrm{d}\,\mathbb{P})$ function which is finite on $\mathcal{O}_f$. Being a countable intersection of sets of measure one, the set

$$\mathcal{O}(\mathcal{D}) = \cap_{f\in\mathcal{D}}\,\mathcal{O}_f$$

has also measure one. Furthermore, for every $\omega \in \mathcal{O}(\mathcal{D})$ fixed, $E_\omega(f)$ is well defined for all $f \in \mathcal{D}$. Without loss of generality, we may assume that $\mathcal{D}$ is a vector space over the countable field $\mathbb{Q} + \mathrm{i}\,\mathbb{Q}$. (If an original $\mathcal{D}$ is not, then we can consider the set of all finite linear combinations of elements of $\mathcal{D}$, with scalars from $\mathbb{Q} + \mathrm{i}\,\mathbb{Q}$. This is again a countable set and we can take that set for $\mathcal{D}$.)

We show in [**11**] that there exists an $\mathcal{O}(\mathcal{D}) \subseteq \mathcal{O}$ of full measure, such that $\forall\omega \in \mathcal{O}(\mathcal{D})$, E_ω is an expectation functional on the Weyl algebra with test functions $f \in \mathcal{D}$, i.e., the Weyl algebra (over the field $\mathbb{C}$ of scalars) generated by all $W(f)$, $f \in \mathcal{D}$.

THEOREM 2.4 (GNS representation [**11**]). *Let $\mathcal{D} \subset L^2(\mathbb{R}^d, \mathrm{d}\,x)$ be a test function subspace with associated $\mathcal{O}(\mathcal{D})$ satisfying $\mathbb{P}(\mathcal{O}(\mathcal{D})) = 1$. For every $\omega \in \mathcal{O}(\mathcal{D})$, the GNS representation of $E_\omega(\cdot)$ (c.f. Theorem 2.3(2)) as a functional of the Weyl algebra with test functions in $\mathcal{D}$, is given by*

$$\begin{aligned}\mathcal{H} &= \mathcal{H}_\mathcal{D} \subseteq \mathcal{F}(L^2(\mathbb{R}^d, \mathrm{d}\,x))\\ \pi_\omega(W(f)) &= W_{\mathrm{Fock}}(f)\,\mathrm{e}^{\mathrm{i}\,\mathrm{Re}\,\chi_\omega(f)}\\ \Psi &= \Omega_{\mathrm{Fock}}.\end{aligned}$$

Recall that a representation π is called regular if $\alpha \mapsto \pi(W(\alpha f))$ is differentiable at $\alpha = 0$, in the strong sense on a dense domain in $\mathcal{H}$. For regular representations, one defines the represented Weyl operators

$$W_\pi(f) = \pi(W(f))$$

and the represented field operators by

$$\Phi_\pi(f) = -i\partial_\alpha|_{\alpha=0}\,\pi(W(\alpha f)).$$

Similarly, the creation and annihilation operators are defined by

$$\begin{aligned}a^*_\pi(f) &= 2^{-1/2}\big[\Phi_\pi(f) - \mathrm{i}\,\Phi_\pi(\mathrm{i}\,f)\big],\\ a_\pi(f) &= 2^{-1/2}\big[\Phi_\pi(f) + \mathrm{i}\,\Phi_\pi(\mathrm{i}\,f)\big] = (a^*_\pi(f))^*.\end{aligned} \tag{2.18}$$

The representation of Theorem 2.4 is actually regular, so that the field and creation operators associated to E_ω, for all $\omega \in \mathcal{O}(\mathcal{D})$ and all $f \in \mathcal{D}$, are given by

$$\begin{aligned}\Phi_\omega(f) &= \Phi_{\mathrm{Fock}}(f) + \mathrm{Re}\,\chi_\omega(f)\\ a^*_\omega(f) &= a^*_{\mathrm{Fock}}(f) + \tfrac{1}{\sqrt{2}}\,\chi_\omega(f).\end{aligned}$$

2.5. Reservoir dynamics. We can now address the dynamical aspects of our construction. The dynamics on the Weyl algebra is given by a Bogoliubov transformation on the functions $f \in L^2(\mathbb{R}^d, \mathrm{d}\,k)$, given by $f \mapsto \mathrm{e}^{\mathrm{i}\,t\varepsilon} f$, where $\varepsilon = \varepsilon(k)$ is a real function of $k \in \mathbb{R}^d$. For photons, $\varepsilon(k) = |k|$. Thus, the dynamics of the random phase expectation functional E_ω, (2.14) satisfies

$$E_\omega(\mathrm{e}^{\mathrm{i}\,t\varepsilon} f) = E_{\mathrm{Fock}}(f)\;\mathrm{e}^{\mathrm{i}\,\mathcal{N}_\omega(0,\sigma_\mu(\mathrm{e}^{\mathrm{i}\,t\varepsilon} f)^2)}\,. \tag{2.19}$$

Now, the Riemann-Lebesgue Lemma and (2.13) show that

$$\lim_{t\to\infty} \sigma_\mu(\mathrm{e}^{\mathrm{i}\,t\varepsilon} f)^2 = \|\sqrt{\rho}\widehat{f}\|_2^2 \tag{2.20}$$

and therefore,

$$\mathcal{N}_\omega(0,\sigma_\mu(\mathrm{e}^{\mathrm{i}\,t\varepsilon}\,f)^2)\ \xrightarrow{\mathcal{D}}\ \mathcal{N}_\omega(0,\|\sqrt{\mu}\widehat{f}\|_2^2),\quad t\to\infty.$$

Consequently, we reach the following conclusion:

PROPOSITION 2.5 (Phase uniformization under reservoir dynamics, [**11**]). *Let μ be a phase distribution satisfying $\widehat{\mu}(1)=0$. Given any $f\in L^2(\mathbb{R}^d,\mathrm{d}\,x)$, we have, as $t\to\infty$,*

$$E_\omega(\mathrm{e}^{\mathrm{i}\,t\varepsilon}\,f)\ \xrightarrow{\mathcal{D}}\ E_{\omega,\mathrm{unif}}(f). \tag{2.21}$$

The convergence in (2.21) *is in distribution of random variables. Here, $E_{\omega,\mathrm{unif}}(\cdot)$ is the expectation functional in which the phase distribution is uniform, $\mathrm{d}\,\mu(\theta)=\frac{\mathrm{d}\,\theta}{2\pi}$.*

2.6. Coupling to an open quantum system. We are now in a position allowing us to consider an N-dimensional quantum system in contact with the reservoir of coherent states. We consider that the phases are uniformly randomly distributed. The Hilbert space of pure states of the system is $\mathbb{C}^N$, that of the reservoir is the GNS space given in Theorem 2.4. The system dynamics is generated by a self-adjoint Hamiltonian with energy levels $e_1,\dots,e_N$,

$$H_{\mathrm{S}}=\mathrm{diag}(e_1,\dots,e_N),$$

and the state of the reservoir is invariant under its own dynamics by Proposition 2.5. The dynamics is implemented as

$$\pi_\omega(W(\mathrm{e}^{\mathrm{i}\,t\varepsilon}\,f))=W_{\mathrm{Fock}}(\mathrm{e}^{\mathrm{i}\,t\varepsilon}\,f)\,\mathrm{e}^{\mathrm{i}\,\mathrm{Re}\chi_\omega(f)}=\mathrm{e}^{\mathrm{i}\,tH_{\mathrm{R}}}\,\pi_\omega(W(f))\,\mathrm{e}^{-\,\mathrm{i}\,tH_{\mathrm{R}}},$$

where the reservoir Hamiltonian is

$$H_{\mathrm{R}}=\mathrm{d}\,\Gamma(\varepsilon).$$

Therefore, the uncoupled dynamics is given by the Hamiltonian

$$H_0=H_{\mathrm{S}}\otimes\mathbb{1}_{\mathrm{R}}+\mathbb{1}_{\mathrm{S}}\otimes H_{\mathrm{R}}.$$

To define a coupled dynamics between the system and the reservoir, one makes use of Dyson's expansion. The free dynamics is given by the group of $*$automorphisms α_0^t on the algebra of observables $\mathfrak{A}=\mathcal{B}(\mathbb{C}^N)\otimes\mathcal{W}$ (where $\mathcal{W}$ is the Weyl algebra), defined by

$$\alpha_0^t(A_{\mathrm{S}}\otimes W(f))=\mathrm{e}^{\mathrm{i}\,tH_{\mathrm{S}}}\,A_{\mathrm{S}}\,\mathrm{e}^{-\,\mathrm{i}\,tH_{\mathrm{S}}}\otimes W(\mathrm{e}^{\mathrm{i}\,t\varepsilon}\,f).$$

Then one defines a coupled dynamics by specifying an interaction operator $V\in\mathfrak{A}$ and using the Dyson series

$$\alpha^t(A)=\alpha_0^t(A)+\sum_{n\geq1}\int_0^t\mathrm{d}\,t_1\cdots\int_0^{t_n-1}\mathrm{d}\,t_n\ [\alpha_0^{t_n}(V),[\cdots[\alpha_0^{t_1}(V),\alpha_0^t(A)]\cdots]]. \tag{2.22}$$

It turns out the series converges in the topology of $\mathfrak{A}$ and thus defines the interacting dynamics α^t, again a group of $*$automorphisms on $\mathfrak{A}$. Applying the representation map π_ω (more precisely, $\mathbb{1}_{\mathrm{S}}\otimes\pi_\omega$) to (2.22), we obtain

$$\begin{aligned}\pi_\omega(\alpha^t(A))&=\tau_0^t(\pi_\omega(A))\\&\quad+\sum_{n\geq1}\int_0^t\mathrm{d}\,t_1\cdots\int_0^{t_n-1}\mathrm{d}\,t_n\ [\tau_0^{t_n}(\pi_\omega(V)),[\cdots[\tau_0^{t_1}(\pi_\omega(V)),\tau_0^t(\pi_\omega(A))]\cdots]],\end{aligned}$$

where

$$\tau_0^t(\cdot) = \mathrm{e}^{\mathrm{i}tH_0}(\cdot)\,\mathrm{e}^{-\mathrm{i}tH_0}.$$

The right side of (2.23) defines a group of $*$automorphisms on the represented algebra of observables which is generated by the self-adjoint operator

$$H = H_0 + \pi_\omega(V),$$

acting on $\mathbb{C}^N \otimes \mathcal{F}(L^2(\mathbb{R}^d, \mathrm{d}\,x))$. Physical considerations would lead us to take $V = G \otimes \Phi(g)$, where G is selfadjoint and $\Phi(g)$ is a field operator. However, this V does not belong to $\mathfrak{A}$ and the above construction cannot be carried out. Nevertheless, one can "regularize" the interaction by introducing V_η, depending on a small parameter η, such that $V_\eta \in \mathfrak{A}$ and in any regular representation π of the algebra $\mathfrak{A}$, $\pi(V_\eta) \to G \otimes \Phi_\pi$, as $\eta \to 0$ (strongly on a dense domain). One can then, for $\eta > 0$, carry out the above construction and finally remove η once placed in a representation. Such a procedure is decribed in [**5**] – and other approaches are possible. This shows we can consider that the dynamics of the coupled system is generated by the Hamiltonian

$$H = H_0 + G \otimes \Phi_\omega(g) = H_0 + G \otimes \big(\Phi_{\mathrm{Fock}}(g) + \mathrm{Re}\chi_\omega(g)\big), \tag{2.23}$$

acting on $\mathbb{C}^N \otimes \mathcal{F}(L^2(\mathbb{R}^d, \mathrm{d}\,x))$.

We take an *energy conserving* (non-demolition) interaction [**10, 19**] between the system and the reservoir, which consists in taking an operator G that commutes with H_{S},

$$G = \mathrm{diag}(g_1, \ldots, g_N).$$

Such models are used to investigate "phase decoherence" of the small system.

The initial system-reservoir state is disentangled, given by a density matrix

$$P_0 = \rho_{\mathrm{S}} \otimes |\Omega\rangle\langle\Omega|,$$

acting on the Hilbert space $\mathbb{C}^N \otimes \mathcal{F}(L^2(\mathbb{R}^d, \mathrm{d}\,x))$. Here, ρ_{S} is an arbitrary intial system density matrix and the reservoir is in the state Ω, which represents the infinitely extended continuous mode coherent state with uniformly distributed phases. The state of the coupled system at any later time t is given by

$$P(t) = \mathrm{e}^{-\mathrm{i}tH}\,P_0\,\mathrm{e}^{\mathrm{i}tH}.$$

Taking the partial trace over the reservoir Hilbert space yields the reduced system density matrix,

$$\rho_{\mathrm{S}}(t) = \mathrm{Tr}_{\mathrm{R}} P(t).$$

We denote its matrix elements in the energy eigenbasis $\{\varphi_j\}_{j=1}^N$ (with $H_{\mathrm{S}}\varphi_j = e_j\varphi_j$) by

$$\rho_{k,l}(t) = \langle \varphi_k, \rho_{\mathrm{S}}(t)\varphi_l\rangle = \mathrm{Tr}\,P(t)|\varphi_l\rangle\langle\varphi_k|. \tag{2.24}$$

As $[H_{\mathrm{S}}, G] = 0$ the populations (diagonal matrix elements) are time-independent. The off-diagonal ones exhibit time decay ("phase decoherence"). For the energy conserving model at hand, the matrix elements (2.24) can be evaluated exactly. The calculation yields (see Appendix D of [**16**])

$$\rho_{k,l}(t) = \mathrm{e}^{-\mathrm{i}t(e_k-e_l)}\,\mathrm{e}^{-\mathrm{i}t(g_k-g_l)\mathrm{Re}\chi_\omega(g)} \tag{2.25}$$

$$\times\,\mathrm{e}^{\frac{\mathrm{i}}{2}(g_k^2-g_l^2)\langle g, \frac{\sin(\varepsilon t)-\varepsilon t}{\varepsilon} g\rangle}\,\mathrm{e}^{-\frac{1}{2}(g_k-g_l)^2\langle g, \frac{1-\cos(\varepsilon t)}{\varepsilon^2} g\rangle} \tag{2.26}$$

$$\times \rho_{k,l}(0).$$

The contribution on the right side of (2.25) is given by the free dynamics and by a random "renormalization" of the system energy due to the interaction with the coherent bath (coming from the term $G\otimes \mathrm{Re}\chi_\omega(g)\mathbb{1}_{\mathrm{R}}$ in the Hamiltonian (2.23)). The two factors (2.26) are the same as in the case of a system coupled to a free bose gas in equilibrium at zero temperature. Therefore, the coherent states character of the reservoir is encoded entirely in the part $\mathrm{e}^{-\mathrm{i}t(g_k-g_l)\mathrm{Re}\chi_\omega(g)}$. The expectation of this oscillating factor is the characteristic function of the random variable $\mathrm{Re}\chi_\omega(g)\sim\mathcal{N}(0,\|\sqrt{\rho}g\|_2^2)$,

$$\mathbb{E}\left[\mathrm{e}^{-\mathrm{i}t(g_k-g_l)\mathrm{Re}\chi_\omega(g)}\right]=\mathrm{e}^{-\frac{t^2}{2}(g_k-g_l)^2\|\sqrt{\rho}g\|_2^2}.$$

This shows that the averaged (reduced system) density matrix $\mathbb{E}[\rho_{\mathrm{S}}(t)]$ acquires *Gaussian time-decay* of off-diagonals at all times, due to the coupling with the coherent reservoir, namely

$$\big|\mathbb{E}[\rho_{k,l}(t)]\big|=\mathrm{e}^{-\frac{t^2}{2}(g_k-g_l)^2\|\sqrt{\rho}g\|_2^2}\,\mathrm{e}^{-\frac{1}{2}(g_k-g_l)^2\Gamma(t)}\,|\rho_{k,l}(0)|, \tag{2.27}$$

with

$$\Gamma(t)=\left\langle g,\frac{1-\cos(\varepsilon t)}{\varepsilon^2}g\right\rangle=2\left\langle g,\frac{\sin^2(\varepsilon t/2)}{\varepsilon^2}g\right\rangle. \tag{2.28}$$

For small times, $\Gamma(t)\sim\frac{1}{2}t^2\|g\|_2^2$ is quadratic in time, but for large t, its behaviour as a function of t depends on the infrared behaviour of the form factor g, $|g(k)|\sim|k|^p$ for $|k|\sim 0$. For instance, in $d=3$ dimensions and for $\varepsilon(k)=|k|$,

$$\Gamma(t)=2\int_0^\infty |k|^2\,\mathrm{d}\,|k|\int_{S^2}\mathrm{d}\,\Sigma|g(|k|,\Sigma)|^2\frac{\sin^2(|k|t/2)}{|k|^2}\sim\frac{\pi t}{2}\lim_{r\to0_+}r^2\int_{S^2}\mathrm{d}\,\Sigma|g(r,\Sigma)|^2,$$

assuming that the latter limit exists and is non-vanishing, meaning that $|g(r,\Sigma)|\sim r^{-1}$ for small r. Note also that for $p>-1/2$, we have $\lim_{t\to\infty}\langle g,\cos(\varepsilon t)/\varepsilon^2 g\rangle=0$ by the Riemann-Lebesgue lemma, so that $\lim_{t\to\infty}\Gamma(t)=\|g/\varepsilon\|_2^2$. For this infra-red behaviour of the form factor, the coupling to the (zero temperature) reservoir does not induce (complete) decoherence, but the coupling to the coherent reservoir does.

References

[1] H. Araki and E. J. Woods, *Representations of the canonical commutation relations describing a nonrelativistic infinite free Bose gas*, J. Mathematical Phys. **4** (1963), 637–662, DOI 10.1063/1.1704002. MR0152295

[2] Huzihiro Araki and Walter Wyss, *Representations of canonical anticommutation relations*, Helv. Phys. Acta **37** (1964), 136–159. MR0171521

[3] Ola Bratteli and Derek W. Robinson, *Operator algebras and quantum statistical mechanics. 2*, 2nd ed., Texts and Monographs in Physics, Springer-Verlag, Berlin, 1997. Equilibrium states. Models in quantum statistical mechanics. MR1441540

[4] Monique Combescure and Didier Robert, *Coherent states and applications in mathematical physics*, Theoretical and Mathematical Physics, Springer, Dordrecht, 2012. MR2952171

[5] Jürg Fröhlich and Marco Merkli, *Thermal ionization*, Math. Phys. Anal. Geom. **7** (2004), no. 3, 239–287, DOI 10.1023/B:MPAG.0000034613.13746.8a. MR2081715

[6] C. W. Gardiner and P. Zoller, *Quantum noise*, 3rd ed., Springer Series in Synergetics, Springer-Verlag, Berlin, 2004. A handbook of Markovian and non-Markovian quantum stochastic methods with applications to quantum optics. MR2096935

[7] Roy J. Glauber, *The quantum theory of optical coherence*, Phys. Rev. (2) **130** (1963), 2529–2539. MR0156576

[8] Roy J. Glauber, *Coherent and incoherent states of the radiation field*, Phys. Rev. (2) **131** (1963), 2766–2788. MR0161585

[9] F. Grosshans, G. Van Assche, J. Wenger, R. Brouri, N. J. Cerg, Ph. Grangier: "Quantum key distribution using gaussian-modulated coherent states", Letters to Nature, Nature **421**, 238-241 (2003)

[10] E. Joos, H. D. Zeh, C. Kiefer, D. Giulini, J. Kupsch, and I.-O. Stamatescu, *Decoherence and the appearance of a classical world in quantum theory*, 2nd ed., Springer-Verlag, Berlin, 2003. MR2148270

[11] Alain Joye and Marco Merkli, *Representations of canonical commutation relations describing infinite coherent states*, Comm. Math. Phys. **347** (2016), no. 2, 421–448, DOI 10.1007/s00220-016-2611-1. MR3545512

[12] John R. Klauder and Bo-Sture Skagerstam (eds.), *Coherent states*, World Scientific Publishing Co., Singapore, 1985. Applications in physics and mathematical physics. MR826247

[13] L. Mandel, E. Wolf: Optical coherence and quantum optics, Cambridge University Press 1995

[14] Philippe A. Martin and François Rothen, *Many-body problems and quantum field theory*, 2nd ed., Texts and Monographs in Physics, Springer-Verlag, Berlin, 2004. An introduction; Translated from the 1990 French original by Steven Goldfarb, Andrew Jordan and Samuel Leach. MR2097344

[15] M. Merkli, G. P. Berman, R. T. Sayre, X. Wang, and A. I. Nesterov, *Production of Entanglement Entropy by Decoherence*, Open Syst. Inf. Dyn. **25** (2018), no. 1, 1850001, 45, DOI 10.1142/S1230161218500014. MR3786871

[16] M. Merkli, I. M. Sigal, and G. P. Berman, *Resonance theory of decoherence and thermalization*, Ann. Physics **323** (2008), no. 2, 373–412, DOI 10.1016/j.aop.2007.04.013. MR2387032

[17] Marco Merkli, *The ideal quantum gas*, Open quantum systems. I, Lecture Notes in Math., vol. 1880, Springer, Berlin, 2006, pp. 183–233, DOI 10.1007/3-540-33922-1_5. MR2248984

[18] Bernt Øksendal, *Stochastic differential equations*, 6th ed., Universitext, Springer-Verlag, Berlin, 2003. An introduction with applications. MR2001996

[19] G. Massimo Palma, Kalle-Antti Suominen, and Artur K. Ekert, *Quantum computers and dissipation*, Proc. Roy. Soc. London Ser. A **452** (1996), no. 1946, 567–584, DOI 10.1098/rspa.1996.0029. MR1378845

[20] E. Schrödinger: "Der stetige Übergang von der Mikro- zur Makromechanik", *Naturwissenschaften* **14**, 664-666 (1926)

[21] M. Schlosshauer, *Classicality, the ensemble interpretation, and decoherence: resolving the Hyperion dispute*, Found. Phys. **38** (2008), no. 9, 796–803, DOI 10.1007/s10701-008-9237-x. MR2447849

CNRS Institut Fourier, Université Grenoble Alpes, 38000 Grenoble, France
Email address: alain.joye@univ-grenoble-alpes.fr

Department of Mathematics and Statistics, Memorial University of Newfoundland, St. John's, Newfoundland, Canada, A1C 5S7
Email address: merkli@mun.ca

Contemporary Mathematics
Volume **717**, 2018
https://doi.org/10.1090/conm/717/14456

Effective evolution of open dimers

Marco Merkli

This paper is based on several collaborations with
G.P. Berman, R.T. Sayre, S. Gnanakaran,
M. Könenberg, A.I. Nesterov and H. Song

ABSTRACT. We analyze the exciton transfer and decoherence processes in a photosynthetic dimer in contact with collective (correlated) and local (uncorrelated) protein-solvent environments. Our approach, based on the framework of the spin-boson model, is mathematically rigorous. We derive explicitly the relaxation and decoherence rates of the exciton transfer process. We establish a generalization of the Marcus formula, yielding the reaction rates for dimer levels which are individually and asymmetrically coupled to environments. Our results hold for arbitrary temperatures and arbitrary strenghts of the dimer-enviornment interaction.

1. Physical Motivation

When a molecule is excited electronically by absorbing a photon, it luminesces by emitting another photon (or the excitation can be lost in a thermal enviornment). For systems we have in mind, the time scale of this fluorescence is ~ 1 nanosecond.

Fluorescence

However, when another molecule with similar excitation energy is present within $\sim 1 - 10$ nanometers, the excitation can be swapped between the molecules. This process, called excitation transfer, happens on time-scales of ~ 1 picosecond, about one thousand times faster than fluorescence! An excited 'donor' and a non-excited 'acceptor' $D^* + A$ transform into a non-excited donor and an excited acceptor $D + A^*$.

This excitation transfer is observed in *biological systems*, notably in chlorophyll molecules during photosynthesis [**7**]. Similar *charge transfer processes* (electron,

Key words and phrases. Excitation transfer reaction, electron transfer reaction, Marcus formula, effective dynamics of open quantum systems.

This work was supported by a Natural Sciences and Engineering Research Council of Canada (NSERC) Discovery Grant and an NSERC Discovery Accelerator Supplement, as well as by the Center for Nonlinear Studies at the Los Alamos National Laboratory.

proton transfer between molecules) happen in chemical (e.g. redox) reactions: $D + A \to D^- + A^+$, where the donor and acceptor are also called the reactant and product, respectively [**5**].

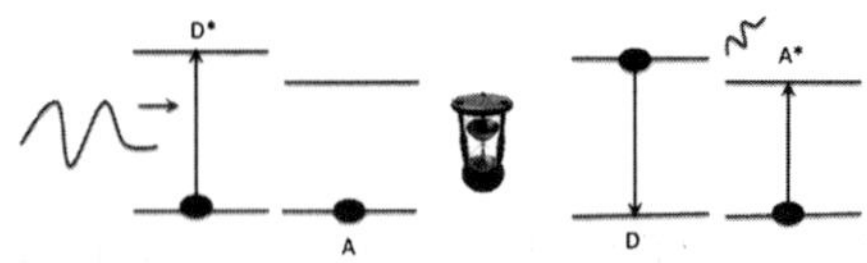

Excitation transfer process: D*+ A→D + A*

The processes in question take place in *noisy environments*, typically thought of as vibrations of surrounding molecules. One can envisage two types of noise interaction. If the donor and acceptor are spatially separated by a distance exceeding the correlation length of the reservoir they are embedded in, then each of D and A may be regarded as interacting with its own, independent reservoir. Otherwise one can model the reservoir as a single one, to which both D and A are coupled.

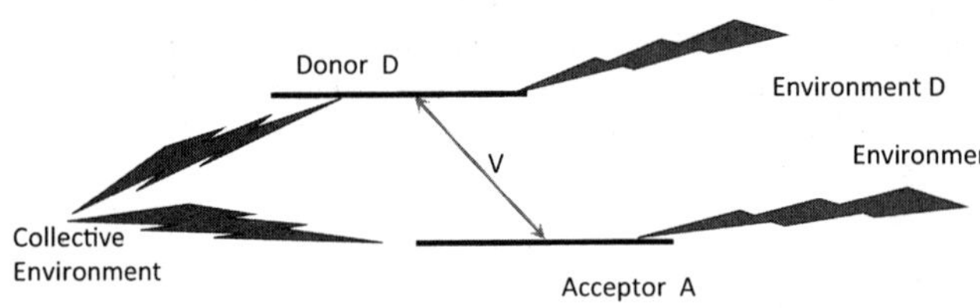

Local model (right arrows) and collective model (left arrows)
V: direct exchange interaction

∘ Local (uncorrelated) model: D, A have individual environments
∘ Collective (correlated) model: D, A have a common environment

Historically, the electronic excitation energy transfer theory is called *Förster theory*, while the charge transfer theory is *Marcus theory*. The main goal of those theories is to predict the *transfer rate*, i.e., the rate at which an initially populated donor depletes by transferring occupation probability to the acceptor. Accordingly, the two theories bring about the rate formulas:

- Förster formula (1948)

$$\gamma_F = \frac{9000\,(\ln 10)\,\kappa^2}{128\,\pi^5 N_A\,\tau_D\,n_r^4\,R^6}\int_0^\infty \frac{f_D(\nu)\epsilon_A(\nu)}{\nu^4}d\nu$$

κ^2 = orientation factor, N_A = Avogadro's number, τ_D = spontaneous decay lifetime of excited donor, n_r = refractive index of medium, R = donor-acceptor distance, $f_D(\nu)$ = normalized donor emission spectrum, $\epsilon_A(\nu)$ = acceptor molar extinction coefficient

- Marcus formula (1956)

$$\gamma_M = \frac{2\pi}{\hbar}|V|^2\frac{1}{\sqrt{4\pi\epsilon_{\rm rec}k_BT}}\exp\left[-\frac{(\Delta G+\lambda)^2}{4\epsilon_{\rm rec}k_BT}\right]$$

V = electronic coupling, $\epsilon_{\rm rec}$ = reconstruction energy, ΔG = Gibbs free energy change in reaction, T = temperature

While γ_F and γ_M look very different indeed, they are obtained by the *same* mathematical perturbation theory, simply expressed in different physical parameters.

Marcus based his considerations on a Hamiltonian

$$\begin{aligned} H_{\text{Marcus}} &= |R\rangle E_R \langle R| + |P\rangle E_P \langle P| + |R\rangle V \langle P| + |P\rangle V \langle R| \\ &= \begin{pmatrix} E_R & V \\ V & E_P \end{pmatrix} \end{aligned}$$

where R is the reactant (donor) level and P is the product (acceptor) level, both considered to be quantum states. The (collective) reservoir is taken to consist of classical harmonic degrees of freedom (with the usual notation),

$$\begin{aligned} E_R &= \sum_\alpha \Big(\tfrac{p_\alpha^2}{2m_\alpha} + \tfrac{1}{2} m_\alpha \omega_\alpha^2 q_\alpha^2 \Big), \\ E_P &= \sum_\alpha \Big(\tfrac{p_\alpha^2}{2m_\alpha} + \tfrac{1}{2} m_\alpha \omega_\alpha^2 (q_\alpha^2 - q_{0,\alpha})^2 - \epsilon_{0,\alpha} \Big). \end{aligned}$$

In a quantum mechanical treatment, E_R and E_P become *operators* H_R and H_P. It is not surprising that the quantum model can be expressed as the famous *spin-boson system* [**4**], defined by the Hamiltonian

$$H_{\text{SB}} = V\sigma_x + \epsilon\,\sigma_z + H_R + \lambda \sigma_z \otimes \varphi(h),$$

where the $\sigma_{x,z}$ are Pauli matrices and

$$\begin{aligned} H_R &= \sum_\alpha \omega_\alpha (a_\alpha^\dagger a_\alpha + 1/2) \\ \varphi(h) &= \tfrac{1}{\sqrt{2}} \sum_\alpha h_\alpha a_\alpha^\dagger + \text{h.c.}, \qquad h_\alpha = \text{form factor} \end{aligned}$$

(free bose field). In [**10**], a dictionary is established, translating the Marcus theory parameters into the spin-boson parameters. In particular, we have $\lambda^2 \propto \epsilon_{\text{rec}}$.

One may then use the heuristic 'time-dependent perturbation theory' for the spin-boson model, as given in [**4**], to obtain the decay law "$p_{donor}(t) = e^{-\gamma t}$" for the donor population, with a specific relaxation rate γ. Experimental investigations show that for the photosystem we are considering (chlorophyll in light harvesting complexes), the ratio of direct donor-acceptor interaction energy V to donor-acceptor excitation energy difference ϵ is $V/\epsilon \approx 0.1$. This parameter regime is called the *weakly coupled dimer.* If in addition to $V \ll \epsilon$, one considers *high temperatures* $k_B T \gg \hbar\omega_c$ (with ω_c a characteristic ultra-violet cutoff frequency), then the rate γ of Leggett et al. [**4**] reduces correctly to the *Marcus formula rate*

$$\gamma_{\text{Marcus}} = \frac{V^2}{4} \sqrt{\frac{\pi}{T\epsilon_{\text{rec}}}}\, e^{-\frac{(\epsilon-\epsilon_{\text{rec}})^2}{4T\epsilon_{\text{rec}}}}.$$

It is also known from experiments that $\epsilon_{\text{rec}} \approx \epsilon$, and since $\epsilon_{\text{rec}} \propto \lambda^2$, we are faced with a system-environment energy of the same order as the Bohr-energy of the system alone. This means that the donor-acceptor pair is *strongly coupled to the reservoir.* We point out that

- Marcus theory is designed to work for large interaction strength with environment (ϵ_{rec}) and for 'high' temperatures (room temperature of biological, living systems).

• The 'usual' theory of open quantum systems is the *Bloch-Redfield* theory, it is designed for small interactions with the environment and hence does not apply to the situation at hand.

• Rudolph A. Marcus received the 1992 Nobel Prize in Chemistry "for his contributions to the theory of electron transfer reactions in chemical systems" [**6**].

Our main contributions are:

1. We develop a "*dynamical resonance theory*," a controlled perturbation theory for the dynamics of the weakly coupled dimer ($V \ll \epsilon$), valid for **all times** ($t \geq 0$ arbitrarily large) and **any reservoir coupling strength** ($\lambda \in \mathbb{R}$).

2. We extract from the dynamical resonance theory the precise meaning and the validity of Leggett's exponential decay law and the rates of relaxation as well as decoherence.

3. We consider individual, possibly different coupling strenghts λ_D, λ_A of the donor and acceptor to the environment(s), in contrast to the Marcus case, where $\lambda_D = -\lambda_A$. We then show that an aysmmetric coupling can significantly speed up the excitation process.

2. Resonance expansion

Spectral deformation ... or not! Consider the propagator e^{itL} of a quantm system, where L is the self-adjoint generator (the 'Liouville operator', or Hamiltonian). The map $z \mapsto f(z) \equiv \langle\psi, (L-z)^{-1}\phi\rangle$ is analytic in $z \in \mathbb{C}_-$. In the *spectral deformation technique*, one assumes that $f(z)$ has a meromorphic extension, across the cut $\mathbb{R}$, into $\mathbb{C}_+$, with poles lying on the 2nd Riemann sheet. A contour defor-

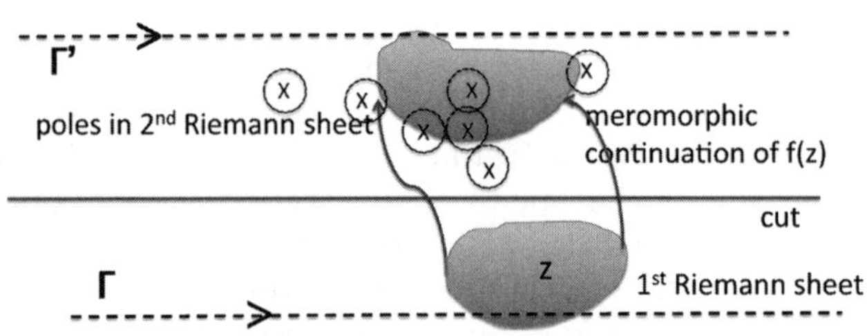

mation ($\Gamma \to \Gamma'$) in the usual contour integral expressing the propagator as a line integral over the resolvent then reveals the dynamics as

$$\langle\psi, e^{itL}\phi\rangle = \frac{-1}{2\pi i}\int_\Gamma e^{itz} f(z)dz = \sum_{\text{poles } a} e^{ita} f(a) + O(e^{-\alpha t}),$$

where $\alpha > \text{Im}a$ for all poles a (see the picture). Poles a with $\text{Im}a > 0$ then drive temporal decay, while poles $a \in \mathbb{R}$ cause oscillations in time.

How does one construct the meromorphic continuation of f? Let U_θ, $\theta \in \mathbb{R}$ be a unitary group acting on the Hilbert space of pure states. Then we have

$$f(z) = \langle\psi, (L-z)^{-1}\phi\rangle = \langle\psi_\theta, (L_\theta - z)^{-1}\phi_\theta\rangle,$$

where $L_\theta = U_\theta L U_\theta^*$ and $\psi_\theta = U_\theta\psi$. Suppose now that, under some regularity assumptions, the above right hand side representing $f(z)$ extends meromorphically in the variable θ to some domain of $\theta \in \mathbb{C}$. For $\theta \in \mathbb{C}\backslash\mathbb{R}$ fixed, the operator L_θ is *not* self-ajdoint any longer, and has eigenvalues in $\mathbb{C}_+$ ('deformed spectrum'!), which are precisely the poles of the function $z \mapsto f(z)$. Then $z \mapsto f(z)$ for $\mathrm{Im}\theta \neq 0$ fixed is the desired meromorphic extension of f. In examples where this scheme is implementable, the eigenvalues of L_θ can be calculated by analytic perturbation theory in some small parameter in L_θ, say the system-environment coupling strength (for $\mathrm{Im}\theta \neq 0$ fixed) and the decay rates are thus explicitly accessible.

So here is the question: What if L_θ does not have a meromorphic extension (due to a lack of some regularity)? Of course, this happens precisely in the kinds of systems of interest here!

- Then we cannot extend to the second Riemann sheet to access the poles.
- How can we recover decay times and directions?

Our task is to develop a method using only a mild regularity condition (as opposed to the existence of a meromorphic continuation), like

$$z \mapsto \langle \psi (L - z)^{-1} \phi \rangle \qquad \text{stays bounded as } \mathrm{Im} z \uparrow 0,$$

for a sufficiently rich set of vectors ψ, ϕ, which is called the *limiting absorption principle* (LAP).

2.1. An example where spectral deformation works. The method works for the spin-boson model (with the Hamiltonian H_{SB} given above) at weak coupling (λ small).

- ∘ $\lambda = 0$: system and reservoir uncoupled, dynamics factorizes
- ∘ $\lambda \neq 0$: U_θ = spectral translation applies and gives $\mathrm{spec}(L_\theta)$:

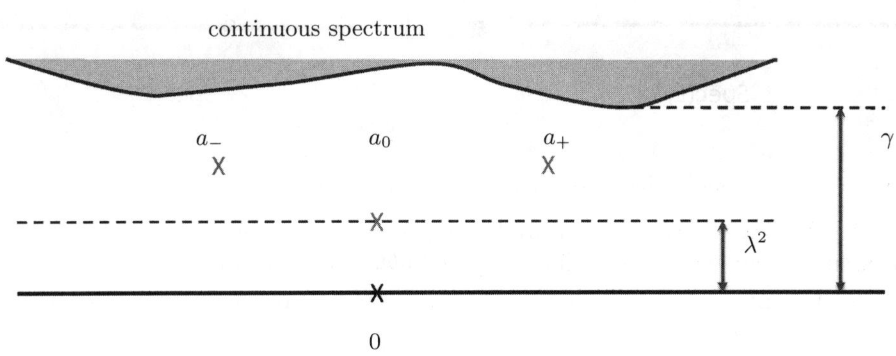

Eigenvalues of L_θ are disjoint from the continuous spectrum and one obtains the resonance expansion

$$e^{itL_\theta} = \sum_j e^{ita_j} \Pi_j + O(e^{-\gamma t}).$$

The regularity requirement for this to work is that $\varphi(h)$, the interaction operator between the spin and the bose field, must be 'translation analytic', a condition which is easily satisfied.

2.2. Why the spectral deformation technique fails for the spin-boson model at weak tunneling (V small).

∘ The system (spin) and reservoir (free bose gas) is already interacting in the unperturbed dynamics $V=0$. One can 'undo the interaction' by the unitary *polaron transformation* $U=U(\lambda)$, [**3**, **4**]. Namely,

$$\mathcal{L}_0 \equiv U\, L_0\, U^* = L_S + L_R \qquad\qquad (\epsilon = 0)$$

∘ When the perturbation $V \neq 0$ is switched on, the transformed Liouville operator becomes

$$\begin{aligned} \mathcal{L} &= \mathcal{L}_0 + V\mathcal{I} \\ \mathcal{I} &= \mathcal{I}(\lambda) \;=\; \sigma_+ \otimes W_\beta(\lambda h) + \text{adj.}, \end{aligned}$$

where σ_+ is the spin raising operator and $W_\beta(\lambda h) = e^{i\varphi_\beta(\lambda h)}$ is the thermal Weyl operator.

∘ The perturbation $\mathcal{I}$ is a bounded operator, but unfortunately, it behaves badly under spectral deformation. Indeed, the spectral deformation of $W_\beta(\lambda h)$ is $\sim e^{V\sqrt{N}}$, where N is the boson number operator. The operator $e^{V\sqrt{N}}$ is 'hugely unbounded' and we do not know how to deal with it as a perturbation of $\mathcal{L}_0$. This means we do not know how to implement the spectral deformation technique!

However, using *Mourre theory* instead of spectral deformation, one can prove the following result.

Theorem ([**3**]) *$\forall\lambda \in \mathbb{R}$, if $V \neq 0$ is small enough, then $\mathcal{L}$ has absolutely continuous spectrum covering $\mathbb{R}$ and a single simple eigenvalue at the origin. The eigenvector is the coupled equilibrium state Ω.*

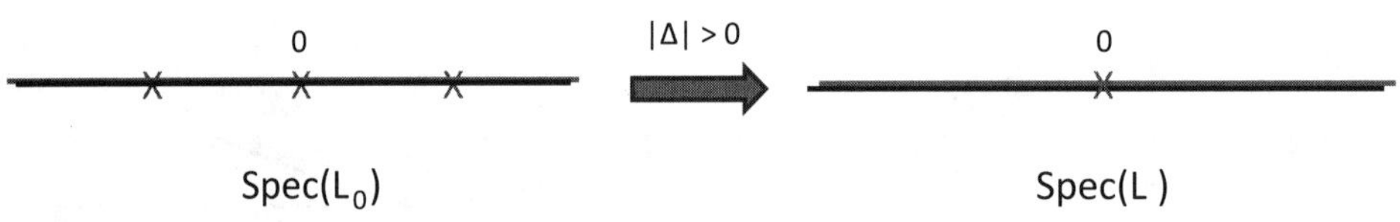

Two main technical tools in proof of this theorem are:

1. Positive commutator methods, used to show instability of eigenvalues by assuming an effective coupling "*Fermi Golden Rule*" condition.

2. The **Limiting Absorption Principle** (LAP) to show absolute continuity of the spectrum. Namely, it is well known that if

$$\sup_{x\in(a,b);y>0} \left|\langle\psi, (\mathcal{L} - x + iy)^{-1}\psi\rangle\right| \le C(\psi)$$

for all ψ in a dense set, then the spectrum of $\mathcal{L}$ in the inverval (a,b) is purely absolutely continuous (AC).

The *dynamical consequences of the Theorem* are:

$$\mathcal{L} = 0\cdot P_\Omega \oplus \bar{\mathcal{L}}P_\Omega^\perp \quad \& \quad \text{spec}(\bar{\mathcal{L}}) \text{ is purely AC, so}$$

$$e^{it\mathcal{L}} = P_\Omega \oplus e^{it\bar{\mathcal{L}}}P_\Omega^\perp \longrightarrow P_\Omega, \qquad \text{weakly, as } t\to\infty.$$

This is rather *incomplete information* compared to the spectral deformation case, where decay rates and directions are obtained as resonance energies and corresponding projections.

How can we recover the full dynamical information

$$e^{it\mathcal{L}} \sim \sum_j e^{ita_j}\Pi_j + \text{remainder}$$

using Mourre theory ?

2.3. Result on Resonance Expansion via Mourre Theory. We consider a family of self-adjoint operators on Hilbert space $\mathcal{H}$,

$$L = L_0 + V\,I, \qquad V \in \mathbb{R} \text{ is a perturbation parameter.}$$

The eigenvalues of L_0 are supposed to be embedded in the continuous spectrum of L_0. An eigenvalue e of L_0 can behave in different ways under perturbation. We call e

- *unstable* if for $V \neq 0$ small, L does not have eigenvalues in a neighbourhood of e
- *partially stable* if for $V \neq 0$ small, L has eigenvalues in a neighbourhood of e with summed multiplicity $< \text{mult}(e)$

We suppose that all eigenvalues of L_0 are either unstable or partially stable with a reduction to dimension one.

If e was an *isolated* eigenvalue of L_0 (which it is not!), with associated spectral projection P_e, then by analytic perturbation theory, the eigenvalues of L near e would be those of the operator

$$eP_e + VP_eIP_e - V^2P_eIP_e^{\perp}(L_0 - e)^{-1}IP_e + O(V^3).$$

We assume that $P_eIP_e = 0$ for all eigenvalues e of L_0. Since e is actually an *embedded eigenvalue* of L_0, the resolvent $P_e^{\perp}(L_0 - e)^{-1}$ does not exist, but we can expect the 2nd order corrections still to be linked to the **level shift operator**

$$\Lambda_e = -P_eIP_e^{\perp}(L_0 - e + i0_+)^{-1}IP_e,$$

where $(L_0 - e + i0_+)^{-1} = \lim_{\epsilon\to 0_+}(L_0 - e + i\epsilon)^{-1}$.

We make an assumption that the instability of eigenvalues is visible at order V^2 in the perturbation already. This assumption on the effectiveness of the coupling is often referred to as the *Fermi Golden Rule Condition.* It is expressed in terms of the level shift operators as follows. We assume that

(1) The eigenvalues of all the level shift operators Λ_e are simple.
(2) e unstable $\Longrightarrow$ all the eigenvalues $\lambda_{e,0}, \dots, \lambda_{e,m_e-1}$ of Λ_e have strictly positive imaginary part.
(3) e partially stable $\Longrightarrow$ Λ_e has a single real eigenvalue $\lambda_{e,0}$. All other eigenvalues $\lambda_{e,1}, \dots, \lambda_{e,m_e-1}$ have imaginary part > 0.

In particular, Λ_e is diagonalizable, so

$$\Lambda_e = \sum_{j=0}^{m_e-1} \lambda_{e,j}P_{e,j},$$

where $P_{e,j}$ are the (rank one) spectral projections.

To state our second main assumption, let P_e denote the eigenprojection associated to eigenvalue e of L_0 and set

$$R_z = (L-z)^{-1} \quad \text{and} \quad R_z^{P_e} = (P_e^\perp L P_e^\perp - z)^{-1}|_{\mathrm{Ran} P_e^\perp}.$$

We assume that the Limiting Absorption Principle holds. Namely, for every eigenvalue e of L_0,

$$\sup_{y<0,\, x\approx e} |\langle \phi, R_{x+iy}^{P_e} \psi\rangle| \le C(\phi,\psi) < \infty$$

and

$$\sup_{y<0,\ x \text{ away from all } e} |\langle \phi, R_{x+iy} \psi\rangle| \le C(\phi,\psi) < \infty$$

for all vectors ϕ, ψ in a dense set $\mathcal{D}$.

Denote by Π_{E_e} the spectral projection of L associated to the eigenvalue E_e (near e). Then we have the following result.

Theorem (Resonance expansion, [2]) *$\exists V_0 > 0$ s.t. if $0 < |V| < V_0$ then $\forall t > 0$,*

$$\begin{aligned} e^{itL} &= \sum_{e \text{ partially stable}} \left\{ e^{itE_e}\, \Pi_{E_e} + \sum_{j=1}^{m_e-1} e^{it(e+V^2 a_{e,j})}\, \Pi'_{e,j} \right\} \\ &\quad + \sum_{e \text{ unstable}} \sum_{j=0}^{m_e-1} e^{it(e+V^2 a_{e,j})}\, \Pi'_{e,j} \ + O(1/t) \end{aligned}$$

(weakly on $\mathcal{D}$). The exponents $a_{e,j}$ and the operators $\Pi'_{e,j}$ are close to the spectral data of the level shift operator Λ_e, namely,

$$a_{e,j} = \lambda_{e,j} + O(V), \qquad \Pi'_{e,j} = P_{e,j} + O(V).$$

3. Application: dynamics of a dimer

We present results for the collective environment model,

$$\begin{aligned} H &= \frac{1}{2}\begin{pmatrix} \epsilon & V \\ V & -\epsilon \end{pmatrix} + H_R + \begin{pmatrix} \lambda_D & 0 \\ 0 & \lambda_A \end{pmatrix} \otimes \phi(g) \\ H_R &= \int_{\mathbb{R}^3} \omega(k)\, a^*(k) a(k) d^3k \\ \phi(g) &= \frac{1}{\sqrt{2}} \int_{\mathbb{R}^3} (g(k) a^*(k) + \text{adj.}) d^3k, \end{aligned}$$

the reservoir being a free bosonic quantum field. We take the intital state to be disentangled,

$$\rho_{\rm in} = \rho_S \otimes \rho_R,$$

where ρ_S is an arbitrary dimer (two-level) density matrix and ρ_R is the reservoir equilibrium state at temperature $T = 1/\beta > 0$. The reduced dimer density matrix is

$$\rho_S(t) = \mathrm{Tr}_{\rm Reservoir}\left(e^{-itH} \rho_{\rm in} e^{itH}\right)$$

and we denote the dimer site basis by $\varphi_1 = \begin{pmatrix}1\\0\end{pmatrix}$ and $\varphi_2 = \begin{pmatrix}0\\1\end{pmatrix}$. The donor population is given by

$$p(t) = \langle \varphi_1, \rho_S(t)\varphi_1\rangle = [\rho_S(t)]_{11}, \qquad p(t) \in [0,1].$$

The evolution $t \mapsto p(t)$ is called the *relaxation* dynamics while *decoherence* is the evolution of the off-diagonal matrix element

$$t \mapsto [\rho_S(t)]_{12} = \langle \varphi_1, \rho_S(t)\varphi_2\rangle.$$

3.1. Dynamics for $V = 0$.

- The populations (diagonal of density matrix) are constant in time, since H_S commutes with the interaction operator.
- The total system has a 2-dimensional manifold of stationary states and one equilibrium (KMS) state given by

$$\rho_{\beta,\lambda} = \frac{e^{-\frac{\beta}{2}(\epsilon-\alpha_D)}}{Z_{\text{ren}}} |\varphi_1\rangle\langle\varphi_1| \otimes \rho_{R,1} + \frac{e^{-\frac{\beta}{2}(-\epsilon-\alpha_A)}}{Z_{\text{ren}}} |\varphi_2\rangle\langle\varphi_2| \otimes \rho_{R,2},$$

  where the *effective energy shifts* $\alpha_{D,A}$ are $\propto \lambda_{D,A}^2$ and $\rho_{R,j}$ are two explicit reservoir states.
- The reduced dimer equilibrium state equals

$$\rho_{\text{ren}} = \text{Tr}_R\ \rho_{\beta,\lambda} = \frac{e^{-\beta H_{\text{ren}}}}{Z_{\text{ren}}}, \qquad H_{\text{ren}} = \tfrac{1}{2}\begin{pmatrix}\epsilon-\alpha_D & 0\\ 0 & -\epsilon-\alpha_A\end{pmatrix}$$

- We will show that for $V \neq 0$ (small), the total system has a unique stationary state (=KMS) and all initial states converge to it as $t \to \infty$.

3.2. Dynamics for $V \neq 0$. An important quantity is the *reservoir spectral function*, defined by

$$J(\omega) = \sqrt{2\pi}\,\tanh(\beta\omega/2)\,\widehat{C}(\omega), \qquad \omega \geq 0,$$

where $\widehat{C}(\omega)$ is the Fourier transform of reservoir correlation function. Reservoir effects are encoded in $J(\omega)$ and our mathematical theory requires the **regularity conditions**

$$J(\omega) \sim \omega^s \qquad \text{with } s \geq 3 \ \text{ as } \omega \to 0$$

$$J(\omega) \sim \omega^{-\sigma} \qquad \text{with } \sigma > 3/2 \ \text{ as } \omega \to \infty.$$

We point out that a minimal a 'priori condition' is $s > 1$ (super-ohmic), but the range $1 < s < 3$ is not treatable up to now with our (or any other rigorous) method. This minimal condition is needed to guarantee that the polaron transformation can be applied to the Hamiltonian, which is the starting point of our analysis for strong system-environment couplings.

Theorem (Population dynamics, relaxation [8]**)** *Consider the local or collective reservoirs model. Let λ_D, λ_A be arbitrary. There is a $V_0 > 0$ s.t. for $0 < |V| < V_0$:*

$$p(t) = p_\infty + e^{-\gamma t}\,(p(0) - p_\infty) + O(\tfrac{t}{1+t^2}),$$

where

$$p_\infty = \frac{1}{1+e^{-\beta\hat{\epsilon}}} + O(V) \qquad \textit{with} \qquad \hat{\epsilon} = \epsilon - \tfrac{\alpha_D-\alpha_A}{2}$$

and γ is the relaxation rate $\propto V^2$ (taking different values for the local and collective reservoir cases), $\alpha_{D,A}$ are renormalizations of the unperturbed dimer energies $\pm\epsilon$ ($\propto \lambda_{D,A}^2$) and p_∞ is the equilibrium value w.r.t. renormalized dimer energies.

Notes: (1) The remainder in the above expression for $p(t)$ is small on a time-scale $\gamma t \ll 1$, *i.e.*, $t \ll V^{-2}$.

(2) The time-asymptotics is resolved correctly: as $t \to \infty$, $p(t)$ approaches its true value p_∞. It is possible to construct a perturbation theory giving the expansion $p(t) = p_\infty + e^{-\gamma t}(p(0) - p_\infty) + O(V)$, with a remainder small in V and uniform in time, but not decaying in time. This needs a slight modification of the proofs in [**2**].

3.3. Properties of final populations. The final donor population (modulo $O(V)$-correction) is given by

$$p_\infty \approx \frac{1}{2} - \frac{\hat{\epsilon}}{4T}, \qquad \text{for} \quad T \gg |\hat{\epsilon}|,$$

where $\hat{\epsilon} := \epsilon - \frac{\alpha_D - \alpha_A}{2}$ is the effective energy gap. If the *donor is strongly coupled* ($\lambda_D^2 \gg \max\{\lambda_A^2, \epsilon\}$) then $\hat{\epsilon} \propto -\lambda_D^2$, and so we have the following fact: *Increased donor-reservoir coupling increases final donor population.* This effect intensifies at lower temperatures,

$$p_\infty \approx \begin{cases} 1, & \text{if} \quad \lambda_D^2 \gg \max\{\lambda_A^2, \epsilon\} \\ 0, & \text{if} \quad \lambda_A^2 \gg \max\{\lambda_D^2, \epsilon\} \end{cases} \qquad \text{for} \quad T \ll |\hat{\epsilon}|.$$

We conclude that *the acceptor gets entirely populated if it is strongly coupled to reservoir*, yielding a good process efficiency.

3.4. Expression for relaxation rate. The rate γ (for the collective reservoir model) is given by

$$\begin{aligned} \gamma &= V^2 \lim_{r \to 0_+} \int_0^\infty e^{-rt} \cos(\hat{\epsilon} t) \cos\left[\frac{(\lambda_D - \lambda_A)^2}{\pi} Q_1(t)\right] \\ &\qquad \times \exp\left[-\frac{(\lambda_D - \lambda_A)^2}{\pi} Q_2(t)\right] dt \end{aligned}$$

where

$$\begin{aligned} Q_1(t) &= \int_0^\infty \frac{J(\omega)}{\omega^2} \sin(\omega t)\, d\omega, \\ Q_2(t) &= \int_0^\infty \frac{J(\omega)(1 - \cos(\omega t))}{\omega^2} \coth(\beta\omega/2)\, d\omega \end{aligned}$$

This expression for γ is a **Generalized Marcus Formula** – in the symmetric case $\lambda_D = -\lambda_A$ and at high temperatures, $k_B T \gg \hbar\omega_c$, it reduces to the usual Marcus Formula

$$\gamma_{\text{Marcus}} = \left(\frac{V}{2}\right)^2 \sqrt{\frac{\pi}{T\epsilon_{\text{rec}}}}\, e^{-\frac{(\epsilon - \epsilon_{\text{rec}})^2}{4T\epsilon_{\text{rec}}}} \qquad (0 < \epsilon_{\text{rec}} \propto \lambda^2).$$

3.5. Some numerical results.

- Accuracy of generalized Marcus formula:
  – For ω_c/T less than about 0.1, the numerics show that the rates given by the generalized Marcus formula coincide extremely well ($\sim \pm 1\%$) with the true value, γ given by the formula above. As the temperature decreases, the Marcus formula is not applicable any more (it is only the high-termperature limit of the true value for the relaxation rate).
  – For ω_c/T larger than about 1, the numerics show serious deviations ($\sim 30\%$) between the Marcus formula and the true value.
- An asymmetric coupling can *significantly increase the transfer* rate:

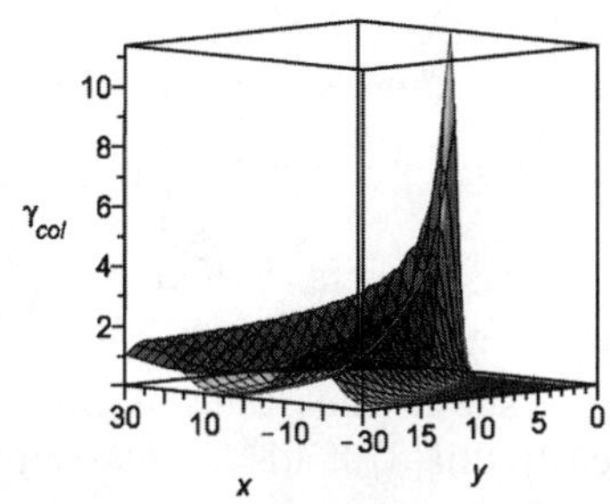

The surface shows the (true) collective relaxation rate $\gamma \equiv \gamma_{\rm col}$ versus the parameters $x \propto \lambda_D^2 - \lambda_A^2$ and $y \propto (\lambda_D - \lambda_A)^2$. The curve (red color in online version) on the surface corresponds to the usual Marcus formula, where $\lambda_A = -\lambda_D$. We see from the graph that a significant increase in γ can be achieved for a suitable asymmetric coupling, $\lambda_D \neq \lambda_A$.

3.6. Decoherece of the dimer. For the non-interacting dimer ($V = 0$), the populations are constant in time and the off-diagonal matrix element evolves as

$$[\rho_S(t)]_{12} = e^{-it\hat{\epsilon}}\,\mathcal{D}(t)\ [\rho_S(0)]_{12}.$$

If $\lim_{t\to\infty}[\rho_S(t)]_{12} = 0$ we say the dimer undergoes *full phase decoherence.* If that limit is not zero, we say it undergoes *partial phase decoherence.* One can show that the low frequency modes of the reservoir are responsible for full decoherence. Namely, we have full decoherence if and only if the low frequency modes are 'well coupled' to the dimer, as the following results explains.

Lemma ([8]) *Suppose the reservoir spectral function has the form $J(\omega) \sim \omega^s$ for ω small. Then the dimer exhibits full phase decoherence if and only if $s \le 2$.*

The following graph shows the decoherence function $\mathcal{D}(t)$ as a function of the (rescaled) time, for different values of the temperature.

Graph: $\mathcal{D}(t) = e^{-\Gamma(t)}$ for $s = 3$

Top curve:	$\beta\omega_c = 0.1$
Middle curve:	$\beta\omega_c = 1$
Bottom curve:	$\beta\omega_c = 5$

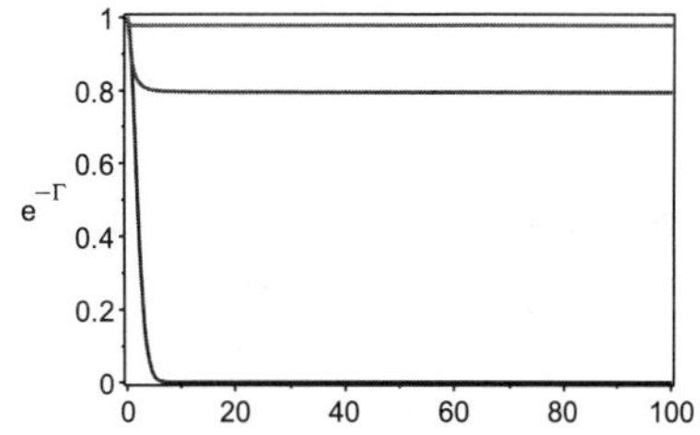

We see that the higher the temperature (i.e., the smaller β), the *lower* the degree of long-time decoherence. This can be explained by the fact that at higher temperatures, the proportion of low energy modes in the reservoir is smaller than at lower temperatures. Then, since decoherence is driven significantly by low frequency modes (see the Lemma), less decoherence results at high temperatures.

Decoherence of the interacting dimer

For $s > 2$, we have residual asymptotic coherence, namely $\lim_{t\to\infty} \mathcal{D}(t) = e^{-\Gamma_\infty} > 0$. The following result examines the decay to the final coherences.

Theorem (Decoherence, [8]) *Consider the local or collective reservoirs model with λ_D, λ_A arbitrary and $s \geq 3$. There is a $V_0 > 0$ such that if $0 < |V| < V_0$, then*

$$[\rho_S(t)]_{12} = e^{-\Gamma_\infty}\, e^{-\gamma t/2}\, e^{-it(\hat{\epsilon}+x_{\rm LS})}\; [\rho_S(0)]_{12} + O(V) + O(\tfrac{1}{1+t}),$$

where γ is the relaxation rate, $x_{\rm LS} \in \mathbb{R}$ is the Lamb shift.

Notes. (1) The theorem shows that the following well-known relation from weak coupling theory (Bloch-Redfield) holds for *all* coupling strengths:

$$\gamma_{\rm decoherence} = \gamma_{\rm relaxation}/2.$$

(2) The theorem holds for $s \geq 3$, which is in the regime of partial decoherence, where $\Gamma_\infty < \infty$. We expect to get a rigorous result in the larger region $s > 1$. But for $s \leq 2$ we will have $\Gamma_\infty = \infty$ (Lemma above!) and so the above expansion not useful. Our analysis needs then to be modified.

4. Outline of the proof of the resonance expansion

We have a self-adjoint operator L_0 which is perturbed into $L_0 + VI$ s.t.

- all eigenvalues of L_0 are embedded
- all eigenvalues of L_0 are either unstable or reduce to dimension one under perturbation
- the Limiting Absorption Principle holds

We want to show an expansion

$$\begin{aligned} e^{itL} &= \sum_{e \text{ part. stable}} \Big\{ e^{itE_e}\, \Pi_{E_e} + \sum_{j=1}^{m_e-1} e^{it(e+V^2 a_{e,j})}\, \Pi'_{e,j} \Big\} \\ &\quad + \sum_{e \text{ unstable}} \sum_{j=0}^{m_e-1} e^{it(e+V^2 a_{e,j})}\, \Pi'_{e,j} \; + \; O(1/t), \end{aligned}$$

in which the resonance data $a_{e,j}$, $\Pi'_{e,j}$ are obtained by perturbation theory in V.

Decomposing the resolvent using the Feshbach map. Let P_e be the spectral projection associated to the eigenvalue e of L_0. According to the Feshbach map decomposition [**2**, **3**], the resolvent $(L - z)^{-1}$ can be written as

$$R_z \equiv (L-z)^{-1} = F_z^{-1} + \bar{R}_z + B_z,$$

where

$$\begin{aligned} F_z &:= P_e\big(e - z - V^2 I\bar{R}_z I\big)P_e \\ \bar{R}_z &:= (P_e^\perp L P_e^\perp - z)^{-1}|_{\mathrm{Ran}P_e^\perp} \\ B_z &= -VF_z^{-1}I\bar{R}_z - V\bar{R}_z I F_z^{-1} + V^2\bar{R}_z I F_z^{-1} I\bar{R}_z \end{aligned}$$

The operator F_z^{-1} is finite-dimensional. It acts on $\mathrm{Ran}P_e$. The operator $\bar{R}_z$ 'is dispersive', namely a LAP holds for it, away from the eigenvalues of L_0. This implies purely AC spectrum of L and hence time-decay in the propagation it generates. The operator B_z is of higher order in V and is dispersive as well.

By the standard resolvent representation of the propagator, we have (for any $w > 0$ [and we will take $w \to 0_+$])

$$e^{itL}\psi = \frac{-1}{2\pi i}\int_{\mathbb{R}-iw} e^{itz}\, R_z\psi\, dz$$

We now subdivide the integration into regions close to e and away from e. For the purpose of the present exposition, let us focus on a partially stable eigenvalue e of L_0 and assume $e = 0$ and that L has a single, simple eigenvalue $E = 0$ (no shift) for small V. Let J be interval around 0 (containing no eigenvalue of L_0 but 0). The contribution to $\langle\varphi, e^{itL}\psi\rangle$ from the integral over J is

$$\int_{J-iw} e^{itz}\langle\varphi, R_z\psi\rangle dz = \int_{J-iw} e^{itz}\left[\langle\varphi, F_z^{-1}\psi\rangle + \langle\varphi, \bar{R}_z\psi\rangle + \langle\varphi, B_z\psi\rangle\right] dz$$

(A) Contribution of F_z^{-1}. Set $P \equiv P_0$ (here $e = 0$). The Feshbach map equals

$$F_z = P(-z - V^2 I\bar{R}_z I)P \equiv -z + V^2 A_z, \qquad z \in \mathbb{C}_-$$

defining the operator A_z. The latter can be diagonalized,

$$A_z = \sum_{j=0}^{d-1} a_j(z)Q_j(z) \quad \text{so that} \quad F_z^{-1} = \sum_{j=0}^{d-1} \frac{Q_j(z)}{-z + V^2 a_j(z)}.$$

Therefore,

$$\int_{J-iw} e^{itz}\langle\varphi, F_z^{-1}\psi\rangle dz = \sum_{j=0}^{d-1}\int_{J-iw} \frac{e^{itz}}{-z + V^2 a_j(z)}\langle\varphi, Q_j(z)\psi\rangle dz$$

Now A_z is approximately the level shift operator Λ_0, namely

$$A_z = -PI(\bar{L} - z)^{-1}IP = \underbrace{-PI(\bar{L}_0 + i0_+)^{-1}IP}_{\text{Level Shift Operator } \Lambda_0} + O(V) + O(z).$$

Therefore, the eigenvalues a_j of A_z are approximately the eigenvalues λ_j of Λ_0, namely

$$\begin{aligned} a_j(z) &= \lambda_j + O(V) + O(z), \qquad j = 1, \ldots, d-1 \\ a_0(z) &= O(z) \end{aligned}$$

(L has a simple eigenvalue at 0, which by the "isospectrality of Feshbach map" implies that $a_0(0) = 0$ for all V.)

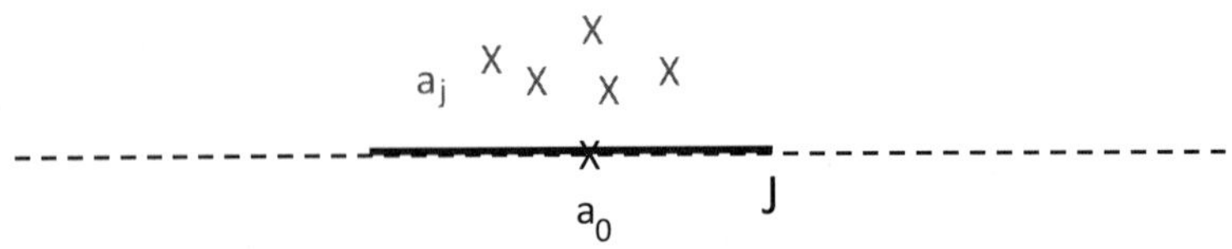

The resonances $a_0 \in \mathbb{R}$ and a_j with $\mathrm{Im} a_j > 0$ close to the interval J.

We thus obtain

$$\begin{aligned}\int_{J-iw} e^{itz}\langle\varphi, F_z^{-1}\psi\rangle dz &= \sum_{j=0}^{d-1}\int_{J-iw}\frac{e^{itz}}{-z+V^2a_j(z)}\langle\varphi, Q_j(z)\psi\rangle\, dz\\ &\approx \sum_{j=0}^{d-1}\langle\varphi, Q_j(0)\psi\rangle\int_{J-iw}\frac{e^{itz}}{-z+V^2a_j(0)}\,dz\end{aligned}$$

where the approximation is for small intervals J around zero. To estimate the last integral in the summand above, we use standard complex analysis. Namely, we complete the line integral into one over a closed contour as depicted in the diagram below, adding two sides S and a top T (which will be moved to $i\infty$). The integrals over the parts of the contour are easily estimated as indicated:

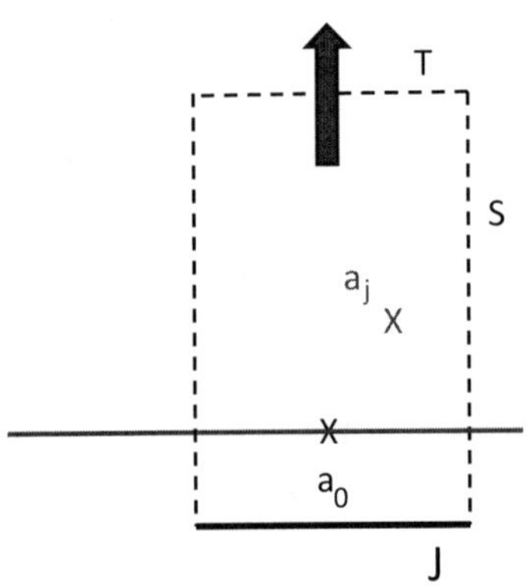

$$\begin{aligned}\int_S\frac{e^{itz}}{-z+V^2a_j(0)}dz &\sim \int_0^\infty e^{-yt}dy = O(1/t)\\ \int_T\frac{e^{itz}}{-z+V^2a_j(0)}dz &\sim 0 \quad (\text{since } T\to i\infty)\\ \oint\frac{e^{itz}}{-z+V^2a_j(0)}dz &\sim e^{itV^2a_j(0)} \quad (\text{residue})\end{aligned}$$

It follows that

$$\frac{1}{2\pi i}\int_{J-iw} e^{itz}\langle\varphi, F_z^{-1}\psi\rangle dz = \langle\varphi, Q_0(0)\psi\rangle + \sum_{j=1}^{d-1} e^{itV^2a_j(0)}\langle\varphi, Q_j(0)\psi\rangle + O(1/t).$$

To analyze the part which is constant in time on the right side, we notice that the spectral projection of L for the (embedded) eigenvalue 0 is

$$\Pi_0 = \lim_{V\to 0_+}(iV)(L-0+iV)^{-1}.$$

Then, using the Feshbach decomposition of the resolvent, one identifies

$$\langle\varphi, Q_0(0)\psi\rangle = \langle\varphi, \Pi_0\psi\rangle + O(V).$$

The contributions coming from B_z will add up precisely to give the remainder term $O(V)$ on the right side. The decaying parts reveal a rate given by the "Fermi Golden Rule",

$$e^{itV^2a_j(0)} = e^{itV^2[\lambda_j+O(V)]}.$$

(B) Contributions of B_z and $\bar{R}_z$. We use the LAP for $\bar{R}_z$ to show that

$$\sup_{z\in\mathbb{C}_-} \left|\tfrac{d}{dz}\langle\varphi, \bar{R}_z\psi\rangle\right| \le C,$$

then we integrate by parts w.r.t. z and obtain

$$\int_{J-iw} e^{itz}\langle\varphi, \bar{R}_z\psi\rangle dz = O(1/t).$$

To treat

$$\int_{J-iw} e^{itz}\langle\varphi, B_z\psi\rangle dz$$

where $B_z = -VF_z^{-1}I\bar{R}_z - V\bar{R}_zIF_z^{-1} + V^2\bar{R}_zIF_z^{-1}I\bar{R}_z$, we use again the spectral representation of F_z^{-1} and get *corrections* (to all orders in V) of the contributions coming from F_z^{-1} explained above.

Summary

• We develop a resonance expansion for the dynamics of a dimer strongly coupled to reservoirs. The expansion is valid for arbitrarily large dimer-reservoir interactions and for all times $t \ge 0$.

• Since analytic spectral deformation theory does not apply to the systems in question (due to the large coupling), we develop a Mourre theory which allows to treat 'singular perturbations' (strong coupling) and we extract from it decay times and directions. This is our main technical novelty, given in [**2**]. The only work we are aware of, which is similar in spirit to ours, is [**1**]. But the method developed there is valid only for zero temperature systems and weak coupling between the system and environment.

• We stablish a generalized Marcus formula for donor-acceptor reaction rates, uncovering new physical properties (e.g. population values), not visible in the previous, usual formula, [**8**].

References

[1] Laura Cattaneo, Gian Michele Graf, and Walter Hunziker, *A general resonance theory based on Mourre's inequality*, Ann. Henri Poincaré **7** (2006), no. 3, 583–601, DOI 10.1007/s00023-005-0261-5. MR2226749

[2] M. Könenberg and M. Merkli, *On the irreversible dynamics emerging from quantum resonances*, J. Math. Phys. **57** (2016), no. 3, 033302, 26, DOI 10.1063/1.4944614. MR3479147

[3] M. Könenberg, M. Merkli, and H. Song, *Ergodicity of the spin-boson model for arbitrary coupling strength*, Comm. Math. Phys. **336** (2015), no. 1, 261–285, DOI 10.1007/s00220-014-2242-3. MR3322374

[4] A.J. Leggett, S. Chakravarty, A.T. Dorsey, M.P. A. Fisher, A. Garg, W. Zwerger, *Dynamics of the dissipative two-state system*, Rev. Mod. Phys. **59**(1), 1-85 (1987)

[5] R.A. Marcus, *On the Theory of Oxidation-Reduction Reactions Involving Electron Transfer I*, J. Chem. Phys. **24**, no.5, 966-978 (1956)

[6] R.A. Marcus, nobel prize lecture,
http://www.nobelprize.org/nobel prizes/chemistry/laureates/1992/marcus-lecture.pdf

[7] M. Mohseni, Y. Omar, G.S. Engel, and M.B. Plenio (Eds), *Quantum Effects in Biology*, Cambridge University Press, 2014

[8] M. Merkli, G. P. Berman, R. T. Sayre, S. Gnanakaran, M. Könenberg, A. I. Nesterov, and H. Song, *Dynamics of a chlorophyll dimer in collective and local thermal environments*, J. Math. Chem. **54** (2016), no. 4, 866–917, DOI 10.1007/s10910-016-0593-z. MR3473683

[9] F. Müh, D. Lindorfer, M. Schmidt am Busch, T. Renger, *Towards a structure-based excitation Hamiltonian for the CP29 antenna of photosystem II*, Phys. Chem. Chem. Phys. **16**, 11848 (2014)

[10] D. Xu, K. Schulten, *Coupling of protein motion to electron transfer in a photosynthetic reaction center: investigating the low temperature behavior in the framework of the spinboson model*, Chem. Phys. **182**, 91-117 (1994)

Department of Mathematics and Statistics, Memorial University of Newfoundland, St. John's, Newfoundland A1C 5S7, Canada

Email address: `merkli@mun.ca`